Indian Statistical Institute Series

Editors-in-chief

B. V. Rajarama Bhat, Indian Statistical Institute, Bengaluru, Karnataka, India
Abhay G. Bhatt, Indian Statistical Institute, New Delhi, India
Joydeb Chattopadhyay, Indian Statistical Institute, Kolkata, West Bengal, India
S. Ponnusamy, Indian Statistical Institute, Chennai, Tamil Nadu, India

The *Indian Statistical Institute Series* publishes high-quality content in the domain of mathematical sciences, bio-mathematics, financial mathematics, pure and applied mathematics, operations research, applied statistics and computer science and applications with primary focus on mathematics and statistics. Editorial board comprises of active researchers from major centres of Indian Statistical Institutes. Launched at the 125th birth Anniversary of P.C. Mahalanobis, the series will publish textbooks, monographs, lecture notes and contributed volumes. Literature in this series will appeal to a wide audience of students, researchers, educators, and professionals across mathematics, statistics and computer science disciplines.

More information about this series at http://www.springer.com/series/15910

Rajeeva L. Karandikar · B. V. Rao

Introduction to Stochastic Calculus

 Springer

Rajeeva L. Karandikar
Chennai Mathematical Institute
Siruseri, Tamil Nadu
India

B. V. Rao
Chennai Mathematical Institute
Siruseri, Tamil Nadu
India

ISSN 2523-3114   ISSN 2523-3122   (electronic)
Indian Statistical Institute Series
ISBN 978-981-13-4121-2   ISBN 978-981-10-8318-1   (eBook)
https://doi.org/10.1007/978-981-10-8318-1

The print edition is not for sale in India. Customers from India please order the print book from: Hindustan Book Agency, P-19 Green Park Extension, New Delhi 110016, India.

Cover photo: Reprography & Photography Unit, Indian Statistical Institute, Kolkata

Printed on acid-free paper

This Springer imprint is published by Springer Nature
The registered company is Springer Nature Singapore Pte Ltd.
The registered company address is: 152 Beach Road, #21-01/04 Gateway East, Singapore 189721, Singapore

Dedicated to the memory of G. Kallianpur

Preface

This book is a comprehensive textbook on stochastic calculus—the branch of mathematics that is most widely applied in financial engineering and mathematical finance. It will be useful for a two-semester graduate-level course on stochastic calculus, where the background required is a course on measure-theoretic probability.

This book begins with conditional expectation and martingales, and basic results on martingales are included with proofs (in discrete time as well as in continuous time). Then a chapter on Brownian motion and Ito's integration with respect to the Brownian motion follows, which includes stochastic differential equations. These three chapters give a soft landing to a reader to the more complex results that follow. The first three chapters form the introductory material.

Taking a cue from the Ito's integral, a *stochastic integrator* is defined and its properties, as well as the properties of the integral, are discussed. In most treatments, one starts by defining the integral for a square integrable martingale and where integrands themselves are in suitable Hilbert space. Then over several stages, the integral is extended, and at each step, one has to reaffirm its properties. We avoid this. Various results including quadratic variation and Ito's formula follow from the definition. Then Emery topology is defined and studied.

We then show that for a square integrable martingale M, the quadratic variation $[M, M]$ exists, and using this, we show that square integrable martingales are stochastic integrators. This approach to stochastic integration is different from the standard approach as we do not use Doob–Meyer decomposition. Instead of using the predictable quadratic variation $\langle M, M \rangle$ of a square integrable martingale M, we use the quadratic variation $[M, M]$. Using an inequality by Burkholder, we show that all martingales and local martingales are stochastic integrators and thus semimartingales are stochastic integrators. We then show that stochastic integrators are semimartingales and obtain various results such as a description of the class of integrands for the stochastic integral. We complete the chapter by giving a proof of the Bichteler–Dellacherie–Meyer–Mokobodzky theorem.

These two chapters form the basic material. We have avoided invoking results from functional analysis but rather included the required steps. Thus, instead of saying that the integral is a continuous linear functional on a dense subset of a Banach space and hence can be extended to the Banach space, we explicitly construct the extension.

Next, we introduce *Pathwise formulae* for the quadratic variation and the stochastic integral. These have not found a place in any textbook on stochastic integration. We briefly specialize in continuous semimartingales and obtain growth estimates and study the solution of a stochastic differential equation (SDE) using the technique of random time change. We also prove pathwise formulae for the solution of an SDE driven by continuous semimartingales.

Then, we move on to a study of predictable increasing processes, introduce predictable stopping times and prove the Doob–Meyer decomposition theorem.

The Davis inequality ($p = 1$ case of the Burkholder–Davis–Gundy inequality) plays an important role in the integral representation of martingales and hence is taken up next. We also introduce the notion of a sigma-martingale.

We then give a comprehensive treatment of integral representation of martingales and its connection with the uniqueness of equivalent martingale measure. This connection is important from the point of view of mathematical finance. Here, we consider the multivariate case and also include the case when the underlying process is a sigma-martingale.

In order to study stochastic differential equations driven by a general semimartingale, we introduce the Metivier–Pellaumail inequality and, using it, introduce a notion of the dominating process of a semimartingale. We then obtain existence and uniqueness of solutions to the SDE and also obtain a pathwise formula by showing that modified Euler–Peano approximations converge almost surely.

We conclude this book by discussing the Girsanov theorem and its role in the construction of weak solutions to SDEs.

We would like to add that this book includes various techniques that we have learnt over the last four decades from different sources. This includes, in addition to books and articles given in the references, the *Séminaire de Probabilités* volumes and various books on stochastic processes and research articles. We must also mention the blog https://almostsure.wordpress.com/stochastic-calculus/ by George Lowther, which brought some of these techniques to our attention.

Siruseri, India   Rajeeva L. Karandikar
                                                                                                        B. V. Rao

Contents

About the Authors

Rajeeva L. Karandikar has been Professor and Director of Chennai Mathematical Institute, Tamil Nadu, India, since 2010. An Indian mathematician, statistician and psephologist, he is a Fellow of the Indian Academy of Sciences, Bengaluru, India, and the Indian National Science Academy, New Delhi, India. He received his M. Stat. and Ph.D. degrees from the Indian Statistical Institute, Kolkata, India, in 1978 and 1981, respectively. He spent 2 years as Visiting Professor at the University of North Carolina at Chapel Hill, USA, and worked with Prof. Gopinath Kallianpur. He returned to the Indian Statistical Institute, New Delhi, India, in 1984. In 2006, he moved to Cranes Software International Limited, where he was Executive Vice President for analytics until 2010. His research interests include stochastic calculus, filtering theory, option pricing theory, psephology in the context of Indian elections and cryptography.

B. V. Rao is Adjunct Professor at Chennai Mathematical Institute, Tamil Nadu, India. He received his M.Sc. degree in Statistics from Osmania University, Hyderabad, India, in 1965 and his doctoral degree from the Indian Statistical Institute, Kolkata, India, in 1970. His research interests include descriptive set theory, analysis, probability theory and stochastic calculus. He was Professor and later Distinguished Scientist at the Indian Statistical Institute, Kolkata. Generations of Indian probabilists have benefitted from his teaching, where he taught from 1973 until 2009.

Chapter 1
Discrete Parameter Martingales

In this chapter, we will discuss martingales indexed by integers (mostly positive integers) and obtain basic inequalities on martingales and other results which are the basis of most of the developments in later chapters on stochastic integration. We will begin with a discussion on *conditional expectation* and then on *filtration*—two notions central to martingales.

1.1 Notations

For an integer $d \geq 1$, $\mathbb{R}^d$ denotes the d-dimensional Euclidean space, and $\mathcal{B}(\mathbb{R}^d)$ will denote the Borel σ-field on $\mathbb{R}^d$. Further, $\mathbb{C}(\mathbb{R}^d)$ and $\mathbb{C}_b(\mathbb{R}^d)$ will denote the classes of continuous functions and bounded continuous functions on $\mathbb{R}^d$, respectively. When $d = 1$, we will write $\mathbb{R}$ in place of $\mathbb{R}^1$. $\mathbb{Q}$ will denote the set of rational numbers in $\mathbb{R}$.

$(\Omega, \mathcal{F}, \mathsf{P})$ will denote a generic probability space, and $\mathbb{B}(\Omega, \mathcal{F})$ will denote the class of real-valued bounded $\mathcal{F}$ measurable functions.

For a collection $\mathcal{A} \subseteq \mathcal{F}$, $\sigma(\mathcal{A})$ will denote the smallest σ-field which contains $\mathcal{A}$ and for a collection $\mathbb{G} \subseteq \mathbb{B}(\Omega, \mathcal{F})$, $\sigma(\mathbb{G})$ will likewise denote the smallest σ-field with respect to which each function in $\mathbb{G}$ is measurable.

It is well known and easy to prove that

$$\sigma(\mathbb{C}_b(\mathbb{R}^d)) = \mathcal{B}(\mathbb{R}^d).$$

An $\mathbb{R}^d$-valued random variable X, on a probability space $(\Omega, \mathcal{F}, \mathsf{P})$, is function from $(\Omega, \mathcal{F})$ to $(\mathbb{R}^d, \mathcal{B}(\mathbb{R}^d))$. For such an X and a function $f \in \mathbb{C}_b(\mathbb{R}^d)$, $\mathsf{E}[f(X)]$ (or $\mathsf{E}_\mathsf{P}[f(X)]$ if there are more than one probability measure under consideration) will denote the integral

© Springer Nature Singapore Pte Ltd. 2018

R. L. Karandikar and B. V. Rao, *Introduction to Stochastic Calculus*,
Indian Statistical Institute Series, https://doi.org/10.1007/978-981-10-8318-1_1

$$E[f(X)] = \int_{\Omega} f(X(\omega)) dP(\omega).$$

For any measure μ on $(\Omega, \mathcal{F})$ and for $1 \le p < \infty$, we will denote by $\mathbb{L}^p(\mu)$ the space $\mathbb{L}^p(\Omega, \mathcal{F}, \mu)$ of real-valued $\mathcal{F}$ measurable functions equipped with the norm

$$\|f\|_p = (\int |f|^p d\mu)^{\frac{1}{p}}.$$

It is well known that $\mathbb{L}^p(\mu)$ is a Banach space under the norm $\|f\|_p$.

For more details and discussions as well as proofs of statements quoted in this chapter, see Billingsley [4], Breiman [5], Ethier and Kurtz [18].

1.2 Conditional Expectation

Let X and Y be random variables. Suppose we are going to observe Y and are required to make a *guess* for the value of X. Of course, we would like to be as close to X as possible. Suppose the penalty function is square of the error. Thus we wish to minimize

$$E[(X - a)^2] \tag{1.2.1}$$

where a is the guess or the *estimate*. For this to be meaningful, we should assume $E[X^2] < \infty$. The value of a that minimizes (1.2.1) is the mean $\mu = E[X]$. On the other hand, if we are allowed to use observations Y while making the guess, then our estimate could be a function of Y. Thus we should choose the function g such that

$$E[(X - g(Y))^2]$$

takes the minimum possible value. It can be shown that there exists a function g (Borel measurable function from $\mathbb{R}$ to $\mathbb{R}$) such that

$$E[(X - g(Y))^2] \le E[(X - f(Y))^2] \tag{1.2.2}$$

for all (Borel measurable) functions f. Further, if g_1, g_2 are two functions satisfying (1.2.2), then $g_1(Y) = g_2(Y)$ almost surely P. Indeed, $\mathbb{A} = \mathbb{L}^2(\Omega, \mathcal{F}, P)$—the space of all square integrable random variables on $(\Omega, \mathcal{F}, P)$ with inner product $\langle X, Y \rangle = E[XY]$, giving rise to the norm $\|Z\| = \sqrt{E[Z^2]}$, is a Hilbert space and

$$\mathbb{K} = \{f(Y) : f \text{ from } \mathbb{R} \text{ to } \mathbb{R} \text{ measurable}, E[(f(Y))^2] < \infty\}$$

is a closed subspace of $\mathbb{A}$. Hence given $X \in \mathbb{A}$, there is a unique element in $\mathbb{K}$, namely the orthogonal projection of X to $\mathbb{K}$, that satisfies (1.2.2). Thus for $X \in \mathbb{A}$, we can define $g(Y)$ to be the conditional expectation of X given Y, written as $E[X \mid Y] =$

$g(Y)$. One can show that for $X, Z \in \mathbb{A}$ and $a, b \in \mathbb{R}$

$$\mathsf{E}[aX + bZ \mid Y] = a\mathsf{E}[X \mid Y] + b\mathsf{E}[Z \mid Y]$$

and

$$X \leq Z \text{ implies } \mathsf{E}[X \mid Y] \leq \mathsf{E}[Z \mid Y].$$

Note that (1.2.2) implies that for all t

$$\mathsf{E}[(X - g(Y))^2] \leq \mathsf{E}[(X - g(Y) + tf(Y))^2]$$

for any $f(Y) \in \mathbb{K}$ and hence that

$$t^2 \mathsf{E}[f(Y)^2] + 2t\mathsf{E}[(X - g(Y))f(Y)] \geq 0, \ \forall t.$$

In particular, $g(Y)$ satisfies

$$\mathsf{E}[(X - g(Y))f(Y)] = 0 \ \forall f \text{ bounded measurable.} \qquad (1.2.3)$$

Indeed, (1.2.3) characterizes $g(Y)$. Also, (1.2.3) is meaningful even when X is not square integrable but only $\mathsf{E}[|X|] < \infty$. With a little work we can show that given X with $\mathsf{E}[|X|] < \infty$, there exists a unique $g(Y)$ such that (1.2.3) is valid. To see this, first consider $X \geq 0$. Take $X_n = \min(X, n)$ and g_n such that $\mathsf{E}[X_n \mid Y] = g_n(Y)$. Clearly, $g_n(Y) \leq g_{n+1}(Y)$ $a.s.$ Define $\tilde{g}(x) = \limsup g_n(x)$, $g(x) = \tilde{g}(x)$ if $\tilde{g}(x) < \infty$ and $g(x) = 0$ otherwise. One can show that

$$\mathsf{E}[(X - g(Y))f(Y)] = 0 \ \forall f \text{ bounded measurable.}$$

The case for general X can be deduced by writing X as difference of two non-negative random variables. It is easy to see that in (1.2.3) it suffices to take f to be $\{0, 1\}$-valued; i.e. indicator function of a Borel set. We are thus led to the following definition: for random variables X, Y with $\mathsf{E}[|X|] < \infty$,

$$\mathsf{E}[X \mid Y] = g(Y)$$

where g is a Borel function satisfying

$$\mathsf{E}[(X - g(Y))1_B(Y)] = 0, \quad \forall B \in \mathcal{B}(\mathbb{R}). \qquad (1.2.4)$$

Now if instead of one random variable Y, we were to observe $Y_1, \ldots, Y_m$, we can similarly define

$$\mathsf{E}[X \mid Y_1, \ldots Y_m] = g(Y_1, \ldots Y_m)$$

where g satisfies

$$\mathsf{E}[(X - g(Y_1, \ldots Y_m))1_B(Y_1, \ldots Y_m)] = 0, \quad \forall B \in \mathcal{B}(\mathbb{R}^m).$$

Also if we were to observe an infinite sequence, we have to proceed similarly, with g being a Borel function on $\mathbb{R}^\infty$. Of course, the random variables could be taking values in $\mathbb{R}^d$. In each case we will have to write down properties and proofs thereof and keep doing the same as the class of observable random variables changes.

Instead, here is a unified way. Let $(\Omega, \mathcal{F}, \mathsf{P})$ be a probability space and Y be a random variable on Ω. The smallest σ-field $\sigma(Y)$ with respect to which Y is measurable (also called the σ-field generated by Y) is given by

$$\sigma(Y) = \{A \in \mathcal{F} : A = \{Y \in B\}, \ B \in \mathcal{B}(\mathbb{R})\}.$$

Likewise, for random variables $Y_1, \dots Y_m$, the σ-field $\sigma(Y_1, \dots Y_m)$ generated by $Y_1, \dots Y_m$ is the smallest σ-field with respect to which $Y_1, \dots Y_m$ are measurable and is given by

$$\sigma(Y_1, \dots Y_m) = \{A \in \mathcal{F} : A = \{(Y_1, \dots Y_m) \in B\}, \ B \in \mathcal{B}(\mathbb{R}^m)\}.$$

Exercise 1.1 Show that

(i) A random variable Z can be written as $Z = g(Y)$ for a measurable function g if and only if Z is measurable with respect to $\sigma(Y)$.
(ii) A random variable Z can be written as $Z = g(Y_1, Y_2, \dots, Y_n)$ for a measurable function g if and only if Z is measurable with respect to $\sigma(Y_1, Y_2, \dots, Y_n)$.

Similar statement is true even for an infinite sequence of random variables. In view of these observations, one can define conditional expectation given a σ-field as follows. It should be remembered that one mostly uses it when the σ-field in question is generated by a collection of observable random variables.

Definition 1.2 Let X be a random variable on $(\Omega, \mathcal{F}, \mathsf{P})$ with $\mathsf{E}[|X|] < \infty$ and let $\mathcal{G}$ be a sub-σ-field of $\mathcal{F}$. Then the *conditional expectation* of X given $\mathcal{G}$ is defined to be the $\mathcal{G}$ measurable random variable Z such that

$$\mathsf{E}[X 1_A] = \mathsf{E}[Z 1_A], \quad \forall A \in \mathcal{G}. \tag{1.2.5}$$

Existence of Z can be proven on the same lines as given above, first for the case when X is square integrable and then for general X. Also, Z is uniquely determined up to null sets—if Z and Z' are $\mathcal{G}$ measurable and both satisfy (1.2.5), then $\mathsf{P}(Z = Z') = 1$. Some properties of conditional expectation are given in the following two propositions.

Throughout the book, we adopt a convention that all statements involving random variables are to be interpreted in almost sure sense—i.e. $X = Y$ means $X = Y$ *a.s.*, $X \leq Y$ means $X \leq Y$ *a.s.*

Proposition 1.3 *Let X, Z be integrable random variables on $(\Omega, \mathcal{F}, \mathsf{P})$ and $\mathcal{G}, \mathcal{H}$ be sub-σ-fields of $\mathcal{F}$ with $\mathcal{G} \subseteq \mathcal{H}$ and $a, b \in \mathbb{R}$. Then we have*

(i) $E[aX + bZ \mid \mathcal{G}] = aE[X \mid \mathcal{G}] + bE[Z \mid \mathcal{G}]$.

(ii) $X \leq Z \Rightarrow E[X \mid \mathcal{G}] \leq E[Z \mid \mathcal{G}]$.

(iii) $|E[X \mid \mathcal{G}]| \leq E[|X| \mid \mathcal{G}]$.

(iv) $E[E[X \mid \mathcal{G}]] = E[X]$.

(v) $E[E[X \mid \mathcal{H}] \mid \mathcal{G}] = E[X \mid \mathcal{G}]$.

(vi) *If Z is $\mathcal{G}$ measurable such that* $E[\,|ZX|\,] < \infty$ *then*

$$E[ZX \mid \mathcal{G}] = ZE[X \mid \mathcal{G}].$$

Of course when X is square integrable, we do have an analogue of (1.2.2):

Proposition 1.4 *Let X be a random variable with $E[X^2] < \infty$ and $\mathcal{H}$ be a sub-σ-field. Then for all $\mathcal{H}$ measurable square integrable random variables U*

$$E[(X - E[X \mid \mathcal{H}])^2] \leq E[(X - U)^2].$$

1.3 Independence

Two events A, B in a probability space $(\Omega, \mathcal{F}, P)$, i.e. $A, B \in \mathcal{F}$, are said to be independent if

$$P(A \cap B) = P(A)P(B).$$

For $j = 1, 2, \ldots m$, let X_j be an $\mathbb{R}^d$-valued random variable on a probability space $(\Omega, \mathcal{F}, P)$. Then $X_1, X_2, \ldots X_m$ are said to be independent if for all $A_j \in \mathcal{B}(\mathbb{R}^d)$, $1 \leq j \leq m$

$$P(X_j \in A_j, \, 1 \leq j \leq m) = \prod_{j=1}^{m} P(X_j \in A_j).$$

A sequence $\{X_n\}$ of random variables is said to be a sequence of independent random variables if $X_1, X_2, \ldots, X_m$ are independent for every $m \geq 2$.

Let $\mathcal{G}$ be a sub-σ-field of $\mathcal{F}$. An $\mathbb{R}^d$-valued random variable X is said to be independent of the σ-field $\mathcal{G}$ if for all $A \in \mathcal{B}(\mathbb{R}^d)$, $D \in \mathcal{G}$,

$$P(\{X \in A\} \cap D) = P(\{X \in A\})P(D).$$

Exercise 1.5 Let X, Y be real-valued random variables. Show that

(i) X, Y are independent if and only if for all bounded Borel measurable functions f, g on $\mathbb{R}$, one has

$$E[f(X)g(Y)] = E[f(X)]E[g(Y)]. \tag{1.3.1}$$

(ii) X, Y are independent if and only if for all bounded Borel measurable functions f, on $\mathbb{R}$, one has

$$\mathsf{E}[f(X) \mid \sigma(Y)] = \mathsf{E}[f(X)]. \tag{1.3.2}$$

(iii) X, Y are independent if and only if for all $t \in \mathbb{R}$, one has

$$\mathsf{E}[\exp\{itX\} \mid \sigma(Y)] = \mathsf{E}[\exp\{itX\}]. \tag{1.3.3}$$

Exercise 1.6 Let U be an $\mathbb{R}^d$-valued random variable, and let $\mathcal{G}$ be a σ-field. Show that U is independent of $\mathcal{G}$ if and only if for all $\lambda \in \mathbb{R}^d$

$$\mathsf{E}[\exp\{i\lambda \cdot U\} \mid \mathcal{G}] = \mathsf{E}[\exp\{i\lambda \cdot U\}]. \tag{1.3.4}$$

1.4 Filtration

Suppose X_n denotes a signal being transmitted at time n over a noisy channel (such as voice over telephone lines), and let N_n denote the noise at time n and Y_n denote the noise-corrupted signal that is observed. Under the assumption of additive noise, we get

$$Y_n = X_n + N_n, \quad n \geq 0.$$

Now the interest typically is in *estimating* the signal X_n at time n. Since the noise as well the true signal is not observed, we must require that the estimate $\widehat{X}_n$ of the signal at time n be a function of only observations up to time n, i.e. $\widehat{X}_n$ must only be a function of $\{Y_k : 0 \leq k \leq n\}$, or $\widehat{X}_n$ is measurable with respect to the σ-field $\mathcal{G}_n = \sigma\{Y_k : 0 \leq k \leq n\}$. A sequence of random variables $X = \{X_n\}$ will also be referred to as a process. Usually, the index n is interpreted as time. This is the framework for filtering theory. See Kallianpur [31] for more on filtering theory.

Consider a situation from finance. Let S_n be the market price of shares of a company UVW at time n. Let A_n denote the value of the assets of the company, B_n denote the value of contracts that the company has bid and C_n denote the value of contracts that the company is about to sign. The process S is observed by the public, but the processes A, B, C are not observed by the public at large. Hence, while making a decision on investing in shares of the company UVW, on the nth day, an investor can only use information $\{S_k : 0 \leq k < n\}$ (we assume that the investment decision is to be made before the price on nth day is revealed). Indeed, in trying to find an optimal investment policy $\pi = (\pi_n)$ (optimum under some criterion), the class of all investment strategies must be taken as all processes π such that for each n, π_n is a (measurable) function of $\{S_k : 0 \leq k < n\}$. In particular, the strategy cannot be a function of the unobserved processes A, B, C.

Let $\mathcal{G}_n$ be the σ-field generated by all the random variables observable before time n, namely $S_0, S_1, S_2, \ldots, S_{n-1}$. It is reasonable to require that any action to be taken at time n (say investment decision) is measurable with respect to $\mathcal{G}_n$. These observations lead to the following definitions.

Definition 1.7 A *filtration* on a probability space $(\Omega, \mathcal{F}, \mathsf{P})$ is an increasing family of sub-σ-fields $(\mathcal{F}_.) = \{\mathcal{F}_n : n \geq 0\}$ of $\mathcal{F}$ indexed by $n \in \{0, 1, 2, \ldots, m, \ldots\}$.

Definition 1.8 A stochastic process X, i.e. a sequence $X = \{X_n\}$ of random variables, is said to be *adapted* to a filtration $(\mathcal{F}_.)$ if for all $n \geq 0$, X_n is $\mathcal{F}_n$ measurable.

In this chapter, we will only consider discrete-time stochastic processes. We will assume that the underlying probability is complete, i.e. $N \in \mathcal{F}, \mathsf{P}(N) = 0$ and $N_1 \subseteq N$ implies $N_1 \in \mathcal{F}$ and that $\mathcal{F}_0$ contains all sets $N \in \mathcal{F}$ with $\mathsf{P}(N) = 0$. We will refer to a stochastic process as a process. Let $\mathcal{N}$ be the class of all null sets (sets with $\mathsf{P}(N) = 0$), and for a process Z, possibly vector-valued, let

$$\mathcal{F}_n^Z = \sigma(Z_k : 0 \leq k \leq n) \tag{1.4.1}$$

and

$$\widetilde{\mathcal{F}}_n^Z = \sigma(\mathcal{N} \cup \mathcal{F}_n^Z). \tag{1.4.2}$$

While it is not required in the definition, in most situations, the filtration $(\mathcal{F}_.)$ under consideration would be chosen to be (for a suitable process Z)

$$(\mathcal{F}_.^Z) = \{\mathcal{F}_n^Z : n \geq 0\}$$

or

$$(\widetilde{\mathcal{F}}_.^Z) = \{\widetilde{\mathcal{F}}_n^Z : n \geq 0\}.$$

Sometimes, a filtration is treated as a mere technicality. We would like to stress that it is not so. It is a technical concept, but a very important ingredient of the analysis. For example, in the estimation problem, one could consider the filtration $(\widetilde{\mathcal{F}}_.^{X,N})$ (recall, X is signal, N is noise and $Y = X + N$ is the observable) as well as $(\mathcal{F}_.^Y)$. While one can use $(\widetilde{\mathcal{F}}_.^{X,N})$ for technical reasons say in a proof, but when it comes to estimating the signal, the estimate at time n has to be measurable with respect to $\widetilde{\mathcal{F}}_n^Y$. If $\widehat{X}_n$ represents the estimate of the signal at time n, then the process $\widehat{X}$ must be required to be adapted to $(\widetilde{\mathcal{F}}_.^Y)$. Requiring it to be adapted to $(\widetilde{\mathcal{F}}_.^{X,N})$ is not meaningful. Indeed, if we can take $\widehat{X}_n$ to be adapted to $(\widetilde{\mathcal{F}}_.^{X,N})$, then we can as well take $\widehat{X}_n = X_n$ which is of course not meaningful. Thus, here $(\widetilde{\mathcal{F}}_.^{X,N})$ is a mere technicality while $(\widetilde{\mathcal{F}}_.^Y)$ is more than a technicality.

1.5 Martingales

In this section, we will fix a probability space $(\Omega, \mathcal{F}, \mathsf{P})$ and a filtration $(\mathcal{F}_.)$. We will only be considering $\mathbb{R}$-valued processes in this section.

Definition 1.9 A sequence $M = \{M_n\}$ of random variables is said to be a *martingale* if M is $(\mathcal{F}_.)$ adapted and for $n \geq 0$ one has $\mathsf{E}[|M_n|] < \infty$ and

$$\mathsf{E}_\mathsf{P}[M_{n+1} \mid \mathcal{F}_n] = M_n.$$

Definition 1.10 A sequence $M = \{M_n\}$ of random variables is said to be a *submartingale* if M is $(\mathcal{F}_.)$ adapted and for $n \geq 0$ one has $\mathsf{E}[|M_n|] < \infty$ and

$$\mathsf{E}_\mathsf{P}[M_{n+1} \mid \mathcal{F}_n] \geq M_n.$$

When there are more than one filtration in consideration, we will call it a $(\mathcal{F}_.)$-martingale or martingale w.r.t. $(\mathcal{F}_.)$. Alternatively, we will say that $\{(M_n, \mathcal{F}_n) : n \geq 0\}$ is a martingale. It is easy to see that for a martingale M, for any $m < n$, one has

$$\mathsf{E}_\mathsf{P}[M_n \mid \mathcal{F}_m] = M_m$$

and similar statement is also true for submartingales. Indeed, one can define martingales and submartingales indexed by an arbitrary partially ordered set. We do not discuss these in this book.

If M is a martingale and ϕ is a convex function on $\mathbb{R}$, then Jensen's inequality implies that the process $X = \{X_n\}$ defined by $X_n = \phi(M_n)$ is a submartingale provided X_n is integrable for all n. If M is a submartingale and ϕ is an increasing convex function then X is also a submartingale provided X_n is integrable for each n. In particular, if M is a martingale or a positive submartingale with $\mathsf{E}[M_n^2] < \infty$ for all n, then Y defined by $Y_n = M_n^2$ is a submartingale.

When we are having only one filtration under consideration, we will drop reference to it and simply say M is a martingale. It is easy to see also that sum of two martingales with respect to the same underlying filtration is also a martingale. We note here an important property of martingales that would be used later.

Theorem 1.11 *Let M^m be a sequence of martingales on some probability space $(\Omega, \mathcal{F}, \mathsf{P})$ w.r.t. a fixed filtration $(\mathcal{F}_.)$. Suppose that*

$$M_n^m \to M_n \text{ in } \mathbb{L}^1(\mathsf{P}), \quad \forall n \geq 0.$$

Then M is also a martingale w.r.t. the filtration $(\mathcal{F}_.)$.

Proof Note that for any X^m converging to X in $\mathbb{L}^1(\mathsf{P})$, for any σ-field $\mathcal{G}$, using (i), (ii), (iii) and (iv) in Proposition 1.3, one has

$$
\begin{aligned}
\mathsf{E}[|\mathsf{E}[X^m \mid \mathcal{G}] - \mathsf{E}[X \mid \mathcal{G}]|] &= \mathsf{E}[|\mathsf{E}[(X^m - X) \mid \mathcal{G}]|] \\
&\leq \mathsf{E}[\mathsf{E}[|X^m - X| \mid \mathcal{G}]] \\
&= \mathsf{E}[|X^m - X|] \\
&\to 0.
\end{aligned}
$$

For $n \geq 0$, applying this to $X^m = M_n^m$, one gets

$$M_n^m = \mathsf{E}[M_{n+1}^m \mid \mathcal{F}_n] \to \mathsf{E}[M_{n+1} \mid \mathcal{F}_n] \text{ in } \mathbb{L}^1(\mathsf{P}).$$

But $M_n^m \to M_n$ in $\mathbb{L}^1(\mathsf{P})$ so that $\mathsf{E}[M_{n+1} \mid \mathcal{F}_n] = M_n$. It follows that M is a martingale. $\qquad\square$

The following decomposition result, called the Doob decomposition, is simple to prove but its analogue in continuous time was the beginning of the theory of stochastic integration.

Theorem 1.12 *Let X be a submartingale. Let $A = \{A_n\}$ be defined by $A_0 = 0$ and for $n \geq 1$,*

$$A_n = \sum_{k=1}^{n} \mathsf{E}[(X_k - X_{k-1}) \mid \mathcal{F}_{k-1}].$$

Then A is an increasing process (i.e. $A_n \leq A_{n+1}$ for $n \geq 0$) such that $A_0 = 0$, A_n is $\mathcal{F}_{n-1}$ measurable for each n and $M = \{M_n\}$ defined by

$$M_n = X_n - A_n$$

is a martingale. Further, if $B = \{B_n\}$ is a process such that $B_0 = 0$, B_n is $\mathcal{F}_{n-1}$ measurable for each n and $N = \{N_n\}$ defined by $N_n = X_n - B_n$ is a martingale, then

$$\mathsf{P}(A_n = B_n \; \forall n \geq 1) = 1.$$

Proof Since X is a submartingale, each summand in the definition of A_n is nonnegative and hence A is an increasing process. Clearly, A_n is $\mathcal{F}_{n-1}$ measurable. By the definition of A_n, M_n we can see that

$$M_n - M_{n-1} = X_n - X_{n-1} - \mathsf{E}[X_n - X_{n-1} \mid \mathcal{F}_{n-1}]$$

and hence that

$$\mathsf{E}[M_n - M_{n-1} \mid \mathcal{F}_{n-1}] = 0$$

showing that M is a martingale. If B_n is as in the statement, we can see that for $n \geq 1$,

$$\begin{aligned}
\mathsf{E}[(X_n - X_{n-1}) \mid \mathcal{F}_{n-1}] &= \mathsf{E}[(N_n - N_{n-1}) \mid \mathcal{F}_{n-1}] + \mathsf{E}[(B_n - B_{n-1}) \mid \mathcal{F}_{n-1}] \\
&= \mathsf{E}[(B_n - B_{n-1}) \mid \mathcal{F}_{n-1}] \\
&= B_n - B_{n-1}.
\end{aligned}$$

Now $B_0 = 0$ implies

$$B_n = \sum_{k=1}^{n} \mathsf{E}[(X_k - X_{k-1}) \mid \mathcal{F}_{k-1}]$$

completing the proof. $\qquad\square$

The uniqueness in the result above depends strongly on the assumption that B_n is $\mathcal{F}_{n-1}$ measurable. The process A is called the compensator of the submartingale X.

Let $M = \{M_n\}$ be a martingale. The sequence D defined by $D_n = M_n - M_{n-1}$, for $n \geq 1$ and $D_0 = M_0$ clearly satisfies

$$\mathsf{E}[D_n \mid \mathcal{F}_{n-1}] = 0 \ \forall n \geq 1. \tag{1.5.1}$$

An adapted sequence $\{D_n : n \geq 0\}$ satisfying (1.5.1) is called a martingale difference sequence.

Definition 1.13 A sequence of random variables $U = \{U_n\}$ is said to be predictable if for all $n \geq 1$, U_n is $\mathcal{F}_{n-1}$ measurable and U_0 is $\mathcal{F}_0$ measurable.

The compensator A appearing in the Doob decomposition of a submartingale M is predictable.

Consider a gambling house, where a sequence of games are being played, at say one hour interval. If an amount a is bet on the nth game, the reward on the nth game is $a D_n$. Since a gambler can use the outcomes of the games that have been played earlier, U_n — the amount she bets on nth game can be random instead of a fixed number. However, $\{U_n : n \geq 1\}$ has to be predictable with respect to the underlying filtration since the gambler has to decide how much to bet before the nth game is played. If the game is fair, i.e. $\mathsf{E}[D_n \mid \mathcal{F}_{n-1}] = 0$, then the partial sums $M_n = D_0 + \cdots + D_n$ is a martingale and the total reward R_n at time n is then given by $R_n = \sum_{k=0}^{n} U_k D_k$. One can see that it is also a martingale, if say U_n is bounded. This leads to the following definition.

Definition 1.14 Let $M = \{M_n\}$ be a martingale and $U = \{U_n\}$ be a predictable sequence of random variables. The process $Z = \{Z_n\}$ defined by $Z_0 = 0$ and for $n \geq 1$

$$Z_n = \sum_{k=1}^{n} U_k (M_k - M_{k-1}) \tag{1.5.2}$$

is called the martingale transform of M by the sequence U.

The following result gives conditions under which the transformed sequence is a martingale.

Theorem 1.15 *Suppose $M = \{M_n\}$ is a martingale and $U = \{U_n\}$ is a predictable sequence of random variables such that*

$$\mathsf{E}[|M_n U_n|] < \infty \ \textit{for all } n \geq 1. \tag{1.5.3}$$

Then the martingale transform Z defined by (1.5.2) is a martingale.

Proof Let us note that

$$\begin{aligned} E[|M_n U_n|] &= E[E[|M_n U_n| \,|\, \mathcal{F}_{n-1}]] \\ &\geq E[|E[M_n U_n \,|\, \mathcal{F}_{n-1}]|\,] \\ &= E[|U_n E[M_n \,|\, \mathcal{F}_{n-1}]|\,] \\ &= E[|U_n M_{n-1}|\,] \end{aligned} \tag{1.5.4}$$

where we have used properties of conditional expectation and the fact that U_n is $\mathcal{F}_{n-1}$ measurable and that M is a martingale. Thus, (1.5.3) implies $E[|U_n M_{n-1}|] < \infty$. This is needed to justify splitting the expression in the next step.

$$\begin{aligned} E[U_n(M_n - M_{n-1}) \,|\, \mathcal{F}_{n-1}] &= E[U_n M_n \,|\, \mathcal{F}_{n-1}] - E[U_n M_{n-1} \,|\, \mathcal{F}_{n-1}] \\ &= U_n E[M_n \,|\, \mathcal{F}_{n-1}] - U_n M_{n-1} \\ &= U_n M_{n-1} - U_n M_{n-1} \\ &= 0. \end{aligned} \tag{1.5.5}$$

This implies $C_n = U_n(M_n - M_{n-1})$ is a martingale difference sequence, and thus Z defined by (1.5.2) is a martingale. $\qquad\square$

The proof given above essentially also yields the following:

Theorem 1.16 *Suppose $M = \{M_n\}$ is a submartingale and $U = \{U_n\}$ is a predictable sequence of random variables such that $U_n \geq 0$ and*

$$E[|M_{n-1}U_n|] < \infty, \quad E[|M_n U_n|] < \infty \text{ for all } n \geq 1. \tag{1.5.6}$$

Then the transform Z defined by (1.5.2) is a submartingale.

Exercise 1.17 Let $(M_n, \mathcal{F}_n)$ be a martingale. Show that M is a $(\mathcal{F}_{\cdot}^M)$ martingale.

1.6 Stopping Times

We continue to work with a fixed probability space $(\Omega, \mathcal{F}, P)$ and a filtration $(\mathcal{F}_{\cdot})$.

Definition 1.18 A stopping time τ is a function from Ω to $\{0, 1, 2, \ldots, \} \cup \{\infty\}$ such that

$$\{\tau = n\} \in \mathcal{F}_n, \quad \forall n < \infty.$$

Equivalently, τ is a stopping time if $\{\tau \leq n\} \in \mathcal{F}_n$ for all $n \geq 1$. Stopping times were introduced in the context of Markov Chains by Doob. Martingales and stopping times together are very important tools in the theory of stochastic process in general and stochastic calculus in particular.

Definition 1.19 Let τ be a stopping time and X be an adapted process. The *stopped random variable* X_τ is defined by

$$X_\tau(\omega) = \sum_{n=0}^{\infty} X_n(\omega) 1_{\{\tau=n\}}.$$

Note that by definition, $X_\tau = X_\tau 1_{\{\tau < \infty\}}$. The following results connecting martingales and submartingales and stopping times (and their counterparts in continuous time) play a very important role in the theory of stochastic processes.

Exercise 1.20 Let σ and τ be two stopping times and let

$$\xi = \tau \vee \sigma \text{ and } \eta = \tau \wedge \sigma.$$

Show that ξ and η are also stopping times. Here and in the rest of this book, $a \vee b = \max(a, b)$ and $a \wedge b = \min(a, b)$.

Exercise 1.21 Let τ be a random variable taking values in $\{0, 1, 2, \ldots\}$, and for $n \geq 0$, let $\mathcal{F}_n$ be the σ-field generated by $\tau \wedge n$. Characterize all the stopping times w.r.t. this filtration.

Theorem 1.22 *Let $M = \{M_n\}$ be a submartingale and τ be a stopping time. Then the process $N = \{N_n\}$ defined by*

$$N_n = M_{n \wedge \tau}$$

is a submartingale. Further, if M is a martingale then so is N.

Proof Without loss of generality, we assume that $M_0 = 0$. Let $U_n = 1_{\{\tau < n\}}$ and $V_n = 1_{\{\tau \geq n\}}$. Since

$$\{U_n = 1\} = \cup_{k=0}^{n-1}\{\tau = k\}$$

it follows that U_n is $\mathcal{F}_{n-1}$ measurable and hence U is a predictable sequence, and since $U_n + V_n = 1$, it follows that V is also a predictable sequence. Noting that $N_0 = 0$ and for $n \geq 1$

$$N_n = \sum_{k=1}^{n} V_k(M_k - M_{k-1})$$

the result follows from Theorem 1.16. $\square$

The following version of this result is also useful.

Theorem 1.23 *Let $M = \{M_n\}$ be a submartingale and σ, τ be stopping times such that $\sigma \leq \tau$. Let $R = \{R_n\}$, $S = \{S_n\}$ be defined as follows: $R_0 = S_0 = 0$ and for $n \geq 1$*

$$S_n = M_n - M_{n \wedge \tau},$$
$$R_n = M_{n \wedge \tau} - M_{n \wedge \sigma}.$$

Then R, S are submartingales. Further, if M is a martingale then so are S, R.

Proof To proceed as earlier, let $U_n = 1_{\{\tau < n\}}$, $V_n = 1_{\{\sigma < n \leq \tau\}} = 1_{\{\tau \geq n\}} - 1_{\{\sigma \geq n\}}$. Once again U, V are predictable and note that $S_0 = R_0 = 0$ and for $n \geq 1$

$$S_n = \sum_{k=1}^{n} U_k(M_k - M_{k-1})$$

$$R_n = \sum_{k=1}^{n} V_k(M_k - M_{k-1}).$$

The result follows from Theorems 1.16. □

The N in Theorem 1.22 is the submartingale M stopped at τ. S in Theorem 1.23 is the increment of M after τ.

Corollary 1.24 *Let $M = \{M_n\}$ be a submartingale and σ, τ be stopping times such that $\sigma \leq \tau$. Then for all $n \geq 1$*

$$\mathsf{E}[M_{n \wedge \sigma}] \leq \mathsf{E}[M_{n \wedge \tau}].$$

It is easy to see that for a martingale M, $\mathsf{E}[M_n] = \mathsf{E}[M_0]$ for all $n \geq 1$. Of course, this property does not characterize martingales. However, we do have the following result.

Theorem 1.25 *Let $M = \{M_n\}$ be an adapted process such that $\mathsf{E}[|M_n|] < \infty$ for all $n \geq 0$. Then M is a martingale if and only if for all bounded stopping times τ,*

$$\mathsf{E}[M_\tau] = \mathsf{E}[M_0]. \tag{1.6.1}$$

Proof If M is a martingale, Theorem 1.22 implies that

$$\mathsf{E}[M_{\tau \wedge n}] = \mathsf{E}[M_0].$$

Thus taking n such that $\tau \leq n$, it follows that (1.6.1) holds.

Conversely, suppose (1.6.1) is true. To show that M is a martingale, suffices to show that for $n \geq 0$, $A \in \mathcal{F}_n$,

$$\mathsf{E}[M_{n+1}1_A] = \mathsf{E}[M_n 1_A]. \tag{1.6.2}$$

Let τ be defined by $\tau = (n+1)1_A + n1_{A^c}$. Since $A \in \mathcal{F}_n$, it is easy to check that τ is a stopping time. Using $\mathsf{E}[M_\tau] = \mathsf{E}[M_0]$, it follows that

$$\mathsf{E}[M_{n+1}1_A + M_n 1_{A^c}] = \mathsf{E}[M_0]. \tag{1.6.3}$$

Likewise, using (1.6.1) for $\tau = n$, we get $\mathsf{E}[M_n] = \mathsf{E}[M_0]$, or equivalently

$$E[M_n 1_A + M_n 1_{A^c}] = E[M_0]. \tag{1.6.4}$$

Now (1.6.2) follows from (1.6.3) and (1.6.4) completing the proof. $\qquad\square$

1.7 Doob's Maximal Inequality

We will now derive an inequality for martingales known as Doob's maximal inequality. It plays a major role in stochastic calculus as we will see later.

Theorem 1.26 *Let M be a martingale or a positive submartingale. Then, for $\lambda > 0$, $n \geq 1$ one has*

$$P(\max_{0 \leq k \leq n} |M_k| > \lambda) \leq \frac{1}{\lambda} E[|M_n| 1_{\{\max_{0 \leq k \leq n} |M_k| > \lambda\}}]. \tag{1.7.1}$$

Further, for $1 < p < \infty$, there exists a universal constant C_p depending only on p such that

$$E[(\max_{0 \leq k \leq n} |M_k|)^p] \leq C_p E[|M_n|^p]. \tag{1.7.2}$$

Proof Under the assumptions, the process N defined by $N_k = |M_k|$ is a positive submartingale. Let

$$\tau = \inf\{k : N_k > \lambda\}.$$

Here, and in what follows, we take *infimum* of the empty set as ∞. Then τ is a stopping time, and further, $\{\max_{0 \leq k \leq n} N_k \leq \lambda\} \subseteq \{\tau > n\}$. By Theorem 1.23, the process S defined by $S_k = N_k - N_{k \wedge \tau}$ is a submartingale. Clearly $S_0 = 0$ and hence $E[S_n] \geq 0$. Note that $\tau \geq n$ implies $S_n = 0$. Thus, $S_n = S_n 1_{\{\max_{0 \leq k \leq n} N_k > \lambda\}}$. Hence

$$E[S_n 1_{\{\max_{0 \leq k \leq n} N_k > \lambda\}}] \geq 0.$$

This yields

$$E[N_{\tau \wedge n} 1_{\{\max_{0 \leq k \leq n} N_k > \lambda\}}] \leq E[N_n 1_{\{\max_{0 \leq k \leq n} N_k > \lambda\}}]. \tag{1.7.3}$$

Noting that $\max_{0 \leq k \leq n} N_k > \lambda$ implies $\tau \leq n$ and $N_\tau > \lambda$, it follows that

$$N_{\tau \wedge n} 1_{\{\max_{0 \leq k \leq n} N_k > \lambda\}} \geq \lambda 1_{\{\max_{0 \leq k \leq n} N_k > \lambda\}}, \tag{1.7.4}$$

combining (1.7.3) and (1.7.4) we conclude that (1.7.1) is valid.

The conclusion (1.7.2) follows from (1.7.1). To see this, fix $0 < \alpha < \infty$ and let us write $f = (\max_{0 \leq k \leq n} |M_k|) \wedge \alpha$ and $g = |M_n|$. Then the inequality (1.7.1) can be rewritten as

$$P(f > \lambda) \leq \frac{1}{\lambda} E[g 1_{\{f > \lambda\}}]. \tag{1.7.5}$$

Now consider the product space $(\Omega, \mathcal{F}, P) \otimes ((0, \alpha], \mathcal{B}((0, \alpha]), \mu)$ where μ is the Lebesgue measure on $(0, \alpha]$. Consider the function $h : \Omega \times (0, \alpha] \mapsto (0, \infty)$ defined by

$$h(\omega, t) = pt^{p-1} 1_{\{t < f(\omega)\}}, \quad (\omega, t) \in \Omega \times (0, \alpha].$$

First, note that

$$\int_{\Omega} [\int_{(0,\alpha]} h(\omega, t)dt]dP(\omega) = \int_{\Omega} [f(\omega)]^p dP(\omega) \tag{1.7.6}$$

$$= E[f^p].$$

On the other hand, using Fubini's theorem in the first and fourth step below and using the estimate (1.7.5) in the second step, we get

$$\int_{\Omega} [\int_{(0,\alpha]} h(\omega, t)dt]dP(\omega) = \int_{(0,\alpha]} [\int_{\Omega} h(\omega, t)dP(\omega)]dt$$

$$= \int_{(0,\alpha]} pt^{(p-1)}P(f > t)dt$$

$$\leq \int_{(0,\alpha]} pt^{(p-1)} \frac{1}{t} E[g 1_{\{f > t\}}]dt$$

$$= \int_{(0,\alpha]} pt^{(p-2)}[\int_{\Omega} g(\omega) 1_{\{f(\omega) > t\}} dP(\omega)]dt \tag{1.7.7}$$

$$= \int_{\Omega} [\int_{(0,\alpha]} pt^{(p-2)} 1_{\{f(\omega) > t\}} dt] g(\omega) dP(\omega)$$

$$= \frac{p}{(p-1)} \int_{\Omega} g(\omega) f^{(p-1)}(\omega) dP(\omega)$$

$$= \frac{p}{(p-1)} E[g f^{(p-1)}].$$

The first step in (1.7.8) below follows from the relations (1.7.6)–(1.7.7) and the next one from Holder's inequality, where $q = \frac{p}{(p-1)}$ so that $\frac{1}{p} + \frac{1}{q} = 1$:

$$E[f^p] \leq \frac{p}{(p-1)} E[g f^{(p-1)}]$$

$$= \frac{p}{(p-1)} (E[g^p])^{\frac{1}{p}} (E[f^{(p-1)q}])^{\frac{1}{q}} \tag{1.7.8}$$

$$= \frac{p}{(p-1)} (E[g^p])^{\frac{1}{p}} (E[f^p])^{\frac{1}{q}}.$$

Since f is bounded, (1.7.8) implies

$$E[f^p]^{(1-\frac{1}{q})} \leq \frac{p}{(p-1)} (E[g^p])^{\frac{1}{p}}$$

which in turn implies (recalling definitions of f, g), writing $C_p = (\frac{p}{(p-1)})^p$

$$E[((\max_{0 \leq k \leq n} |M_k|) \wedge \alpha)^p] \leq C_p E[|M_n|^p]. \qquad (1.7.9)$$

Since (1.7.9) holds for all α, taking limit as $\alpha \uparrow \infty$ (via integers) and using monotone convergence theorem, we get that (1.7.2) is true. $\square$

1.8 Martingale Convergence Theorem

Martingale convergence theorem is one of the main results on martingales. We begin this section with an upcrossings inequality—a key step in its proof. Let $\{a_n : 1 \leq n \leq m\}$ be a sequence of real numbers and $\alpha < \beta$ be real numbers. Let s_k, t_k be defined (inductively) as follows: $s_0 = 0, t_0 = 0$, and for $k = 1, 2, \ldots m$

$$s_k = \inf\{n > t_{k-1} : a_n \leq \alpha\}, \quad t_k = \inf\{n \geq s_k : a_n \geq \beta\}.$$

Recall our convention—infimum of an empty set is taken to be ∞. It is easy to see that if $t_k = j < \infty$, then

$$0 \leq s_1 < t_1 < \ldots < s_k < t_k = j < \infty \qquad (1.8.1)$$

and for $i \leq j$

$$a_{s_i} \leq \alpha, \ a_{t_i} \geq \beta. \qquad (1.8.2)$$

We define

$$U_m(\{a_j : 1 \leq j \leq m\}, \alpha, \beta) = \max\{k : t_k \leq m\}.$$

$U_m(\{a_j : 1 \leq j \leq m\}, \alpha, \beta)$ is the number of upcrossings of the interval (α, β) by the sequence $\{a_j : 1 \leq j \leq m\}$. Eqs. (1.8.1) and (1.8.2) together also imply $t_k > (2k - 1)$ and also writing $c_j = \max(\alpha, a_j)$ that

$$\sum_{j=1}^{m} (c_{t_j \wedge m} - c_{s_j \wedge m}) \geq (\beta - \alpha) U_m(\{a_j\}, \alpha, \beta). \qquad (1.8.3)$$

This inequality follows because each completed upcrossings contributes at least $\beta - \alpha$ to the sum, one term could be non-negative and rest of the terms are zero.

Lemma 1.27 *For a sequence* $\{a_n : n \geq 1\}$ *of real numbers,*

$$\liminf_{n \to \infty} a_n = \limsup_{n \to \infty} a_n \qquad (1.8.4)$$

if and only if

$$\lim_{m \to \infty} U_m(\{a_j : 1 \le j \le m\}, \alpha, \beta) < \infty, \quad \forall \alpha < \beta, \ \alpha, \beta \text{ rationals.} \tag{1.8.5}$$

Proof If $\liminf_{n \to \infty} a_n < \alpha < \beta < \limsup_{n \to \infty} a_n$ then

$$\lim_{m \to \infty} U_m(\{a_j : 1 \le j \le m\}, \alpha, \beta) = \infty.$$

Thus (1.8.5) implies (1.8.4). The other implication follows easily.   □

It follows that if (1.8.5) holds, then $\lim_{n \to \infty} a_n$ exists in $\bar{\mathbb{R}} = \mathbb{R} \cup \{-\infty, \infty\}$. The next result gives an estimate on expected value of

$$U_m(\{X_j : 1 \le j \le m\}, \alpha, \beta)$$

for a submartingale X.

Theorem 1.28 (Doob's upcrossings inequality) . *Let X be a submartingale. Then for $\alpha < \beta$*

$$\mathsf{E}[U_m(\{X_j : 1 \le j \le m\}, \alpha, \beta)] \le \frac{\mathsf{E}[|X_m| + |\alpha|]}{\beta - \alpha}. \tag{1.8.6}$$

Proof Fix $\alpha < \beta$ and define $\sigma_0 = 0 = \tau_0$ and for $k \ge 1$

$$\sigma_k = \inf\{n > \tau_{k-1} : X_n \le \alpha\}, \quad \tau_k = \inf\{n \ge \sigma_k : X_n \ge \beta\}.$$

Then for each k, σ_k and τ_k are stopping times. Writing $Y_k = (X_k - \alpha)^+$, we see that Y is also submartingale and as noted in (1.8.3)

$$\sum_{j=1}^{m} (Y_{\tau_j \wedge m} - Y_{\sigma_j \wedge m}) \ge (\beta - \alpha) U_m(\{X_j : 1 \le j \le m\}, \alpha, \beta). \tag{1.8.7}$$

On the other hand, using $0 \le (\sigma_1 \wedge m) \le (\tau_1 \wedge m) \le \ldots \le (\sigma_m \wedge m) \le (\tau_m \wedge m) = m$ we have

$$Y_m - Y_{\sigma_1 \wedge m} = \sum_{j=1}^{m} (Y_{\tau_j \wedge m} - Y_{\sigma_j \wedge m}) + \sum_{j=1}^{m-1} (Y_{\sigma_{j+1} \wedge m} - Y_{\tau_j \wedge m}). \tag{1.8.8}$$

Since $\sigma_{j+1} \wedge m \ge \tau_j \wedge m$ are stopping times and Y is a submartingale, using Corollary 1.24 we have

$$\mathsf{E}[Y_{\sigma_{j+1} \wedge m} - Y_{\tau_j \wedge m}] \ge 0. \tag{1.8.9}$$

Putting together (1.8.7), (1.8.8) and (1.8.9) we get

$$\mathsf{E}[Y_m - Y_{\sigma_1 \wedge m}] \ge \mathsf{E}[(\beta - \alpha) U_m(\{X_j : 1 \le j \le m\}, \alpha, \beta)]. \tag{1.8.10}$$

Since $Y_k \ge 0$ we get

$$E[U_m(\{X_j : 1 \leq j \leq m\}, \alpha, \beta)] \leq \frac{E[Y_m]}{\beta - \alpha}. \qquad (1.8.11)$$

The inequality (1.8.6) now follows using $Y_m = (X_m - \alpha)^+ \leq |X_m| + |\alpha|$.   □

We recall the notion of uniform integrability of a class of random variables and related results.

A collection $\{Z_\alpha : \alpha \in \Delta\}$ of random variables is said to be *uniformly integrable* if

$$\lim_{K \to \infty} [\sup_{\alpha \in \Delta} E[|Z_\alpha| 1_{\{|Z_\alpha| \geq K\}}]] = 0. \qquad (1.8.12)$$

Here are some exercises on uniform integrability.

Exercise 1.29 If $\{X_n : n \geq 1\}$ is uniformly integrable and $|Y_n| \leq |X_n|$ for each n then show that $\{Y_n : n \geq 1\}$ is also uniformly integrable.

Exercise 1.30 Let $\{Z_\alpha : \alpha \in \Delta\}$ be such that for some $p > 1$

$$\sup_\alpha E[|Z_\alpha|^p] < \infty.$$

show that $\{Z_\alpha : \alpha \in \Delta\}$ is uniformly integrable.

Exercise 1.31 Let $\{Z_\alpha : \alpha \in \Delta\}$ be uniformly integrable. Show that

(i) $\sup_\alpha E[|Z_\alpha|] < \infty$.
(ii) $\forall \epsilon > 0 \; \exists \delta > 0$ such that $P(A) < \delta$ implies

$$E[1_A |Z_\alpha|] < \epsilon \qquad (1.8.13)$$

HINT: For (ii), observe that for any $K > 0$,

$$E[1_A |Z_\alpha|] \leq K P(A) + E[|Z_\alpha| 1_{\{|Z_\alpha| \geq K\}}].$$

Exercise 1.32 Show that $\{Z_\alpha : \alpha \in \Delta\}$ satisfies (i), (ii) in Exercise 1.31 if and only if $\{Z_\alpha : \alpha \in \Delta\}$ is uniformly integrable.

Exercise 1.33 Suppose $\{X_\alpha : \alpha \in \Delta\}$ and $\{Y_\alpha : \alpha \in \Delta\}$ are uniformly integrable and for each $\alpha \in \Delta$, let $Z_\alpha = X_\alpha + Y_\alpha$. Show that $\{Z_\alpha : \alpha \in \Delta\}$ is also uniformly integrable.

The following result on uniform integrability is standard.

Lemma 1.34 *Let* $Z, Z_n \in \mathbb{L}^1(P)$ *for* $n \geq 1$.

(i) Z_n *converges in* $\mathbb{L}^1(P)$ *to* Z *if and only if it converges to* Z *in probability and* $\{Z_n\}$ *is uniformly integrable.*

(ii) Let $\{\mathcal{G}_\alpha : \alpha \in \Delta\}$ be a collection of sub-σ-fields of $\mathcal{F}$. Then

$$\{\mathsf{E}[Z \mid \mathcal{G}_\alpha] : \alpha \in \Delta\} \text{ is uniformly integrable.}$$

Exercise 1.35 Prove Lemma 1.34. For (i), use Exercise 1.32.

We are now in a position to prove the basic martingale convergence theorem.

Theorem 1.36 (Martingale Convergence Theorem) *Let $\{X_n : n \geq 0\}$ be a sub-martingale such that*

$$\sup_n \mathsf{E}[|X_n|] = K_1 < \infty. \tag{1.8.14}$$

Then the sequence of random variables X_n converges a.e. to a random variable ξ with $\mathsf{E}[|\xi|] < \infty$. Further if $\{X_n\}$ is uniformly integrable, then X_n converges to ξ in $\mathbb{L}^1(\mathsf{P})$. If $\{X_n\}$ is a martingale or a positive submartingale and if for some p, $1 < p < \infty$

$$\sup_n \mathsf{E}[|X_n|^p] = K_p < \infty, \tag{1.8.15}$$

then X_n converges to ξ in $\mathbb{L}^p(\mathsf{P})$.

Proof The upcrossings inequality Theorem 1.28 gives

$$\mathsf{E}[U_m(\{X_j\}, \alpha, \beta)] \leq \frac{K_1 + |\alpha|}{\beta - \alpha}$$

for any $\alpha < \beta$ and hence by monotone convergence theorem

$$\mathsf{E}[\sup_{m \geq 1} U_m(\{X_j\}, \alpha, \beta)] < \infty.$$

Let $N_{\alpha\beta} = \{\sup_{m \geq 1} U_m(\{X_j\}, \alpha, \beta) = \infty\}$. Then $\mathsf{P}(N_{\alpha\beta}) = 0$ and hence if

$$N^* = \cup\{N_{\alpha\beta} : \alpha < \beta, \ \alpha, \beta \text{ rationals}\}$$

then $\mathsf{P}(N^*) = 0$. Clearly, for $\omega \notin N^*$ one has

$$\sup_{m \geq 1} U_m(\{X_j(\omega)\}, \alpha, \beta) < \infty \ \forall \alpha < \beta, \ \alpha, \beta \text{ rationals.}$$

Hence by Lemma 1.27, for $\omega \notin N^*$

$$\xi^*(\omega) = \liminf_{n \to \infty} X_n(\omega) = \limsup_{n \to \infty} X_n(\omega).$$

Defining $\xi^*(\omega) = 0$ for $\omega \in N^*$, by Fatou's lemma we get

$$\mathsf{E}[|\xi^*|] < \infty$$

so that $P(\xi^* \in \mathbb{R}) = 1$. Then defining $\xi(\omega) = \xi^*(\omega)1_{\{|\xi^*(\omega)|<\infty\}}$ we get

$$X_n \to \xi \, a.e.$$

If $\{X_n\}$ is uniformly integrable, the *a.e.* convergence also implies $\mathbb{L}^1(P)$ convergence.

If $\{X_n\}$ is a martingale or a positive submartingale, then by Doob's maximal inequality,

$$E[(\max_{1 \le k \le n} |X_k|)^p] \le C_p E[|X_n|^p] \le C_p K_p < \infty$$

and hence by monotone convergence theorem $Z = (\sup_{k\ge1} |X_k|)^p$ is integrable. Now the convergence in $\mathbb{L}^p(P)$ follows from the dominated convergence theorem. $\square$

Theorem 1.37 *Let $\{\mathcal{F}_m\}$ be an increasing family of sub-σ-fields of $\mathcal{F}$, and let $\mathcal{F}_\infty = \sigma(\cup_{m=1}^\infty \mathcal{F}_m)$. Let $Z \in \mathbb{L}^1(P)$ and for $1 \le n < \infty$, let*

$$Z_n = E[Z \mid \mathcal{F}_n].$$

Then $Z_n \to Z^* = E[Z \mid \mathcal{F}_\infty]$ in $\mathbb{L}^1(P)$ and a.s.

Proof From the definition it is clear that Z is a martingale and it is uniformly integrable by Lemma 1.34. Thus Z_n converges in $\mathbb{L}^1(P)$ and *a.s.* to say Z^*. To complete the proof, we will show that

$$Z^* = E[Z \mid \mathcal{F}_\infty]. \tag{1.8.16}$$

Since each Z_n is $\mathcal{F}_\infty$ measurable and Z^* is limit of $\{Z_n\}$ (*a.s.*), it follows that Z^* is $\mathcal{F}_\infty$ measurable. Fix $m \ge 1$ and $A \in \mathcal{F}_m$. Then

$$E[Z_n 1_A] = E[Z 1_A], \quad \forall n \ge m \tag{1.8.17}$$

since $\mathcal{F}_m \subseteq \mathcal{F}_n$ for $n \ge m$. Taking limit in (1.8.17) and using that $Z_n \to Z^*$ in $\mathbb{L}^1(P)$, we conclude that

$$E[Z^* 1_A] = E[Z 1_A] \tag{1.8.18}$$

$\forall A \in \cup_{n=1}^\infty \mathcal{F}_n$, which is a field that generates the σ-field $\mathcal{F}_\infty$. The monotone class theorem implies that (1.8.18) holds for all $A \in \mathcal{F}_\infty$ and hence (1.8.16) holds. $\square$

The previous result has an analogue when the σ-fields are decreasing. Usually one introduces a reverse martingale (martingale indexed by negative integers) to prove this result. We avoid it by incorporating the same in the proof.

Theorem 1.38 *Let $\{\mathcal{G}_m\}$ be a decreasing family of sub-σ-fields of $\mathcal{F}$, i.e. $\mathcal{G}_m \supseteq \mathcal{G}_{m+1}$ for all $m \ge 1$. Let $\mathcal{G}_\infty = \cap_{m=1}^\infty \mathcal{G}_m$. Let $Z_0 \in \mathbb{L}^1(P)$, and for $1 \le n \le \infty$, let*

$$Z_n = E[Z_0 \mid \mathcal{G}_n].$$

Then $Z_n \to Z_\infty$ in $\mathbb{L}^1(P)$ and a.s.

Proof Fix m. Then $\{(Z_{m-j}, \mathcal{F}_{m-j}) : 1 \leq j \leq m\}$ is a martingale where $\mathcal{G}_0 = \mathcal{F}$ and the upcrossings inequality Theorem 1.28 gives

$$E[U_m(\{Z_{m-j} : 1 \leq j \leq m\}, \alpha, \beta)] \leq \frac{E[|Z_0| + |\alpha|]}{\beta - \alpha}.$$

and hence proceeding as in Theorem 1.36, we can conclude that there exists $N^* \subseteq \Omega$ with $P(N^*) = 0$ and for $\omega \notin N^*$ one has

$$\sup_{m \geq 1} U_m(\{Z_{m-j}(\omega)\}, \alpha, \beta) < \infty \ \forall \alpha < \beta, \ \alpha, \beta \text{ rationals.}$$

Now arguments as in Lemma 1.27 imply that

$$\liminf_{n \to \infty} Z_n(\omega) = \limsup_{n \to \infty} Z_n(\omega).$$

By Jensen's inequality $|Z_n| \leq E[|Z_0| \mid \mathcal{G}_n]$. It follows that $\{Z_n\}$ is uniformly integrable (by Lemma 1.34) and thus Z_n converge *a.e.* and in $L^1(P)$ to a real-valued random variable Z^*. Since Z_n for $n \geq m$ is $\mathcal{G}_m$ measurable, it follows that Z^* is $\mathcal{G}_m$ measurable for every m and hence Z^* is $\mathcal{G}_\infty$ measurable.

Also for all $A \in \mathcal{G}_\infty$, for all $n \geq 1$ we have

$$\int Z_n 1_A dP = \int Z_0 1_A dP$$

since $\mathcal{G}_\infty \subseteq \mathcal{G}_n$. Now $L^1(P)$ convergence of Z_n to Z^* implies that for all $A \in \mathcal{G}_\infty$

$$\int Z^* 1_A dP = \int Z_0 1_A dP.$$

Thus $E[Z_0 \mid \mathcal{G}_\infty] = Z^*$. $\square$

Exercise 1.39 Let $\Omega = [0, 1]$, $\mathcal{F}$ be the Borel σ-field on $[0, 1]$, and let P denote the Lebesgue measure on $(\Omega, \mathcal{F})$. Let Q be another probability measure on $(\Omega, \mathcal{F})$ absolutely continuous with respect to P, i.e. satisfying $P(A) = 0$ implies $Q(A) = 0$. For $n \geq 1$ let

$$X^n(\omega) = \sum_{j=1}^{2^n} 2^n Q((2^{-n}(j-1), 2^{-n} j]) 1_{(2^{-n}(j-1), 2^{-n} j]}(\omega).$$

Let $\mathcal{F}_n$ be the field generated by the intervals $\{(2^{-n}(j-1), 2^{-n} j] : 1 \leq j \leq 2^n\}$. Show that $(X_n, \mathcal{F}_n)$ is a uniformly integrable martingale and thus converges to a random variable ξ in $\mathbb{L}^1(P)$. Show that ξ satisfies

$$\int_A \xi \, dP = Q(A) \quad \forall A \in \mathcal{F}. \tag{1.8.19}$$

HINT: To show uniform integrability of $\{X_n : n \geq 1\}$ use *Exercise* 1.32 along with the fact that absolute continuity of Q w.r.t. P implies that for all $\epsilon > 0$, $\exists \delta > 0$ such that $P(A) < \delta$ implies $Q(A) < \epsilon$.

Exercise 1.40 Let $\mathcal{F}$ be a countably generated σ-field on a set Ω, i.e. there exists a sequence of sets $\{B_n : n \geq 1\}$ such that $\mathcal{F} = \sigma(\{B_n : n \geq 1\})$. Let P, Q be probability measures on $(\Omega, \mathcal{F})$ such that Q absolutely continuous with respect to P. For $n \geq 1$, let $\mathcal{F}_n = \sigma(\{B_k : k \leq n\})$. Then show the following.

(i) For each $n \geq 1$, $\exists$ a partition $\{C_1, C_2 \ldots, C_{k_n}\}$ of Ω such that

$$\mathcal{F}_n = \sigma(C_1, C_2 \ldots, C_{k_n}).$$

(ii) For $n \geq 1$ let

$$X_n(\omega) = \sum_{j=1}^{k_n} 1_{\{P(C_j)>0\}} \frac{Q(C_j)}{P(C_j)} 1_{C_j}(\omega). \tag{1.8.20}$$

Show that $(X_n, \mathcal{F}_n)$ is a uniformly integrable martingale on $(\Omega, \mathcal{F}, P)$.

(iii) X_n converges in $L^1(P)$ and also P almost surely to X satisfying

$$\int_A X \, dP = Q(A) \quad \forall A \in \mathcal{F}. \tag{1.8.21}$$

The random variable X in (1.8.21) is called the Radon–Nikodym derivative of Q w.r.t. P.

Exercise 1.41 Let $\mathcal{F}$ be a countably generated σ-field on a set Ω. Let Γ be a non-empty set and $\mathcal{A}$ be a σ-field on Γ. For each $\alpha \in \Gamma$ let P_α and Q_α be probability measures on $(\Omega, \mathcal{F})$ such that Q_α is absolutely continuous with respect to P_α. Suppose that for each $A \in \mathcal{F}$, $\alpha \mapsto P_\alpha(A)$ and $\alpha \mapsto Q_\alpha(A)$ are measurable. Show that there exists $\xi : \Omega \times \Gamma \mapsto [0, \infty)$ such that ξ is measurable w.r.t. $\mathcal{F} \otimes \mathcal{A}$ and

$$\int_A \xi(\cdot, \alpha) \, dP_\alpha = Q_\alpha(A) \quad \forall A \in \mathcal{F}, \ \forall \alpha \in \Gamma \tag{1.8.22}$$

1.9 Square Integrable Martingales

Martingales M such that

$$E[|M_n|^2] < \infty, \quad n \geq 0 \tag{1.9.1}$$

are called square integrable martingales, and they play a special role in the theory of stochastic integration as we will see later. Let us note that for $p = 2$, the constant C_p appearing in (1.7.2) equals 4. Thus for a square integrable martingale M, we have

$$\mathsf{E}[(\max_{0 \le k \le n} |M_k|)^2] \le 4\mathsf{E}[|M_n|^2]. \tag{1.9.2}$$

As seen earlier, $X_n = M_n^2$ is a submartingale and the compensator of X—namely the predictable increasing process A such that $X_n - A_n$ is a martingale, is given by $A_0 = 0$ and for $n \ge 1$,

$$A_n = \sum_{k=1}^{n} \mathsf{E}[(X_k - X_{k-1}) \mid \mathcal{F}_{k-1}].$$

The compensator A is denoted as $\langle M, M \rangle$. Using

$$\begin{aligned} \mathsf{E}[(M_k - M_{k-1})^2 \mid \mathcal{F}_{k-1}] &= \mathsf{E}[(M_k^2 - 2M_k M_{k-1} + M_{k-1}^2) \mid \mathcal{F}_{k-1}] \\ &= \mathsf{E}[(M_k^2 - M_{k-1}^2) \mid \mathcal{F}_{k-1}] \end{aligned} \tag{1.9.3}$$

it follows that the compensator can be described as

$$\langle M, M \rangle_n = \sum_{k=1}^{n} \mathsf{E}[(M_k - M_{k-1})^2 \mid \mathcal{F}_{k-1}] \tag{1.9.4}$$

Thus $\langle M, M \rangle$ is the unique predictable increasing process with $\langle M, M \rangle_0 = 0$ such that $M_n^2 - \langle M, M \rangle_n$ is a martingale. Let us also define another increasing process $[M, M]$ associated with a martingale M: $[M, M]_0 = 0$ and

$$[M, M]_n = \sum_{k=1}^{n} (M_k - M_{k-1})^2. \tag{1.9.5}$$

The process $[M, M]$ is called the quadratic variation of M, and the process $\langle M, M \rangle$ is called the predictable quadratic variation of M. It can be easily checked that

$$M_n^2 - [M, M]_n = M_0^2 + 2 \sum_{k=1}^{n} M_{k-1}(M_k - M_{k-1})$$

and hence using Theorem 1.15 it follows that $M_n^2 - [M, M]_n$ is also a martingale. If $M_0 = 0$, then it follows that

$$\mathsf{E}[M_n^2] = \mathsf{E}[\langle M, M \rangle_n] = \mathsf{E}[[M, M]_n]. \tag{1.9.6}$$

We have already seen that if U is a bounded predictable sequence then the transform Z defined by (1.5.2)

$$Z_n = \sum_{k=1}^{n} U_k(M_k - M_{k-1})$$

is itself a martingale. The next result includes an estimate on the $\mathbb{L}^2(P)$ norm of Z_n.

Theorem 1.42 *Let M be a square integrable martingale and U be a bounded predictable process. Let $Z_0 = 0$ and for $n \geq 1$, let*

$$Z_n = \sum_{k=1}^{n} U_k(M_k - M_{k-1}).$$

Then Z is itself a square integrable martingale and further

$$\langle Z, Z \rangle_n = \sum_{k=1}^{n} U_k^2(\langle M, M \rangle_k - \langle M, M \rangle_{k-1}) \tag{1.9.7}$$

$$[Z, Z]_n = \sum_{k=1}^{n} U_k^2([M, M]_k - [M, M]_{k-1}). \tag{1.9.8}$$

As a consequence

$$\mathsf{E}[\max_{1 \leq n \leq N} |\sum_{k=1}^{n} U_k(M_k - M_{k-1})|^2] \leq 4\mathsf{E}[\sum_{k=1}^{N} U_k^2([M, M]_k - [M, M]_{k-1})]. \tag{1.9.9}$$

Proof Since U is bounded and predictable, Z_n is square integrable for each n. That Z is square integrable martingale follows from Theorem 1.15. Since

$$(Z_k - Z_{k-1})^2 = U_k^2(M_k - M_{k-1})^2 \tag{1.9.10}$$

the relation (1.9.8) follows from the definition of the quadratic variation. Further, taking conditional expectation given $\mathcal{F}_{k-1}$ in (1.9.10) and using that U_k is $\mathcal{F}_{k-1}$ measurable, one concludes

$$\begin{aligned}
\mathsf{E}[(Z_k - Z_{k-1})^2 \mid \mathcal{F}_{k-1}] &= U_k^2 \mathsf{E}[(M_k - M_{k-1})^2 \mid \mathcal{F}_{k-1}] \\
&= U_k^2(\langle M, M \rangle_k - \langle M, M \rangle_{k-1})
\end{aligned} \tag{1.9.11}$$

by (1.9.4). This proves (1.9.7). Now (1.9.9) follows from (1.9.8) and Doob's maximal inequality, Theorem 1.26. $\qquad\square$

Corollary 1.43 *For a square integrable martingale M and a predictable process U bounded by 1, defining $Z_n = \sum_{k=1}^{n} U_k(M_k - M_{k-1})$ we have*

$$\mathsf{E}[\max_{1 \leq n \leq N} |\sum_{k=1}^{n} U_k(M_k - M_{k-1})|^2] \leq 4\mathsf{E}[[M, M]_N]. \tag{1.9.12}$$

Further,

$$[Z, Z]_n \leq [M, M]_n, \ n \geq 1. \tag{1.9.13}$$

The inequality (1.9.12) plays a very important role in the theory of stochastic integration as later chapters will reveal. We will obtain another estimate due to Burkholder [6] on martingale transform which is valid even when the martingale is not square integrable.

Theorem 1.44 (Burkholder's inequality) *Let M be a martingale and U be a bounded predictable process, bounded by 1. Then*

$$P(\max_{1 \leq n \leq N} |\sum_{k=1}^{n} U_k(M_k - M_{k-1})| \geq \lambda) \leq \frac{9}{\lambda} E[|M_N|]. \qquad (1.9.14)$$

Proof Let

$$\tau = \inf\{n \geq 0 : |M_n| \geq \frac{\lambda}{4}\} \wedge (N+1)$$

so that τ is a stopping time. Since $\{\tau \leq N\} = \{\max_{0 \leq n \leq N} |M_n| \geq \frac{\lambda}{4}\}$, using Doob's maximal inequality (Theorem 1.26) for the positive submartingale $\{|M_k| : 0 \leq k \leq N\}$, we have

$$P(\tau \leq N) \leq \frac{4}{\lambda} E(|M_N|). \qquad (1.9.15)$$

For $0 \leq n \leq N$, let $\xi_n = M_n 1_{\{n < \tau\}}$. By definition of τ, $|\xi_n| \leq \frac{\lambda}{4}$ for $0 \leq n \leq N$. Since on the set $\{\tau = N+1\}$, $\xi_n = M_n$, $\forall n \leq N$, it follows that

$$P(\max_{1 \leq n \leq N} |\sum_{k=1}^{n} U_k(M_k - M_{k-1})| \geq \lambda)$$

$$\leq P(\max_{1 \leq n \leq N} |\sum_{k=1}^{n} U_k(\xi_k - \xi_{k-1})| \geq \lambda) + P(\tau \leq N) \qquad (1.9.16)$$

$$\leq P(\max_{1 \leq n \leq N} |\sum_{k=1}^{n} U_k(\xi_k - \xi_{k-1})| \geq \lambda) + \frac{4}{\lambda} E(|M_N|).$$

We will now prove that for any predictable sequence of random variables $\{V_n\}$ bounded by 1, we have

$$|E[\sum_{k=1}^{N} V_k(\xi_k - \xi_{k-1})]| \leq E[|M_N|]. \qquad (1.9.17)$$

Let us define $\tilde{M}_k = M_{k \wedge \tau}$ for $k \geq 0$. Since $\tilde{M}$ is a martingale, $E[\sum_{k=1}^{N} V_k(\tilde{M}_k - \tilde{M}_{k-1})] = 0$. Writing $Y_n = \xi_n - \tilde{M}_n$, it follows that

$$E[\sum_{k=1}^{N} V_k(\xi_k - \xi_{k-1})] = E[\sum_{k=1}^{N} V_k(Y_k - Y_{k-1})]. \qquad (1.9.18)$$

Since $Y_n = M_n 1_{\{n < \tau\}} - M_{n \wedge \tau} = -M_{n \wedge \tau} 1_{\{\tau \le n\}} = -M_\tau 1_{\{\tau \le n\}}$, it follows that $Y_k - Y_{k-1} = M_\tau 1_{\{\tau \le (k-1)\}} - M_\tau 1_{\{\tau \le k\}}$ and thus $Y_k - Y_{k-1} = -M_\tau 1_{\{\tau = k\}}$. Using (1.9.18), we get

$$|\mathsf{E}[\sum_{k=1}^{N} V_k(\xi_k - \xi_{k-1})]| = |\mathsf{E}[\sum_{k=1}^{N} V_k(Y_k - Y_{k-1})]|$$

$$\le \mathsf{E}[\sum_{k=1}^{N} |Y_k - Y_{k-1}|] \qquad (1.9.19)$$

$$\le \mathsf{E}[\sum_{k=1}^{N} |M_\tau| 1_{\{\tau = k\}}]$$

$$\le \mathsf{E}[|M_\tau| 1_{\{\tau \le N\}}]$$

$$\le \mathsf{E}[|M_N|]$$

where for the last step we have used that $|M_n|$ is a submartingale. We will decompose $\{\xi_n : 0 \le n \le N\}$ into a martingale $\{R_n : 0 \le n \le N\}$ and a predictable process $\{B_n : 0 \le n \le N\}$ as follows: $B_0 = 0$, $R_0 = \xi_0$ and for $0 < n \le N$

$$B_n = B_{n-1} + (\mathsf{E}[\xi_n \mid \mathcal{F}_{n-1}] - \xi_{n-1})$$

$$R_n = \xi_n - B_n.$$

By construction, $\{R_n\}$ is a martingale and $\{B_n\}$ is predictable. Note that

$$B_n - B_{n-1} = \mathsf{E}[\xi_n \mid \mathcal{F}_{n-1}] - \xi_{n-1}$$

and hence

$$R_n - R_{n-1} = (\xi_n - \xi_{n-1}) - \mathsf{E}[\xi_n - \xi_{n-1} \mid \mathcal{F}_{n-1}].$$

As a consequence

$$\mathsf{E}[(R_n - R_{n-1})^2] \le \mathsf{E}[(\xi_n - \xi_{n-1})^2]. \qquad (1.9.20)$$

For $x \in \mathbb{R}$, let $\mathsf{sgn}(x) = 1$ for $x \ge 0$ and $\mathsf{sgn}(x) = -1$ for $x < 0$, so that $|x| = \mathsf{sgn}(x)x$. Now taking $V_k = \mathsf{sgn}(B_k - B_{k-1})$ and noting that V_k is $\mathcal{G}_{k-1}$ measurable, we have $\mathsf{E}[V_k(B_k - B_{k-1})] = \mathsf{E}[V_k(\xi_k - \xi_{k-1})]$ (since $\xi = R + B$ and R is a martingale) and hence

$$\mathsf{E}[\max_{1 \le n \le N} |\sum_{k=1}^{n} U_k(B_k - B_{k-1})|] \le \mathsf{E}[\sum_{k=1}^{N} |(B_k - B_{k-1})|]$$

$$= \mathsf{E}[\sum_{k=1}^{N} V_k(B_k - B_{k-1})] \qquad (1.9.21)$$

$$= E[\sum_{k=1}^{N} V_k(\xi_k - \xi_{k-1})]$$

$$\leq E[|M_N|]$$

where we have used (1.9.19). Thus

$$P(\max_{1 \leq n \leq N} |\sum_{k=1}^{n} U_k(B_k - B_{k-1})| \geq \frac{\lambda}{2}) \leq \frac{2}{\lambda} E[|M_N|]. \qquad (1.9.22)$$

Since R is a martingale and U is predictable, $X_n = \sum_{k=1}^{n} U_k(R_k - R_{k-1})$ is a martingale and hence X_n^2 is a submartingale. Thus

$$P(\max_{1 \leq n \leq N} |\sum_{k=1}^{n} U_k(R_k - R_{k-1})| \geq \frac{\lambda}{2}) = P(\max_{1 \leq n \leq N} |\sum_{k=1}^{n} U_k(R_k - R_{k-1})|^2 \geq \frac{\lambda^2}{4})$$

$$\leq \frac{4}{\lambda^2} E[X_N^2].$$

$$(1.9.23)$$

Since X is a martingale transform of the martingale R, with U bounded by one, we have

$$E[X_N^2] = E[[X, X]_N]$$

$$\leq E[[R, R]_N]$$

$$\leq E[\sum_{k=1}^{N} (R_k - R_{k-1})^2] \qquad (1.9.24)$$

$$\leq E[\sum_{k=1}^{N} (\xi_k - \xi_{k-1})^2],$$

where we have used (1.9.20). The estimates (1.9.23) and (1.9.24) together yield

$$P(\max_{1 \leq n \leq N} |\sum_{k=1}^{n} U_k(R_k - R_{k-1})| \geq \frac{\lambda}{2}) \leq \frac{4}{\lambda^2} E[\sum_{k=1}^{N} (\xi_k - \xi_{k-1})^2]. \qquad (1.9.25)$$

Using the identity $(y - x)^2 = y^2 - x^2 - 2x(y - x)$ for $y = \xi_k$ and $x = \xi_{k-1}$ and summing over k, one gets

$$\sum_{k=1}^{N} (\xi_k - \xi_{k-1})^2 = \xi_N^2 - \xi_0^2 - 2\sum_{k=1}^{N} \xi_{k-1}(\xi_k - \xi_{k-1})$$

$$(1.9.26)$$

$$\leq \frac{\lambda}{4}|M_N| + 2\frac{\lambda}{4}\sum_{k=1}^{N} W_k(\xi_k - \xi_{k-1})$$

where $W_k = -\frac{4}{\lambda}\xi_{k-1}$. Here we have used $\xi_N^2 = M_N^2 1_{\{N < \tau\}}$. Note that $|W_k| \le 1$ and W_k is $\mathcal{G}_{k-1}$ measurable. Using (1.9.19) and (1.9.26), we get

$$\mathsf{E}[\sum_{k=1}^{N}(\xi_k - \xi_{k-1})^2] \le \frac{3\lambda}{4}\mathsf{E}[|M_N|]. \tag{1.9.27}$$

Now (1.9.25) and (1.9.27) together yield

$$\mathsf{P}(\max_{1 \le n \le N}|\sum_{k=1}^{n} U_k(R_k - R_{k-1})| \ge \frac{\lambda}{2}) \le \frac{3}{\lambda}\mathsf{E}[|M_N|]. \tag{1.9.28}$$

Since $\xi_k = R_k + B_k$, (1.9.22) and (1.9.28) give

$$\mathsf{P}(\max_{1 \le n \le N}|\sum_{k=1}^{n} U_k(\xi_k - \xi_{k-1})| \ge \lambda) \le \frac{5}{\lambda}\mathsf{E}[|M_N|]. \tag{1.9.29}$$

Finally, (1.9.16) and (1.9.29) together imply the required estimate (1.9.14). $\square$

1.10   Burkholder–Davis–Gundy Inequality

If M is a square integrable martingale with $M_0 = 0$, then we have

$$\mathsf{E}[M_n^2] = \mathsf{E}[\langle M, M \rangle_n] = \mathsf{E}[[M, M]_n].$$

And Doob's maximal inequality (1.9.2) yields

$$\begin{aligned}
\mathsf{E}[[M, M]_n] &= \mathsf{E}[M_n^2] \\
&\le \mathsf{E}[(\max_{1 \le k \le n}|M_k|)^2] \\
&\le 4\mathsf{E}[[M, M]_n].
\end{aligned} \tag{1.10.1}$$

Burkholder and Gundy proved that indeed for $1 < p < \infty$, there exist constants c_p^1, c_p^2 such that

$$c_p^1\mathsf{E}[([M, M]_n)^{\frac{p}{2}}] \le \mathsf{E}[(\max_{1 \le k \le n}|M_k|)^p] \le c_p^2\mathsf{E}[([M, M]_n)^{\frac{p}{2}}].$$

Note that for $p = 2$, this reduces to (1.10.1). Davis went on to prove the above inequality for $p = 1$. This case plays an important role in the result on integral representation of martingales that we will later consider. Hence we include a proof for the case $p = 1$—essentially this is the proof given by Davis [14]

Theorem 1.45 *Let M be a martingale with $M_0 = 0$. Then there exist universal constants c^1, c^2 such that for all $N \geq 1$*

$$c^1 E[([M, M]_N)^{\frac{1}{2}}] \leq E[\max_{1 \leq k \leq N} |M_k|] \leq c^2 E[([M, M]_N)^{\frac{1}{2}}]. \qquad (1.10.2)$$

Proof Let us define for $n \geq 1$

$$U_n^1 = \max_{1 \leq k \leq n} |M_k|$$

$$U_n^2 = (\sum_{k=1}^{n} (M_k - M_{k-1})^2)^{\frac{1}{2}}$$

$$W_n = \max_{1 \leq k \leq n} |M_k - M_{k-1}|$$

and $V_n^1 = U_n^2$, $V_n^2 = U_n^1$. The reason for unusual notation is that we will prove

$$E[U_n^t] \leq 130 E[V_n^t], \quad t = 1, 2 \qquad (1.10.3)$$

and this will prove both the inequalities in (1.10.2). Note that by definition, for all $n \geq 1$,

$$W_n \leq 2V_n^t, \quad W_n \leq 2U_n^t \quad t = 1, 2. \qquad (1.10.4)$$

We begin by decomposing M as $M_n = X_n + Y_n$, where X and Y are also martingales defined as follows. For $n \geq 1$ let

$$R_n = (M_n - M_{n-1})1_{\{|(M_n - M_{n-1})| > 2W_{n-1}\}}$$

$$S_n = E[R_n \mid \mathcal{F}_{n-1}]$$

$$X_n = \sum_{k=1}^{n} (R_k - S_k)$$

$$T_n = (M_n - M_{n-1})1_{\{|(M_n - M_{n-1})| \leq 2W_{n-1}\}}$$

$$Y_n = M_n - X_n$$

and $X_0 = Y_0 = 0$. Let us note that X is a martingale by definition and hence so is Y. Also that

$$Y_n = \sum_{k=1}^{n} (T_k + S_k).$$

If $|R_n| > 0$ then $|(M_n - M_{n-1})| > 2W_{n-1}$ and so $W_n = |(M_n - M_{n-1})|$ and, in turn, $|R_n| = W_n < 2(W_n - W_{n-1})$. Thus (noting $W_n \geq W_{n-1}$ for all n)

$$\sum_{k=1}^{n} |R_k| \leq 2W_n \qquad (1.10.5)$$

and using that $E[|S_n|] \leq E[|R_n|]$, it also follows that

$$\sum_{k=1}^{n} \mathsf{E}[\,|S_k|\,] \leq 2\mathsf{E}(W_n).$$ (1.10.6)

Thus (using (1.10.4)) we have

$$\mathsf{E}[\sum_{k=1}^{n}|R_k| + \sum_{k=1}^{n}|S_k|\,] \leq 4\mathsf{E}(W_n) \leq 8\mathsf{E}[V_n^t], \quad t = 1, 2.$$ (1.10.7)

Since $X_k - X_{k-1} = R_k - S_k$, (1.10.7) gives us for all $n \geq 1$,

$$\sum_{k=1}^{n} \mathsf{E}[\,|X_k - X_{k-1}|\,] \leq 8\mathsf{E}[V_n^t], \quad t = 1, 2.$$ (1.10.8)

Let us define for $n \geq 1$,

$$A_n^1 = \max_{1 \leq k \leq n} |X_k|$$

$$A_n^2 = (\sum_{k=1}^{n}(X_k - X_{k-1})^2)^{\frac{1}{2}}$$

$$F_n^1 = \max_{1 \leq k \leq n} |Y_k|$$

$$F_n^2 = (\sum_{k=1}^{n}(Y_k - Y_{k-1})^2)^{\frac{1}{2}}$$

and $B_n^1 = A_n^2, B_n^2 = A_n^1, G_n^1 = F_n^2, G_n^2 = F_n^1$. Since for $1 \leq j \leq n, |X_j| \leq \sum_{k=1}^{n}|R_k| + \sum_{k=1}^{n}|S_k|$, the estimates (1.10.7)–(1.10.8) immediately give us

$$\mathsf{E}[A_n^t] \leq 8\mathsf{E}[V_n^t], \quad \mathsf{E}[B_n^t] \leq 8\mathsf{E}[V_n^t], \quad t = 1, 2.$$ (1.10.9)

Also we note here that for all $n \geq 1, t = 1, 2$

$$U_n^t \leq A_n^t + F_n^t$$ (1.10.10)

and

$$G_n^t \leq B_n^t + V_n^t.$$ (1.10.11)

Thus, for all $n \geq 1, t = 1, 2$ using (1.10.9) we conclude

$$\mathsf{E}[U_n^t] \leq 8\mathsf{E}[V_n^t] + \mathsf{E}[F_n^t].$$ (1.10.12)

Now using Fubini's theorem (as used in proof of Theorem 1.26) it follows that

$$\mathsf{E}[F_n^t] = \int_0^\infty \mathsf{P}(F_n^t > x)\,dx.$$ (1.10.13)

We now will estimate $P(F_N^t > x)$. For this we fix $t \in \{1, 2\}$, and for $x \in (0, \infty)$, define a stopping time

$$\sigma_x = \inf\{n \geq 1 : V_n^t > x \text{ or } G_n^t > x \text{ or } |S_{n+1}| > x\} \wedge N.$$

Since S_{n+1} is $\mathcal{F}_n$ measurable, it follows that σ_x is a stopping time. If $\sigma_x < N$ then either $V_N^t > x$ or $G_N^t > x$ or $\sum_{k=1}^N |S_k| > x$. Thus

$$P(\sigma_x < N) \leq P(V_N^t > x) + P(G_N^t > x) + P(\sum_{k=1}^N |S_k| > x)$$

and hence

$$\int_0^\infty P(\sigma_x < N)dx \leq \int_0^\infty P(V_N^t > x)dx + \int_0^\infty P(G_N^t > x)dx$$
$$+ \int_0^\infty P(\sum_{k=1}^N |S_n| > x)dx$$
$$= E[V_N^t] + E[G_N^t] + E[\sum_{k=1}^N |S_n|]$$
$$\leq E[V_N^t] + E[V_N^t] + E[B_N^t] + 2E[W_N]$$

where we have used (1.10.11) and (1.10.6) in the last step. Now using (1.10.9), (1.10.4) gives us

$$\int_0^\infty P(\sigma_x < N)dx \leq 14E[V_N^t]. \tag{1.10.14}$$

Note that $S_{\sigma_x} \leq x$ and hence

$$|G_{\sigma_x}^t - G_{\sigma_x-1}^t| \leq |T_{\sigma_x} + S_{\sigma_x}|$$
$$\leq 2W_{\sigma_x-1} + x$$
$$\leq 4V_{\sigma_x-1}^t + x$$
$$\leq 5x$$

and as a result, using $G_{\sigma_x-1}^t \leq x$, we conclude

$$G_{\sigma_x}^t \leq 6x.$$

Hence in view of (1.10.11), we have

$$G_{\sigma_x}^t \leq \min\{B_N^t + V_N^t, 6x\}. \tag{1.10.15}$$

Let $Z_n = Y_{n \wedge \sigma_x}$. Then Z is a martingale with $Z_0 = 0$ and

$$E[Z_N^2] = \sum_{k=1}^{N} E[(Z_k - Z_{k-1})^2] = E[(G_{\sigma_x}^1)^2].$$

Further, Z_n^2 is a positive submartingale and hence

$$P(F_{\sigma_x}^1 > x) = P(\max_{1 \le k \le N} |Z_k| > x)$$

$$\le \frac{1}{x^2} E[Z_N^2]$$

$$= \frac{1}{x^2} E[(G_{\sigma_x}^1)^2].$$

On the other hand

$$P(F_{\sigma_x}^2 > x) = P((\sum_{k=1}^{N}(Z_k - Z_{k-1})^2)^{\frac{1}{2}} > x)$$

$$\le \frac{1}{x^2} \sum_{k=1}^{N} E[(Z_k - Z_{k-1})^2]$$

$$= \frac{1}{x^2} E[(Z_N)^2]$$

$$\le \frac{1}{x^2} E[(G_{\sigma_x}^2)^2].$$

Thus, we have for $t = 1, 2$, for $n \ge 1$

$$P(F_{\sigma_x}^t > x) \le \frac{1}{x^2} E[(G_{\sigma_x}^t)^2]. \tag{1.10.16}$$

Now, writing $Q^t = \frac{1}{6}(B_N^t + V_N^t)$, we have

$$E[F_N^t] = \int_0^\infty P(F_N^t > x)dx$$

$$\le \int_0^\infty P(\sigma_x < N)dx + \int_0^\infty P(F_{\sigma_x}^t > x)dx$$

$$\le 14E[V_N^t] + \int_0^\infty \frac{1}{x^2} E[(G_{\sigma_x}^t)^2]dx$$

$$\le 14E[V_N^t] + \int_0^\infty \frac{1}{x^2} E[(\min\{B_N^t + V_N^t, 6x\})^2]dx \tag{1.10.17}$$

$$\le 14E[V_N^t] + E[\int_0^\infty \frac{36}{x^2} (\min\{Q^t, x\})^2 dx]$$

$$\le 14E[V_N^t] + E[\int_0^{Q^t} \frac{36}{x^2}(x)^2 dx] + E[\int_{Q^t}^\infty \frac{36}{x^2}(Q^t)^2 dx]$$

$$\le 14E[V_N^t] + E[36Q^t] + E[36Q^t].$$

Using the estimate (1.10.9), we have

$$
\begin{aligned}
\mathsf{E}[Q^t] &= \frac{1}{6}(\mathsf{E}[B_N^t] + \mathsf{E}[V_N^t]) \\
&\le \frac{9}{6}\mathsf{E}[V_N^t]
\end{aligned}
\tag{1.10.18}
$$

and putting together (1.10.17)–(1.10.18) we conclude

$$
\mathsf{E}[F_N^t] \le 122\mathsf{E}[V_N^t]
$$

and along with (1.10.12) we finally conclude, for $t = 1, 2$

$$
\mathsf{E}[U_N^t] \le 130\mathsf{E}[V_N^t].
$$

$\square$

Chapter 2
Continuous-Time Processes

In this chapter, we will give definitions, set up notations that will be used in the rest of the book and give some basic results. While some proofs are included, several results are stated without proof. The proofs of these results can be found in standard books on stochastic processes.

2.1 Notations and Basic Facts

E will denote a complete separable metric space, $\mathbb{C}(E)$ will denote the space of real-valued continuous functions on E, $\mathbb{C}_b(E)$ will denote bounded functions in $\mathbb{C}(E)$, and $\mathcal{B}(E)$ will denote the Borel σ-field on E. $\mathbb{R}^d$ will denote the d-dimensional Euclidean space and $\mathsf{L}(m, d)$ will denote the space of $m \times d$ matrices with real entries. For $x \in \mathbb{R}^d$ and $A \in \mathsf{L}(m, d)$, $|x|$ and $\|A\|$ will denote the Euclidean norms of x, A, respectively.

$(\Omega, \mathcal{F}, \mathsf{P})$ will denote a generic probability space, and $\mathbb{B}(\Omega, \mathcal{F})$ will denote the class of real-valued bounded $\mathcal{F}$ measurable functions. When $\Omega = E$ and $\mathcal{F} = \mathcal{B}(E)$, we will write $\mathbb{B}(E)$ for real-valued bounded Borel measurable functions.

Recall that for a collection $\mathcal{A} \subseteq \mathcal{F}$, $\sigma(\mathcal{A})$ will denote the smallest σ-field which contains $\mathcal{A}$ and for a collection $\mathbb{G}$ of measurable functions on $(\Omega, \mathcal{F})$, $\sigma(\mathbb{G})$ will likewise denote the smallest σ-field on Ω with respect to which each function in $\mathbb{G}$ is measurable.

It is well known and easy to prove that for a complete separable metric space E,

$$\sigma(\mathbb{C}_b(E)) = \mathcal{B}(E)$$

For an integer $d \geq 1$, let $\mathbb{C}_d = \mathbb{C}([0, \infty), \mathbb{R}^d)$ with the ucc topology, i.e. uniform convergence on compact subsets of $[0, \infty)$. With this topology, $\mathbb{C}_d$ is itself a complete separable metric space. We will denote a generic element in $\mathbb{C}_d$ by ζ. Denoting the coordinate mappings on $\mathbb{C}_d$ by

© Springer Nature Singapore Pte Ltd. 2018
R. L. Karandikar and B. V. Rao, *Introduction to Stochastic Calculus*,
Indian Statistical Institute Series, https://doi.org/10.1007/978-981-10-8318-1_2

$$\beta_t(\zeta) = \zeta(t), \quad \zeta \in \mathbb{C}_d \text{ and } 0 \le t < \infty$$

it can be shown that

$$\mathcal{B}(\mathbb{C}_d) = \sigma(\beta_t \; : \; 0 \le t < \infty).$$

A function γ from $[0, \infty)$ to $\mathbb{R}^d$ is said to be r.c.l.l. (right continuous with left limits) if γ is right continuous everywhere ($\gamma(t) = \lim_{u \downarrow t} \gamma(u)$ for all $0 \le t < \infty$) and such that the left limit $\gamma(t-) = \lim_{u \uparrow t} \gamma(u)$ exists for all $0 < t < \infty$. We define $\gamma(0-) = 0$ and for $t \ge 0$, $\Delta\gamma(t) = \gamma(t) - \gamma(t-)$.

For an integer $d \ge 1$, let $\mathbb{D}_d = \mathbb{D}([0, \infty), \mathbb{R}^d)$ be the space of all r.c.l.l. functions γ from $[0, \infty)$ to $\mathbb{R}^d$ with the topology of uniform convergence on compact subsets, abbreviated it as ucc. Thus γ^n converges to γ in ucc topology if

$$\sup_{t \le T} |\gamma^n(t) - \gamma(t)| \to 0 \quad \forall T < \infty.$$

Exercise 2.1 Let $\gamma \in \mathbb{D}_d$.

(i) Show that for any $\epsilon > 0$, and $T < \infty$, the set

$$\{t \in [0, T] \; : \; |(\Delta\gamma)(t)| > \epsilon\}$$

is a finite set.

(ii) Show that the set $\{t \in [0, \infty) \; : \; |(\Delta\gamma)(t)| > 0\}$ is a countable set.

(iii) Let $K = \{\gamma(t) : 0 \le t \le T\} \cup \{\gamma(t-) : 0 \le t \le T\}$. Show that K is compact.

The space $\mathbb{D}_d$ is equipped with the σ-field $\sigma(\theta_t \; : \; 0 \le t < \infty)$ where θ_t are coordinate mappings on $\mathbb{D}_d$ defined by

$$\theta_t(\gamma) = \gamma(t), \quad \gamma \in \mathbb{D}_d \text{ and } 0 \le t < \infty.$$

We will denote this σ-field as $\mathcal{B}(\mathbb{D}_d)$. It can be shown that this σ-field is same as the Borel σ-field for the Skorokhod topology (see Ethier and Kurtz [18]). However, we do not need this fact.

An E-valued random variable X defined on a probability space $(\Omega, \mathcal{F}, \mathsf{P})$ is a measurable function from $(\Omega, \mathcal{F})$ to $(E, \mathcal{B}(E))$. For such an X and a function $f \in \mathbb{C}_b(E)$, $\mathsf{E}[f(X)]$ will denote the integral

$$\mathsf{E}[f(X)] = \int_{\Omega} f(X(\omega))d\mathsf{P}(\omega).$$

If there are more than one probability measures under consideration, we will denote it as $\mathsf{E}_{\mathsf{P}}[f(X)]$.

An E-valued stochastic process X is a collection $\{X_t : 0 \le t < \infty\}$ of E-valued random variables. While one can consider families of random variables indexed by sets other than $[0, \infty)$, say $[0, 1]$ or even $[0, 1] \times [0, 1]$, unless stated otherwise we will take the index set to be $[0, \infty)$. Sometimes for notational clarity we will also use $X(t)$ to denote X_t.

From now on unless otherwise stated, a process will mean a continuous-time stochastic process $X = (X_t)$ with $t \in [0, \infty)$ or $t \in [0, T]$. For more details and discussions as well as proofs of statements given without proof in this chapter, see Breiman [5], Ethier and Kurtz [18], Ikeda and Watanabe [24], Jacod [26], Karatzas and Shreve [43], Metivier [50], Meyer [51], Protter [52], Revuz-Yor [53], Stroock and Varadhan [60], Williams [59].

Definition 2.2 Two processes X, Y, defined on the same probability space $(\Omega, \mathcal{F}, \mathsf{P})$ are said to be equal (written as $X = Y$) if

$$\mathsf{P}(X_t = Y_t \text{ for all } t \geq 0) = 1.$$

In other words, $X = Y$ if $\exists \Omega_0 \in \mathcal{F}$ such that $\mathsf{P}(\Omega_0) = 1$ and

$$\forall t \geq 0, \ \forall \omega \in \Omega_0, \ X_t(\omega) = Y_t(\omega).$$

Definition 2.3 A process Y is said to be a version of another process X (written as $X \stackrel{v}{\longleftrightarrow} Y$) if both are defined on the same probability space and if

$$\mathsf{P}(X_t = Y_t) = 1 \quad \text{for all } t \geq 0.$$

It should be noted that in general, $X \stackrel{v}{\longleftrightarrow} Y$ does not imply $X = Y$. Take $\Omega = [0, 1]$ with $\mathcal{F}$ to be the Borel σ-field and P to be the Lebesgue measure. Let $X_t(\omega) = 0$ for all $t, \omega \in [0, 1]$. For $t, \omega \in [0, 1]$, $Y_t(\omega) = 0$ for $\omega \in [0, 1]$, $\omega \neq t$ and $Y_t(\omega) = 1$ for $\omega = t$. Easy to see that $X \stackrel{v}{\longleftrightarrow} Y$ but $\mathsf{P}(X_t = Y_t \text{ for all } t \geq 0) = 0$.

Definition 2.4 Two E-valued processes X, Y are said to have same distribution (written as $X \stackrel{d}{=} Y$), where X is defined on $(\Omega_1, \mathcal{F}_1, \mathsf{P}_1)$ and Y is defined on $(\Omega_2, \mathcal{F}_2, \mathsf{P}_2)$, if for all $m \geq 1, 0 \leq t_1 < t_2 < \ldots < t_m, A_1, A_2, \ldots, A_m \in \mathcal{B}(E)$

$$\mathsf{P}_1(X_{t_1} \in A_1, X_{t_2} \in A_2, \ldots, X_{t_m} \in A_m)$$
$$= \mathsf{P}_2(Y_{t_1} \in A_1, Y_{t_2} \in A_2, \ldots, Y_{t_m} \in A_m).$$

Thus, two processes have same distribution if their finite-dimensional distributions are the same. It is easy to see that if $X \stackrel{v}{\longleftrightarrow} Y$ then $X \stackrel{d}{=} Y$.

Definition 2.5 Let X be an E-valued process and D be a Borel subset of E. X is said to be D-valued if

$$\mathsf{P}(X_t \in D \ \forall t) = 1.$$

Definition 2.6 An E-valued process X is said to be a *continuous process* (or a process with continuous paths) if for all $\omega \in \Omega$, the *path* $t \mapsto X_t(\omega)$ is continuous.

Definition 2.7 An E-valued process X is said to be an *r.c.l.l. process* (or a process with right continuous paths with left limits) if for all $\omega \in \Omega$, the *path* $t \mapsto X_t(\omega)$ is right continuous and admits left limits for all $t > 0$.

Definition 2.8 An E-valued process X is said to be an *l.c.r.l. process* (or a process with left continuous paths with right limits) if for all $\omega \in \Omega$, the *path* $t \mapsto X_t(\omega)$ is left continuous on $(0, \infty)$ and admits right limits for all $t \geq 0$.

For an r.c.l.l. process X, X^- will denote the r.c.l.l. process defined by $X_t^- = X(t-)$, i.e. the left limit at t, with the convention $X(0-) = 0$ and let

$$\Delta X = X - X^-$$

so that $(\Delta X)_t = 0$ at each continuity point and equals the jump otherwise. Note that by the above convention

$$(\Delta X)_0 = X_0.$$

Let X, Y be r.c.l.l. processes such that $X \overset{v}{\longleftrightarrow} Y$. Then it is easy to see that $X = Y$. The same is true if both X, Y are l.c.r.l. processes.

Exercise 2.9 Prove the statements in the previous paragraph.

It is easy to see that if X is an $\mathbb{R}^d$-valued r.c.l.l. process or an l.c.r.l. process then

$$\sup_{0 \leq t \leq T} |X_t(\omega)| < \infty \ \forall T < \infty, \ \forall \omega.$$

Exercise 2.10 If X is an $\mathbb{R}^d$-valued r.c.l.l. process then show that the process Z defined by

$$Z_t(\omega) = \sup_{0 \leq s \leq t} |X_s(\omega)|$$

is an r.c.l.l. process.

When X is an $\mathbb{R}^d$-valued continuous process, the mapping $\omega \mapsto X_\cdot(\omega)$ from Ω into $\mathbb{C}_d$ is measurable and induces a measure $\mathsf{P} \circ X^{-1}$ on $(\mathbb{C}_d, \mathcal{B}(\mathbb{C}_d))$. This is so because the Borel σ-field $\mathcal{B}(\mathbb{C}_d)$ is also the smallest σ-field with respect to which the coordinate process is measurable. Likewise, when X is an r.c.l.l. process, the mapping $\omega \mapsto X_\cdot(\omega)$ from Ω into $\mathbb{D}_d$ is measurable and induces a measure $\mathsf{P} \circ X^{-1}$ on $(\mathbb{D}_d, \mathcal{B}(\mathbb{D}_d))$. In both cases, the probability measure $\mathsf{P} \circ X^{-1}$ is called the distribution of the process X.

Definition 2.11 A d-dimensional *Brownian motion* (also called a Wiener process) is an $\mathbb{R}^d$-valued continuous process X such that

(i) $P(X_0 = 0) = 1$.
(ii) For $0 \leq s < t < \infty$, the distribution of $X_t - X_s$ is normal (Gaussian) with mean 0 and co-variance matrix $(t - s)I$, i.e. for $u \in \mathbb{R}^d$

$$E[\exp\{iu \cdot (X_t - X_s)\}] = \exp\{-(t - s)|u|^2\}$$

(iii) For $m \geq 1$, $0 = t_0 < t_1 < \ldots < t_m$, the random variables $Y_1, \ldots, Y_m$ are independent, where $Y_j = X_{t_j} - X_{t_{j-1}}$.

Equivalently, it can be seen that a $\mathbb{R}^d$-valued continuous process X is a Brownian motion if and only if for $m \geq 1$, $0 \leq t_1 < \ldots < t_m$, $u_1, u_2, \ldots, u_m \in \mathbb{R}^d$,

$$E[\exp\{i \sum_{j=1}^{m} u_j \cdot X_{t_j}\}] = \exp\{-\frac{1}{2} \sum_{j,k=1}^{m} \min(t_j, t_k) u_j \cdot u_k\}.$$

Remark 2.12 The process X in the definition above is sometimes called a standard Brownian motion and Y given by

$$Y_t = \mu t + \sigma X_t$$

where $\mu \in \mathbb{R}^d$ and σ is a positive constant and is also called a Brownian motion for any μ and σ.

When X is a d-dimensional Brownian motion, its distribution, i.e. the induced measure $\mu = P \circ X^{-1}$ on $(\mathbb{C}_d, \mathcal{B}(\mathbb{C}_d))$, is known as the Wiener measure. The Wiener measure was constructed by Wiener before Kolmogorov's axiomatic formulation of probability theory. The existence of Brownian motion or the Wiener measure can be proved in many different ways. One method is to use Kolmogorov's consistency theorem to construct a process X satisfying (i), (ii) and (iii) in the definition, which determine the finite-dimensional distributions of X, and then to invoke the following criterion for existence of a continuous process $\tilde{X}$ such that $X \overset{v}{\longleftrightarrow} \tilde{X}$.

Theorem 2.13 *Let X be an $\mathbb{R}^d$-valued process. Suppose that for each $T < \infty$, $\exists m > 0$, $K < \infty$, $\beta > 0$ such that*

$$E[|X_t - X_s|^m] \leq K|t - s|^{1+\beta}, \quad 0 \leq s \leq t \leq T.$$

Then there exists a continuous process $\tilde{X}$ such that $X \overset{v}{\longleftrightarrow} \tilde{X}$.

Exercise 2.14 Let X be a Brownian motion and let Y be defined as follows: $Y_0 = 0$ and for $0 < t < \infty$, $Y_t = t X_s$ where $s = \frac{1}{t}$. Show that Y is also a Brownian motion.

Definition 2.15 A *Poisson Process* (with rate parameter λ) is an r.c.l.l. non-negative integer-valued process N such that

(i) $N_0 = 0$.

(ii) $P(N_t - N_s = n) = \exp\{-\lambda(t - s)\}\frac{(\lambda(t-s))^n}{n!}$.

(iii) For $m \geq 1$, $0 = t_0 < t_1 < \ldots < t_m$, the random variables $Y_1, \ldots, Y_m$ are independent, where $Y_j = N_{t_j} - N_{t_{j-1}}$.

Exercise 2.16 Let N^1, N^2 be Poisson processes with rate parameters λ^1 and λ^2, respectively. Suppose N^1 and N^2 are independent. Show that N defined by $N_t = N_t^1 + N_t^2$ is also a Poisson process with rate parameter $\lambda = \lambda^1 + \lambda^2$.

Brownian motion and Poisson process are the two most important examples of continuous time stochastic processes and arise in modelling of phenomena occurring in nature.

2.2 Filtration

As in the discrete-time case, it is useful to define for $t \geq 0$, $\mathcal{G}_t$ to be the σ-field generated by all the random variables observable up to time t and then require any action to be taken at time t (an estimate of some quantity or investment decision) should be measurable with respect to $\mathcal{G}_t$. These observations lead to the following definitions.

Definition 2.17 A *filtration* on a probability space $(\Omega, \mathcal{F}, P)$ is an increasing family of sub σ-fields $(\mathcal{F}_\cdot) = \{\mathcal{F}_t : t \geq 0\}$ of $\mathcal{F}$ indexed by $t \in [0, \infty)$.

Definition 2.18 A process X is said to be *adapted* to a filtration $(\mathcal{F}_\cdot)$ if for all $t \geq 0$, X_t is $\mathcal{F}_t$ measurable.

We will always assume that the underlying probability space is complete, i.e. $N \in \mathcal{F}$, $P(N) = 0$, and $N_1 \subseteq N$ implies $N \in \mathcal{F}$ and (ii) that $\mathcal{F}_0$ contains all sets $N \in \mathcal{F}$ with $P(N) = 0$.

Note that if X is $(\mathcal{F}_\cdot)$ adapted and $Y = X$ (see Definition 2.2) then Y is also $(\mathcal{F}_\cdot)$ adapted in view of the assumption that $\mathcal{F}_0$ contains all null sets.

Given a filtration $(\mathcal{F}_\cdot)$, we will denote by $(\mathcal{F}_\cdot^+)$ the filtration $\{\mathcal{F}_t^+ : t \geq 0\}$ where

$$\mathcal{F}_t^+ = \cap_{u>t} \mathcal{F}_u.$$

Let $\mathcal{N}$ be the class of all null sets (sets with $P(N) = 0$), and for a process Z, possibly vector-valued, let

$$\mathcal{F}_t^Z = \sigma(\mathcal{N} \cup \sigma(Z_u : 0 \leq u \leq t)). \tag{2.2.1}$$

Then $(\mathcal{F}_\cdot^Z)$ is the smallest filtration such that $\mathcal{F}_0$ contains all null sets with respect to which Z is adapted.

While it is not required in the definition, in most situations, the filtration $(\mathcal{F}_\cdot)$ under consideration would be chosen to be (for a suitable process Z)

$$(\mathcal{F}_{\cdot}^{Z}) = \{\mathcal{F}_{t}^{Z} : t \geq 0\}$$

Sometimes, a filtration is treated as a mere technicality, specially in continuous-time setting as a necessary detail just to define stochastic integral. We would like to stress that it is not so. See discussion in Sect. 1.4.

2.3 Martingales and Stopping Times

Let M be a process defined on a probability space $(\Omega, \mathcal{F}, \mathsf{P})$ and $(\mathcal{F}_{\cdot})$ be a filtration.

Definition 2.19 M is said to be $(\mathcal{F}_{\cdot})$-*martingale* if M is $(\mathcal{F}_{\cdot})$ adapted, M_t is integrable for all t and for $0 \leq s < t$ one has

$$\mathsf{E}_{\mathsf{P}}[M_t \mid \mathcal{F}_s] = M_s.$$

Definition 2.20 M is said to be $(\mathcal{F}_{\cdot})$-*submartingale* if M is $(\mathcal{F}_{\cdot})$ adapted, M_t is integrable for all t and for $0 \leq s < t$ one has

$$\mathsf{E}_{\mathsf{P}}[M_t \mid \mathcal{F}_s] \geq M_s.$$

Remark **2.21** Likewise M is said to be a supermartingale if N defined by $N_t = -M_t$ is a submartingale.

When there is only one filtration under consideration, we will drop reference to it and call M to be a martingale (or a submartingale). If M is a martingale and ϕ is a convex function on $\mathbb{R}$, then Jensen's inequality implies that $\phi(M)$ is a submartingale provided each $\phi(M_t)$ is integrable. In particular, if M is a martingale with $\mathsf{E}[M_t^2] < \infty$ for all t then M^2 is a submartingale. We are going to be dealing with martingales that have r.c.l.l. paths. The next result shows that under minimal conditions on the underlying filtration one can assume that every martingale has r.c.l.l. paths.

Theorem 2.22 *Suppose that the filtration* $(\mathcal{F}_{\cdot})$ *satisfies*

$$N_0 \subseteq N, \ N \in \mathcal{F}, \ \mathsf{P}(N) = 0 \ \Rightarrow \ N_0 \in \mathcal{F}_0 \qquad (2.3.1)$$

$$\cap_{t>s} \mathcal{F}_t = \mathcal{F}_s \ \forall s \geq 0. \qquad (2.3.2)$$

Then every martingale M *admits an r.c.l.l. version* $\tilde{M}$, *i.e. there exists an r.c.l.l. process* $\tilde{M}$ *such that*

$$\mathsf{P}(M_t = \tilde{M}_t) = 1 \ \forall t \geq 0. \qquad (2.3.3)$$

Proof For $k, n \geq 1$, let $s_k^n = k2^{-n}$ and $X_k^n = M_{s_k^n}$ and $\mathcal{G}_k^n = \mathcal{F}_{s_k^n}$. Then for fixed n, $\{X_k^n : k \geq 0\}$ is a martingale w.r.t. the filtration $\{\mathcal{G}_k^n : k \geq 0\}$. Fix rational numbers

$\alpha < \beta$ and an integer T. Let $T_n = T2^n$. Doob's upcrossings inequality Theorem 1.28 yields

$$\mathsf{E}[U_{T_n}(\{X_k^n \ : \ 0 \le k \le T_n\}, \alpha, \beta)] \le \frac{\mathsf{E}[|M_T|] + |\alpha|}{\beta - \alpha}.$$

Thus, using the observation that

$$U_{T_n}(\{X_k^n \ : \ 0 \le k \le T_n\}, \alpha, \beta) \le U_{T_{n+1}}(\{X_k^{n+1} \ : \ 0 \le k \le T_{n+1}\}, \alpha, \beta)$$

we get by the monotone convergence theorem,

$$\mathsf{E}[\sup_{n \ge 1} U_{T_n}(\{X_k^n \ : \ 0 \le k \le T_n\}, \alpha, \beta)] \le \frac{\mathsf{E}[|M_T|] + |\alpha|}{\beta - \alpha}.$$

Thus we conclude that there exists a set N with $\mathsf{P}(N) = 0$ such that for $\omega \notin N$,

$$\sup_{n \ge 1} U_{T_n}(\{X_k^n \ : \ 0 \le k \le T_n\}, \alpha, \beta) < \infty \qquad (2.3.4)$$

for all rational numbers $\alpha < \beta$ and integers T. Thus if $\{t_k : k \ge 1\}$ are dyadic rational numbers increasing or decreasing to t, then for $\omega \notin N$, $M_{t_k}(\omega)$ converges. Define $\theta_k(t) = ([t2^k] + 1)2^{-k}$, where $[r]$ is the largest integer less than or equal to r (the integer part of r). For $\omega \notin N$, letting

$$\tilde{M}_t(\omega) = \lim_{k \to \infty} M_{\theta_k(t)}(\omega), \ \ 0 \le t < \infty \qquad (2.3.5)$$

it follows that $t \mapsto \tilde{M}_t(\omega)$ is a r.c.l.l. function. Fix t and let $t_n = \theta_n(t)$. Then $\cap_n \mathcal{F}_{t_n} = \mathcal{F}_t$ so that

$$\begin{aligned}
M_t &= \mathsf{E}[M_{t+1} \mid \mathcal{F}_t] \\
&= \lim_{n \to \infty} \mathsf{E}[M_{t+1} \mid \mathcal{F}_{t_n}] \\
&= \lim_{n \to \infty} M_{t_n} \\
&= \tilde{M}_t.
\end{aligned}$$

Here has used Theorem 1.38. Hence (2.3.3) follows. $\square$

The conditions (2.3.1) and (2.3.2) together are known as *usual hypothesis* in the literature. We will assume (2.3.1) but will not assume (2.3.2). We will mostly consider martingales with r.c.l.l. paths, and refer to it as an r.c.l.l. martingale.

When we are having only one filtration under consideration, we will drop reference to it and simply say M is a martingale. We note here an important property of martingales that would be used later.

Theorem 2.23 *Let M^n be a sequence of martingales on some probability space $(\Omega, \mathcal{F}, \mathsf{P})$ w.r.t. a fixed filtration $(\mathcal{F}_.)$. Suppose that*

$$M_t^n \to M_t \text{ in } \mathbb{L}^1(P), \quad \forall t \geq 0.$$

Then M is also a martingale w.r.t. the filtration $(\mathcal{F}_.)$.

Proof Note that for any X^n converging to X in $\mathbb{L}^1(P)$, for any σ-field $\mathcal{G}$, using (i), (ii), (iii) and (iv) in Proposition 1.3, one has

$$\begin{aligned}
E[|E[X^n \mid \mathcal{G}] - E[X \mid \mathcal{G}]|] &= E[|E[(X^n - X) \mid \mathcal{G}]|] \\
&\leq E[E[|X^n - X| \mid \mathcal{G}]] \\
&= E[|X^n - X|] \\
&\to 0.
\end{aligned}$$

For $s < t$, applying this to $X^n = M_t^n$, one gets

$$M_s^n = E[M_t^n \mid \mathcal{F}_s] \to E[M_t \mid \mathcal{F}_s] \text{ in } \mathbb{L}^1(P).$$

Since $M_s^n \to M_s$ in $\mathbb{L}^1(P)$, we conclude that $E[M_t \mid \mathcal{F}_s] = M_s$ and hence M is a martingale. $\square$

Remark **2.24** It may be noted that in view of our assumption that $\mathcal{F}_0$ contains all null sets, M as in the statement of the previous theorem is adapted.

Here is a consequence of Theorem 1.36 in continuous time.

Theorem 2.25 *Let M be a martingale such that $\{M_t : t \geq 0\}$ is uniformly integrable. Then, there exists a random variable ξ such that $M_t \to \xi$ in $\mathbb{L}^1(P)$. Moreover,*

$$M_t = E[\xi \mid \mathcal{F}_t].$$

Proof Here $\{M_n : n \geq 1\}$ is a uniformly integrable martingale, and Theorem 1.36 yields that $M_n \to \xi$ in $\mathbb{L}^1(P)$. Similarly, for any sequence $t_n \uparrow \infty$, M_{t_n} converges to say η in $\mathbb{L}^1(P)$. Interlacing argument gives $\eta = \xi$ and hence $M_t \to \xi$ in $\mathbb{L}^1(P)$. $\square$

Doob's maximal inequality for martingales in continuous time follows from the discrete version easily. We do not need the L^p version for $p \neq 2$ in the sequel, and hence we will state only $p = 2$ version here.

Theorem 2.26 *Let M be an r.c.l.l. process. If M is a martingale or a positive submartingale then one has*

(i) For $\lambda > 0$, $0 < t < \infty$,

$$P[\sup_{0 \leq s \leq t} |M_s| > \lambda] \leq \frac{1}{\lambda} E[|M_t|]. \tag{2.3.6}$$

(ii) For $0 < t < \infty$,

$$E[\sup_{0 \leq s \leq t} |M_s|^2] \leq 4E[|M_t|^2]. \tag{2.3.7}$$

Example **2.27** Let (W_t) denote a one-dimensional Brownian motion on a probability space $(\Omega, \mathcal{F}, P)$. Then by the definition of Brownian motion, for $0 \leq s \leq t$, $W_t - W_s$ has mean zero and is independent of $\{W_u, \, 0 \leq u \leq s\}$. Hence (see (1.3.2)) we have

$$E[W_t - W_s \mid \mathcal{F}_s^W] = 0$$

and hence W is an $(\mathcal{F}_{\cdot}^W)$-martingale. Likewise,

$$E[(W_t - W_s)^2 \mid \mathcal{F}_s^W] = E[(W_t - W_s)^2] = t - s$$

and using this and the fact that (W_t) is a martingale, it is easy to see that

$$E[W_t^2 - W_s^2 \mid \mathcal{F}_s^W] = t - s$$

and so defining $M_t = W_t^2 - t$, it follows that M is also an $(\mathcal{F}_{\cdot}^W)$-martingale.

Example **2.28** Let (N_t) denote a Poisson process with rate parameter $\lambda = 1$ on a probability space $(\Omega, \mathcal{F}, P)$. Then by the definition of Poisson process, for $0 \leq s \leq t$, $N_t - N_s$ has mean $(t - s)$ and is independent of $\{N_u, \, 0 \leq u \leq s\}$. Writing $M_t = N_t - t$, and using (1.3.2), we can see that

$$E[M_t - M_s \mid \mathcal{F}_s^N] = 0$$

and hence M is an $(\mathcal{F}_{\cdot}^N)$-martingale. Likewise,

$$E[(M_t - M_s)^2 \mid \mathcal{F}_s^M] = E[(M_t - M_s)^2] = t - s$$

and using this and the fact that (M_t) is a martingale, it is easy to see that

$$E[M_t^2 - M_s^2 \mid \mathcal{F}_s^N] = t - s$$

and so defining $U_t = M_t^2 - t$, it follows that U is also an $(\mathcal{F}_{\cdot}^N)$-martingale.

The notion of stopping time, as mentioned in Sect. 1.6, was first introduced in the context of Markov chains. Martingales and stopping times together are very important tools in the theory of stochastic process in general and stochastic calculus in particular.

Definition 2.29 A *stopping time* with respect to a filtration $(\mathcal{F}_{\cdot})$ is a mapping τ from Ω into $[0, \infty]$ such that for all $t < \infty$,

$$\{\tau \leq t\} \in \mathcal{F}_t.$$

If the filtration under consideration is fixed, we will only refer to it as a stopping time. Of course, for a stopping time, $\{\tau < t\} = \cup_n\{\tau \le t - \frac{1}{n}\} \in \mathcal{F}_t$. For stopping times τ and σ, it is easy to see that $\tau \wedge \sigma$ and $\tau \vee \sigma$ are stopping times. In particular, $\tau \wedge t$ and $\tau \vee t$ are stopping times for any $t \ge 0$.

Example **2.30** Let X be a $\mathbb{R}^d$-valued process with continuous paths and adapted to a filtration $(\mathcal{F}_.)$. Let C be a closed set in $\mathbb{R}$. Then

$$\tau_C = \inf\{t \ge 0 : X_t \in C\} \tag{2.3.8}$$

is a stopping time. To see this, define open sets $U_k = \{x : d(x, C) < \frac{1}{k}\}$. Writing $A_{r,k} = \{X_r \in U_k\}$ and $Q_t = \{r : r \text{ rational}, 0 \le r \le t\}$,

$$\{\tau_C \le t\} = \cap_{k=1}^\infty [\cup_{r \in Q_t} A_{r,k}]. \tag{2.3.9}$$

τ_C is called the *hitting time* of C.

If X is an r.c.l.l. process, then the hitting time τ_C may not be a random variable and hence may not be a stopping time. Let us define the *contact time* σ_C by

$$\sigma_C = \min\{\inf\{t \ge 0 : X_t \in C\}, \inf\{t > 0 : X_{t-} \in C\}\}. \tag{2.3.10}$$

With same notations as above, we now have

$$\{\sigma_C \le t\} = [\cap_{k=1}^\infty (\cup_{r \in Q_t} A_{r,k})] \cup \{X_t \in C\} \tag{2.3.11}$$

and thus σ_C is a stopping time. If $0 \notin C$, then σ_C can also be described as

$$\sigma_C = \inf\{t \ge 0 : X_t \in C \text{ or } X_{t-} \in C\}.$$

Exercise 2.31 Construct an example to show that this alternate description may be incorrect when $0 \in C$.

If τ is a $[0, \infty)$-valued stopping time, then for integers $m \ge 1$, τ^m defined via

$$\tau^m = 2^{-m}([2^m \tau] + 1) \tag{2.3.12}$$

is also a stopping time since $\{\tau^m \le t\} = \{\tau^m \le 2^{-m}([2^m t])\} = \{\tau < 2^{-m}([2^m t])\}$ and it follows that $\{\tau^m \le t\} \in (\mathcal{F}_.)$. Clearly, $\tau^m \downarrow \tau$.

One can see that if σ_k is an increasing sequence of stopping times ($\sigma_k \le \sigma_{k+1}$ for all $k \ge 1$), then $\sigma = \lim_{k\uparrow\infty} \sigma_k$ is also a stopping time. However, if $\sigma_k \downarrow \sigma$, σ may not be a stopping time as seen in the next exercise.

Exercise 2.32 Let $\Omega = \{-1, 1\}$ and P be such that $0 < P(1) < 1$. Let $X_t(1) = X_t(-1) = 0$ for $0 \le t \le 1$. For $t > 1$, let $X_t(1) = \sin(t - 1)$ and $X_t(-1) = -\sin(t - 1)$. Let

$$\sigma_k = \inf\{t \geq 0 : X_t \geq 2^{-k}\},$$

$$\sigma = \inf\{t \geq 0 : X_t > 0\}.$$

Show that $\mathcal{F}_t^X = \{\phi, \Omega\}$ for $0 \leq t \leq 1$ and $\mathcal{F}_t^X = \{\phi, \{1\}, \{-1\}, \Omega\}$ for $t > 1$, σ_k are stopping times w.r.t. $(\mathcal{F}_{\cdot}^X)$ and $\sigma_k \downarrow \sigma$ but σ is not a stopping time w.r.t. $(\mathcal{F}_{\cdot}^X)$. Note that $\{\sigma < s\} \in \mathcal{F}_s^X$ for all s and yet σ is not a stopping time.

Exercise 2.33 Let $\Omega = [0, \infty)$ and $\mathcal{F}$ be the Borel σ-field on Ω. For $t \geq 0$, let $\mathcal{F}_t$ be the σ-field generated by Borel subsets of $[0, t]$ along with the set (t, ∞). Let τ be a $[0, \infty)$-valued measurable function on $(\Omega, \mathcal{F})$. Show that τ is a stopping time w.r.t. $(\mathcal{F}_{\cdot})$ if and only if there exists α, $0 \leq \alpha \leq \infty$, such that $\tau(t) \geq t$ for $t \in [0, \alpha]$ and $\tau(t) = \alpha$ for $t \in (\alpha, \infty)$. Note that here, $[0, \infty]$ is to be taken as $[0, \infty)$ and (∞, ∞) as the empty set.

Definition 2.34 For a stochastic process X and a stopping time τ, X_τ is defined via

$$X_\tau(\omega) = X_{\tau(\omega)}(\omega) 1_{\{\tau(\omega) < \infty\}}.$$

Remark **2.35** A random variable τ taking countably many values $\{s_j\}$, $0 \leq s_j < \infty$ $\forall j$ is a stopping time if and only if $\{\tau = s_j\} \in \mathcal{F}_{s_j}$ for all j. For such a stopping time τ and a stochastic process X,

$$X_\tau = \sum_{j=1}^{\infty} 1_{\{\tau = s_j\}} X_{s_j}$$

and thus X_τ is a random variable (i.e. a measurable function).

In general, X_τ may not be a random variable, i.e. may not be a measurable function. However, if X has right continuous paths X_τ is a random variable. To see this, given τ that is $[0, \infty)$-valued, X_{τ^m} is measurable where τ^m is defined via (2.3.12) and $X_{\tau^m} \to X_\tau$ and hence X_τ is a random variable. Finally, for a general stopping time τ, $\xi_n = X_{\tau \wedge n}$ is a random variable and hence

$$X_\tau = (\limsup_{n \to \infty} \xi_n) 1_{\{\tau < \infty\}}$$

is also a random variable.

Lemma 2.36 *Let Z be an r.c.l.l. adapted process and τ be a stopping time with $\tau < \infty$. Let $U_t = Z_{\tau \wedge t}$. Then U is an adapted process (with respect to the same filtration $(\mathcal{F}_{\cdot})$ that Z is adapted to).*

Proof When τ takes finitely many values it is easy to see that U is adapted and for the general case, the proof is by approximating τ by τ^m defined via (2.3.12). □

Definition 2.37 For a stopping time τ with respect to a filtration $(\mathcal{F}_.)$, the stopped σ-field is defined by

$$\mathcal{F}_\tau = \{A \in \sigma(\cup_t \mathcal{F}_t) : A \cap \{\tau \le t\} \in \mathcal{F}_t \; \forall t\}.$$

We have seen that for a right continuous process Z, Z_τ is a random variable, i.e. measurable. The next lemma shows that indeed Z_τ is $\mathcal{F}_\tau$ measurable.

Lemma 2.38 *Let X be a right continuous $(\mathcal{F}_.)$ adapted process and τ be a stopping time. Then X_τ is $\mathcal{F}_\tau$ measurable.*

Proof For a $t < \infty$, we need to show that $\{X_\tau \le a\} \cap \{\tau \le t\} \in \mathcal{F}_t$ for all $a \in \mathbb{R}$. Note that

$$\{X_\tau \le a\} \cap \{\tau \le t\} = \{X_{t \wedge \tau} \le a\} \cap \{\tau \le t\}.$$

As seen in Lemma 2.36, $\{X_{t \wedge \tau} \le a\} \in \mathcal{F}_t$ and of course, $\{\tau \le t\} \in \mathcal{F}_t$. □

Applying the previous result to the (deterministic) process $X_t = t$ we get

Corollary 2.39 *Every stopping time τ is $\mathcal{F}_\tau$ measurable.*

Also, we have

Corollary 2.40 *Let σ, τ be two stopping times. Then $\{\sigma \le \tau\} \in \mathcal{F}_\tau$.*

Proof Let $X_t = 1_{[\sigma(\omega), \infty)}(t)$. We have seen that X is r.c.l.l. adapted and hence X_τ is $\mathcal{F}_\tau$ measurable. Note that $X_\tau = 1_{\{\sigma \le \tau\}}$. This completes the proof. □

Here is another observation.

Lemma 2.41 *Let τ be a stopping time and ξ be $\mathcal{F}_\tau$ measurable random variable. Then $Z = \xi 1_{[\tau, \infty)}$ is an adapted r.c.l.l. process.*

Proof Note that for any t, $\{Z_t \le a\} = \{\xi \le a\} \cap \{\tau \le t\}$ if $a < 0$ and $\{Z_t \le a\} = (\{\xi \le a\} \cap \{\tau \le t\}) \cup \{\tau > t\}$ if $a \ge 0$. Since $\{\xi \le a\} \in \mathcal{F}_\tau$, we have $\{\xi \le a\} \cap \{\tau \le t\} \in \mathcal{F}_t$ and also $\{\tau > t\} = \{\tau \le t\}^c \in \mathcal{F}_t$. This shows Z is adapted. Of course it is r.c.l.l. by definition. □

The following result will be needed repeatedly in later chapters.

Lemma 2.42 *Let X be any real-valued r.c.l.l. process adapted to a filtration $(\mathcal{F}_.)$. Let σ be any stopping time. Then for any $a > 0$, τ defined as follows is a stopping time : if $\sigma = \infty$ then $\tau = \infty$ and if $\sigma < \infty$ then*

$$\tau = \inf\{t > \sigma : |X_t - X_\sigma| \ge a \text{ or } |X_{t-} - X_\sigma| \ge a\}. \tag{2.3.13}$$

If $\tau < \infty$ then $\tau > \sigma$ and either $|X_{\tau-} - X_\sigma| \ge a$ or $|X_\tau - X_\sigma| \ge a$. In other words, when the infimum is finite, it is actually minimum.

Proof Let a process Y be defined by $Y_t = X_t - X_{t \wedge \sigma}$. Then as seen in Lemma 2.36, Y is an r.c.l.l. adapted process. For any $t > 0$, let $\mathbb{Q}^t = (\mathbb{Q} \cap [0, t]) \cup \{t\}$. Note that since Y has r.c.l.l. paths one has

$$\{\omega : \tau(\omega) \le t\} = \cap_{n=1}^{\infty} (\cup_{r \in \mathbb{Q}^t} \{|Y_r(\omega)| \ge a - \tfrac{1}{n}\}). \qquad (2.3.14)$$

To see this, let Ω_1 denote the left-hand side and Ω_2 denote the right-hand side.

Let $\omega \in \Omega_1$. If $\tau(\omega) < t$, then there is an $s \in [0, t)$ such that $|Y_s(\omega)| \ge a$ or $|Y_{s-}(\omega)| \ge a$ (recall the convention that $Y_{0-} = 0$ and here $a > 0$). In either case, there exists a sequence of rational numbers $\{r_k\}$ in $(0, t)$ converging to s and $\lim_k |Y_{r_k}(\omega)| \ge a$ and thus $\omega \in \Omega_2$. If $\tau(\omega) = t$ and if $|Y_{t-}(\omega)| \ge a$ then also $\omega \in \Omega_2$. To complete the proof of $\Omega_1 \subseteq \Omega_2$ (recall $t \in \mathbb{Q}^t$), we will show that $\tau(\omega) = t$ and $|Y_{t-}(\omega)| < a$ implies $|Y_t(\omega)| \ge a$. Observe that $\tau(\omega) = t$ implies that there exist $s_m > t$, $s_m \downarrow t$ such that for each m, $|Y_{s_m-}(\omega)| \ge a$ or $|Y_{s_m}(\omega)| \ge a$. This implies $|Y_t(\omega)| \ge a$, and this proves $\Omega_1 \subseteq \Omega_2$.

For the other part, if $\omega \in \Omega_2$, then there exist $\{r_n \in \mathbb{Q}^t : n \ge 1\}$ such that $|Y_{r_n}(\omega)| \ge a - \tfrac{1}{n}$. Since $\mathbb{Q}^t \subseteq [0, t]$, it follows that we can extract a subsequence r_{n_k} that converges to $s \in [0, t]$. By taking a further subsequence if necessary, we can assume that either $r_{n_k} \ge s \; \forall k$ or $r_{n_k} < s \; \forall k$ and thus $|Y_{r_{n_k}}(\omega)| \to |Y_s(\omega)| \ge a$ in the first case and $|Y_{r_{n_k}}(\omega)| \to |Y_{s-}(\omega)| \ge a$ in the second case. Also, $Y_0(\omega) = 0$ and $Y_{0-}(\omega) = 0$ and hence $s \in (0, t]$. This shows $\tau(\omega) \le s \le t$ and thus $\omega \in \Omega_1$. This proves (2.3.14).

In each of the cases, we see that either $|Y_\tau| \ge a$ or $|Y_{\tau-}| \ge a$. Since $Y_t = 0$ for $t \le \sigma$, it follows that $\tau > \sigma$. This completes the proof. $\qquad \square$

Remark 2.43 Note that τ is the contact time for the set $C = [a, \infty)$ for the process Y.

Remark 2.44 If X is a continuous process, the definition (2.3.13) is same as

$$\tau = \inf\{t > \sigma : |X_t - X_\sigma| \ge a\}. \qquad (2.3.15)$$

Exercise 2.45 Construct an example to convince yourself that in the definition (2.3.13) of τ, $t > \sigma$ cannot be replaced by $t \ge \sigma$. However, when X is continuous, in (2.3.15), $t > \sigma$ can be replaced by $t \ge \sigma$.

Here is another result that will be used in the sequel.

Theorem 2.46 *Let X be an r.c.l.l. adapted process with $X_0 = 0$. For $a > 0$, let $\{\sigma_i, : i \ge 0\}$ be defined inductively as follows: $\sigma_0 = 0$ and having defined $\sigma_j : j \le i$, let $\sigma_{i+1} = \infty$ if $\sigma_i = \infty$ and otherwise*

$$\sigma_{i+1} = \inf\{t > \sigma_i : |X_t - X_{\sigma_i}| \ge a \text{ or } |X_{t-} - X_{\sigma_i}| \ge a\}. \qquad (2.3.16)$$

Then each σ_i is a stopping time. Further, (i) if $\sigma_i < \infty$, then $\sigma_i < \sigma_{i+1}$ and (ii) $\lim_{i \uparrow \infty} \sigma_i = \infty$.

Proof That each σ_i is a stopping time and observation (i) follows from Lemma 2.42. Remains to prove (ii). If for some $\omega \in \Omega$,

$$\lim_{i \uparrow \infty} \sigma_i(\omega) = t_0 < \infty$$

then for such an ω, the left limit of the mapping $s \rightarrow X_s(\omega)$ at t_0 does not exist, a contradiction. This proves (ii). □

Exercise 2.47 Let X be the Poisson process with rate parameter λ. Let $\sigma_0 = 0$ and for $i \geq 0$, σ_{i+1} be defined by (2.3.16) with $a = 1$. For $n \geq 1$ let

$$\tau_n = \sigma_n - \sigma_{n-1}.$$

Show that

(i) $N_t(\omega) = k$ if and only if $\sigma_k(\omega) \leq t < \sigma_{k+1}(\omega)$.
(ii) $\{\tau_n : n \geq 1\}$ are independent random variables with $P(\tau_n > t) = \exp\{-\lambda t\}$ for all $n \geq 1$ and $t \geq 0$.

Recall that $(\mathcal{F}_\cdot^+)$ denotes the right continuous filtration corresponding to the filtration $(\mathcal{F}_\cdot)$. We now show that the hitting time of the interval $(-\infty, a)$ by an r.c.l.l. adapted process is a stopping time for the filtration $(\mathcal{F}_\cdot^+)$.

Lemma 2.48 *Let Y be an r.c.l.l. adapted process. Let $a \in \mathbb{R}$ and let*

$$\tau = \inf\{t \geq 0 : Y_t < a\}. \tag{2.3.17}$$

Then τ is a stopping time for the filtration $(\mathcal{F}_\cdot^+)$.

Proof Note that for any $s > 0$, right continuity of $t \mapsto Y_t(\omega)$ implies that

$$\{\omega : \tau(\omega) < s\} = \{\omega : Y_r(\omega) < a \text{ for some } r \text{ rational}, \ r < s\}$$

and hence $\{\tau < s\} \in \mathcal{F}_s$.
Now for any t, for all k

$$\{\tau \leq t\} = \cap_{n=k}^\infty \{\tau < t + \frac{1}{n}\} \in \mathcal{F}_{t+\frac{1}{k}}$$

and hence $\{\tau \leq t\} \in \mathcal{F}_t^+$. □

Remark **2.49** Similarly, it can be shown that the hitting time of an open set by an r.c.l.l. adapted process is a stopping time for the filtration $(\mathcal{F}_\cdot^+)$.

Here is an observation about $(\mathcal{F}_\cdot^+)$ stopping times.

Lemma 2.50 *Let $\sigma : \Omega \mapsto [0, \infty]$ be such that*

$$\{\sigma < t\} \subset \mathcal{F}_t \quad \forall t \geq 0.$$

Then σ is a $(\mathcal{F}_\cdot^+)$ stopping time.

Proof Let t be fixed and let $t_m = t + 2^{-m}$. Note that for all $m \geq 1$,

$$\{\sigma \leq t\} = \cap_{n=m}^\infty \{\sigma < t_n\} \in \mathcal{F}_{t_m}.$$

Thus $\{\sigma \leq t\} \in \cap_{m=1}^\infty \mathcal{F}_{t_m} = \mathcal{F}_t^+$. $\qquad\qquad\qquad\qquad\qquad\qquad\qquad\qquad$ □

Corollary 2.51 *Let $\{\tau_m : m \geq 1\}$ be a sequence of $(\mathcal{F}_\cdot)$ stopping times. Let $\sigma = \inf\{\tau_m : m \geq 1\}$ and $\theta = \sup\{\tau_m : m \geq 1\}$. Then θ is a $(\mathcal{F}_\cdot)$ stopping time whereas σ is $(\mathcal{F}_\cdot^+)$ stopping time.*

Proof The result follows by observing that

$$\{\theta \leq t\} = \cap_m \{\tau_m \leq t\}$$

and

$$\{\sigma < t\} = \cup_m \{\tau_m < t\}.$$

$\qquad\qquad\qquad\qquad\qquad\qquad\qquad\qquad\qquad\qquad\qquad\qquad\qquad\qquad\qquad\qquad$ □

Exercise 2.52 Show that $A \in \mathcal{F}_\tau$ if and only if the process X defined by

$$X_t(\omega) = 1_A(\omega) 1_{[\tau(\omega), \infty)}(t)$$

is $(\mathcal{F}_\cdot)$ adapted.

Lemma 2.53 *Let $\sigma \leq \tau$ be two stopping times. Then*

$$\mathcal{F}_\sigma \subseteq \mathcal{F}_\tau.$$

Proof Let $A \in \mathcal{F}_\sigma$. Then for $t \geq 0$, $A \cap \{\sigma \leq t\} \in \mathcal{F}_t$ and $\{\tau \leq t\} \in \mathcal{F}_t$ and thus $A \cap \{\sigma \leq t\} \cap \{\tau \leq t\} = A \cap \{\tau \leq t\} \in \mathcal{F}_t$. Hence $A \in \mathcal{F}_\tau$. $\qquad\qquad$ □

Here is another result on the family of σ-fields $\{\mathcal{F}_\tau : \tau$ a stopping time$\}$.

Theorem 2.54 *Let σ, τ be stopping times. Then*

$$\{\sigma = \tau\} \in \mathcal{F}_\sigma \cap \mathcal{F}_\tau.$$

Proof Fix $0 \leq t < \infty$. For $n \geq 1$ and $1 \leq k \leq 2^n$ let $A_{k,n} = \{\frac{k-1}{2^n}t \leq \sigma \leq \frac{k}{2^n}t\}$ and $B_{k,n} = \{\frac{k-1}{2^n}t \leq \tau \leq \frac{k}{2^n}t\}$. Note that

$$\{\sigma = \tau\} \cap \{\sigma \le t\} = \cap_{n=1}^{\infty} \cup_{k=1}^{2^n} [A_{k,n} \cap B_{k,n}].$$

This shows $\{\sigma = \tau\} \cap \{\sigma \le t\} \in \mathcal{F}_t$ for all t and thus $\{\sigma = \tau\} \in \mathcal{F}_\sigma$. Symmetry now implies $\{\sigma = \tau\} \in \mathcal{F}_\tau$ as well. □

Martingales and stopping times are intricately related as the next result shows.

Theorem 2.55 *Let M be a r.c.l.l. martingale and τ be a bounded stopping time, $\tau \le T$. Then*

$$\mathsf{E}[M_T \mid \mathcal{F}_\tau] = M_\tau. \tag{2.3.18}$$

Proof We have observed that M_τ is $\mathcal{F}_\tau$ measurable. First let us consider the case when τ is a stopping time taking finitely many values, $s_1 < s_2 < \ldots < s_m \le T$. Let $A_j = \{\tau = s_j\}$. Since τ is a stopping time, $A_j \in \mathcal{F}_{s_j}$. Clearly, $\{A_1, \ldots, A_m\}$ forms a partition of Ω. Let $B \in \mathcal{F}_\tau$. Then, by definition of $\mathcal{F}_\tau$ it follows that $C_j = B \cap A_j \in \mathcal{F}_{s_j}$. Since M is a martingale, $\mathsf{E}[M_T \mid \mathcal{F}_{s_j}] = M_{s_j}$ and hence

$$\mathsf{E}[M_T 1_{C_j}] = \mathsf{E}[M_{s_j} 1_{C_j}] = \mathsf{E}[M_\tau 1_{C_j}].$$

Summing over j we get

$$\mathsf{E}[M_T 1_B] = \mathsf{E}[M_\tau 1_B].$$

This proves (2.3.18) when τ takes finitely many values. For the general case, given $\tau \le T$, let

$$\tau_n = \frac{([2^n \tau] + 1)}{2^n} \wedge T.$$

Then for each n, τ_n is a stopping time that takes only finitely many values and further the sequence $\{\tau_n\}$ decreases to τ. By the part proven above we have

$$\mathsf{E}[M_T \mid \mathcal{F}_{\tau_n}] = M_{\tau_n}. \tag{2.3.19}$$

Now given $B \in \mathcal{F}_\tau$, using $\tau \le \tau_n$ and Lemma 2.53, we have $B \in \mathcal{F}_{\tau_n}$ and hence using (2.3.19), we have

$$\mathsf{E}[M_T 1_B] = \mathsf{E}[M_{\tau_n} 1_B]. \tag{2.3.20}$$

Now M_{τ_n} converges to M_τ (pointwise). Further, in view of Lemma 1.34, (2.3.19) implies that $\{M_{\tau_n} : n \ge 1\}$ is uniformly integrable and hence M_{τ_n} converges to M_τ in $\mathbb{L}^1(\mathsf{P})$. Thus taking limit as $n \to \infty$ in (2.3.20) we get

$$\mathsf{E}[M_T 1_B] = \mathsf{E}[M_\tau 1_B].$$

This holds for all $B \in \mathcal{F}_\tau$ and hence we conclude that (2.3.18) is true. □

It should be noted that we have not assumed that the underlying filtration is right continuous. This result leads to the following:

Corollary 2.56 *Let* M *be an r.c.l.l. martingale. Let* $\sigma \le \tau \le T$ *be two bounded stopping times. Then*

$$\mathsf{E}[M_\tau \mid \mathcal{F}_\sigma] = M_\sigma. \qquad (2.3.21)$$

Proof Taking conditional expectation given $\mathcal{F}_\sigma$ in (2.3.18) and using $\mathcal{F}_\sigma \subseteq \mathcal{F}_\tau$ we get (2.3.21). □

As in the discrete case, here too we have the following characterization of martingales via stopping times.

Theorem 2.57 *Let* X *be an r.c.l.l.* $(\mathcal{F}_.)$ *adapted process with* $\mathsf{E}[|X_t|] < \infty$ *for all* $t < \infty$. *Then* X *is an* $(\mathcal{F}_.)$-*martingale if and only if for all* $(\mathcal{F}_.)$ *stopping times* τ *taking finitely many values in* $[0, \infty)$, *one has*

$$\mathsf{E}[X_\tau] = \mathsf{E}[X_0]. \qquad (2.3.22)$$

Further if X *is a martingale then for all bounded stopping times* σ *one has*

$$\mathsf{E}[X_\sigma] = \mathsf{E}[X_0]. \qquad (2.3.23)$$

Proof Suppose X is a martingale. Then (2.3.22) and (2.3.23) follow from (2.3.18).

On the other hand suppose (2.3.22) is true for stopping times taking finitely many values. Fix $s < t$ and $A \in \mathcal{F}_s$. To show $\mathsf{E}[X_t \mid \mathcal{F}_s] = X_s$, suffices to prove that $\mathsf{E}[X_t 1_A] = \mathsf{E}[X_s 1_A]$. Now take

$$\tau = s1_A + t1_{A^c}$$

to get

$$\mathsf{E}[X_0] = \mathsf{E}[X_\tau] = \mathsf{E}[X_s 1_A] + \mathsf{E}[X_t 1_{A^c}] \qquad (2.3.24)$$

and of course taking the constant stopping time t, one has

$$\mathsf{E}[X_0] = \mathsf{E}[X_t] = \mathsf{E}[X_t 1_A] + \mathsf{E}[X_t 1_{A^c}]. \qquad (2.3.25)$$

Now using (2.3.24) and (2.3.25) it follows that $\mathsf{E}[X_s 1_A] = \mathsf{E}[X_t 1_A]$ and hence X is a martingale. □

Corollary 2.58 *For an r.c.l.l. martingale* M *and a stopping time* σ, N *defined by*

$$N_t = M_{t \wedge \sigma}$$

is a martingale.

We also have the following result about two stopping times.

Theorem 2.59 *Suppose* M *is an r.c.l.l.* $(\mathcal{F}_.)$-*martingale and* σ *and* τ *are* $(\mathcal{F}_.)$ *stopping times with* $\sigma \le \tau$. *Suppose* X *is an r.c.l.l.* $(\mathcal{F}_.)$ *adapted process. Let*

$$N_t = X_{\sigma \wedge t}(M_{\tau \wedge t} - M_{\sigma \wedge t}).$$

Then N is a $(\mathcal{F}_.)$-martingale if either (i) X is bounded or if (ii) $\mathsf{E}[X_\sigma^2] < \infty$ and M is square integrable.

Proof Clearly, N is adapted. First consider the case when X is bounded. We will show that for any bounded stopping time θ (bounded by T),

$$\mathsf{E}[N_\theta] = 0$$

and then invoke Theorem 2.55. Note that

$$N_\theta = X_{\sigma \wedge \theta}(M_{\tau \wedge \theta} - M_{\sigma \wedge \theta})$$
$$= X_{\tilde{\sigma}}(M_{\tilde{\tau}} - M_{\tilde{\sigma}})$$

where $\tilde{\sigma} = \sigma \wedge \theta \leq \tilde{\tau} = \tau \wedge \theta$ are also bounded stopping times. Now

$$\mathsf{E}[N_\theta] = \mathsf{E}[\mathsf{E}[N_\theta \mid \mathcal{F}_{\tilde{\sigma}}]]$$
$$= \mathsf{E}[\mathsf{E}[X_{\tilde{\sigma}}(M_{\tilde{\tau}} - M_{\tilde{\sigma}}) \mid \mathcal{F}_{\tilde{\sigma}}]]$$
$$= \mathsf{E}[X_{\tilde{\sigma}}(\mathsf{E}[M_{\tilde{\tau}} \mid \mathcal{F}_{\tilde{\sigma}}] - M_{\tilde{\sigma}})]$$
$$= 0$$

as $\mathsf{E}[M_{\tilde{\tau}} \mid \mathcal{F}_{\tilde{\sigma}}] = M_{\tilde{\sigma}}$ by part (ii) of Corollary 2.56. This proves the result when X is bounded. For (ii), approximating X by X^n, where $X_t^n = \max\{\min\{X_t, n\}, -n\}$ and using (i) we conclude

$$\mathsf{E}[X_{\sigma \wedge \theta}^n(M_{\tau \wedge \theta} - M_{\sigma \wedge \theta})] = 0.$$

Since $\sigma \leq \tau$, we can check that

$$X_{\sigma \wedge \theta}^n(M_{\tau \wedge \theta} - M_{\sigma \wedge \theta}) = X_\sigma^n(M_{\tau \wedge \theta} - M_{\sigma \wedge \theta})$$

and hence that

$$\mathsf{E}[X_\sigma^n(M_{\tau \wedge \theta} - M_{\sigma \wedge \theta})] = 0.$$

Using Doob's maximal inequality Theorem 2.26 we have $(\sup_{s \leq T}|M_s|)$ is square integrable. Since X_σ is square integrable, we conclude that

$$\mathsf{E}[|X_\sigma|(\sup_{s \leq T}|M_s|)] < \infty.$$

The required result follows using dominated convergence theorem. $\qquad\square$

Corollary 2.60 *Suppose M is a r.c.l.l. $(\mathcal{F}_.)$-martingale and σ and τ are stopping times with $\sigma \leq \tau$. Suppose U is a $\mathcal{F}_\sigma$ measurable random variable. Let*

$$N_t = U(M_{\tau \wedge t} - M_{\sigma \wedge t}).$$

Then N is a $(\mathcal{F}_.)$-martingale if either (i) U is bounded or if (ii) $\mathsf{E}[U^2] < \infty$ and M is square integrable.

Proof Let us define a process X as follows:

$$X_t(\omega) = U(\omega) 1_{[\sigma(\omega), \infty)}(t).$$

Then X is adapted by Lemma 2.41 and $X_\sigma = U$. Now the result follows from Theorem 2.59.   □

Here is another variant that will be useful later.

Theorem 2.61 *Suppose M, N are r.c.l.l. $(\mathcal{F}_.)$-martingales with $\mathsf{E}(M_t^2) < \infty$, $\mathsf{E}(N_t^2)$ $< \infty$ for all t. Let σ and τ be $(\mathcal{F}_.)$ stopping times with $\sigma \le \tau$. Suppose B, X are r.c.l.l. $(\mathcal{F}_.)$ adapted processes such that $\mathsf{E}(|B_t|) < \infty$ for all t and X is bounded. Suppose that Z is also a martingale where $Z_t = M_t N_t - B_t$. Let*

$$Y_t = X_{\sigma \wedge t}[(M_{\tau \wedge t} - M_{\sigma \wedge t})(N_{\tau \wedge t} - N_{\sigma \wedge t}) - (B_{\tau \wedge t} - B_{\sigma \wedge t})].$$

Then Y is a $(\mathcal{F}_.)$-martingale.

Proof Since X is assumed to be bounded, it follows that Y_t is integrable for all t. Once again we will show that for all bounded stopping times θ, $\mathsf{E}[Y_\theta] = 0$. Let $\tilde{\sigma} = \sigma \wedge \theta \le \tilde{\tau} = \tau \wedge \theta$. Note that

$$\begin{aligned}
\mathsf{E}[Y_\theta] &= \mathsf{E}[X_{\tilde{\sigma}}[(M_{\tilde{\tau}} - M_{\tilde{\sigma}})(N_{\tilde{\tau}} - N_{\tilde{\sigma}}) - (B_{\tilde{\tau}} - B_{\tilde{\sigma}})]] \\
&= \mathsf{E}[X_{\tilde{\sigma}}[(M_{\tilde{\tau}} N_{\tilde{\tau}} - M_{\tilde{\sigma}} N_{\tilde{\sigma}}) - (B_{\tilde{\tau}} - B_{\tilde{\sigma}}) \\
&\qquad - M_{\tilde{\sigma}}(N_{\tilde{\tau}} - N_{\tilde{\sigma}}) - N_{\tilde{\sigma}}(M_{\tilde{\tau}} - M_{\tilde{\sigma}})]] \\
&= \mathsf{E}[X_{\tilde{\sigma}}(Z_{\tilde{\tau}} - Z_{\tilde{\sigma}})] - \mathsf{E}[X_{\tilde{\sigma}} M_{\tilde{\sigma}}(N_{\tilde{\tau}} - N_{\tilde{\sigma}}) - \mathsf{E}[X_{\tilde{\sigma}} N_{\tilde{\sigma}}(M_{\tilde{\tau}} - M_{\tilde{\sigma}})]] \\
&= 0
\end{aligned}$$

by Theorem 2.59 since M, N, Z are martingales. This completes the proof.   □

2.4   A Version of Monotone Class Theorem

The following functional version of the usual monotone class theorem is very useful in dealing with integrals and their extension. Its proof is on the lines of the standard version of monotone class theorem for sets. We will include a proof; because of the central role, this result will play in our development of stochastic integrals.

Definition 2.62 A subset $\mathbb{A} \subseteq \mathbb{B}(\Omega, \mathcal{F})$ is said to be closed under uniform and monotone convergence if $f, g, h \in \mathbb{B}(\Omega, \mathcal{F})$, $f^n, g^n, h^n \in \mathbb{A}$ for $n \ge 1$ are such that

(i) $f^n \le f^{n+1}$ for all $n \ge 1$ and f^n converges to f pointwise
(ii) $g^n \ge g^{n+1}$ for all $n \ge 1$ and g^n converges to g pointwise
(iii) h^n converges to h uniformly

then $f, g, h \in \mathbb{A}$.

Here is a functional version of the monotone class theorem:

Theorem 2.63 *Let* $\mathbb{A} \subseteq \mathbb{B}(\Omega, \mathcal{F})$ *be closed under uniform and monotone conver-gence. Suppose* $\mathbb{G} \subseteq \mathbb{A}$ *is an algebra such that*

(i) $\sigma(\mathbb{G}) = \mathcal{F}$.
(ii) $\exists f^n \in \mathbb{G}$ *such that* $f^n \le f^{n+1}$ *and* f^n *converges to 1 pointwise.*

Then $\mathbb{A} = \mathbb{B}(\Omega, \mathcal{F})$.

Proof Let $\mathbb{K} \subseteq \mathbb{B}(\Omega, \mathcal{F})$ be the smallest class that contains $\mathbb{G}$ and is closed under uniform and monotone convergence. Clearly, $\mathbb{K}$ contains constants and $\mathbb{K} \subseteq \mathbb{A}$. Using arguments similar to the usual (version for sets) monotone class theorem we will first prove that $\mathbb{K}$ itself is an algebra. First we show that $\mathbb{K}$ is a vector space. For $f \in \mathbb{B}(\Omega, \mathcal{F})$, let

$$\mathbb{K}_0(f) = \{g \in \mathbb{K} : \alpha f + \beta g \in \mathbb{K}, \ \forall \alpha, \beta \in \mathbb{R}\}.$$

Note that $\mathbb{K}_0(f)$ is closed under uniform and monotone convergence. First fix $f \in \mathbb{G}$. Since $\mathbb{G}$ is an algebra, it follows that $\mathbb{G} \subseteq \mathbb{K}_0(f)$ and hence $\mathbb{K} = \mathbb{K}_0(f)$. Now fix $f \in \mathbb{K}$. The statement proven above implies that $\mathbb{G} \subseteq \mathbb{K}_0(f)$, and since $\mathbb{K}_0(f)$ is closed under uniform and monotone convergence, it follows that $\mathbb{K} = \mathbb{K}_0(f)$. Thus $\mathbb{K}$ is a vector space.

To show that $\mathbb{K}$ is an algebra, for $f \in \mathbb{B}(\Omega, \mathcal{F})$ let

$$\mathbb{K}_1(f) = \{g \in \mathbb{K} : fg \in \mathbb{K}\}.$$

Since we have shown that $\mathbb{K}$ is a vector space containing constants, it follows that

$$\mathbb{K}_1(f) = \mathbb{K}_1(f + c)$$

for any $c \in \mathbb{R}$. Clearly, if $f \ge 0$, $\mathbb{K}_1(f)$ is closed under monotone and uniform con-vergence. Given $g \in \mathbb{B}(\Omega, \mathcal{F})$, choosing $c \in \mathbb{R}$ such that $f = g + c \ge 0$, it follows that $\mathbb{K}_1(g) = \mathbb{K}_1(g + c)$ is closed under monotone and uniform convergence. Now proceeding as in the proof of $\mathbb{K}$ being vector space, we can show that $\mathbb{K}$ is closed under multiplication and thus $\mathbb{K}$ is an algebra.

Let $\mathcal{C} = \{A \in \mathcal{F} : 1_A \in \mathbb{K}\}$. In view of the assumption (ii), $\Omega \in \mathcal{C}$. It is clearly a σ-field and also $\mathbb{B}(\Omega, \mathcal{C}) \subseteq \mathbb{K}$. Since $\sigma(\mathbb{G}) = \mathcal{F}$, one has $\sigma(\mathcal{A}) = \mathcal{F}$ where

$$\mathcal{A} = \{\, \{f \le \alpha\} : f \in \mathbb{G}, \alpha \in \mathbb{R}\}.$$

Hence to complete the proof, suffices to show $\mathcal{A} \subseteq \mathcal{C}$ as that would imply $\mathcal{F} = \mathcal{C}$ and in turn

$$\mathbb{B}(\Omega, \mathcal{F}) \subseteq \mathbb{K} \subseteq \mathbb{A} \subseteq \mathbb{B}(\Omega, \mathcal{F})$$

implying $\mathbb{K} = \mathbb{A} = \mathbb{B}(\Omega, \mathcal{F})$. Now to show that $\mathcal{A} \subseteq \mathcal{C}$, fix $f \in \mathbb{G}$ and $\alpha \in \mathbb{R}$ and let $A = \{f \leq \alpha\} \in \mathcal{A}$. Let $|f| \leq M$. Let $\psi(x) = 1$ for $x \leq \alpha$ and $\psi(x) = 0$ for $x \geq \alpha + 1$ and $\psi(x) = 1 - x + \alpha$ for $\alpha < x < \alpha + 1$. Using Weierstrass's approximation theorem, we can get polynomials p_n that converge to ψ uniformly on $[-M, M]$. Now $p_n(f) \in \mathbb{G}$ as $\mathbb{G}$ is an algebra and $p_n(f)$ converges uniformly to $\psi(f)$. Thus $\psi(f) \in \mathbb{K}$. Since $\mathbb{K}$ is an algebra, it follows that $(\psi(f))^m \in \mathbb{K}$ for all $m \geq 1$. Clearly $(\psi(f))^m$ converges monotonically to 1_A. Thus $1_A \in \mathbb{K}$ i.e. $A \in \mathcal{C}$ completing the proof. $\square$

Here is a useful variant of the monotone class theorem.

Definition 2.64 A sequence of real-valued functions f_n (on a set S) is said to converge *boundedly pointwise* to a function f (written as $f_n \xrightarrow{bp} f$) if there exists number K such that $|f_n(u)| \leq K$ for all n, u and $f_n(u) \to f(u)$ for all $u \in S$.

Definition 2.65 A class $\mathbb{A} \subseteq \mathbb{B}(\Omega, \mathcal{F})$ is said to be *bp-closed* if

$$f_n \in \mathbb{A} \ \forall n \geq 1, \ f_n \xrightarrow{bp} f \text{ implies } f \in \mathbb{A}.$$

If a set $\mathbb{A}$ is bp-closed then it is also closed under monotone and uniform limits and thus we can deduce the following useful variant of the monotone class theorem from Theorem 2.63.

Theorem 2.66 *Let $\mathbb{A} \subseteq \mathbb{B}(\Omega, \mathcal{F})$ be bp-closed. Suppose $\mathbb{G} \subseteq \mathbb{A}$ is an algebra such that*

(i) $\sigma(\mathbb{G}) = \mathcal{F}$.
(ii) $\exists f^n \in \mathbb{G}$ such that $f^n \leq f^{n+1}$ and f^n converges to 1 pointwise.

Then $\mathbb{A} = \mathbb{B}(\Omega, \mathcal{F})$.

Here is an important consequence of Theorem 2.66.

Theorem 2.67 *Let $\mathcal{F}$ be a σ-field on Ω, and let $\mathbf{Q}$ be a probability measure on $(\Omega, \mathcal{F})$. Suppose $\mathbb{G} \subseteq \mathbb{B}(\Omega, \mathcal{F})$ be an algebra such that $\sigma(\mathbb{G}) = \mathcal{F}$. Further, $\exists f^n \in \mathbb{G}$ such that $f^n \leq f^{n+1}$ and f^n converges to 1 pointwise. Then $\mathbb{G}$ is dense in $\mathbb{L}^2(\Omega, \mathcal{F}, \mathbf{Q})$.*

Proof Let $\mathbb{K}$ denote the closure of $\mathbb{G}$ in $\mathbb{L}^2(\Omega, \mathcal{F}, \mathbf{Q})$ and let $\mathbb{A}$ be the set of bounded functions in $\mathbb{K}$. Then $\mathbb{G} \subseteq \mathbb{A}$, and hence by Theorem 2.66 it follows that $\mathbb{A} = \mathbb{B}(\Omega, \mathcal{F})$. Hence it follows that $\mathbb{K} = \mathbb{L}^2(\Omega, \mathcal{F}, \mathbf{Q})$ as every function in $\mathbb{L}^2(\mathbf{Q})$ can be approximated by bounded functions. $\square$

Exercise 2.68 Show that Theorem 2.66 remains true if the condition (ii) in the theorem is replaced by $\exists f^n \in \mathbb{G}$ such that $f^n \xrightarrow{bp} 1$.

2.5 The UCP Metric

Let $\mathbb{R}^0(\Omega, (\mathcal{F}_.), P)$ denote the class of all r.c.l.l. $(\mathcal{F}_.)$ adapted processes. For processes $X, Y \in \mathbb{R}^0(\Omega, (\mathcal{F}_.), P)$, let

$$\mathbf{d}_{ucp}(X, Y) = \sum_{m=1}^{\infty} 2^{-m} E[\min(1, \sup_{0 \le t \le m} |X_t - Y_t|)]. \qquad (2.5.1)$$

Noting that $\mathbf{d}_{ucp}(X, Y) = 0$ if and only if $X = Y$ (i.e. $P(X_t = Y_t \, \forall t) = 1$), it follows that $\mathbf{d}_{ucp}$ is a metric on $\mathbb{R}^0(\Omega, (\mathcal{F}_.), P)$. Now $\mathbf{d}_{ucp}(X^n, X) \to 0$ is equivalent to

$$\sup_{0 \le t \le T} |X_t^n - X_t| \text{ converges to 0 in probability } \forall T < \infty,$$

also called uniform convergence in probability, written as $X^n \xrightarrow{ucp} X$.

Remark 2.69 We have defined $\mathbf{d}_{ucp}(X, Y)$ when X, Y are real-valued r.c.l.l. processes. We can similarly define $\mathbf{d}_{ucp}(X, Y)$ when X, Y are $\mathbb{R}^d$-valued r.c.l.l. or l.c.r.l. processes. We will use the same notation $\mathbf{d}_{ucp}$ in each of these cases.

In the rest of this section, d is a fixed integer, we will be talking about $\mathbb{R}^d$-valued processes, and $|\cdot|$ will be the Euclidean norm on $\mathbb{R}^d$.

When $\mathbf{d}_{ucp}(X^n, X) \to 0$, sometimes we will write it as $X^n \xrightarrow{ucp} X$ (and thus the two mean the same thing). Let $X, Y \in \mathbb{R}^0(\Omega, (\mathcal{F}_.), P)$. Then for $\delta > 0$ and integers $N \ge 1$, observe that

$$\mathbf{d}_{ucp}(X, Y) \le 2^{-N} + \delta + P(\sup_{0 \le t \le N} |X_t - Y_t| > \delta) \qquad (2.5.2)$$

and

$$P(\sup_{0 \le t \le N} |X_t - Y_t| > \delta) \le \frac{2^N}{\delta} \mathbf{d}_{ucp}(X, Y). \qquad (2.5.3)$$

The following observation, stated here as a remark, follows from (2.5.2)–(2.5.3).

Remark 2.70 $Z^n \xrightarrow{ucp} Z$ if and only if for all integers $T < \infty$, $\varepsilon > 0, \delta > 0$, $\exists n_0$ such that for $n \ge n_0$

$$P[\sup_{t \le T} |Z_t^n - Z_t| > \delta] < \varepsilon. \qquad (2.5.4)$$

For an r.c.l.l. process Y, given $T < \infty$, $\varepsilon > 0$ one can choose $K < \infty$ such that

$$P[\sup_{t \le T} |Y_t| > K] < \varepsilon \qquad (2.5.5)$$

and hence using (2.5.4) it follows that if $Z^n \xrightarrow{ucp} Z$, then given $T < \infty$, $\varepsilon > 0$ there exists $K < \infty$ such that

$$\sup_{n \geq 1} P[\ \sup_{t \leq T} |Z_t^n| > K\ ] < \varepsilon. \tag{2.5.6}$$

The following result uses standard techniques from measure theory and functional analysis, but a proof is included as it plays an important part in subsequent chapters.

Theorem 2.71 *The space* $\mathbb{R}^0(\Omega, (\mathcal{F}_.), P)$ *is complete under the metric* $\mathbf{d}_{ucp}$.

Proof Let $\{X^n : n \geq 1\}$ be a Cauchy sequence in $\mathbf{d}_{ucp}$ metric. By taking a subsequence if necessary, we can assume without loss of generality that $\mathbf{d}_{ucp}(X^{n+1}, X^n) < 2^{-n}$ and hence

$$\sum_{n=1}^{\infty} \sum_{m=1}^{\infty} 2^{-m} \mathsf{E}[\min(1, \sup_{0 \leq t \leq m} |X_t^{n+1} - X_t^n|)] < \infty$$

or equivalently

$$\sum_{m=1}^{\infty} 2^{-m} \sum_{n=1}^{\infty} \mathsf{E}[\min(1, \sup_{0 \leq t \leq m} |X_t^{n+1} - X_t^n|)] < \infty$$

and thus for all $m \geq 1$ one has

$$\sum_{n=1}^{\infty} [\min(1, \sup_{0 \leq t \leq m} |X_t^{n+1} - X_t^n|)] < \infty \ \text{ a.s.}$$

because its expectation is finite. Note that for a sequence $\{a_n\}$, $\sum_n |a_n| < \infty$ if and only if $\sum_n \min(1, |a_n|) < \infty$. Hence for all $m \geq 1$ we have

$$\sum_{n=1}^{\infty} \sup_{0 \leq t \leq m} |X_t^{n+1} - X_t^n| < \infty \text{ a.s.}$$

Again, noting that for a real-valued sequence $\{b_n\}$, $\sum_n |b_{n+1} - b_n| < \infty$ implies that $\{b_n\}$ is Cauchy and hence converges, we conclude that outside a fixed null set (say N), X_t^n converges uniformly on $[0, m]$ for every m. So we define the limit to be X_t, which is an r.c.l.l. process. On the exceptional null set N, X_t is defined to be zero. Note that

$$\mathbf{d}_{ucp}(X^{n+k}, X^n) \leq \sum_{j=n}^{n+k-1} \mathbf{d}_{ucp}(X^{j+1}, X^j)$$

$$\leq \sum_{j=n}^{n+k-1} 2^{-j}$$

$$\leq 2^{-n+1}.$$

As a consequence, (using definition of $\mathbf{d}_{ucp}$) we get that for all integers m,

$$E[\min(1, \sup_{0 \le t \le m} |X_t^n - X_t^{n+k}|)] \le 2^m 2^{-n+1}.$$

Taking limit as $k \to \infty$ and invoking dominated convergence theorem we conclude

$$E[\min(1, \sup_{0 \le t \le m} |X_t^n - X_t|)] \le 2^m 2^{-n+1}.$$

It follows that for any $T < \infty$, $0 < \delta < 1$ we have

$$P(\sup_{0 \le t \le T} |X_t^n - X_t|) > \delta) \le \frac{1}{\delta} 2^{(T+1)} 2^{-n+1}. \tag{2.5.7}$$

and hence, invoking Remark 2.70, it follows that X^n converges in *ucp* metric to X. Thus every Cauchy sequence converges and so the space is complete under the metric $\mathbf{d}_{ucp}$. $\qquad\square$

Here is a result that will be useful later.

Theorem 2.72 *Suppose Z^n, Z are r.c.l.l. adapted processes such that*

$$Z^n \xrightarrow{ucp} Z.$$

Then there exists a subsequence $\{n^k\}$ such that $Y^k = Z^{n^k}$ satisfies

(i) $\sup_{0 \le t \le T} |Y_t^k - Z_t| \to 0$ a.s. $\forall T < \infty$.
(ii) *There exists an r.c.l.l. adapted increasing process H such that*

$$|Y_t^k| \le H_t \ \forall t < \infty, \ \forall k \ge 1, \tag{2.5.8}$$

Proof Since $\mathbf{d}_{ucp}(Z^n, Z) \to 0$, for $k \ge 1$, we can choose n^k with such that $\mathbf{d}_{ucp}(Z^{n^k}, Z) \le 2^{-k}$ and $n^{k+1} > n^k$. Let $Y^k = Z^{n^k}$. Then as seen in the proof of Theorem 2.71, this implies

$$\sum_{k=1}^{\infty} [\sup_{t \le T} |Y_t^k - Z_t|] < \infty, \ \forall T < \infty \ a.s.$$

Hence (i) above holds for this choice of $\{n^k\}$. Further, $U_s = \sum_{k=1}^{\infty} |Y_s^k - Z_s| + |Z_s|$ is an r.c.l.l. process as the series converges uniformly on $[0, T]$ for every $T < \infty$. Thus defining

$$H_t = \sup_{0 \le s \le t} [\sum_{k=1}^{\infty} |Y_s^k - Z_s| + |Z_s|]$$

it follows that H is an r.c.l.l. adapted increasing process. See Exercise 2.10. Clearly, $|Y_t^k| \le H_t$ for all $k \ge 1$. $\qquad\square$

Remark **2.73** If we have two, or finitely many sequences $\{Z^{i,n}\}$ converging to Z^i in $\mathbf{d}_{ucp}$, $i = 1, 2, \ldots, p$ then we can get one common subsequence $\{n^k\}$ and a process H such that (i), (ii) above hold for $i = 1, 2, \ldots p$. All we need to do is to choose $\{n^k\}$ such that

$$\mathbf{d}_{ucp}(Z^{i,n^k}, Z^i) \leq 2^{-k}, \quad i = 1, 2 \ldots p.$$

Exercise 2.74 An alternative way of obtaining the conclusion in Remark 2.73 is to apply Theorem 2.72 to an appropriately defined $\mathbb{R}^{dp}$-valued processes.

The following lemma will play an important role in the theory of stochastic integration.

Lemma 2.75 *Let Z^n, Z be adapted processes and let τ^m be a sequence of stopping times such that $\tau^m \uparrow \infty$. Suppose that for each m*

$$Z^n_{t \wedge \tau^m} \xrightarrow{ucp} Z_{t \wedge \tau^m} \text{ as } n \uparrow \infty. \tag{2.5.9}$$

Then

$$Z^n \xrightarrow{ucp} Z \text{ as } n \uparrow \infty.$$

Proof Fix $T < \infty$, $\varepsilon > 0$ and $\eta > 0$. We need to show that $\exists n_0$ such that for $n \geq n_0$

$$P[\sup_{t \leq T} |Z^n_t - Z_t| > \eta] < \varepsilon. \tag{2.5.10}$$

First, using $\tau^m \uparrow \infty$, fix m such that

$$P[\tau^m < T] < \varepsilon/2. \tag{2.5.11}$$

Using $Z^n_{t \wedge \tau^m} \xrightarrow{ucp} Z_{t \wedge \tau^m}$, get n_0 such that for $n \geq n_0$

$$P[\sup_{t \leq T} |Z^n_{t \wedge \tau^m} - Z_{t \wedge \tau^m}| > \eta] < \varepsilon/2. \tag{2.5.12}$$

Now,

$$\{\sup_{t \leq T} |Z^n_t - Z_t| > \eta\} \subseteq \{\sup_{t \leq T} |Z^n_{t \wedge \tau^m} - Z_{t \wedge \tau^m}| > \eta\} \cup \{\tau^m < T\}$$

and hence for $n \geq n_0$, (2.5.10) follows from (2.5.11) and (2.5.12). $\square$

The same argument as above also yields the following.

Corollary 2.76 *Let Z^n be r.c.l.l. adapted processes and let τ^m be a sequence of stopping times such that $\tau^m \uparrow \infty$. For $n \geq 1, m \geq 1$ let*

$$Y^{n,m}_t = Z^n_{t \wedge \tau^m}.$$

Suppose that for each m, $\{Y^{n,m} : n \geq 1\}$ is Cauchy in $\mathbf{d}_{ucp}$ metric then Z^n is Cauchy in $\mathbf{d}_{ucp}$ metric.

2.6 The Lebesgue–Stieltjes Integral

Let $G : [0, \infty) \mapsto \mathbb{R}$ be an r.c.l.l. function. For $0 \leq a < b < \infty$ the *total variation* $\text{VAR}_{[a,b]}$ of $G(s)$ over $[a, b]$ and $\text{VAR}_{(a,b]}$ over $(a, b]$ are defined as follows:

$$\text{VAR}_{(a,b]}(G) = \sup\{\sum_{j=1}^{m} |G(t_j) - G(t_{j-1})| : a = t_0 < t_1 < \ldots < t_m = b, \, m \geq 1\}. \quad (2.6.1)$$

$$\text{VAR}_{[a,b]}(G) = |G(a)| + \text{VAR}_{(a,b]}(G).$$

If $\text{VAR}_{[0,t]}(G) < \infty$ for all t, then G will be called a function with *finite variation*. It is well known that a function has finite variation paths if and only if it can be expressed as difference of two increasing functions.

If $\text{VAR}_{[0,t]}(G) < \infty$ for all t, the function $|G|_t = \text{VAR}_{[0,t]}(G)$ is then an increasing $[0, \infty)$-valued function. Let us fix such a function G.

For any T fixed, there exists a unique countably additive measure ν and a countably additive signed measure μ on the Borel σ-field of $[0, T]$ such that

$$\nu([0, t]) = |G|(t) \quad \forall t \leq T \quad (2.6.2)$$

$$\mu([0, t]) = G(t) \quad \forall t \leq T. \quad (2.6.3)$$

Here, ν is the total variation measure of the signed measure μ.

For measurable function h on $[0, T]$, if $\int h d\nu < \infty$, then we define

$$\int_0^t |h|_s d|G|_s = \int |h| 1_{[0,t]} d\nu \quad (2.6.4)$$

and

$$\int_0^t h_s dG_s = \int h 1_{[0,t]} d\mu. \quad (2.6.5)$$

Note that we have

$$|\int_0^t h_s dG_s| \leq \int_0^t |h|_s d|G|_s. \quad (2.6.6)$$

It follows that if h is a bounded measurable function on $[0, \infty)$, then $\int_0^t h dG$ is defined and further if $h^n \xrightarrow{bp} h$, then the dominated convergence theorem yields that $H_t^n = \int_0^t h^n dG$ converges to $H(t) = \int_0^t h dG$ uniformly on compact subsets.

Exercise 2.77 Let G be an r.c.l.l. function on $[0, \infty)$ such that $|G|_t < \infty$ for all $t < \infty$. Show that for all $T < \infty$

$$\sum_{t \leq T} |(\Delta G)_t| < \infty. \tag{2.6.7}$$

Note that as seen in Exercise 2.1, $\{t : |(\Delta G)_t| > 0\}$ is a countable set and thus the sum appearing above is a sum of countably many terms.

HINT: Observe that the left-hand side in (2.6.7) is less than or equal to $|G|_T$.

Exercise 2.78 Let B be an r.c.l.l. adapted processes such that $A_t(\omega) = \text{VAR}_{[0,t]}$ $(B.(\omega)) < \infty$ for all t, ω. Show that A is an r.c.l.l. adapted process.

Let us denote by $\mathbb{V}^+ = \mathbb{V}^+(\Omega, (\mathcal{F}.), P)$ the class of $(\mathcal{F}.)$ adapted r.c.l.l. increasing processes A with $A_0 \geq 0$ and by $\mathbb{V} = \mathbb{V}(\Omega, (\mathcal{F}.), P)$ the class of r.c.l.l. adapted processes B such that

$$A_t(\omega) = \text{VAR}_{[0,t]}(B.(\omega)) < \infty \quad \forall t \geq 0, \forall \omega \in \Omega. \tag{2.6.8}$$

As seen above, $A \in \mathbb{V}^+$. A process $B \in \mathbb{V}$ will be called process with *finite variation paths*. It is easy to see that $B \in \mathbb{V}$ if and only if B can be written as difference of two processes in $\mathbb{V}^+$: indeed, if A is defined by (2.6.8), we have $B = D - C$ where $D = \frac{1}{2}(A + B)$ and $C = \frac{1}{2}(A - B)$ and $C, D \in \mathbb{V}^+$. Let $\mathbb{V}_0$ and $\mathbb{V}_0^+$ denote the class of processes A in $\mathbb{V}$ and $\mathbb{V}^+$, respectively, such that $A_0 = 0$. For $B \in \mathbb{V}$, we will denote the process A defined by (2.6.8) as $A = \text{VAR}(B)$.

A process $A \in \mathbb{V}$ will be said to be purely discontinuous if

$$A_t = A_0 + \sum_{0 < s \leq t} (\Delta A)_s.$$

Exercise 2.79 Show that every $A \in \mathbb{V}$ can be written uniquely as $A = B + C$ with $B, C \in \mathbb{V}$, B being a continuous process with $B_0 = 0$ and C being a purely discontinuous process.

Lemma 2.80 *Let $B \in \mathbb{V}$ and let X be a bounded l.c.r.l. adapted process. Then*

$$C_t(\omega) = \int_0^t X_s(\omega) d B_s(\omega) \tag{2.6.9}$$

is well defined and is an r.c.l.l. adapted process. Further, $C \in \mathbb{V}$.

Proof For every $\omega \in \Omega, t \mapsto X_t(\omega)$ is a bounded measurable function and hence C is well defined. For $n \geq 1$ and $i \geq 0$ let $t_i^n = \frac{i}{n}$. Let

$$X_t^n = X_{t_i^n} \quad \text{for } t_i^n \leq t < t_{i+1}^n.$$

Then $X^n \xrightarrow{bp} X$. Clearly

$$C_t^n(\omega) = \int_0^t X_s^n(\omega)\, dB_s(\omega) = \sum_{i:t_i^n \le t} X_{t_i^n}(B_{t_{i+1}^n \wedge t} - B_{t_i^n \wedge t})$$

is r.c.l.l. adapted and further for every $\omega \in \Omega$, $C_t^n(\omega) \to C_t(\omega)$ uniformly in $t \in [0, T]$ for every $T < \infty$. Thus C is also an r.c.l.l. adapted process. Let $A = \mathrm{VAR}(B)$. Let K be a bound for the process X. For any $s < t$,

$$|C_t(\omega) - C_s(\omega)| \le \int_s^t |X_u(\omega)|\, dA_s(\omega)$$
$$\le K(A_t(\omega) - A_s(\omega)).$$

Since A is an increasing process, it follows that $\mathrm{VAR}_{[0,T]}(C)(\omega) \le K A_T(\omega) < \infty$ for all $T < \infty$ and for all ω. $\qquad\square$

Exercise 2.81 Show that the conclusion in Lemma 2.80 is true even when X is a bounded r.c.l.l. adapted process. In fact show that D_t^n defined by

$$D_t^n(\omega) = \sum_{i:t_i^n \le t} X_{t_{i+1}^n \wedge t}(B_{t_{i+1}^n \wedge t} - B_{t_i^n \wedge t})$$

converges to $\int_0^t X\, dB$.

Exercise 2.82 Show that the assumption of boundedness of X in Lemma 2.80 can be dropped.

Chapter 3
The Ito's Integral

We begin this chapter with the quadratic variation and Levy's characterization of the Brownian motion. Later, we will outline the basic development of the Ito's Integral w.r.t. Brownian motion. We also discuss existence and uniqueness of solutions to the classical stochastic differential equations driven by Brownian motion.

3.1   Quadratic Variation of Brownian Motion

Let (W_t) denote a one-dimensional Brownian motion on a probability space $(\Omega, \mathcal{F}, \mathsf{P})$. We have seen that W is a martingale w.r.t. its natural filtration $(\mathcal{F}_{\cdot}^W)$ and with $M_t = W_t^2 - t$, M is also $(\mathcal{F}_{\cdot}^W)$-martingale. These properties are easy consequence of the independent increment property of Brownian motion.

Wiener and Ito's realized the need to give a meaning to limit of what appeared to be Riemann–Stieltjes sums for the integral

$$\int_0^t f_s \, dW_s \tag{3.1.1}$$

in different contexts—while in case of Wiener, the integrand was a deterministic function, Ito's needed to consider a random process (f_s) that was a non-anticipating function of W—i.e. f is adapted to $(\mathcal{F}_{\cdot}^W)$.

It is well known that paths $s \mapsto W_s(\omega)$ are nowhere differentiable for almost all ω, and hence we cannot interpret the integral in (3.1.1) as a path-by-path Riemann–Stieltjes integral. We will deduce the later from the following result that is relevant for stochastic integration.

© Springer Nature Singapore Pte Ltd. 2018
R. L. Karandikar and B. V. Rao, *Introduction to Stochastic Calculus*,
Indian Statistical Institute Series, https://doi.org/10.1007/978-981-10-8318-1_3

Theorem 3.1 *Let* (W_t) *be a Brownian motion. Let* $t_i^n = i2^{-n}$, $i \geq 0$, $n \geq 1$. *Let* $V_t^n = \sum_{i=0}^{\infty} |W_{t_{i+1}^n \wedge t} - W_{t_i^n \wedge t}|$, $Q_t^n = \sum_{i=0}^{\infty} (W_{t_{i+1}^n \wedge t} - W_{t_i^n \wedge t})^2$. *Then for all* $t > 0$, *(a)* $V_t^n \to \infty$ *a.s. and (b)* $Q_t^n \to t$ *a.s.*

Proof We will first prove (b). Let us fix $t < \infty$ and let

$$X_i^n = W_{t_{i+1}^n \wedge t} - W_{t_i^n \wedge t}, \quad Z_i^n = (W_{t_{i+1}^n \wedge t} - W_{t_i^n \wedge t})^2 - (t_{i+1}^n \wedge t - t_i^n \wedge t).$$

Then from properties of Brownian motion it follows that $\{X_i^n, \ i \geq 0\}$ are independent random variables with normal distribution and $\mathsf{E}(X_i^n) = 0$, $\mathsf{E}(X_i^n)^2 = (t_{i+1}^n \wedge t - t_i^n \wedge t)$. So, $\{Z_i^n, \ i \geq 0\}$ are independent random variables with $\mathsf{E}(Z_i^n) = 0$ and $\mathsf{E}(Z_i^n)^2 = 2(t_{i+1}^n \wedge t - t_i^n \wedge t)^2$. Now

$$
\begin{aligned}
\mathsf{E}(Q_t^n - t)^2 &= \mathsf{E}(\sum_{i=0}^{\infty} Z_i^n)^2 \\
&= \sum_{i=0}^{\infty} \mathsf{E}(Z_i^n)^2 \\
&= 2 \sum_{i=0}^{\infty} (t_{i+1}^n \wedge t - t_i^n \wedge t)^2 \\
&\leq 2^{-n+1} \sum_{i=0}^{\infty} (t_{i+1}^n \wedge t - t_i^n \wedge t) \\
&= 2^{-n+1} t.
\end{aligned}
\tag{3.1.2}
$$

Note that each of the sum appearing above is actually a finite sum. Thus

$$\mathsf{E} \sum_{n=1}^{\infty} (Q_t^n - t)^2 \leq t < \infty$$

so that $\sum_{n=1}^{\infty} (Q_t^n - t)^2 < \infty$ a.s. and hence $Q_t^n \to t$ a.s.

For (a), let $\alpha(\delta, \omega, t) = \sup\{|W_u(\omega) - W_v(\omega)| : |u - v| \leq \delta, u, v \in [0, t]\}$. Then uniform continuity of $u \mapsto W_u(\omega)$ implies that for all t finite and for each ω,

$$\lim_{\delta \downarrow 0} \alpha(\delta, \omega, t) = 0. \tag{3.1.3}$$

Now note that for any ω,

$$
\begin{aligned}
Q_t^n(\omega) &\leq (\max_{0 \leq i < \infty} |W_{t_{i+1}^n \wedge t}(\omega) - W_{t_i^n \wedge t}(\omega)|)(\sum_{i=0}^{\infty} |W_{t_{i+1}^n \wedge t} - W_{t_i^n \wedge t}|) \\
&= \alpha(2^{-n}, \omega, t) V_t^n(\omega).
\end{aligned}
\tag{3.1.4}
$$

So if $\liminf_n V_t^n(\omega) < \infty$ for some ω, then $\liminf_n Q_t^n(\omega) = 0$ in view of (3.1.3) and (3.1.4).

For $t > 0$, since $Q_t^n \to t$ a.s., we must have $V_t^n \to \infty$ a.s. $\qquad\square$

Exercise 3.2 For any sequence of partitions

$$0 = t_0^m < t_1^m < \ldots < t_n^m < \ldots; \quad t_n^m \uparrow \infty \text{ as } n \uparrow \infty \tag{3.1.5}$$

of $[0, \infty)$ such that for all $T < \infty$,

$$\delta_m(T) = (\sup_{\{n : t_n^m \leq T\}} (t_{n+1}^m - t_n^m)) \to 0 \quad \text{as } m \uparrow \infty \tag{3.1.6}$$

let

$$Q_t^m = \sum_{n=0}^{\infty} (W_{t_{n+1}^m \wedge t} - W_{t_n^m \wedge t})^2. \tag{3.1.7}$$

Show that for each t, Q_t^m converges in probability to t.

Remark 3.3 It is well known that the paths of Brownian motion are nowhere differentiable. For this and other path properties of Brownian motion, see Breiman [5], McKean [47], Karatzas and Shreve [43].

Remark 3.4 Since the paths of Brownian motion do not have finite variation on any interval, we cannot invoke Riemann–Stieltjes integration theory for interpreting $\int X \, dW$, where W is Brownian motion. The following calculation shows that the Riemann–Stieltjes sums do not converge in any weaker sense (say in probability) either. Let us consider $\int W \, dW$. Let $t_i^n = i2^{-n}, i \geq 0, n \geq 1$. The question is whether the sums

$$\sum_{i=0}^{\infty} W_{s_i^n \wedge t}(W_{t_{i+1}^n \wedge t} - W_{t_i^n \wedge t}) \tag{3.1.8}$$

converge to some limit for all choices of s_i^n such that $t_i^n \leq s_i^n \leq t_{i+1}^n$. Let us consider two cases $s_i^n = t_{i+1}^n$ and $s_i^n = t_i^n$:

$$A_t^n = \sum_{i=0}^{\infty} W_{t_{i+1}^n \wedge t}(W_{t_{i+1}^n \wedge t} - W_{t_i^n \wedge t}) \tag{3.1.9}$$

$$B_t^n = \sum_{i=0}^{\infty} W_{t_i^n \wedge t}(W_{t_{i+1}^n \wedge t} - W_{t_i^n \wedge t}). \tag{3.1.10}$$

Now A_t^n and B_t^n cannot converge to the same limit as their difference satisfies

$$(A_t^n - B_t^n) = Q_t^n.$$

Thus even in this simple case, the Riemann–Stieltjes sums do not converge to a unique limit. In this case, it is possible to show that A_t^n and B_t^n actually do converge but to two different limits. It is possible to choose $\{s_i^n\}$ so that the Riemann sums in (3.1.8) do not converge.

3.2   Levy's Characterization of Brownian Motion

Definition 3.5   Let (W_t) be d-dimensional Brownian motion adapted to a filtration $(\mathcal{F}_t)$. Then $(W_t, \mathcal{F}_t)_{\{t \geq 0\}}$ is said to be a Wiener martingale if W is a martingale w.r.t. $(\mathcal{F}_\cdot)$ and

$$\{W_t - W_s : t \geq s\} \text{ is independent of } \mathcal{F}_s. \tag{3.2.1}$$

It follows that for a one-dimensional Wiener martingale $(W_t, \mathcal{F}_t)_{\{t \geq 0\}}$ $M_t = W_t^2 - t$ is also a martingale w.r.t $(\mathcal{F}_t)$. Levy had proved that if W is any continuous process such that both W, M are $(\mathcal{F}_\cdot)$-martingales then W is a Brownian motion and (3.2.1) holds. Most proofs available in texts today deduce this as an application of Ito's formula. We will give an elementary proof of this result which uses interplay of martingales and stopping times. The proof is motivated by the proof given in Ito's lecture notes [25], but the same has been simplified using partition via stopping times instead of deterministic partitions.

Exercise 3.6   Show that (3.2.1) is equivalent to

$$W_t - W_s \text{ is independent of } \mathcal{F}_s \text{ for all } t \geq s. \tag{3.2.2}$$

We will use the following inequalities on the exponential function which can be easily proven using Taylor's theorem with remainder. For $a, b \in \mathbb{R}$

$$|e^b - (1 + b)| \leq \frac{1}{2} e^{|b|} |b|^2$$

$$|e^{ia} - 1| \leq |a|$$

$$|e^{ia} - (1 + ia - \frac{1}{2} a^2)| \leq \frac{1}{6} |a|^3.$$

Using these inequalities, we conclude that for a, b such that $|a| \leq \delta$, $|b| \leq \delta$, $\delta < \log_e(2)$, we have

$$|e^{ia+b} - (1 + ia - \frac{1}{2}a^2 + b)|$$

$$\leq |e^{ia}[e^b - (1+b)] + b(e^{ia} - 1) + (e^{ia} - (1 + ia - \frac{1}{2}a^2))| \tag{3.2.3}$$

$$\leq (\frac{1}{2}e^{|b|}|b|^2 + |b||a| + \frac{1}{6}|a|^3)$$

$$\leq \delta(2|b| + |a|^2).$$

Theorem 3.7 *Let X be a continuous process adapted to a filtration $(\mathcal{F}_.)$ and let $M_t = X_t^2 - t$ for $t \geq 0$. Suppose that (i) $X_0 = 0$, (ii) X is a $(\mathcal{F}_.)$-martingale and (iii) M is a $(\mathcal{F}_.)$-martingale. Then X is a Brownian motion and further, for all s*

$$\{(X_t - X_s) : t \geq s\} \text{ is independent of } \mathcal{F}_s.$$

Proof We will prove this in a series of steps.
step 1: For bounded stopping times $\sigma \leq \tau$, say $\tau \leq T$,

$$\mathsf{E}[(X_\tau - X_\sigma)^2] = \mathsf{E}[(\tau - \sigma)]. \tag{3.2.4}$$

To see this, Corollary 2.60 and the hypothesis that M is a martingale imply that

$$Y_t = X_{\tau \wedge t}^2 - X_{\sigma \wedge t}^2 - (\tau \wedge t - \sigma \wedge t)$$

is a martingale and hence $\mathsf{E}[Y_T] = \mathsf{E}[Y_0] = 0$. This proves step 1.
Let us fix $\lambda \in \mathbb{R}$ and let

$$Z_t^\lambda = \exp\{i\lambda X_t + \frac{1}{2}\lambda^2 t\}.$$

step 2: For each bounded stopping time σ, $\mathsf{E}[Z_\sigma^\lambda] = 1$. This would show that Z^λ is a $(\mathcal{F}_.)$-martingale. To prove this claim, fix λ and let $\sigma \leq T$. Let δ be sufficiently small such that $(|\lambda| + \lambda^2)\delta < \log_e(2)$. Let us define a sequence of stopping times $\{\tau_i : i \geq 1\}$ inductively as follows: $\tau_0 = 0$ and for $i \geq 0$,

$$\tau_{i+1} = \inf\{t \geq \tau_i : |X_t - X_{\tau_i}| \geq \delta \text{ or } |t - \tau_i| \geq \delta \text{ or } t \geq \sigma\} \tag{3.2.5}$$

Inductively, using Theorem 2.46 one can prove that each τ_i is a stopping time and that for each ω, $\tau_j(\omega) = \sigma(\omega)$ for sufficiently large j. Further, continuity of the process implies

$$|X_{\tau_{i+1}} - X_{\tau_i}| \leq \delta, \quad |\tau_{i+1} - \tau_i| \leq \delta. \tag{3.2.6}$$

Let us write

$$Z_{\tau_m}^\lambda - 1 = \sum_{k=0}^{m-1}(Z_{\tau_{k+1}}^\lambda - Z_{\tau_k}^\lambda). \tag{3.2.7}$$

Note that

$$\mathsf{E}[(Z^\lambda_{\tau_{k+1}} - Z^\lambda_{\tau_k})] = \mathsf{E}[Z^\lambda_{\tau_k}(e^{i\lambda(X_{\tau_{k+1}} - X_{\tau_k}) + \frac{1}{2}\lambda^2(\tau_{k+1} - \tau_k)} - 1)]$$

$$= \mathsf{E}[Z^\lambda_{\tau_k}\mathsf{E}(e^{i\lambda(X_{\tau_{k+1}} - X_{\tau_k}) + \frac{1}{2}\lambda^2(\tau_{k+1} - \tau_k)} - 1 \mid \mathcal{F}_{\tau_k})]$$

Since $\mathsf{E}[X_{\tau_{k+1}} - X_{\tau_k} \mid \mathcal{F}_{\tau_k}] = 0$ as X is a martingale and

$$\mathsf{E}[(X_{\tau_{k+1}} - X_{\tau_k})^2 - (\tau_{k+1} - \tau_k) \mid \mathcal{F}_{\tau_k}] = 0 \qquad (3.2.8)$$

as seen in step 1 above, we have

$$\mathsf{E}[e^{i\lambda(X_{\tau_{k+1}} - X_{\tau_k}) + \frac{1}{2}\lambda^2(\tau_{k+1} - \tau_k)} - 1 \mid \mathcal{F}_{\tau_k}]$$

$$= \mathsf{E}[e^{i\lambda(X_{\tau_{k+1}} - X_{\tau_k}) + \frac{1}{2}\lambda^2(\tau_{k+1} - \tau_k)} - \{1 + i\lambda(X_{\tau_{k+1}} - X_{\tau_k})$$

$$- \tfrac{1}{2}\lambda^2(X_{\tau_{k+1}} - X_{\tau_k})^2 + \tfrac{1}{2}\lambda^2(\tau_{k+1} - \tau_k)\} \mid \mathcal{F}_{\tau_k}]$$

Using (3.2.3), (3.2.6) and the choice of δ, we can conclude that the expression on the right-hand side inside the conditional expectation is bounded by

$$\delta\lambda^2((X_{\tau_{k+1}} - X_{\tau_k})^2 + (\tau_{k+1} - \tau_k))$$

Putting together these observations and (3.2.8), we conclude that

$$|\mathsf{E}[(Z^\lambda_{\tau_{k+1}} - Z^\lambda_{\tau_k})]| \le 2\delta\lambda^2 e^{\frac{1}{2}\lambda^2 T}\mathsf{E}[(\tau_{k+1} - \tau_k)].$$

As a consequence

$$|\mathsf{E}[Z^\lambda_{\tau_m} - 1]| \le 2\delta\lambda^2 e^{\frac{1}{2}\lambda^2 T}\mathsf{E}[\tau_m].$$

Now $Z^\lambda_{\tau_m}$ is bounded by $e^{\frac{1}{2}\lambda^2 T}$ and converges to Z^λ_σ, we conclude

$$|\mathsf{E}[Z^\lambda_\sigma - 1]| \le 2\delta\lambda^2 T e^{\frac{1}{2}\lambda^2 T}.$$

Since this holds for all small $\delta > 0$ it follows that

$$\mathsf{E}[Z^\lambda_\sigma] = 1$$

and this completes the proof of step 2.
step 3: For $s < t, \lambda \in \mathbb{R}$,

$$\mathsf{E}[e^{(i\lambda(X_t - X_s))} \mid \mathcal{F}_s] = e^{-\frac{1}{2}\lambda^2(t-s)}. \qquad (3.2.9)$$

We have seen that Z^λ_t is a martingale and (3.2.9) follows from it since X_s is $\mathcal{F}_s$ measurable.

As a consequence

$$\mathsf{E}[e^{(i\lambda(X_t-X_s))}] = e^{-\frac{1}{2}\lambda^2(t-s)}, \tag{3.2.10}$$

and so the distribution of $X_t - X_s$ is Gaussian with mean 0 and variance $(t - s)$.
step 4:
For $s < t$, $\lambda, \theta \in \mathbb{R}$, a $\mathcal{F}_s$ measurable random variable Y we have

$$\mathsf{E}[e^{(i\lambda(X_t-X_s)+i\theta Y)}] = \mathsf{E}[e^{(i\lambda(X_t-X_s))}]\mathsf{E}[e^{(i\theta Y)}]. \tag{3.2.11}$$

The relation (3.2.11) is an immediate consequence of (3.2.9).

We have already seen that $X_t - X_s$ has Gaussian distribution with mean 0 and variance $t - s$ and (3.2.11) implies that $X_t - X_s$ is independent of $\mathcal{F}_s$, in particular, $X_t - X_s$ is independent of $\{X_u : u \leq s\}$. This completes the proof. $\qquad\square$

Let $W = (W^1, W^2, \ldots, W^d)$ be d-dimensional Brownian motion, where W^j is the jth component, i.e. each W^j is a real-valued Brownian motion and $W^1, W^2, \ldots, W^d$ are independent.

For any $\theta = (\theta^1, \theta^2, \ldots, \theta^d) \in \mathbb{R}^d$ with $|\theta| = \sum_{j=1}^d (\theta^j)^2 = 1$,

$$X_t^\theta = \sum_{j=1}^d \theta^j W_t^j \tag{3.2.12}$$

is itself a Brownian motion. Defining

$$M_t^\theta = (X_t^\theta)^2 - t, \tag{3.2.13}$$

we have

$$\forall \theta \in \mathbb{R}^d \text{ with } |\theta| = 1; \quad X^\theta, M^\theta \text{ are } (\mathcal{F}_\cdot^W)\text{-martingales}. \tag{3.2.14}$$

Indeed, using Theorem 3.7, we will show that (3.2.14) characterizes multidimensional Brownian motion.

Theorem 3.8 *Let W be an $\mathbb{R}^d$-valued continuous process such that $W_0 = 0$. Suppose $(\mathcal{F}_\cdot)$ is a filtration such that W is $(\mathcal{F}_\cdot)$ adapted. Suppose W satisfies*

$$\forall \theta \in \mathbb{R}^d \text{ with } |\theta| = 1; \quad X^\theta, M^\theta \text{ are } (\mathcal{F}_\cdot)\text{-martingales}, \tag{3.2.15}$$

where X^θ and M^θ are defined via (3.2.12) and (3.2.13). Then W is a d-dimensional Brownian motion and further, for $0 \leq s \leq t$

$$(W_t - W_s) \text{ is independent of } \mathcal{F}_s.$$

Proof Theorem 3.7 implies that for $\theta \in \mathbb{R}^d$ with $|\theta| = 1$ and $\lambda \in \mathbb{R}$

$$\mathsf{E}[\exp\{i\lambda(\theta \cdot W_t - \theta \cdot W_s)\} \mid \mathcal{F}_s] = \exp\{-\tfrac{1}{2}\lambda^2(t-s)\}.$$

This implies that W is a Brownian motion. Independence of $\{W_t - W_s : t \geq s\}$ and $\mathcal{F}_s$ also follows as in Theorem 3.7. $\square$

Theorems 3.7 and 3.8 are called Levy's characterization of Brownian motion (one-dimensional and multidimensional cases, respectively).

3.3   The Ito's Integral

Let $\mathbb{S}$ be the class of stochastic processes f of the form

$$f_s(\omega) = a_0(\omega)1_{\{0\}}(s) + \sum_{j=0}^{m} a_{j+1}(\omega)1_{(s_j, s_{j+1}]}(s) \tag{3.3.1}$$

where $0 = s_0 < s_1 < s_2 < \ldots < s_{m+1} < \infty$, a_j is bounded $\mathcal{F}_{s_{j-1}}$ measurable random variable for $1 \leq j \leq (m+1)$, and a_0 is bounded $\mathcal{F}_0$ measurable. Elements of $\mathbb{S}$ will be called simple processes. For an f given by (3.3.1), we define $X = \int f \, dW$ by

$$X_t(\omega) = \sum_{j=0}^{m} a_{j+1}(\omega)(W_{s_{j+1} \wedge t}(\omega) - W_{s_j \wedge t}(\omega)). \tag{3.3.2}$$

a_0 does not appear on the right side because $W_0 = 0$. It can be easily seen that $\int f \, dW$ defined via (3.3.1) and (3.3.2) for $f \in \mathbb{S}$ does not depend upon the representation (3.3.1). In other words, if g is given by

$$g_t(\omega) = b_0(\omega)1_{\{0\}}(s) + \sum_{j=0}^{n} b_{j+1}(\omega)1_{(r_j, r_{j+1}]}(t) \tag{3.3.3}$$

where $0 = r_0 < r_1 < \ldots < r_{n+1}$ and b_j is $\mathcal{F}_{r_{j-1}}$ measurable bounded random variable, $1 \leq j \leq (n+1)$, and b_0 is bounded $\mathcal{F}_0$ measurable and $f = g$, then $\int f \, dW = \int g \, dW$, i.e.

$$\sum_{j=0}^{m} a_{j+1}(\omega)(W_{s_{j+1} \wedge t}(\omega) - W_{s_j \wedge t}(\omega))$$

$$= \sum_{j=0}^{n} b_{j+1}(\omega)(W_{r_{j+1} \wedge t}(\omega) - W_{r_j \wedge t}(\omega)). \tag{3.3.4}$$

By definition, X is a continuous adapted process. We will denote X_t as $\int_0^t f \, dW$. We will obtain an estimate on the growth of the integral defined above for simple

$f \in \mathbb{S}$ and then extend the integral to an appropriate class of integrands—those that can be obtained as limits of simple processes. This approach is different from the one adopted by Ito's, and we have adopted this approach with an aim to generalize the same to martingales.

We first note some properties of $\int f\, dW$ for $f \in \mathbb{S}$ and obtain an estimate.

Lemma 3.9 *Let* $f, g \in \mathbb{S}$ *and let* $a, b \in \mathbb{R}$. *Then*

$$\int_0^t (af + bg)\, dW = a \int_0^t f\, dW + b \int_0^t g\, dW. \qquad (3.3.5)$$

Proof Let f, g have representations (3.3.1) and (3.3.3), respectively. Easy to see that we can get $0 = t_0 < t_1 < \ldots < t_k$ such that

$$\{t_j : 0 \le j \le k\} = \{s_j : 0 \le j \le m\} \cup \{r_j : 0 \le j \le n\}$$

and then represent both f, g over common time partition. Then the result (3.3.5) follows easily. $\qquad \square$

Lemma 3.10 *Let* $f, g \in \mathbb{S}$, *and let* $Y_t = \int_0^t f\, dW$, $Z_t = \int_0^t g\, dW$ *and* $A_t = \int_0^t f_s g_s\, ds$, $M_t = Y_t Z_t - A_t$. *Then* Y, Z, M *are* $(\mathcal{F}_{\cdot})$-*martingales.*

Proof By linearity property (3.3.5) and the fact that sum of martingales is a martingale, suffices to prove the lemma in the following two cases:
Case 1: $0 \le s < r$ and

$$f_t = a 1_{(s,r]}(t), \quad g_t = b 1_{(s,r]}(t), \quad a, b \text{ are } \mathcal{F}_s \text{ measurable}.$$

Case 2: $0 \le s < r \le u < v$ and

$$f_t = a 1_{(s,r]}(t), \quad g_t = b 1_{(u,v]}(t), \quad a \text{ is } \mathcal{F}_s \text{ measurable and } b \text{ is } \mathcal{F}_u \text{ measurable}.$$

Here in both cases, a, b are assumed to be bounded. In both the cases, $Y_t = a(W_{t \wedge r} - W_{t \wedge s})$. That Y is a martingale follows from Theorem 2.59. Thus in both cases, Y is an $(\mathcal{F}_{\cdot})$-martingale and similarly, so is Z. Remains to show that M is a martingale. In case 1, writing $N_t = W_t^2 - t$

$$\begin{aligned} M_t &= ab((W_{t \wedge r} - W_{t \wedge s})^2 - (t \wedge r - t \wedge s)) \\ &= ab((W_{t \wedge r}^2 - W_{t \wedge s}^2) - (t \wedge r - t \wedge s) - 2W_{t \wedge s}(W_{t \wedge r} - W_{t \wedge s})) \\ &= ab(N_{t \wedge r} - N_{t \wedge s}) - 2ab W_{t \wedge s}(W_{t \wedge r} - W_{t \wedge s}). \end{aligned}$$

Recalling that N, W are martingales, it follows from Theorem 2.59 that M is a martingale as

$$ab W_{t \wedge s}(W_{t \wedge r} - W_{t \wedge s}) = ab W_s(W_{t \wedge r} - W_{t \wedge s}).$$

In case 2, recalling $0 \le s \le r \le u \le v$, note that

$$M_t = a(W_{t \wedge r} - W_{t \wedge s})b(W_{t \wedge v} - W_{t \wedge u})$$
$$= a(W_r - W_s)b(W_{t \wedge v} - W_{t \wedge u})$$

as $M_t = 0$ if $t \le u$. Proof is again completed using Theorem 2.59.   □

Theorem 3.11 *Let $f \in \mathbb{S}$, $M_t = \int_0^t f \, dW$ and $N_t = M_t^2 - \int_0^t f_s^2 ds$. Then M and N are $(\mathcal{F}_.)$-martingales. Further, for any $T < \infty$,*

$$\mathsf{E}[\sup_{t \le T} |\int_0^t f \, dW|^2] \le 4\mathsf{E}[\int_0^T f_s^2 ds]. \tag{3.3.6}$$

Proof The fact that M and N are martingales follows from Lemma 3.10. As a consequence $\mathsf{E}[N_T] = 0$ and hence

$$\mathsf{E}[(\int_0^T f \, dW)^2] = \mathsf{E}[\int_0^T f_s^2 ds]. \tag{3.3.7}$$

Now the growth inequality (3.3.6) follows from Doob's maximal inequality, Theorem 2.26 applied to M and using (3.3.7).   □

We will use the growth inequality (3.3.7) to extend the integral to a larger class of functions that can be approximated in the *norm* defined by the right-hand side in (3.3.7).

Each $f \in \mathbb{S}$ can be viewed as a real-valued function on $\widetilde{\Omega} = [0, \infty) \times \Omega$. It is easy to see that $\mathbb{S}$ is an algebra. Let $\mathcal{P}$ be the σ-field generated by $\mathbb{S}$, i.e. the smallest σ-field on $\widetilde{\Omega}$ such that every element of $\mathbb{S}$ is measurable w.r.t. $\mathcal{P}$.

The σ-field $\mathcal{P}$ is called the predictable σ-field. We will discuss the predictable σ-field in the next chapter. We note here that every bounded left continuous adapted process X is $\mathcal{P}$ measurable as it is the pointwise limit of

$$X_t^m = X_0 1_{\{0\}} + \sum_{j=0}^{m 2^m} X_{\frac{j}{2^m}} 1_{(\frac{j}{2^m}, \frac{j+1}{2^m}]}(t).$$

Hence every left continuous adapted process X is $\mathcal{P}$ measurable. Process f which is $\mathcal{P}$ measurable is called a *predictable process*.

Lemma 3.12 *Let f be a predictable process such that*

$$\mathsf{E}[\int_0^T f_s^2 ds] < \infty \quad \forall T < \infty. \tag{3.3.8}$$

Then there exists a continuous adapted process Y such that for all simple predictable processes $h \in \mathbb{S}$,

$$\mathsf{E}[(\sup_{0 \le t \le T} |Y_t - \int_0^t h \, dW|)^2] \le 4\mathsf{E}[\int_0^T (f_s - h_s)^2 ds] \quad \forall T < \infty. \tag{3.3.9}$$

Further, Y and Z are $(\mathcal{F}_{\cdot})$-martingales where

$$Z_t = Y_t^2 - \int_0^t f_s^2 ds.$$

Proof For $r > 0$, let μ_r be the measure on $(\widetilde{\Omega}, \mathcal{P})$ defined as follows: for $\mathcal{P}$ measurable bounded functions g

$$\int_{\widetilde{\Omega}} g d\mu_r = \mathsf{E}[\int_0^r g_s ds]$$

and let us denote the $\mathbb{L}^2$ norm on $\mathbb{L}^2(\mu_r)$ by $\|\cdot\|_{2,\mu_r}$. By Theorem 2.67, $\mathbb{S}$ is dense in $\mathbb{L}^2(\mu_r)$ for every $r > 0$ and hence for integers $m \geq 1$, we can get $f^m \in \mathbb{S}$ such that

$$\|f - f^m\|_{2,\mu_m} \leq 2^{-m-1}. \tag{3.3.10}$$

Using $\|\cdot\|_{2,\mu_r} \leq \|\cdot\|_{2,\mu_s}$ for $r \leq s$ it follows that for $k \geq 1$

$$\|f^{m+k} - f^m\|_{2,\mu_m} \leq 2^{-m}. \tag{3.3.11}$$

Denoting the $\mathbb{L}^2(\Omega, \mathcal{F}, \mathsf{P})$ norm by $\|\cdot\|_{2,\mathsf{P}}$, the growth inequality (3.3.6) can be rewritten as, for $g \in \mathbb{S}$, $m \geq 1$,

$$\| \sup_{0 \leq t \leq m} |\int_0^t g dW| \|_{2,\mathsf{P}} \leq 2\|g\|_{2,\mu_m} \tag{3.3.12}$$

Recall that $f^k \in \mathbb{S}$ and hence $\int_0^t f^k dW$ is already defined. Let $Y_t^k = \int_0^t f^k dW$. Now using (3.3.11) and (3.3.12), we conclude that for $k \geq 1$

$$\|[\sup_{0 \leq t \leq m} |Y_t^{m+k} - Y_t^m|]\|_{2,\mathsf{P}} \leq 2^{-m+1}. \tag{3.3.13}$$

Fix an integer n. For $m \geq n$, using (3.3.13) for $k = 1$ we get

$$\|[\sup_{0 \leq t \leq n} |Y_t^{m+1} - Y_t^m|]\|_{2,\mathsf{P}} \leq 2^{-m+1}. \tag{3.3.14}$$

and hence

$$\|[\sum_{m=n}^{\infty} \sup_{0 \leq t \leq n} |Y_t^{m+1} - Y_t^m|]\|_{2,\mathsf{P}} \leq \sum_{m=n}^{\infty} \|[\sup_{0 \leq t \leq n} |Y_t^{m+1} - Y_t^m|]\|_{2,\mathsf{P}}$$

$$\leq \sum_{m=n}^{\infty} 2^{-m+1}$$

$$< \infty.$$

Hence,

$$\sum_{m=n}^{\infty} [\sup_{0 \le t \le n} |Y_t^{m+1} - Y_t^m|] < \infty \quad a.s. \ \mathsf{P}. \tag{3.3.15}$$

So let

$$N_n = \{\omega : \sum_{m=n}^{\infty} [\sup_{0 \le t \le n} |Y_t^{m+1}(\omega) - Y_t^m(\omega)|] = \infty\}$$

and let $N = \cup_{n=1}^{\infty} N_n$. Then N is a P null set. For $\omega \notin N$, let us define

$$Y_t(\omega) = \lim_{m \to \infty} Y_t^m(\omega)$$

and for $\omega \in N$, let $Y_t(\omega) = 0$. It follows from (3.3.15) that for all $T < \infty$, $\omega \notin N$

$$\sup_{0 \le t \le T} |Y_t^m(\omega) - Y_t(\omega)| \to 0. \tag{3.3.16}$$

Thus Y is a process with continuous paths. Now using (3.3.12) for $f^m - h \in \mathbb{S}$ we get

$$\mathsf{E}[(\sup_{0 \le t \le T} |Y_t^m - \int_0^t h \, dW|)^2] \le 4\mathsf{E}[\int_0^T (f^m - h)_s^2 ds]. \tag{3.3.17}$$

In view of (3.3.10), the right-hand side above converges to

$$\mathsf{E}[\int_0^T (f_s - h_s)^2 ds].$$

Using Fatou's lemma and (3.3.16) along with $\mathsf{P}(N) = 0$, taking lim inf in (3.3.17) we conclude that (3.3.9) is true. From these observations, it follows that Y_t^m converges to Y_t in $\mathbb{L}^2(\mathsf{P})$ for each fixed t. The observation $\|\cdot\|_{2,\mu_r} \le \|\cdot\|_{2,\mu_s}$ for $r \le s$ and (3.3.10) implies that for all r, $\|f - f^m\|_{2,\mu_r} \to 0$ and hence for all t

$$\mathsf{E}[\int_0^t (f_s - f_s^m)^2 ds] \to 0.$$

As a consequence,

$$\mathsf{E}[\int_0^t |(f_s)^2 - (f_s^m)^2| ds] \to 0$$

and hence

$$\mathsf{E}[|\int_0^t (f_s^m)^2 ds - \int_0^t f_s^2 ds|] \to 0.$$

By Theorem 3.11, we have Y^n and Z^n which are martingales where $Z_t^n = (Y_t^n)^2 - \int_0^t (f_s^n)^2 ds$. As observed above Y_t^n converges in $L^1(P)$ to Y_t, and Z_t^n converges in $L^1(P)$ to Z_t for each t and hence Y and Z are martingales. □

Remark 3.13 From the proof of the Lemma it also follows that Y is uniquely determined by the property (3.3.9), for if $\tilde{Y}$ is another process that satisfies (3.3.9), then using it for $h = f^m$ as in the proof above, we conclude that Y^m converges almost surely to $\tilde{Y}$ and hence $Y = \tilde{Y}$.

Definition 3.14 For a predictable process f such that $E[\int_0^T f_s^2 ds] < \infty \ \forall T < \infty$, we define the *Ito's integral* $\int_0^\cdot f dW$ to be the process Y that satisfies (3.3.9).

The next result gives the basic properties of the Ito's integral $\int f dW$; most of them have essentially been proved above.

Theorem 3.15 *Let f, g be predictable processes satisfying (3.3.8).Then*

$$\int_0^t (af + bg) dW = a \int_0^t f dW + b \int_0^t g dW. \tag{3.3.18}$$

Let $M_t = \int_0^t f dW$ and $N_t = M_t^2 - \int_0^t f_s^2 ds$. Then M and N are $(\mathcal{F}_\cdot)$-martingales. Further, for any $T < \infty$,

$$E[\sup_{t \leq T} |\int_0^t f dW|^2] \leq 4E[\int_0^T f_s^2 ds]. \tag{3.3.19}$$

Proof The linearity (3.3.18) follows by linearity for the integral for simple functions observed in Lemma 3.9 and then for general predictable processes via approximation. That M, N are martingales has been observed in Lemma 3.12. The growth inequality (3.3.19) follows from (3.3.9) with $h = 0$. □

Remark 3.16 For a bounded predictable process f, let $I_W(f) = \int f dW$. Then the growth inequality (3.3.19) and linearity of I_W imply that for f_n, f bounded predictable processes

$$f_n \xrightarrow{bp} f \text{ implies } I_W(f_n) \xrightarrow{ucp} I_W(f).$$

Exercise 3.17 Let $t_i^n = i2^{-n}, i \geq 0, n \geq 1$ and let

$$f_s^n = \sum_{i=0}^\infty W_{t_i^n} 1_{(t_i^n, t_{i+1}^n]}(s).$$

Show that

(i) $\int_0^t f^n dW = \sum_{i=0}^\infty W_{t_i^n} (W_{t_{i+1}^n \wedge t} - W_{t_i^n \wedge t}).$

(ii) $E[\int_0^T |f_s^n - W_s|^2 ds] \to 0$.

(iii) $\int_0^t f^n dW \to \int_0^t W dW$.

(iv) $\int_0^t W dW = \frac{1}{2}(W_t^2 - t)$.

Hint: For (iii) using notations as in (3.1.9) and (3.1.10) we have $\int_0^t f^n dW = B_t^n$, $(A_t^n - B_t^n) = Q_t^n$ along with $(A_t^n + B_t^n) = W_t^2$ and $Q_t^n \to t$ (see Exercise 3.2). Thus A_t^n and B_t^n converge, as mentioned in Remark 3.4.

Exercise 3.18 Let $f_t = t$. Show that $\int_0^t f dW = t W_t - \int_0^t W_s ds$.

Exercise 3.19 Let $f \in \mathbb{L}^2([0, \infty)$ be a deterministic function. Show that $Z_t = \int_0^t f dW$ is a Gaussian process, i.e. for any $t_1 < t_2 < \ldots < t_m < \infty$, $(Z_{t_1}, Z_{t_2}, \ldots, Z_{t_m})$ has multivariate normal (Gaussian) distribution.

For deterministic f the integral $\int f dW$ had been defined and studied by Wiener and is also called the Wiener integral.

Exercise 3.20 Let $A \in V$ be a bounded r.c.l.l. adapted process with finite variation paths. Show that

$$\int_0^t A^- dW = A_t W_t - \int_0^t W dA. \tag{3.3.20}$$

HINT: Let $t_i^n = i2^{-n}$, $i \geq 0$, $n \geq 1$. Observe that

$$\sum_{i=0}^{\infty} A_{t_i^n \wedge t}(W_{t_{i+1}^n \wedge t} - W_{t_i^n \wedge t}) = A_t W_t - \sum_{i=0}^{\infty} W_{t_{i+1}^n \wedge t}(A_{t_{i+1}^n \wedge t} - A_{t_i^n \wedge t}).$$

The left-hand side converges to $\int_0^t A^- dW$ while the second term on right-hand side converges to $\int_0^t W dA$ as seen in Exercise 2.81.

Remark 3.21 The Ito's integral can be extended to a larger class of predictable integrands f satisfying

$$\int_0^T f_s^2 ds < \infty \ a.s. \ \forall T < \infty.$$

We will outline this later when we discuss integration w.r.t. semimartingales.

3.4 Multidimensional Ito's Integral

Let $W = (W^1, W^2, \ldots, W^d)$ be d-dimensional Brownian motion, where W^j is the jth component. In other words, each W^j is a real-valued Brownian motion and $W^1, W^2, \ldots, W^d$ are independent. Suppose further that $(\mathcal{F}_.)$ is a filtration such that

$(W_t, \mathcal{F}_t)_{\{t \geq 0\}}$ is a Wiener martingale. Thus for each s, $\{W_t - W_s : t \geq s\}$ is independent of $\mathcal{F}_s$. Denoting $\theta = (\theta^1, \ldots, \theta^d) \in \mathbb{R}^d$ and defining

$$X_t^\theta = \sum_{j=1}^d \theta^j W_t^j \tag{3.4.1}$$

$$M_t^\theta = (X_t^\theta)^2 - t, \tag{3.4.2}$$

we have

$$\forall \theta \in \mathbb{R}^d \text{ with } |\theta| = 1; \quad X^\theta, M^\theta \text{ are } (\mathcal{F}_\cdot)\text{-martingales} . \tag{3.4.3}$$

The argument given in the proof of next lemma is interesting. Throughout this section, the filtration $(\mathcal{F}_\cdot)$ will remain fixed.

Lemma 3.22 *For $j \neq k$, $W^j W^k$ is also a martingale.*

Proof Let $X_t = \frac{1}{\sqrt{2}}(W_t^j + W_t^k)$. Then, as seen above, X is a Brownian motion and hence $X_t^2 - t$ is a martingale. Note that

$$X_t^2 - t = \frac{1}{2}[(W_t^j)^2 + (W_t^k)^2 + 2W_t^j W_t^k] - t$$
$$= \frac{1}{2}[(W_t^j)^2 - t] + \frac{1}{2}[(W_t^k)^2 - t] + W_t^j W_t^k.$$

Since the left-hand side above as well as the first two terms of right-hand side above are martingales, it follows that so is the third term. □

Suppose that for $1 \leq j \leq m$ and $1 \leq k \leq d$, f^{jk} are $(\mathcal{F}_\cdot)$ predictable processes that satisfy (3.3.8). Let

$$X_t^j = \sum_{k=1}^d \int_0^t f^{jk} dW^k.$$

Let $X = (X^1, \ldots X^m)$ denote the m-dimensional process. It is natural to define X_t to be $\int_0^t f\, dX$ where we interpret f to be $L(m, d)$ ($m \times d$-matrix-valued) predictable process. We will obtain a growth estimate on the stochastic integral $\int_0^t f\, dW$ which in turn would be crucial in the study of stochastic differential equations. The following lemma is a first step towards it.

Lemma 3.23 *Let h be a predictable process satisfying (3.3.8). Let $Y_t^k = \int_0^t h\, dW^k$. Then for $j \neq k$, $Y_t^k Y_t^j$ is a martingale.*

Proof Let $X_t = \frac{1}{\sqrt{2}}(W_t^j + W_t^k)$. Then, as seen above, X is a Brownian motion. For simple functions f it is easy to check that

$$\int_0^t f \, dX = \frac{1}{\sqrt{2}} \left(\int_0^t f \, dW^j + \int_0^t f \, dW^k \right)$$

and hence for all f satisfying (3.3.8) via approximation. Thus,

$$\int_0^t h \, dX = \frac{1}{\sqrt{2}} (Y_t^j + Y_t^k)$$

and so $(\int_0^t h \, dX)^2 - \int_0^t h_s^2 ds$ is a martingale. Now as in the proof of Lemma 3.22

$$\left(\int_0^t h \, dX \right)^2 - \int_0^t h_s^2 ds = \frac{1}{2}[(Y_t^j)^2 + (Y_t^k)^2 + 2Y_t^j Y_t^k] - \int_0^t h_s^2 ds$$

$$= \frac{1}{2}[(Y_t^j)^2 - \int_0^t h_s^2 ds] + \frac{1}{2}[(Y_t^k)^2 - \int_0^t h_s^2 ds] + Y_t^j Y_t^k.$$

and once again the left-hand side as well as the first two terms on the right-hand side are martingales and hence so is the last term completing the proof.   □

Lemma 3.24 *Let f, g be predictable processes satisfying (3.3.8). Let $Y_t^k = \int_0^t f \, dW^k$, $Z_t^k = \int_0^t g \, dW^k$. Then for $j \neq k$, $Y_t^k Z_t^j$ is a martingale.*

Proof Let $X_t = Y_t^k Z_t^j$. We will first prove that X is a martingale when f, g are simple, the general case follows by approximation. The argument is similar to the proof of Theorem 3.10. By linearity, suffices to prove the required result in the following cases.
Case 1: $0 \leq s < r$ and

$$f_t = a \mathbf{1}_{(s,r]}(t), \quad g_t = b \mathbf{1}_{(s,r]}(t), \quad a, b \text{ are } \mathcal{F}_s \text{ measurable}, a \geq 0, b \geq 0.$$

Case 2: $0 \leq s < r \leq u < v$ and

$$f_t = a \mathbf{1}_{(s,r]}(t), \quad g_t = b \mathbf{1}_{(u,v]}(t), \quad a \text{ is } \mathcal{F}_s \text{ measurable and } b \text{ is } \mathcal{F}_u \text{ measurable}.$$

In case 1,

$$X_t = Y_t^k Z_t^j = ab(W_{r \wedge t}^k - W_{s \wedge t}^k)(W_{r \wedge t}^j - W_{s \wedge t}^j) = \left(\int_0^t h \, dW^k \right) \left(\int_0^t h \, dW^j \right)$$

where $h = \sqrt{(ab)} \mathbf{1}_{(s,r]}(t)$ and hence by Lemma 3.23, X is a martingale.
   In case 2,

$$X_t = Y_t^k Z_t^j = ab(W_{r \wedge t}^k - W_{s \wedge t}^k)(W_{v \wedge t}^j - W_{u \wedge t}^j).$$

Here, $X_t = 0$ for $t \leq r$ and

$$X_t = \xi(W_{v \wedge t}^j - W_{u \wedge t}^j)$$

with $\xi = ab(W_{r\wedge t}^k - W_{s\wedge t}^k)$ is $\mathcal{F}_u$ measurable and hence by Corollary 2.60, X is a martingale.

This proves the required result for simple predictable processes f, g. The general case follows by approximating f, g by simple predictable processes $\{f^n\}$, $\{g^n\}$ such that for all $T < \infty$

$$\int_0^T [|f_s^n - f_s|^2 + |g_s^n - g_s|^2]ds \to 0.$$

Then it follows from (3.3.19) that for each $t < \infty$,

$$\int_0^t f^n dW^k \to \int_0^t f dW^k \text{ in } \mathbb{L}^2(P),$$

$$\int_0^t g^n dW^j \to \int_0^t g dW^j \text{ in } \mathbb{L}^2(P)$$

and hence the martingale $(\int_0^t f^n dW^k)(\int_0^t g^n dW^j)$ converges to $Y_t^k Z_t^j = (\int_0^t f dW^k)(\int_0^t g dW^j)$ in $\mathbb{L}^1(P)$ and thus $Y_t^k Z_t^j$ is a martingale. □

We are now ready to prove the main growth inequality. Recall that $L(m, d)$ denotes the space of $m \times d$ matrices and for $x = (x_1, x_2, \ldots, x_d) \in \mathbb{R}^d$, $|x| = \sqrt{\sum_{j=1}^d x_j^2}$ denotes the Euclidean norm on $\mathbb{R}^m$ and for $a = (a_{jk}) \in L(m, d)$, $\|a\| = \sqrt{\sum_{j=1}^d \sum_{k=1}^m a_{jk}^2}$ is the Euclidean norm on $L(m, d)$.

Let $f = (f^{jk})$ be $L(m, d)$-valued process. For $\mathbb{R}^d$-valued Brownian motion W, we have seen that $X = \int f dW$ is an $\mathbb{R}^m$-valued process.

Theorem 3.25 *Let W be an $\mathbb{R}^d$-valued Brownian motion. Then for $m \times d$-matrix-valued predictable process f with*

$$E[\int_0^T \|f_s\|^2 ds] < \infty$$

we have

$$E[\sup_{0 \le t \le T} |\int_0^t f dW|^2] \le 4E[\int_0^T \|f_s\|^2 ds]. \tag{3.4.4}$$

Proof Let $X_t^j = \sum_{k=1}^d \int f^{jk} dW^k$. Then

$$(X_t^j)^2 - \sum_{k=1}^d \int_0^t (f^{jk})^2 ds = \sum_{k=1}^d [(\int_0^t f^{jk} dW^k)^2 - \int_0^t (f^{jk})^2 ds]$$

$$+ \sum_{l=1}^d \sum_{k=1}^d 1_{\{k \ne l\}} \int_0^t f^{jk} dW^k \int_0^t f^{jl} dW^l$$

and each term on the right-hand side above is a martingale and thus summing over j we conclude

$$| \int_0^t f\, dW|^2 - \int_0^t \|f_s\|^2 ds$$

is a martingale. Thus

$$E[| \int_0^t f\, dW|^2] = E[\int_0^t \|f_s\|^2 ds] \tag{3.4.5}$$

and $| \int_0^t f\, dW|^2$ is a submartingale. The required estimate (3.4.4) now follows from Doob's maximal inequality (2.3.7) and (3.4.5). $\square$

Exercise 3.26 Let W be an $\mathbb{R}^d$-valued Brownian motion. Let $f \in \mathbb{L}^2([0, \infty)$ be an $\mathsf{L}(m, d)$-valued deterministic function. Show that $Z_t = \int_0^t f\, dW$ is a $\mathbb{R}^d$-valued Gaussian process, i.e. for any $t_1 < t_2 < \ldots < t_n < \infty$, $(Z_{t_1}, Z_{t_2}, \ldots, Z_{t_n})$ considered as a dn-dimensional vector has multivariate normal (Gaussian) distribution.

3.5   Stochastic Differential Equations

We are going to consider stochastic differential equations (SDE) of the type

$$dX_t = \sigma(t, X_t)dW_t + b(t, X_t)dt. \tag{3.5.1}$$

Equation (3.5.1) is to be interpreted as an integral equation:

$$X_t = X_0 + \int_0^t \sigma(s, X_s)dW_s + \int_0^t b(s, X_s)ds. \tag{3.5.2}$$

Here W is an $\mathbb{R}^d$-valued Brownian motion, X_0 is an $\mathbb{R}^d$-valued $\mathcal{F}_0$ measurable random variable, $\sigma : [0, \infty) \times \mathbb{R}^m \mapsto \mathsf{L}(m, d)$ and $b : [0, \infty) \times \mathbb{R}^m \mapsto \mathbb{R}^m$ are given functions, and one is seeking a process X such that (3.5.2) is true. The solution X to the SDE (3.5.1), when it exists, is called a diffusion process with diffusion coefficient $\sigma\sigma^*$ and drift coefficient b.

We shall impose the following conditions on σ, b:

$$\sigma : [0, \infty) \times \mathbb{R}^m \mapsto \mathsf{L}(m, d) \text{ is a continuous function}$$
$$b : [0, \infty) \times \mathbb{R}^m \mapsto \mathbb{R}^m \text{ is a continuous function} \tag{3.5.3}$$

$$\forall T < \infty \ \exists C_T < \infty \text{ such that for all } t \in [0, T], \ x^1, x^2 \in \mathbb{R}^d$$
$$\|\sigma(t, x^1) - \sigma(t, x^2)\| \le C_T |x^1 - x^2|,$$
$$|b(t, x^1) - b(t, x^2)| \le C_T |x^1 - x^2|. \tag{3.5.4}$$

Since $t \mapsto \sigma(t, 0)$ and $t \mapsto b(t, 0)$ are continuous and hence bounded on $[0, T]$ for every $T < \infty$, using the Lipschitz conditions (3.5.4), we can conclude that for each $T < \infty, \exists K_T < \infty$ such that

$$\|\sigma(t, x)\| \leq K_T(1 + |x|),$$
$$|b(t, x)| \leq K_T(1 + |x|). \tag{3.5.5}$$

We will need the following lemma, known as Gronwall's lemma, for proving uniqueness of solution to (3.5.2) under the Lipschitz conditions.

Lemma 3.27 *Let $\beta(t)$ be a bounded measurable function on $[0, T]$ satisfying, for some $0 \leq a < \infty, 0 < b < \infty$,*

$$\beta(t) \leq a + b \int_0^t \beta(s) \, ds, \ 0 \leq t \leq T. \tag{3.5.6}$$

Then

$$\beta(t) \leq ae^{bt}. \tag{3.5.7}$$

Proof Let

$$g(t) = e^{-bt} \int_0^t \beta(s) \, ds.$$

Then by definition, g is absolutely continuous and

$$g'(t) = e^{-bt} \beta(t) - be^{-bt} \int_0^t \beta(s) \, ds \ a.e.$$

where almost everywhere refers to the Lebesgue measure on $\mathbb{R}$. Using (3.5.6), it follows that

$$g'(t) \leq ae^{-bt} \ a.e.$$

Hence (using $g(0) = 0$ and that g is absolutely continuous) $g(t) \leq \frac{a}{b}(1 - e^{-bt})$ from which we get

$$\int_0^t \beta(s) \, ds \leq \frac{a}{b}(e^{bt} - 1).$$

The conclusion $\beta(t) \leq ae^{bt}$ follows immediately from (3.5.6). $\qquad \square$

So now let $(\mathcal{F}_.)$ be a filtration on $(\Omega, \mathcal{F}, P)$ and W be a d-dimensional Brownian motion adapted to $(\mathcal{F}_.)$ and such that $(W_t, \mathcal{F}_t)_{\{t \geq 0\}}$ is a Wiener martingale. Without loss of generality, let us assume that $(\Omega, \mathcal{F}, P)$ is complete and that $\mathcal{F}_0$ contains all P null sets in $\mathcal{F}$. Let $\mathbb{K}_m$ denote the class of $\mathbb{R}^m$-valued continuous $(\mathcal{F}_.)$ adapted process Z such that $E[\int_0^T |Z_s|^2 ds] < \infty \ \forall T < \infty$. For $Y \in \mathbb{K}_m$ let

$$\xi_t = Y_0 + \int_0^t \sigma(s, Y_s)dW_s + \int_0^t b(s, Y_s)ds. \tag{3.5.8}$$

Note that in view of the growth condition (3.5.5) the Ito's integral above is defined. Using the growth estimate (3.4.4) we see that

$$\mathsf{E}[\sup_{0 \le t \le T} |\xi_t|^2] \le 3[\mathsf{E}[|Y_0|^2] + 4\mathsf{E}[\int_0^T \|\sigma(s, Y_s)\|^2 ds]$$

$$+ \mathsf{E}[(\int_0^T |b(s, Y_s)|ds)^2]]$$

$$\le 3\mathsf{E}[|Y_0|^2] + 3K_T^2(4 + T)\int_0^T (1 + \mathsf{E}[|Y_s|^2])ds$$

and hence $\xi \in \mathbb{K}_m$. Let us define a mapping Λ from $\mathbb{K}_m$ into itself as follows: $\Lambda(Y) = \xi$ where ξ is defined by (3.5.8). Thus solving the SDE (3.5.2) amounts to finding a fixed point Z of the functional Λ with $Z_0 = X_0$, where X_0 is pre-specified. We are going to prove that given X_0, there exists a unique solution (or a unique fixed point of Λ) with the given initial condition. The following lemma is an important step in that direction.

Lemma 3.28 *Let $Y, Z \in \mathbb{K}_m$ and let $\xi = \Lambda(Y)$ and $\eta = \Lambda(Z)$. Then for $0 \le t \le T$ one has*

$$\mathsf{E}[\sup_{0 \le s \le t} |\xi_s - \eta_s|^2] \le 3\mathsf{E}[|Y_0 - Z_0|^2] + 3C_T^2(4 + T)\int_0^t \mathsf{E}[|Y_s - Z_s|^2]ds$$

Proof Let us note that

$$\xi_t - \eta_t = Y_0 - Z_0 + \int_0^t [\sigma(s, Y_s) - \sigma(s, Z_s)]dW_s + \int_0^t [b(s, Y_s) - b(s, Z_s)]ds$$

and hence this time using the Lipschitz condition (3.5.4) along with the growth inequality (3.4.4) we now have

$$\mathsf{E}[\sup_{0 \le s \le t} |\xi_s - \eta_s|^2] \le 3[\mathsf{E}[|Y_0 - Z_0|^2] + 4\mathsf{E}[\int_0^t \|\sigma(s, Y_s) - \sigma(s, Z_s)\|^2 ds]$$

$$+ \mathsf{E}[(\int_0^t |b(s, Y_s) - b(s, Z_s)|ds)^2]]$$

$$\le 3\mathsf{E}[|Y_0 - Z_0|^2] + 3C_T^2(4 + T)\int_0^t \mathsf{E}[|Y_s - Z_s|^2]ds.$$

$$\square$$

Corollary 3.29 *Suppose $Y, Z \in \mathbb{K}_m$ be such that $Y_0 = Z_0$. Then for $0 \le t \le T$*

$$E[\sup_{0 \le s \le t} |A(Y)_s - A(Z)_s|^2] \le 3C_T^2(4+T) \int_0^t E[|Y_s - Z_s|^2]ds$$

We are now in a position to prove the main result of this section.

Theorem 3.30 *Suppose σ, b satisfy conditions (3.5.3) and (3.5.4) and X_0 is a $\mathcal{F}_0$ measurable $\mathbb{R}^m$-valued random variable with $E[|X_0|^2] < \infty$. Then there exists a process X such that $E[\int_0^T |X_s|^2 ds] < \infty \ \forall T < \infty$ and*

$$X_t = X_0 + \int_0^t \sigma(s, X_s)dW_s + \int_0^t b(s, X_s)ds. \tag{3.5.9}$$

Further if $\tilde{X}$ is another process such that $\tilde{X}_0 = X_0$, $E[\int_0^T |\tilde{X}_s|^2 ds] < \infty$ for all $T < \infty$ and

$$\tilde{X}_t = \tilde{X}_0 + \int_0^t \sigma(s, \tilde{X}_s)dW_s + \int_0^t b(s, \tilde{X}_s)ds$$

then $X = \tilde{X}$, i.e. $P(X_t = Y_t \ \forall t) = 1$.

Proof Let us first prove uniqueness. Let X and $\tilde{X}$ be as in the statement of the theorem. Then, using Corollary 3.29 it follows that

$$u(t) = E[\sup_{s \le t} |X_s - \tilde{X}_s|^2]$$

satisfies for $0 \le t \le T$ (recalling $X_0 = \tilde{X}_0$)

$$u(t) \le 3C_T^2(4+T) \int_0^t E[|X_s - \tilde{X}_s|^2]ds.$$

Hence u is bounded and satisfies

$$u(t) \le 3C_T^2(4+T) \int_0^t u(s)ds, \quad 0 \le t \le T.$$

By (Gronwall's) Lemma 3.27, it follows that $u(t) = 0, 0 \le t \le T$ for every $T < \infty$. Hence $X = \tilde{X}$.

We will now construct a solution. Let $X_t^1 = X_0$ for all $t \ge 0$. Note that $X^1 \in \mathbb{K}_m$. Now define X^n inductively by

$$X^{n+1} = A(X^n).$$

Since $X_t^1 = X_0$ for all t and $X_0^2 = X_0^1$,

$$X_t^2 - X_t^1 = \int_0^t \sigma(s, X_0)dW_s + \int_0^t b(s, X_0)ds$$

and hence

$$E[\sup_{s \leq t}|X_s^2 - X_s^1|^2] \leq 2K_T^2(4 + T)(1 + E[|X_0|^2])t. \qquad (3.5.10)$$

Note that $X_0^n = X_0^1 = X_0$ for all $n \geq 1$ and hence from Lemma 3.28 it follows that for $n \geq 2$, for $0 \leq t \leq T$,

$$E[\sup_{s \leq t}|X_s^{n+1} - X_s^n|^2] \leq 3C_T^2(4 + T)\int_0^t E[|X_s^n - X_s^{n-1}|^2]ds$$

Thus defining for $n \geq 1$, $u_n = E[\sup_{s \leq t}|X_s^{n+1} - X_s^n|^2]$ we have for $n \geq 2$, for $0 \leq t \leq T$,

$$u_n(t) \leq 3C_T^2(4 + T)\int_0^t u_{n-1}(s)ds. \qquad (3.5.11)$$

As seen in (3.5.10),

$$u_1(t) \leq 2K_T^2(4 + T)(1 + E[|X_0|^2])t$$

and hence using (3.5.11), which is true for $n \geq 2$, we can deduce by induction on n that for a constant $\tilde{C}_T = 3(C_T^2 + K_T^2)(4 + T)(1 + E[|X_0|^2])$

$$u_n(t) \leq \frac{(\tilde{C}_T)^n t^n}{n!}, \quad 0 \leq t \leq T.$$

Thus $\sum_n \sqrt{u_n(T)} < \infty$ for every $T < \infty$ which is same as

$$\sum_{n=1}^{\infty} \|\sup_{s \leq T}|X_s^{n+1} - X_s^n|\|_2 < \infty \qquad (3.5.12)$$

$\|Z\|_2$ denoting the $\mathbb{L}^2(P)$ norm here. The relation (3.5.12) implies

$$\|[\sum_{n=1}^{\infty} \sup_{s \leq T}|X_s^{n+1} - X_s^n|]\|_2 < \infty \qquad (3.5.13)$$

as well as

$$\sup_{k \geq 1}\|[\sup_{s \leq T}|X_s^{n+k} - X_s^n|]\|_2 \leq \sup_{k \geq 1}\|[\sum_{j=n}^{n+k} \sup_{s \leq T}|X_s^{j+1} - X_s^j|]\|_2$$

$$\leq [\sum_{j=n}^{\infty} \|\sup_{s \leq T}|X_s^{j+1} - X_s^j|\|_2] \qquad (3.5.14)$$

$$\to 0 \text{ as } n \text{ tends to } \infty.$$

Let $N = \cup_{T=1}^{\infty} \{\omega : \sum_{n=1}^{\infty} \sup_{s \leq T} |X_s^{n+1}(\omega) - X_s^n(\omega)| = \infty\}$. Then by (3.5.13), $P(N)$ $= 0$ and for $\omega \notin N$, $X_s^n(\omega)$ converges uniformly on $[0, T]$ for every $T < \infty$. So let us define X as follows:

$$X_t(\omega) = \begin{cases} \lim_{n \to \infty} X_t^n(\omega) & \text{if } \omega \in N^c \\ 0 & \text{if } \omega \in N. \end{cases}$$

By definition, X is a continuous adapted process (since by assumption $N \in \mathcal{F}_0$) and X^n converges to X uniformly in $[0, T]$ for every T almost surely. Using Fatou's lemma and (3.5.14) we get

$$\|[\sup_{s \leq T} |X_s - X_s^n|]\|_2 \leq \liminf_{k \to \infty} \|[\sum_{j=n}^{n+k} \sup_{s \leq T} |X_s^{j+1} - X_s^j|]\|_2$$

$$\leq [\sum_{j=n}^{\infty} \|\sup_{s \leq T} |X_s^{j+1} - X_s^j|\|_2] \tag{3.5.15}$$

$$\to 0 \text{ as } n \text{ tends to } \infty.$$

In particular, $X \in \mathbb{K}_m$. Since $\Lambda(X^n) = X^{n+1}$ by definition, (3.5.15) also implies that

$$\lim_{n \to \infty} \|[\sup_{s \leq T} |X_s - \Lambda(X^n)_s|]\|_2 = 0 \tag{3.5.16}$$

while (3.5.15) and Corollary 3.29 (remembering that $X_0^n = X_0$ for all n) imply that

$$\lim_{n \to \infty} \|[\sup_{s \leq T} |\Lambda(X)_s - \Lambda(X^n)_s|]\|_2 = 0. \tag{3.5.17}$$

From (3.5.16) and (3.5.17) it follows that $X = \Lambda(X)$ or that X is a solution to the SDE (3.5.9). $\qquad\square$

Chapter 4
Stochastic Integration

In this chapter we consider processes X that are good integrators: i.e.

$$J_X(f)(t) = \int_0^t f dX$$

can be defined for a suitable class of integrands f and the integral has some natural continuity properties. We will call such a process a *stochastic integrator*. In this chapter, we will prove basic properties of the stochastic integral $\int_0^t f dX$ for a stochastic integrator X.

In the rest of the book, $(\Omega, \mathcal{F}, \mathsf{P})$ will denote a complete probability space and $(\mathcal{F}_\cdot)$ will denote a filtration such that $\mathcal{F}_0$ contains all null sets in $\mathcal{F}$. All notions such as adapted, stopping time, martingale will refer to this filtration unless otherwise stated explicitly.

For some of the auxiliary results, we need to consider the corresponding right continuous filtration $(\mathcal{F}_\cdot^+) = \{\mathcal{F}_t^+ : t \geq 0\}$ where

$$\mathcal{F}_t^+ = \cap_{s>t} \mathcal{F}_s.$$

We begin with a discussion on the predictable σ-field.

4.1 The Predictable σ-Field

Recall our convention that a process $X = (X_t)$ is viewed as a function on $\widetilde{\Omega} = [0, \infty) \times \Omega$ and the predictable σ-field $\mathcal{P}$ has been defined as the σ-field on $\widetilde{\Omega}$ generated by $\mathbb{S}$. Here $\mathbb{S}$ consists of simple adapted processes:

$$f(s) = a_0 1_{\{0\}}(s) + \sum_{k=0}^m a_{k+1} 1_{(s_k, s_{k+1}]}(s) \tag{4.1.1}$$

© Springer Nature Singapore Pte Ltd. 2018
R. L. Karandikar and B. V. Rao, *Introduction to Stochastic Calculus*,
Indian Statistical Institute Series, https://doi.org/10.1007/978-981-10-8318-1_4

where $0 = s_0 < s_1 < s_2 < \ldots < s_{m+1} < \infty$, a_k is bounded $\mathcal{F}_{s_{k-1}}$ measurable random variable, $1 \leq k \leq (m+1)$, and a_0 is bounded $\mathcal{F}_0$ measurable. $\mathcal{P}$ measurable processes have appeared naturally in the definition of the stochastic integral w.r.t. Brownian motion and play a very significant role in the theory of stochastic integration with respect to general semimartingales as we will see. A process f will be called a *predictable process* if it is $\mathcal{P}$ measurable. Of course, $\mathcal{P}$ depends upon the underlying filtration and would refer to the filtration that we have fixed. If there are more than one filtration under consideration, we will state it explicitly. For example $\mathcal{P}(\mathcal{G}_.)$ denotes the predictable σ-field corresponding to a filtration $(\mathcal{G}_.)$ and $\mathbb{S}(\mathcal{G}_.)$ denotes simple predictable process for the filtration $(\mathcal{G}_.)$.

The following proposition lists various facts about the σ-field $\mathcal{P}$.

Proposition 4.1 *Let* $(\mathcal{F}_.)$ *be a filtration and* $\mathcal{P} = \mathcal{P}(\mathcal{F}_.)$.

(i) *Let* f *be* $\mathcal{P}$ *measurable. Then* f *is* $(\mathcal{F}_.)$ *adapted. Moreover, for every* $t < \infty$, f_t *is* $\sigma(\cup_{s<t}\mathcal{F}_s)$ *measurable.*

(ii) *Let* Y *be a left continuous adapted process. Then* Y *is* $\mathcal{P}$ *measurable.*

(iii) *Let* $\mathbb{A}$ *be the class of all bounded adapted continuous processes. Then* $\mathcal{P} = \sigma(\mathbb{A})$ *and the smallest* bp*-closed class that contains* $\mathbb{A}$ *is* $\mathbb{B}(\widetilde{\Omega}, \mathcal{P})$.

(iv) *For any stopping time* τ, $U = 1_{[0,\tau]}$ *(i.e.* $U_t = 1_{[0,\tau]}(t)$) *is* $\mathcal{P}$ *measurable.*

(v) *For an r.c.l.l. adapted process* Z *and a stopping time* τ, *the process* X *defined by*

$$X_t = Z_\tau 1_{(\tau,\infty)}(t) \tag{4.1.2}$$

is predictable.

(vi) *For a predictable process* g *and a stopping time* τ, g_τ *is a random variable and* h *defined by*

$$h_t = g_\tau 1_{(\tau,\infty)}(t) \tag{4.1.3}$$

is itself predictable.

Proof It suffices to prove the assertions assuming that the processes f, Y, g are bounded (by making a $\tan^{-1}$ transformation, if necessary). Now, for (i) let

$$\mathbb{K}_1 = \{f \in \mathbb{B}(\widetilde{\Omega}, \mathcal{P}) : f_t \text{ is } \sigma(\cup_{s<t}\mathcal{F}_s)\text{- measurable}\}.$$

It is easily seen that $\mathbb{K}_1$ is *bp*-closed and contains $\mathbb{S}$ and thus by Theorem 2.66 equals $\mathbb{B}(\widetilde{\Omega}, \mathcal{P})$ proving (i).

For (ii), given a left continuous bounded adapted process Y, let Y^n be defined by

$$Y_t^n = Y_0 1_{\{0\}}(t) + \sum_{k=0}^{n2^n} Y_{\frac{k}{2^n}} 1_{(\frac{k}{2^n}, \frac{k+1}{2^n}]}(t). \tag{4.1.4}$$

Then $Y^n \in \mathbb{S}$ and $Y^n \xrightarrow{bp} Y$ and this proves (ii).

For (iii), Let $\mathbb{K}_2$ be the smallest bp-closed class containing $\mathbb{A}$. From part (ii) above, it follows that $\mathbb{A} \subseteq \mathbb{B}(\widetilde{\Omega}, \mathcal{P})$ and hence $\mathbb{K}_2 \subseteq \mathbb{B}(\widetilde{\Omega}, \mathcal{P})$. For $t \in [0, \infty), n \geq 1$ let

$$\phi^n(t) = (1 - nt)1_{[0,\frac{1}{n}]}(t),$$

$$\psi^n(t) = nt1_{(0,\frac{1}{n}]}(t) + 1_{(\frac{1}{n},1]}(t) + (1 - n(t-1))1_{(1,1+\frac{1}{n}]}(t).$$

Then ϕ^n and ψ^n are continuous functions, bounded by 1 and $\phi^n \xrightarrow{bp} 1_{\{0\}}$ and $\psi^n \xrightarrow{bp} 1_{(0,1]}$.

For $f \in \mathbb{S}$ given by

$$f(s) = a_0 1_{\{0\}}(s) + \sum_{j=0}^{m} a_{j+1} 1_{(s_j, s_{j+1}]}(s)$$

where $0 = s_0 < s_1 < s_2 \ldots < s_{m+1} < \infty$, a_{j+1} is bounded $\mathcal{F}_{s_j}$ measurable random variable, $0 \leq j \leq m$, and a_0 is bounded $\mathcal{F}_0$ measurable random variable. Let

$$Y_s^n = a_0 \phi^n(s) + \sum_{j=0}^{m} a_{j+1} \psi^n\left(\frac{s-s_j}{s_{j+1}-s_j}\right).$$

Then it follows that $Y^n \in \mathbb{A}$ and $Y^n \xrightarrow{bp} Y$. Thus $\mathbb{S} \subseteq \mathbb{K}_2$ and hence $\mathbb{B}(\widetilde{\Omega}, \mathcal{P}) \subseteq \mathbb{K}_2$ completing the proof of (iii).

For part (iv) note that U is adapted left continuous process and hence $\mathcal{P}$ measurable by part (ii).

For (v), suffices to prove that X is adapted since X is left continuous by construction. Using Lemmas 2.38 and 2.41 it follows that that Z_τ is $\mathcal{F}_\tau$ measurable and W defined by $W_t = Z_\tau 1_{[\tau,\infty)}$ is an r.c.l.l. adapted process. Hence $X = W^-$ is predictable.

For (vi), the class of processes g for which (vi) is true is closed under bp-convergence and contains the class of continuous adapted processes as shown above. In view of part (iii), this completes the proof. $\qquad \square$

If X is $\mathcal{P}(\mathcal{F}_\cdot^Y)$ measurable, then part (i) of the result proven above says that for every t, X_t is measurable w.r.t. $\sigma(Y_u : 0 \leq u < t)$. Thus having observed Y_u, $u < t$, the value X_t can be known (predicted with certainty) even before observing Y_t. This justifies the name predictable σ-field for $\mathcal{P}$.

Exercise 4.2 Show that (i) $\mathcal{U} = \{A \subseteq \widetilde{\Omega} : 1_A \in \mathbb{S}\}$ is a field and that $\mathcal{U}$ generates $\mathcal{P}$ i.e. $\mathcal{P}$ is the smallest σ-field on $\widetilde{\Omega}$ that contains $\mathcal{U}$. (ii) If μ is a signed-measure on $\mathcal{P}$ such that $f \in \mathbb{S}$, $f \geq 0$ implies $\int f d\mu \geq 0$, then μ is a positive measure.

Exercise 4.3 For $t \geq 0$, let $\mathcal{G}_t = \mathcal{F}_t^+$. Show that for $t > 0$

$$\sigma(\cup_{s<t}\mathcal{F}_s) = \sigma(\cup_{s<t}\mathcal{G}_s).$$

Here is an important observation regarding the predictable σ-field $\mathcal{P}(\mathcal{F}_{\cdot}^+)$ corresponding to the filtration $(\mathcal{F}_{\cdot}^+)$.

Theorem 4.4 *Let f be a $(\mathcal{F}_{\cdot}^+)$ predictable process (i.e. $\mathcal{P}(\mathcal{F}_{\cdot}^+)$ measurable). Then g defined by*

$$g_t(\omega) = f_t(\omega)1_{\{(0,\infty)\times\Omega\}}(t,\omega) \qquad (4.1.5)$$

is a $(\mathcal{F}_{\cdot})$ predictable process.

Proof Let f be a $(\mathcal{F}_{\cdot}^+)$ adapted bounded continuous process. A crucial observation is that for $t > 0$, f_t is $\mathcal{F}_t$ measurable (see Exercise 4.3) and thus defining for $n \geq 1$

$$h_t^n = (nt \wedge 1)f_t$$

it follows that h^n are $(\mathcal{F}_{\cdot})$ adapted continuous processes and $h^n \overset{bp}{\to} g$ where g is defined by (4.1.5) and hence in this case g is $(\mathcal{F}_{\cdot})$ predictable.

Now let $\mathcal{H}$ be the class of bounded $\mathcal{P}(\mathcal{F}_{\cdot}^+)$ measurable f for which the conclusion is true. Then easy to see that $\mathcal{H}$ is an algebra that is bp-closed and contains all $(\mathcal{F}_{\cdot}^+)$ adapted continuous processes. The conclusion follows by the monotone class theorem, Theorem 2.66.   □

The following is an immediate consequence of this.

Corollary 4.5 *Let f be an $(\mathcal{F}_{\cdot}^+)$ predictable process such that f_0 is $\mathcal{F}_0$ measurable. Then f is $(\mathcal{F}_{\cdot})$ predictable.*

Proof Since f_0 is $\mathcal{F}_0$ measurable, $h_t = f_0 1_{\{0\}}(t)$ is $(\mathcal{F}_{\cdot})$ predictable and $f = g + h$ where g is the $(\mathcal{F}_{\cdot})$ predictable process given by (4.1.5). Hence f is $(\mathcal{F}_{\cdot})$ predictable.   □

Corollary 4.6 *Let $A \subseteq (0, \infty) \times \Omega$. Then $A \in \mathcal{P}(\mathcal{F}_{\cdot})$ if and only if $A \in \mathcal{P}(\mathcal{F}_{\cdot}^+)$.*

The following example will show that the result may not be true if $f_0 = 0$ is dropped in Corollary 4.5 above.

Exercise 4.7 Let $\Omega = C([0, \infty)$ and let X_t denote the coordinate process and $\mathcal{F}_t = \sigma(X_s : 0 \leq s \leq t)$. Let A be the set of all $\omega \in \Omega$ that take positive as well as negative values in $(0, \epsilon)$ for every $\epsilon > 0$. Show that $A \in \mathcal{F}_0^+$ but A does not belong to $\mathcal{F}_0$. Use this to show the relevance of the hypothesis on f_0 in the corollary given above.

Exercise 4.8 Let ξ be an $\mathcal{F}_\tau$ measurable random variable. Show that $Y = \xi 1_{(\tau,\infty)}$ is predictable.
HINT: Use Lemma 2.41 along with part (v) in Proposition 4.1.

Exercise 4.9 Let X denote the coordinate mappings on $(\mathbb{D}_d, \mathbb{B}(C_d))$ and let $\mathcal{D}_t = \sigma(X_u : 0 \le u \le t)$ and let $\mathcal{P} = \mathcal{P}(\mathcal{D}.)$. Let $f : [0, \infty) \times \mathbb{D}_d \mapsto \mathbb{R}$ be $\mathcal{P}$ measurable. Let $(\Omega, \mathcal{F}, P)$ be a probability space with a filtration $(\mathcal{F}.)$. Let Y be a r.c.l.l. $(\mathcal{F}.)$ adapted process. Let Z be defined by $Z_t = f(t, Y)$. Show that Z is $\mathcal{P}(\mathcal{F}.)$ measurable, i.e. a predictable process on $(\Omega, \mathcal{F}, P)$.
HINT: Verify for simple f and then use monotone class theorem.

4.2 Stochastic Integrators

Let us fix an r.c.l.l. $(\mathcal{F}.)$ adapted stochastic process X.

Recall, $\mathbb{S}$ consists of the class of processes f of the form

$$f(s) = a_0 1_{\{0\}}(s) + \sum_{j=0}^{m} a_{j+1} 1_{(s_j, s_{j+1}]}(s) \tag{4.2.1}$$

where $0 = s_0 < s_1 < s_2 < \ldots < s_{m+1} < \infty, a_j$ is bounded $\mathcal{F}_{s_{j-1}}$ measurable random variable, $1 \le j \le (m+1)$, and a_0 is bounded $\mathcal{F}_0$ measurable.

For simple predictable $f \in \mathbb{S}$ given by (4.2.1), let $J_X(f)$ be the r.c.l.l. process defined by

$$J_X(f)(t) = a_0 X_0 + \sum_{j=0}^{m} a_{j+1}(X_{s_{j+1} \wedge t} - X_{s_j \wedge t}). \tag{4.2.2}$$

One needs to verify that J_X is unambiguously defined on $\mathbb{S}$. That is, if a given f has two representations of type (4.2.1), then the corresponding expressions in (4.2.2) agree. This as well as linearity of $J_X(f)$ for $f \in \mathbb{S}$ can be verified using elementary algebra. By definition, for $f \in \mathbb{S}$, $J_X(f)$ is an r.c.l.l. adapted process. In analogy with the Ito's integral with respect to Brownian motion discussed in the earlier chapter, we wish to explore if we can extend J_X to the smallest bp-closed class of integrands that contain $\mathbb{S}$. Each $f \in \mathbb{S}$ can be viewed as a real-valued function on $\widetilde{\Omega} = [0, \infty) \times \Omega$. Since $\mathcal{P}$ is the σ-field generated by $\mathbb{S}$, by Theorem 2.66, the smallest class of functions that contains $\mathbb{S}$ and is closed under bp-convergence is $\mathbb{B}(\widetilde{\Omega}, \mathcal{P})$.

When the space, filtration and the probability measure are clear from the context, we will write the class of adapted r.c.l.l. processes $\mathbb{R}^0(\Omega, (\mathcal{F}.), P)$ simply as $\mathbb{R}^0$.

Definition 4.10 An r.c.l.l. adapted process X is said to be a *stochastic integrator* if the mapping J_X from $\mathbb{S}$ to $\mathbb{R}^0(\Omega, (\mathcal{F}.), P)$ has an extension J_X: $\mathbb{B}(\widetilde{\Omega}, \mathcal{P}) \mapsto \mathbb{R}^0(\Omega, (\mathcal{F}.), P)$ satisfying the following continuity property:

$$f^n \xrightarrow{bp} f \text{ implies } J_X(f^n) \xrightarrow{ucp} J_X(f). \tag{4.2.3}$$

It should be noted that for a given r.c.l.l. process X, J_X may not be continuous on $\mathbb{S}$. See the next exercise. So this definition, in particular, requires that J_X is continuous on $\mathbb{S}$ and has a continuous extension to $\mathbb{B}(\widetilde{\Omega}, \mathcal{P})$. Though not easy to prove, continuity of J_X on $\mathbb{S}$ does imply that it has a continuous extension and hence it is a stochastic integrator. We will see this later in Sect. 4.10. The next exercise shows that J_G is not continuous on $\mathbb{S}$.

Exercise 4.11 Let G be a real-valued function from $[0, \infty)$ with $G(0) = 0$ such that for some $T < \infty$, $\text{VAR}_{[0,T]}(G) = \infty$.

(i) Show that there exists a sequence of partitions of $[0, T]$

$$0 = t_0^m < t_1^m < \ldots < t_{n_m}^m = T \tag{4.2.4}$$

such that

$$\alpha_m = \left(\sum_{j=0}^{n_m-1} |G(t_j^m) - G(t_{j-1}^m)| \right) \to \infty. \tag{4.2.5}$$

(ii) Let $\text{sgn} : \mathbb{R} \mapsto \mathbb{R}$ be defined by $\text{sgn}(x) = 1$ for $x \geq 0$ and $\text{sgn}(x) = -1$ for $x < 0$. Let $f^m = (\alpha_m)^{-\frac{1}{2}} \text{sgn}(G(t_j^m) - G(t_{j-1}^m)) 1_{(t_{j-1}^m, t_j^m]}$. Show that f^m converges to 0 uniformly.
(iii) $J_G(f^m)$ converges to ∞.

Conclude that J_G is not continuous on $\mathbb{S}$.

We next observe that the extension, when it exists, is unique.

Theorem 4.12 *Let X be an r.c.l.l. process. Suppose there exist mappings J_X, J_X' from $\mathbb{B}(\widetilde{\Omega}, \mathcal{P})$ into $\mathbb{R}^0(\Omega, (\mathcal{F}_.), P)$, such that for $f \in \mathbb{S}$ (given by (4.2.1)),*

$$J_X(f)(t) = J_X'(f)(t) = a_0 X_0 + \sum_{j=1}^m a_{j+1}(X_{s_{j+1} \wedge t} - X_{s_j \wedge t}). \tag{4.2.6}$$

Further suppose that both J_X, J_X' satisfy (4.2.3). Then

$$P(J_X(f)(t) = J_X'(f)(t) \; \forall t) = 1 \; \forall f \in \mathbb{B}(\widetilde{\Omega}, \mathcal{P}).$$

Proof Since J_X, J_X' both satisfy (4.2.3), the class

$$\mathbb{K}_1 = \{f \in \mathbb{B}(\widetilde{\Omega}, \mathcal{P}) : P(J_X(f)(t) = J_X'(f)(t) \; \forall t) = 1\}$$

is *bp*-closed and by our assumption (4.2.6), contains $\mathbb{S}$. Since $\mathcal{P} = \sigma(\mathbb{S})$, by Theorem 2.66 it follows that $\mathbb{K}_1 = \mathbb{B}(\widetilde{\Omega}, \mathcal{P})$. $\qquad \square$

The following result which is almost obvious in this treatment of stochastic integration is a deep result in the traditional approach to stochastic integration and is known as Stricker's theorem.

Theorem 4.13 *Let X be a stochastic integrator for the filtration $(\mathcal{F}.)$. Let $(\mathcal{G}.)$ be a filtration such that $\mathcal{F}_t \subseteq \mathcal{G}_t$ for all t. Suppose X is a stochastic integrator for the filtration $(\mathcal{G}.)$ as well. Denoting the mapping defined by (4.2.2) for the filtration $(\mathcal{G}.)$ and its extension by H_X, we have*

$$J_X(f) = H_X(f) \quad \forall f \in \mathbb{B}(\widetilde{\Omega}, \mathcal{P}(\mathcal{F}.)). \tag{4.2.7}$$

Proof Let J'_X be the restriction of H_X to $\mathbb{B}(\widetilde{\Omega}, \mathcal{P}(\mathcal{F}.))$. Then J'_X satisfies (4.2.6) as well as (4.2.3). Thus (4.2.7) follows from uniqueness of extension, Theorem 4.12. $\square$

Here is an observation that plays an important role in next result.

Lemma 4.14 *Let X be a stochastic integrator and ξ be a $\mathcal{F}_0$ measurable bounded random variable. Then $\forall f \in \mathbb{B}(\widetilde{\Omega}, \mathcal{P})$*

$$J_X(\xi f) = \xi J_X(f). \tag{4.2.8}$$

Proof Let $\mathbb{K}_2$ consist of all $f \in \mathbb{B}(\widetilde{\Omega}, \mathcal{P})$ such that (4.2.8) is true. Easy to verify that $\mathbb{S} \subseteq \mathbb{K}_2$ and that $\mathbb{K}_2$ is *bp*-closed. Thus, $\mathbb{K}_2 = \mathbb{B}(\widetilde{\Omega}, \mathcal{P})$ by Theorem 2.66. $\square$

The next observation is about the role of P null sets.

Theorem 4.15 *Let X be a stochastic integrator. Then $f, g \in \mathbb{B}(\widetilde{\Omega}, \mathcal{P})$,*

$$\mathsf{P}(\omega \in \Omega \ : \ f_t(\omega) = g_t(\omega) \ \forall t \geq 0) = 1 \tag{4.2.9}$$

implies

$$\mathsf{P}(\omega \in \Omega \ : \ J_X(f)_t(\omega) = J_X(g)_t(\omega) \ \forall t \geq 0) = 1. \tag{4.2.10}$$

In other words, the mapping J_X maps equivalence classes of process under the relation $f = g$ (see Definition 2.2) to equivalence class of processes.

Proof Given f, g such that (4.2.9) holds, let

$$\Omega_0 = \{\omega \in \Omega \ : \ f_t(\omega) = g_t(\omega) \ \forall t \geq 0\}.$$

Then the assumption that $\mathcal{F}_0$ contains all P null sets implies that $\Omega_0 \in \mathcal{F}_0$ and thus $\xi = 1_{\Omega_0}$ is $\mathcal{F}_0$ measurable and $\xi f = \xi g$ in the sense that these are identical processes:

$$\xi(\omega) f_t(\omega) = \xi(\omega) g_t(\omega) \quad \forall \omega \in \Omega, \ t \geq 0.$$

Now we have

$$\begin{aligned}
\xi J_X(f) &= J_X(\xi f) \\
&= J_X(\xi g) \\
&= \xi J_X(g).
\end{aligned}$$

Since $\mathsf{P}(\xi = 1) = 1$, (4.2.10) follows. $\square$

Corollary 4.16 *Let f_n, f be bounded predictable such that there exists $\Omega_0 \subseteq \Omega$ with $\mathrm{P}(\Omega_0) = 1$ and such that*

$$1_{\Omega_0} f_n \xrightarrow{bp} 1_{\Omega_0} f.$$

Then

$$J_X(f_n) \xrightarrow{ucp} J_X(f).$$

For a stochastic integrator X, we will be defining the stochastic integral $Y = \int f \, dX$, for an appropriate class of integrands, given as follows:

Definition 4.17 For a stochastic integrator X, let $\mathbb{L}(X)$ denote the class of predictable processes f such that

$$h^n \in \mathbb{B}(\widetilde{\Omega}, \mathcal{P}), \ h^n \to 0 \text{ pointwise}, \ |h^n| \le |f| \Rightarrow J_X(h^n) \xrightarrow{ucp} 0. \qquad (4.2.11)$$

From the definition, it follows that if $f \in \mathbb{L}(X)$ and g is predictable such that $|g| \le |f|$, then $g \in \mathbb{L}(X)$. Here is an interesting consequence of this definition.

Theorem 4.18 *Let X be a stochastic integrator. Then $f \in \mathbb{L}(X)$ if and only if*

$$g^n \in \mathbb{B}(\widetilde{\Omega}, \mathcal{P}), \ g^n \to g \text{ pointwise} , \ |g^n| \le |f| \Rightarrow J_X(g^n) \text{ is Cauchy in } \mathbf{d}_{ucp}. \qquad (4.2.12)$$

Proof Suppose f satisfies (4.2.11). Let

$$g^n \in \mathbb{B}(\widetilde{\Omega}, \mathcal{P}), \ g^n \to g \text{ pointwise} , \ |g^n| \le |f|.$$

Given any subsequences $\{m^k\}$, $\{n^k\}$ of integers, increasing to ∞, let

$$h^k = \frac{1}{2}(g^{m^k} - g^{n^k}).$$

Then $\{h^k\}$ satisfies (4.2.11) and hence $J_X(h^k) \xrightarrow{ucp} 0$ and as a consequence

$$(J_X(g^{m^k}) - J_X(g^{n^k})) \xrightarrow{ucp} 0. \qquad (4.2.13)$$

Since (4.2.13) holds for all subsequences $\{m^k\}$, $\{n^k\}$ of integers, increasing to ∞, it follows that $J_X(g^n)$ is Cauchy. Conversely, suppose f satisfies (4.2.12). Given $\{h^n\}$ as in (4.2.11), let a sequence g^k be defined as $g^{2k} = h^k$ and $g^{2k-1} = 0$ for $k \ge 1$. Then g^k converges to $g = 0$ and hence $J_X(g^k)$ is Cauchy. Since for odd integers n, $J_X(g^n) = 0$, $J_X(g^{2k}) = J_X(h^k)$ converges to 0. $\qquad\qquad \square$

Remark **4.19** Note that in the previous theorem, each g^n was assumed to be bounded but no such assumption was made about g.

This result enables us to define $\int f \, dX$ for $f \in \mathbb{L}(X)$.

Definition 4.20 Let X be a stochastic integrator and let $f \in \mathbb{L}(X)$. Then

$$\int f \, dX = \lim_{n \to \infty} J_X(f 1_{\{|f| \le n\}}) \tag{4.2.14}$$

where the limit is in the $\mathbf{d}_{ucp}$ metric.

Exercise 4.21 Let $a_n \to \infty$. Show that for $f \in \mathbb{L}(X)$,

$$J_X(f 1_{\{|f| \le a_n\}}) \xrightarrow{ucp} \int f \, dX.$$

When f is bounded, it follows that $\int f \, dX = J_X(f)$ by definition. Here is an important result which is essentially a version of dominated convergence theorem.

Theorem 4.22 Let $f \in \mathbb{L}(X)$ and $g^n \in \mathbb{B}(\widetilde{\Omega}, \mathcal{P})$ be such that $g^n \to g$ pointwise, $|g^n| \le |f|$. Then

$$\int g^n \, dX \to \int g \, dX \text{ in } \mathbf{d}_{ucp} \text{ metric as } n \to \infty.$$

Proof For $n \ge 1$, let $\xi^{2n-1} = g 1_{\{|g| \le n\}}$ and $\xi^{2n} = g^n$. Then $|\xi^m| \le |f|$ and ξ^m converges pointwise to g. Thus, $\int \xi^m \, dX$ is Cauchy in $\mathbf{d}_{ucp}$ metric. On the other hand

$$\int \xi^{2n-1} \, dX \xrightarrow{ucp} \int g \, dX$$

from the definition of $\int g \, dX$. Thus

$$\int \xi^{2n} \, dX = \int g^n \, dX \xrightarrow{ucp} \int g \, dX. \qquad \square$$

Note that in the result given above, we did not require g to be bounded. Even if g were bounded, the convergence was not required to be bounded pointwise.

The process $\int f \, dX$ is called the *stochastic integral* of f with respect to X, and we will also write

$$\left(\int f \, dX \right)_t = \int_0^t f \, dX.$$

We interpret $\int_0^t f \, dX$ as the definite integral of f with respect to X over the interval $[0, t]$. We sometimes need the integral of f w.r.t. X over $(0, t]$ and so we introduce

$$\int_{0+}^t f \, dX = \int_0^t f 1_{(0,\infty)} dX = \int_0^t f \, dX - f_0 X_0.$$

Note that $\int f dX$ is an r.c.l.l. process by definition. We will also write $\int_0^{t-} f dX$ to denote Y_{t-} where $Y_s = \int_0^s f dX$.

A simple example of a stochastic integrator is an r.c.l.l. adapted process X such that the mapping $t \mapsto X_t(\omega)$ satisfies $\text{VAR}_{[0,T]}(X.(\omega)) < \infty$ for all $\omega \in \Omega$, for all $T < \infty$.

Theorem 4.23 *Let $X \in \mathbb{V}$ be a process with finite variation paths, i.e. X be an r.c.l.l. adapted process such that*

$$\text{VAR}_{[0,T]}(X.(\omega)) < \infty \text{ for all } T < \infty. \tag{4.2.15}$$

Then X is a stochastic integrator. Further, for $f \in \mathbb{B}(\tilde{\Omega}, \mathcal{P})$ the stochastic integral $J_X(f) = \int f dX$ is the Lebesgue–Stieltjes integral for every $\omega \in \Omega$:

$$J_X(f)(t)(\omega) = \int_0^t f(s, \omega) dX_s(\omega) \tag{4.2.16}$$

where the integral above is the Lebesgue–Stieltjes integral.

Proof For $f \in \mathbb{S}$, the right-hand side of (4.2.16) agrees with the specification in (4.2.1)–(4.2.2). Further the dominated convergence theorem (for Lebesgue–Stieltjes integral) implies that the right-hand side of (4.2.16) satisfies (4.2.3) and thus X is a stochastic integrator and (4.2.16) is true. $\qquad\qquad\square$

As seen in Remark 3.16, Brownian motion W is also a stochastic integrator.

Remark **4.24** If $X \in \mathbb{V}$ and $A_t = |X|_t$ is the total variation of X on $[0, t]$ and f is predictable such that

$$\int_0^t |f_s| dA_s < \infty \ \forall t < \infty \ a.s. \tag{4.2.17}$$

then $f \in \mathbb{L}(X)$ and the stochastic integral is the same as the Lebesgue–Stieltjes integral. This follows from the dominated convergence theorem and Theorem 4.23. However, $\mathbb{L}(X)$ may include processes f that may not satisfy (4.2.17). We will return to this later (see Exercise 5.77).

Remark **4.25** Suppose X is an r.c.l.l. adapted process such that

$$f^n \to 0 \text{ uniformly} \ \Rightarrow \ J_X(f^n) \xrightarrow{ucp} 0. \tag{4.2.18}$$

Of course every stochastic integrator satisfies this property. Let $\mathbb{S}_1$ denote the class of $f \in \mathbb{S}$ (simple functions) that are bounded by 1. For $T < \infty$ the family of random variables

$$\left\{ \int_0^T f dX : f \in \mathbb{S}_1 \right\} \tag{4.2.19}$$

is *bounded in probability* or *tight* in the sense that

$$\forall \varepsilon > 0 \ \exists \ M < \infty \text{ such that } \sup_{f \in \mathbb{S}_1} P(|\int_0^T f dX| \geq M) \leq \varepsilon. \qquad (4.2.20)$$

To see this, suppose (4.2.18) is true but (4.2.20) is not true. Then we can get an $\varepsilon > 0$ and for each $m \geq 1$, $f^m \in \mathbb{S}$ with $|f^m| \leq 1$ such that

$$P(|\int_0^T f^m dX| \geq m) \geq \varepsilon. \qquad (4.2.21)$$

Writing $g^m = \frac{1}{m} f^m$, it follows that $g^m \to 0$ uniformly but in view of (4.2.21), $\int_0^T g^m dX$ does not converge to zero in probability—which contradicts (4.2.18). Indeed, the apparently weaker property (4.2.18) characterizes stochastic integrators as we will see later. See Theorem 5.89.

Remark **4.26** *Equivalent Probability Measures*: Let Q be a probability measure equivalent to P. In other words, for $A \in \mathcal{F}$,

$$Q(A) = 0 \text{ if and only if } P(A) = 0.$$

Then it is well known and easy to see that (for a sequence of random variables) convergence in P probability implies and is implied by convergence in Q probability and the same is true for *ucp* convergence. Thus, it follows that an r.c.l.l. adapted process X is a stochastic integrator on $(\Omega, \mathcal{F}, P)$ if and only if it is a stochastic integrator on $(\Omega, \mathcal{F}, Q)$. Moreover, the class $\mathbb{L}(X)$ under the two measures is the same and for $f \in \mathbb{L}(X)$, the stochastic integral $\int f dX$ on the two spaces is identical.

4.3 Properties of the Stochastic Integral

First we note linearity of $(f, X) \mapsto \int f dX$.

Theorem 4.27 *Let X, Y be stochastic integrators, f, g be predictable processes and $\alpha, \beta \in \mathbb{R}$.*

(i) *Suppose $f, g \in \mathbb{L}(X)$. Let $h = \alpha f + \beta g$. Then $h \in \mathbb{L}(X)$ and*

$$\int h dX = \alpha \int f dX + \beta \int g dX. \qquad (4.3.1)$$

(ii) *Let $Z = \alpha X + \beta Y$. Then Z is a stochastic integrator. Further, if $f \in \mathbb{L}(X)$ and $f \in \mathbb{L}(Y)$. then $f \in \mathbb{L}(Z)$ and*

$$\int f dZ = \alpha \int f dX + \beta \int f dY. \tag{4.3.2}$$

Proof We will begin by showing that (4.3.1) is true for f, g bounded predictable processes. For a bounded predictable process f, let

$$\mathbb{K}(f) = \{g \in \mathbb{B}(\widetilde{\Omega}, \mathcal{P}) : \int (\alpha f + \beta g) dX = \alpha \int f dX + \beta \int g dX, \ \forall \alpha, \beta \in \mathbb{R}\}.$$

If $f \in \mathbb{S}$, it easy to see that $\mathbb{S} \subseteq \mathbb{K}(f)$ and Theorem 4.22 implies that $K(f)$ is *bp*-closed. Hence invoking Theorem 2.66, it follows that $\mathbb{K}(f) = \mathbb{B}(\widetilde{\Omega}, \mathcal{P})$.

Now we take $f \in \mathbb{B}(\widetilde{\Omega}, \mathcal{P})$ and the part proven above yields $\mathbb{S} \subseteq \mathbb{K}(f)$. Once again, using that $K(f)$ is *bp*-closed we conclude that $\mathbb{K}(f) = \mathbb{B}(\widetilde{\Omega}, \mathcal{P})$. Thus (4.3.1) is true when f, g are bounded predictable process.

Now let us fix $f, g \in \mathbb{L}(X)$. We will show $(|\alpha f| + |\beta g|) \in \mathbb{L}(X)$, let u^n be bounded predictable processes converging to u pointwise and

$$|u^n| \leq (|\alpha f| + |\beta g|); \ \forall n \geq 1.$$

Let $v^n = u^n 1_{\{|\alpha f| \leq |\beta g|\}}$ and $w^n = u^n 1_{\{|\alpha f| > |\beta g|\}}$. Then v^n and w^n converge pointwise to $v = u 1_{\{|\alpha f| \leq |\beta g|\}}$ and $w = u 1_{\{|\alpha f| > |\beta g|\}}$, respectively, and further

$$|v^n| \leq 2|\beta g|$$

$$|w^n| \leq 2|\alpha f|.$$

Note that since v^n, w^n are bounded and $u^n = v^n + w^n$, from the part proven above, we have

$$\int v^n dX + \int w^n dX = \int u^n dX.$$

Since $f, g \in \mathbb{L}(X)$, it follows that $\int v^n dX$ and $\int w^n dX$ are Cauchy in $\mathbf{d}_{ucp}$ metric and hence so is their sum $\int u^n dX$. Thus $(|\alpha f| + |\beta g|) \in \mathbb{L}(X)$ and as a consequence, $(\alpha f + \beta g) \in \mathbb{L}(X)$ as well.

Now let

$$f^n = f 1_{\{|f| \leq n\}}, \ \ g^n = g 1_{\{|g| \leq n\}}.$$

Then by definition, $\int f^n dX$ converges to $\int f dX$ and $\int g^n dX$ converges to $\int g dX$ in $\mathbf{d}_{ucp}$ metric. Also $(\alpha f^n + \beta g^n)$ are bounded predictable processes, converge pointwise to $(\alpha f + \beta g)$ and are dominated by $(|\alpha f| + |\beta g|) \in \mathbb{L}(X)$. Hence by Theorem 4.22 we have

$$\int (\alpha f^n + \beta g^n) dX \xrightarrow{ucp} \int (\alpha f + \beta g) dX.$$

On the other hand, the validity of (4.3.1) for bounded predictable processes yields

$$\int (\alpha f^n + \beta g^n) dX = \int \alpha f^n dX + \int \beta g^n dX$$

$$= \alpha \int f^n dX + \beta \int g^n dX$$

$$\overset{ucp}{\longrightarrow} \alpha \int f dX + \beta \int g dX.$$

This completes proof of (i).

For (ii), we begin by noting that (4.3.2) is true when f is simple i.e.

$$J_Z(f) = \alpha J_X(f) + \beta J_Y(f) \quad f \in \mathbb{S}.$$

Since J_X, J_Y have a continuous extension to $\mathbb{B}(\widetilde{\Omega}, \mathcal{P})$, it follows that so does J_Z and hence Z is also a stochastic integrator and thus (4.3.2) is true for bounded predictable processes.

Now if $f \in \mathbb{L}(X)$ and $f \in \mathbb{L}(Y)$ and $g^n \in \mathbb{B}(\widetilde{\Omega}, \mathcal{P})$, converge pointwise to g, g^n is dominated by $|f|$, then $\alpha \int g^n dX$ and $\beta \int g^n dY$ are Cauchy in $\mathbf{d}_{ucp}$ metric and hence so is their sum, which equals $\int g^n d(\alpha X + \beta Y) = \int g^n dZ$. Thus $f \in \mathbb{L}(Z)$. Equation (4.3.2) follows by using (4.3.2) for the bounded process $f^n = f 1_{\{|f| \le n\}}$ and passing to the limit.   □

Thus the class of stochastic integrators is a linear space. We will see later that it is indeed an Algebra. Let us note that when X is a continuous process, then so is $\int f dX$.

Theorem 4.28 *Let X be a continuous process and further X be a stochastic integrator. Then for $f \in \mathbb{L}(X)$, $\int f dX$ is also a continuous process.*

Proof Let

$$\mathbb{K} = \{ f \in \mathbb{B}(\widetilde{\Omega}, \mathcal{P}) : \int f dX \text{ is a continuous process} \}.$$

By using the definition of $\int f dX$ it is easy to see that $\mathbb{S} \subseteq \mathbb{K}$. Also that $\mathbb{K}$ is *bp*-closed since Z^n continuous, $Z^n \overset{ucp}{\longrightarrow} Z$ implies Z is also continuous. Hence invoking Theorem 2.66 we conclude $\mathbb{K} = \mathbb{B}(\widetilde{\Omega}, \mathcal{P})$. The general case follows by noting that limit in $\mathbf{d}_{ucp}$ metric of continuous process is a continuous process and using that for $f \in \mathbb{L}(X)$, $\int f dX$ is the limit in $\mathbf{d}_{ucp}$-metric of $\int f 1_{\{|f| \le n\}} dX$.   □

We can now prove:

Theorem 4.29 Dominated Convergence Theorem for the Stochastic Integral *Let X be a stochastic integrator. Suppose h^n, h are predictable processes such that*

$$h_t^n(\omega) \to h_t(\omega) \quad \forall t \ge 0, \quad \forall \omega \in \Omega \tag{4.3.3}$$

and there exists $f \in \mathbb{L}(X)$ such that

$$|h^n| \leq |f| \quad \forall n. \tag{4.3.4}$$

Then

$$\int h^n dX \xrightarrow{ucp} \int h dX. \tag{4.3.5}$$

Proof Let $g^n = h^n 1_{\{|h^n| \leq n\}}$ and $f^n = h^n 1_{\{|h^n| > n\}}$

Note that in view of (4.3.4) and the assumption $f \in \mathbb{L}(X)$ it follows that $h^n \in \mathbb{L}(X)$. Pointwise convergence of h^n to h also implies $|h| \leq |f|$ which in turn yields $h \in \mathbb{L}(X)$. Thus $\int h^n dX$, $\int h dX$ are defined. Clearly, $g^n \to h$ pointwise and also $f^n \to 0$ pointwise. Further, $h^n = g^n + f^n$, $|g^n| \leq |f|$ and $|f^n| \leq |f|$.

From Theorem 4.22 it follows that $\int g^n dX \xrightarrow{ucp} \int h dX$ and from the definition of $\mathbb{L}(X)$, it follows that $\int f^n dX \xrightarrow{ucp} 0$. Now linearity of the stochastic integral, Theorem 4.27, shows that (4.3.5) is true.   □

The reader should note the subtle difference between this result and Theorem 4.22.

Remark **4.30** The condition (4.3.3) that $h^n \to h$ pointwise can be replaced by requiring that convergence holds pointwise outside a null set, namely that there exists $\Omega_0 \subseteq \Omega$ with $P(\Omega_0) = 1$ such that

$$h_t^n(\omega) \to h_t(\omega) \quad \forall t \geq 0, \quad \forall \omega \in \Omega_0. \tag{4.3.6}$$

See Theorem 4.15 and Corollary 4.16.

It should be noted that the hypothesis in the dominated convergence theorem given above are exactly the same as in the case of Lebesgue integrals.

Recall, for an r.c.l.l. process X, X^- denotes the l.c.r.l. process defined by $X_t^- = X(t-)$, i.e. the left limit at t with the convention $X(0-) = 0$ and $\Delta X = X - X^-$. Note that $(\Delta X)_t = 0$ at each continuity point $t > 0$ and equals the jump otherwise. Note that by the above convention

$$(\Delta X)_0 = X_0.$$

Exercise 4.31 Let X^n, $n \geq 1$ and X be r.c.l.l. adapted processes such that $X^n \xrightarrow{ucp} X$. Show that $\Delta X^n \xrightarrow{ucp} \Delta X$.

The next result connects the jumps of the stochastic integral with the jumps of the integrator.

Theorem 4.32 *Let X be a stochastic integrator and let $f \in \mathbb{L}(X)$. Then we have*

$$\Delta(\int f dX) = f \cdot (\Delta X). \tag{4.3.7}$$

Equation (4.3.7) *is to be interpreted as follows: if* $Y_t = \int_0^t f dX$ *then* $(\Delta Y)_t = f_t(\Delta X)_t$.

Proof For $f \in \mathbb{S}$, (4.3.7) can be verified from the definition. Now the class $\mathbb{K}$ of f such that (4.3.7) is true can be seen to be *bp*-closed and hence is the class of all bounded predictable processes. The case of general $f \in \mathbb{L}(X)$ can be completed as in the proof of Theorem 4.28. □

We have already seen that the class of stochastic integrators (with respect to a filtration on a given probability space) is a linear space.

Theorem 4.33 *Let X be a stochastic integrator and let $f \in \mathbb{L}(X)$. Let $Y = \int f dX$. Then Y is also a stochastic integrator and for a bounded predictable process g,*

$$\int_0^t g dY = \int_0^t g f dX, \quad \forall t. \tag{4.3.8}$$

Further, for a predictable process g, $g \in \mathbb{L}(Y)$ if and only if $gf \in \mathbb{L}(X)$ and then (4.3.8) holds.

Proof We first prove that Y is a stochastic integrator and that (4.3.8) holds. Let us first assume that $f, g \in \mathbb{S}$ are given by

$$f(s) = a_0 1_{\{0\}}(s) + \sum_{j=0}^m a_{j+1} 1_{(s_j, s_{j+1}]}(s) \tag{4.3.9}$$

$$g(s) = b_0 1_{\{0\}}(s) + \sum_{j=0}^n b_{j+1} 1_{(t_j, t_{j+1}]}(s) \tag{4.3.10}$$

where a_0 and b_0 are bounded $\mathcal{F}_0$ measurable random variables, $0 = s_0 < s_1 < s_2 < \ldots, < s_{m+1} < \infty$, $0 = t_0 < t_1 < t_2 < \ldots, < t_{n+1} < \infty$; a_{j+1} is bounded $\mathcal{F}_{s_j}$ measurable random variable, $0 \le j \le m$; and b_{j+1} is bounded $\mathcal{F}_{t_j}$ measurable random variable, $0 \le j \le n$. Let us put

$$A = \{s_j : 0 \le j \le (m+1)\} \cup \{t_j : 0 \le j \le (n+1)\}.$$

Let us enumerate the set A as

$$A = \{r_i : 0 \le i \le k\}$$

where $0 = r_0 < r_1 < \ldots < r_{k+1}$. Note k may be smaller than $m + n$ as there could be repetitions among the $\{s_j\}$ and $\{t_i\}$. We can then represent f, g as

$$f(s) = c_0 1_{\{0\}}(s) + \sum_{j=0}^k c_{j+1} 1_{(r_j, r_{j+1}]}(s) \tag{4.3.11}$$

$$g(s) = d_0 1_{\{0\}}(s) + \sum_{j=0}^{k} d_{j+1} 1_{(r_j, r_{j+1}]}(s) \tag{4.3.12}$$

where c_{j+1}, d_{j+1} are bounded $\mathcal{F}_{r_j}$ measurable. Then

$$(gf)(s) = d_0 c_0 1_{\{0\}}(s) + \sum_{j=0}^{k} d_{j+1} c_{j+1} 1_{(r_j, r_{j+1}]}(s)$$

and hence

$$\int_0^t (gf) dX = d_0 c_0 X_0 + \sum_{j=0}^{k} d_{j+1} c_{j+1} (X_{r_{j+1} \wedge t} - X_{r_j \wedge t}). \tag{4.3.13}$$

Since

$$Y_t = \int_0^t f dX \tag{4.3.14}$$

we have $Y_0 = c_0 X_0$ and $Y_t = c_0 X_0 + \sum_{j=0}^{k} c_{j+1}(X_{r_{j+1} \wedge t} - X_{r_j \wedge t})$ and hence

$$(Y_{r_{j+1} \wedge t} - Y_{r_j \wedge t}) = c_{j+1}(X_{r_{j+1} \wedge t} - X_{r_j \wedge t}). \tag{4.3.15}$$

Thus, using (4.3.13) and (4.3.15), we conclude

$$\int_0^t (gf) dX = d_0 Y_0 + \sum_{j=0}^{k} d_{j+1} (Y_{r_{j+1} \wedge t} - Y_{r_j \wedge t}). \tag{4.3.16}$$

The right-hand side in (4.3.16) is $\int g dY$ and thus (4.3.8) is true when both $f, g \in \mathbb{S}$.
   Let us note that in view of (4.3.10) we also have

$$\int_0^t (gf) dX = b_0 Y_0 + \sum_{j=0}^{n} b_{j+1} (Y_{t_{j+1} \wedge t} - Y_{t_j \wedge t}) \tag{4.3.17}$$

or in other words,

$$\int_0^t (gf) dX = b_0 Y_0 + \sum_{j=0}^{n} b_{j+1} \left(\int_0^{t_{j+1} \wedge t} f dX - \int_0^{t_j \wedge t} f dX \right). \tag{4.3.18}$$

Now fix $g \in \mathbb{S}$. Note that this fixes n (appearing in (4.3.10)) as well. Let

$$\mathbb{K} = \{ f \in \mathbb{B}(\widetilde{\Omega}, \mathcal{P}) : \ (4.3.18) \text{ holds} \}.$$

We have seen that $\mathbb{S} \subseteq \mathbb{K}$. Easy to see using Theorem 4.29 (dominated convergence theorem) that $\mathbb{K}$ is bp- closed and since it contains $\mathbb{S}$, it equals $\mathbb{B}(\widetilde{\Omega}, \mathcal{P})$. Thus, for all bounded predictable f, (4.3.18) is true.

If $f \in \mathbb{L}(X)$, then approximating f by $f^n = f 1_{\{|f| \leq n\}}$, using (4.3.18) for f^n and taking limits, we conclude (invoking dominated convergence theorem, which is justified as g is bounded) that (4.3.18) is true for $g \in \mathbb{S}$ and $f \in \mathbb{L}(X)$.

Now fix $f \in \mathbb{L}(X)$ and let Y be given by (4.3.14). Note that right-hand side in (4.3.18) is $J_Y(g)(t)$, as defined by (4.2.1)–(4.2.2), so that we have

$$\int_0^t (gf) dX = J_Y(g)(t), \quad \forall g \in \mathbb{S}. \tag{4.3.19}$$

Let us define $J(g) = \int_0^t gf \, dX$ for bounded predictable g, then J is an extension of J_Y as noted above. Theorem 4.29 again yields that if $g^n \xrightarrow{bp} g$ then $J(g^n) \xrightarrow{ucp} J(g)$. Thus, J is the extension of J_Y as required in the definition of stochastic integrator. Thus Y is a stochastic integrator and (4.3.8) holds (for bounded predictable g).

Now we shall prove the last statement of the theorem. Suppose g is predictable such that $fg \in \mathbb{L}(X)$. First, we will prove that $g \in \mathbb{L}(Y)$ and that (4.3.8) holds for such a g.

To prove $g \in \mathbb{L}(Y)$, let h^k be bounded predictable, converging pointwise to h such that h^k are dominated by g. We need to show that $\int h^k dY$ is Cauchy in $\mathbf{d}_{ucp}$ metric.

Let $u^k = h^k f$. Then u^k are dominated by $fg \in \mathbb{L}(X)$ and converge pointwise to hf. Invoking DCT, Theorem 4.29, we conclude that $Z^k = \int u^k dX$ converges to $Z = \int hf dX$ and hence is Cauchy in $\mathbf{d}_{ucp}$ metric. On the other hand, since h^k is bounded, invoking (4.3.8) for h^k, we conclude $\int h^k dY = Z^k$ and hence is Cauchy in $\mathbf{d}_{ucp}$ metric. This shows $g \in \mathbb{L}(Y)$.

Further, with $h^k = g 1_{\{|g| \leq k\}}$ above we conclude that the limit of $\int h^k dY$ is $\int g dY$. On the other hand as seen above, (with $h = g$) $\int h^k dY = \int h^k f dX$ converges to $\int gf dX$. Thus (4.3.8) is true.

To complete the proof, we need to show that if $g \in \mathbb{L}(Y)$ then $fg \in \mathbb{L}(X)$. For this, suppose u^n are bounded predictable, $|u^n| \leq |fg|$ and u^n converges to 0 pointwise. Need to show $\int u^n dX \xrightarrow{ucp} 0$. Let

$$\tilde{f}_s(\omega) = \begin{cases} \frac{1}{f_s(\omega)} & \text{if } f_s(\omega) \neq 0 \\ 0 & \text{if } f_s(\omega) = 0. \end{cases}$$

and $v^n = u^n \tilde{f}$. Now v^n are predictable and are dominated by $|g|$ and v^n converges pointwise to 0. Thus,

$$\int v^n dY \xrightarrow{ucp} 0.$$

Since $|u^n| \leq |fg|$, $f_s(\omega) = 0$ implies $u_s^n(\omega) = 0$. Thus it follows that $v^n f = u^n$ and thus $v^n f$ is bounded and hence in $\mathbb{L}(X)$. Thus by the part proven above, invoking (4.3.8) for v^n we have

$$\int v^n dY = \int v^n f dX = \int u^n dX.$$

This shows $\int u^n dX \xrightarrow{ucp} 0$. Hence $fg \in \mathbb{L}(X)$. This completes the proof. □

We had seen in Theorem 4.4 that the class of predictable processes is essentially the same for the filtrations $(\mathcal{F}_.)$ and $(\mathcal{F}_.^+)$—the only difference being at $t = 0$.
We now observe:

Theorem 4.34 *For an r.c.l.l.* $(\mathcal{F}_.)$ *adapted process X, it is a stochastic integrator w.r.t. the filtration* $(\mathcal{F}_.)$ *if and only if it is a stochastic integrator w.r.t. the filtration* $(\mathcal{F}_.^+)$.

Proof Let X be a stochastic integrator w.r.t. the filtration $(\mathcal{F}_.)$, so that $\int h dX$ is defined for bounded $(\mathcal{F}_.)$ predictable processes h. Given a bounded $(\mathcal{F}_.^+)$ predictable processes f, let g be defined by

$$g_t(\omega) = f_t(\omega) 1_{\{(0,\infty) \times \Omega\}}(t, \omega).$$

Then by Theorem 4.4, g is a bounded $(\mathcal{F}_.)$ predictable processes. So we define

$$J_X(f) = \int g dX + f_0 X_0.$$

It is easy to check that J_X satisfies the required properties for X to be a stochastic integrator.

Conversely if X is a stochastic integrator w.r.t. the filtration $(\mathcal{F}_.^+)$, so that $\int h dX$ is defined for bounded $(\mathcal{F}_.^+)$ predictable processes h, of course for a bounded $(\mathcal{F}_.)$ predictable f we can define by $J_X(f) = \int f dX$ and J_X will have the required continuity properties. However, we need to check that $J_X(f)$ so defined is $(\mathcal{F}_.)$ adapted or in other words, belongs to $\mathbb{R}^0(\Omega, (\mathcal{F}_.), \mathsf{P})$. For $f \in \mathbb{S}$, it is clear that $J_X(f) \in \mathbb{R}^0(\Omega, (\mathcal{F}_.), \mathsf{P})$ since X is $(\mathcal{F}_.)$ adapted. Since the space $\mathbb{R}^0(\Omega, (\mathcal{F}_.), \mathsf{P})$ is a closed subspace of $\mathbb{R}^0(\Omega, (\mathcal{F}_.^+), \mathsf{P})$ in the $\mathbf{d}_{ucp}$ metric, it follows that $J_X(f) \in \mathbb{R}^0(\Omega, (\mathcal{F}_.), \mathsf{P})$ for $f \in \mathbb{B}(\widetilde{\Omega}, \mathcal{P}(\mathcal{F}_.))$ and thus X is a stochastic integrator for the filtration $(\mathcal{F}_.)$. □

Exercise 4.35 Let X be a $(\mathcal{F}_.)$- stochastic integrator. For each t let $\{\mathcal{G}_t : t \geq 0\}$ be a filtration such that for all t, $\mathcal{F}_t \subseteq \mathcal{G}_t \subseteq \mathcal{F}_t^+$. Show that X is a $(\mathcal{G}_.)$- stochastic integrator.

4.4 Locally Bounded Processes

We will introduce an important class of integrands, namely that of locally bounded predictable processes, that is contained in $\mathbb{L}(X)$ for every stochastic integrator X. For a stopping time τ, $[0, \tau]$ will denote the set $\{(t, \omega) \in \widetilde{\Omega} : 0 \le t \le \tau(\omega)\}$ and thus $g = f 1_{[0,\tau]}$ means the following: $g_t(\omega) = f_t(\omega)$ if $t \le \tau(\omega)$ and $g_t(\omega)$ is zero if $t > \tau(\omega)$.

The next result gives interplay between stopping times and stochastic integration.

Lemma 4.36 *Let X be a stochastic integrator and $f \in \mathbb{L}(X)$. Let τ be a stopping time. Let $g = f 1_{[0,\tau]}$. Let*

$$Y_t = \int_0^t f \, dX \tag{4.4.1}$$

and $V = \int g \, dX$. Then $V_t = Y_{t \wedge \tau}$, i.e.

$$Y_{t \wedge \tau} = \int_0^t f 1_{[0,\tau]} dX. \tag{4.4.2}$$

Proof When $f \in \mathbb{S}$ is a simple predictable process and τ is a stopping time taking only finitely many values, then $g \in \mathbb{S}$ and (4.4.1)–(4.4.2) can be checked as in that case, the integrals $\int f \, dX$ and $\int g \, dX$ are both defined directly by (4.2.2). Thus fix $f \in \mathbb{S}$. Approximating a bounded stopping time τ from above by stopping time taking finitely many values (as seen in the proof of Theorem 2.54), it follows that (4.4.2) is true for any bounded stopping time, then any stopping time τ can be approximated by $\tilde{\tau}^n = \tau \wedge n$ and one can check that (4.4.2) continues to be true. Thus we have proven the result for simple integrands.

Now fix a stopping time τ and let

$$\mathbb{K} = \{f \in \mathbb{B}(\widetilde{\Omega}, \mathcal{P}) : \ (4.4.1) - (4.4.2) \text{ is true for all } t \ge 0.\}.$$

Then it is easy to see that $\mathbb{K}$ is closed under *bp*-convergence and as noted above it contains $\mathbb{S}$. Hence by Theorem 2.66, it follows that $\mathbb{K} = \mathbb{B}(\widetilde{\Omega}, \mathcal{P})$. Finally, for a general $f \in \mathbb{L}(X)$, the result follows by approximating f by $f^n = f 1_{\{|f| \le n\}}$ and using dominated convergence theorem. This completes the proof. $\qquad \square$

Exercise 4.37 In the proof given above, we first proved the required result for $f \in \mathbb{S}$ and any stopping time τ. Complete the proof by first fixing a simple stopping time τ and prove it for all $f \in \mathbb{L}(X)$ and subsequently prove it for all stopping times τ.

Remark **4.38** We can denote $Y_{t \wedge \tau}$ as $\int_0^{t \wedge \tau} f \, dX$ so that (4.4.2) can be recast as

$$\int_0^{t \wedge \tau} f \, dX = \int_0^t f 1_{[0,\tau]} dX. \tag{4.4.3}$$

Corollary 4.39 *If X is a stochastic integrator and $f, g \in \mathbb{L}(X)$ and τ is a stopping time such that*

$$f 1_{[0,\tau]} = g 1_{[0,\tau]}$$

then for each t

$$\int_0^{t\wedge\tau} f \, dX = \int_0^{t\wedge\tau} g \, dX. \tag{4.4.4}$$

Definition 4.40 A process f is said to be locally bounded if there exist stopping times τ^n, $0 \le \tau^1 \le \tau^2 \le \ldots$, $\tau^n \uparrow \infty$ such that for every n,

$$f 1_{[0,\tau^n]} \text{ is bounded.}$$

The sequence $\{\tau^n : n \ge 1\}$ is called a localizing sequence.

Thus if a process f is locally bounded, then we can get stopping times τ^n increasing to infinity (can even choose each τ^n to be bounded) and constants C_n such that

$$P(\omega : \sup_{0 \le t \le \tau_n(\omega)} |f_t(\omega)| \le C_n) = 1, \ \forall n \ge 1. \tag{4.4.5}$$

Note that given finitely many locally bounded processes one can choose a common localizing sequence $\{\tau^n : n \ge 1\}$. A continuous adapted process X such that X_0 is bounded is easily seen to be locally bounded. We can take the localizing sequence to be

$$\tau_n = \inf\{t \ge 0 : |X(t)| \ge n \text{ or } t \ge n\}.$$

For an r.c.l.l. adapted process X, recall that X^- is the process defined by $X^-(t) = X(t-)$, where $X(t-)$ is the left limit of $X(s)$ at $s = t$ for $t > 0$ and $X^-(0) = X(0-) = 0$. Let τ_n be the stopping times defined by

$$\tau_n = \inf\{t \ge 0 : |X(t)| \ge n \text{ or } |X(t-)| \ge n \text{ or } t \ge n\}. \tag{4.4.6}$$

Then it can be easily seen that $X^- 1_{[0,\tau_n]}$ is bounded by n and that $\tau_n \uparrow \infty$ and hence X^- is locally bounded. Easy to see that sum of two locally bounded processes is itself locally bounded. Further, if X is a r.c.l.l. process with bounded jumps: $|\Delta X| \le K$, then X is locally bounded, since X^- is locally bounded and (ΔX) is bounded and $X = X^- + (\Delta X)$.

Exercise 4.41 Let X be an r.c.l.l. adapted process. Show that X is locally bounded if and only if ΔX is locally bounded.

For future reference, we record these observations as a lemma.

Lemma 4.42 *Let X be an adapted r.c.l.l. process.*

(i) If X is continuous with X_0 bounded, then X is locally bounded.

(ii) X^- *is locally bounded.*

(iii) If ΔX is locally bounded, then X is locally bounded.

We now prove an important property of the class $\mathbb{L}(X)$.

Theorem 4.43 *Let X be an integrator and f be a predictable process such that there exist stopping times τ_m increasing to ∞ with*

$$f 1_{[0,\tau_m]} \in \mathbb{L}(X) \quad \forall n \geq 1. \tag{4.4.7}$$

Then $f \in \mathbb{L}(X)$.

Proof Let g^n be bounded predictable such that $g^n \to g$ pointwise and $|g^n| \leq |f|$. Let $Z^n = \int g^n dX$. Now for each m,

$$g^n 1_{[0,\tau_m]} \to g 1_{[0,\tau_m]} \text{ pointwise}, \tag{4.4.8}$$

$$|g^n 1_{[0,\tau_m]}| \leq |f 1_{[0,\tau_m]}| \tag{4.4.9}$$

and $Y^{n,m}$ defined by

$$Y^{n,m} = \int g^n 1_{[0,\tau_m]} dX \tag{4.4.10}$$

satisfies

$$Y_t^{n,m} = Z_{t \wedge \tau_m}^n. \tag{4.4.11}$$

In view of (4.4.8), (4.4.9) the assumption (4.4.7) implies that for each m, $\{Y^{n,m} : n \geq 1\}$ is Cauchy in $\mathbf{d}_{ucp}$ metric. Thus invoking Corollary 2.76, we conclude that Z^n is Cauchy in $\mathbf{d}_{ucp}$ metric and hence $f \in \mathbb{L}(X)$. $\qquad\square$

As noted earlier, $f \in \mathbb{L}(X)$, h predictable, $|h| \leq C|f|$ for some constant $C > 0$ implies $h \in \mathbb{L}(X)$. Thus the previous result gives us

Corollary 4.44 *Let X be a stochastic integrator, g be a locally bounded predictable process and $f \in \mathbb{L}(X)$. Then fg belongs to $\mathbb{L}(X)$.*

In particular, we have the following.

Corollary 4.45 *Let g be a locally bounded predictable process. Then g belongs to $\mathbb{L}(X)$ for every stochastic integrator X. As a consequence, if Y is an r.c.l.l. adapted process, then $Y^- \in \mathbb{L}(X)$ for every stochastic integrator X.*

Exercise 4.46 Let X be a stochastic integrator. Let $s_0 = 0 < s_1 < s_2 < \ldots < s_n < \ldots$ with $s_n \uparrow \infty$. Let ξ_j, $j = 1, 2 \ldots$, be such that ξ_j is $\mathcal{F}_{s_{j-1}}$ measurable.

(i) For $n \geq 1$ let $h^n = \sum_{j=1}^n \xi_j 1_{(s_{j-1}, s_j]}$. Show that $h^n \in \mathbb{L}(X)$ and

$$\int_0^t h^n dX = \sum_{j=1}^n \xi_j (X_{s_j \wedge t} - X_{s_{j-1} \wedge t}).$$

(ii) Let $h = \sum_{j=1}^{\infty} \xi_j 1_{(s_{j-1}, s_j]}$. Show that $h \in \mathbb{L}(X)$ and

$$\int_0^t h \, dX = \sum_{j=1}^{\infty} \xi_j (X_{s_j \wedge t} - X_{s_{j-1} \wedge t}).$$

For an r.c.l.l. process X and a stopping time σ, let $X^{[\sigma]}$ denote the process X stopped at σ defined as follows

$$X_t^{[\sigma]} = X_{t \wedge \sigma}. \tag{4.4.12}$$

Next result shows that if X is a stochastic integrator and σ is a stopping time, then $Y = X^{[\sigma]}$ is a stochastic integrator as well.

Lemma 4.47 *Let X be a stochastic integrator and σ be a stopping time. Then $X^{[\sigma]}$ is also a stochastic integrator and for $f \in \mathbb{L}(X)$, writing $Z_t = \int_0^t f \, dX$, one has*

$$Z_{t \wedge \sigma} = \int_0^t f \, dX^{[\sigma]} = \int_0^t (f 1_{[0, \sigma]}) dX. \tag{4.4.13}$$

Proof First one checks that (4.4.13) is true for $f \in \mathbb{S}$. Then for any bounded predictable f, defining the process

$$J_0(f) = \int (f 1_{[0, \sigma]}) dX$$

one can check that $f_n \xrightarrow{bp} f$ implies $J_0(f_n) \xrightarrow{ucp} J_0(f)$ and hence it follows that $X^{[\sigma]}$ is a stochastic integrator. Using (4.4.4), it follows that (4.4.13) is true for all bounded predictable processes f. Finally, for a general $f \in \mathbb{L}(X)$, the result follows by approximating f by $f^n = f 1_{\{|f| \le n\}}$ and using dominated convergence theorem. This completes the proof.   □

Exercise 4.48 Deduce the previous result using Theorem 4.33 by identifying $X^{[\sigma]}$ as $\int g \, dX$ for a suitable $g \in \mathbb{L}(X)$.

The next result shows that if we localize the concept of integrator, we do not get anything new; i.e. if a process is locally a stochastic integrator, then it is already a stochastic integrator.

Theorem 4.49 *Suppose X is an adapted r.c.l.l. process such that there exist stopping times τ^n with $\tau^n \le \tau^{n+1}$ for all n and $\tau^n \uparrow \infty$ and the stopped processes $X^n = X^{[\tau^n]}$ are stochastic integrators. Then X is itself a stochastic integrator.*

Proof Fix $f \in \mathbb{B}(\widetilde{\Omega}, \mathcal{P})$ and for $m \ge 1$ let $U^m = \int f \, dX^m$. Without loss of generality we assume that $\tau^0 = 0$. Then using (4.4.4), it follows that $U_t^m = U_{t \wedge \tau^m}^k$ for $m \le k$. We define $J_0(f)$ by $J_0(f)_0 = f_0 X_0$ and for $m \ge 1$,

$$J_0(f)_t = U_t^m, \qquad \tau^{m-1} < t \le \tau^m. \tag{4.4.14}$$

It follows that

$$J_0(f)_{t \wedge \tau^m} = \int_0^t f dX^m. \tag{4.4.15}$$

Of course, for simple predictable f, $J_0(f) = \int_0^t f dX$ and thus J_0 is an extension of J_X. Now let f^n be bounded predictable such that $f^n \overset{bp}{\to} f$. Using (4.4.15), for $n, m \ge 1$ we have

$$J_0(f^n)_{t \wedge \tau^m} = \int_0^t f^n dX^m.$$

Writing $Z^n = J_0(f^n)$ and $Z = J_0(f)$, it follows using (4.4.15) that Z^n, $n \ge 1$, and Z satisfy(2.5.9) and hence by Lemma 2.75 it follows that $Z^n \overset{ucp}{\to} Z$. We have thus proved that $f^n \overset{bp}{\to} f$ implies $J_0(f^n) \overset{ucp}{\to} J_0(f)$ and since $J_0(f)$ agrees with $J_X(f)$ for simple predictable f, it follows that X is a stochastic integrator. $\qquad\square$

We have seen a version of the dominated convergence theorem for stochastic integrals. Here is another result on convergence that will be needed later.

Theorem 4.50 *Suppose* $Y^n, Y \in \mathbb{R}^0(\Omega, (\mathcal{F}.), \mathsf{P})$, $Y^n \overset{ucp}{\to} Y$ *and* X *is a stochastic integrator. Then*

$$\int (Y^n)^- dX \overset{ucp}{\to} \int Y^- dX.$$

Proof We have noted that Y^- and $(Y^n)^-$ belong to $\mathbb{L}(X)$. Let

$$b_n = \mathbf{d}_{ucp}\Big(\int (Y^n)^- dX, \int Y^- dX\Big).$$

To prove that $b_n \to 0$ suffices to prove the following: *For any subsequence* $\{n_k : k \ge 1\}$, *there exists a further subsequence* $\{m_j : j \ge 1\}$ *of* $\{n_k : k \ge 1\}$ *(i.e.* $\exists$ *subsequence* $\{k_j : j \ge 1\}$ *such that* $m_j = n_{k_j}$) *such that*

$$b_{m_j} \to 0. \tag{4.4.16}$$

So now, given a subsequence $\{n_k : k \ge 1\}$, using $\mathbf{d}_{ucp}(Y^{n_k}, Y) \to 0$, let us choose $m_j = n_{k_j}$ with $k_{j+1} > k_j$ and $\mathbf{d}_{ucp}(Y^{m_j}, Y) \le 2^{-j}$. Then as seen in the proof of Theorem 2.71, this would imply

$$\sum_{j=1}^{\infty} [\sup_{t \le T} |Y_t^{m_j} - Y_t|] < \infty, \quad \forall T < \infty.$$

Thus defining

$$H_t = \sum_{j=1}^{\infty} |Y_t^{m_j} - Y_t| \tag{4.4.17}$$

it follows that (outside a fixed null set) the convergence in (4.4.17) is uniform on $t \in [0, T]$ for all $T < \infty$ and as a result H is an r.c.l.l. adapted process. Thus the processes $(Y^{m_j})^-$ are dominated by $(H + Y)^-$ which is a locally bounded process, as $H + Y$ is an r.c.l.l. adapted process. Thus the dominated convergence Theorem 4.29 yields

$$b_{m_j} = \mathbf{d}_{ucp}(\textstyle\int (Y^{m_j})^- dX, \int Y^- dX) \to 0.$$

This completes the proof as explained above. $\square$

Exercise 4.51 Show that if $Y^n \xrightarrow{ucp} Y$, then there is a subsequence that is dominated by a locally bounded process.

The subsequence technique used in the proof also yields the following result, which will be useful alter.

Proposition 4.52 *Let* $Y^n \xrightarrow{ucp} Y$, *where* Y^n, Y *are* $\mathbb{R}^d$*-valued r.c.l.l. processes and* $g^n, g : [0, \infty) \times \mathbb{R}^d \mapsto \mathbb{R}$ *be continuous functions such that* g^n *converges to* g *uniformly on compact subsets of* $[0, \infty) \times \mathbb{R}^d$. *Let* $Z_t^n = g^n(t, Y_t^n)$ *and* $Z_t = g(t, Y_t)$. *Then* $Z^n \xrightarrow{ucp} Z$.

Proof Like in the previous proof, let $b_n = \mathbf{d}_{ucp}(Z^n, Z)$ and given any subsequence $\{n_k : k \geq 1\}$, using $\mathbf{d}_{ucp}(Y^{n_k}, Y) \to 0$, choose $m_j = n_{k_j}$ with $k_{j+1} > k_j$ and $\mathbf{d}_{ucp}(Y^{m_j}, Y) \leq 2^{-j}$. It follows that

$$\sum_{j=1}^{\infty} [\sup_{t \leq T} |Y_t^{m_j} - Y_t|] < \infty, \quad \forall T < \infty.$$

and so $[\sup_{t \leq T} |Y_t^{m_j} - Y_t|]$ converges to zero for all $T < \infty$ a.s. and now uniform convergence of g^n to g on compact subsets would yield convergence of Z^{m_j} to Z uniformly on $[0, T]$ for all T and thus $b_{m_j} = \mathbf{d}_{ucp}(Z^{m_j}, Z)$ converges to 0. Thus every subsequence of $\{b_n\}$ has a further subsequence converging to zero and hence $\lim_{n \to \infty} b_n = 0$. $\square$

Essentially the same proof as given above for Theorem 4.50 gives the following result, only difference being that if $X \in \mathbb{V}$, i.e. if $\mathrm{VAR}_{[0,t]}(X) < \infty$ for all $t < \infty$, and Y is an r.c.l.l. adapted process, the integral $\int Y dX$ is defined in addition to $\int Y^- dX$, both are defined as Lebesgue–Stieltjes integrals while the later agrees with the stochastic integral (as seen in Theorem 4.23).

Proposition 4.53 *Suppose* $Y^n, Y \in \mathbb{R}^0(\Omega, (\mathcal{F}_.), \mathrm{P})$, $Y^n \xrightarrow{ucp} Y$ *and* $X \in \mathbb{V}$ (*a process with finite variation paths*). *Then*

$$\int (Y^n)dX \xrightarrow{ucp} \int YdX.$$

Exercise 4.54 Let $(\Omega, \mathcal{F}, P)$ be a probability space with a filtration $(\mathcal{F}.)$. Let $W = (W^1, \ldots, W^d)$ be such that W^j is a stochastic integrator on $(\Omega, \mathcal{F}, P)$ for $1 \le j \le d$. Let $\mu = P \circ W^{-1}$. Let X denote the coordinate mappings on $(\mathbb{D}_d, \mathbb{B}(D_d))$ and let $\mathcal{D}_t = \mathcal{F}_t^X$ be the filtration generated by X and let $\mathcal{P} = \mathcal{P}(\mathcal{D}.)$. Let $X = (X^1, \ldots, X^d)$ denote the coordinate mappings on $(\mathbb{D}_d, \mathbb{B}(D_d))$.

(i) Show that X^j is a stochastic integrator on $(\mathbb{D}_d, \mathbb{B}(D_d), \mu)$ with the filtration $(\mathcal{D}.)$ for $1 \le j \le d$.

(ii) Let f^j be $\mathcal{P}(\mathcal{D}.)$ measurable such that $f^j \in \mathbb{L}(X^j)$. Let Z^j be defined by $Z_t^j = f_t^j(W)$. Show that $Z^j \in \mathbb{L}(W^j)$ for $1 \le j \le d$. (See Exercise 4.9).

(iii) Let $h : \mathbb{D}_d \mapsto \mathbb{R}$ be a measurable mapping and $g : \mathbb{R}^{d+1} \mapsto \mathbb{R}$ be a bounded continuous function, $A \in \mathcal{B}(\mathbb{D}_d)$ and let $T < \infty$. Show that

$$E_P[1_A(W)g(\int_0^T f^1(W)dW^1, \ldots, \int_0^T f^d(W)dW^d, h(W))]$$
$$= \int_A g(\int_0^T f^1 dX^1, \ldots, \int_0^T f^d dX^d, h)d\mu. \tag{4.4.18}$$

(iv) Show that (4.4.18) is true if g is a bounded measurable function or $[0, \infty)$-valued measurable function.

HINT: For simple processes u, v on $(\mathbb{D}_d, \mathbb{B}(D_d), \mu)$, observe that

$$\mu(|J_{X^j}(u) - J_{X^j}(v)| > \varepsilon) = P(|J_{W^j}(u) - J_{W^j}(v)| > \varepsilon).$$

(i) and (ii) follow from this. For (iii), note that it holds for simple processes $f^1, \ldots, f^d$ and that the class of processes $f^1, \ldots, f^d$ such that (4.4.18) is true is closed under bounded pointwise convergence. Using monotone class theorem, deduce that (4.4.18) is true for all bounded predictable $f^1, \ldots, f^d$. The general case follows by truncation. For (iv), the validity for bounded measurable g follows from yet another application of monotone class theorem.

4.5 Approximation by Riemann Sums

The next result shows that for an r.c.l.l. process Y and a stochastic integrator X, the stochastic integral $\int Y^- dX$ can be approximated by Riemann-like sums. The difference is that the integrand must be evaluated at the lower end point of the interval as opposed to any point in the interval in the Riemann–Stieltjes integral.

Theorem 4.55 *Let Y be an r.c.l.l. adapted process and X be a stochastic integrator. Let*

$$0 = t_0^m < t_1^m < \ldots < t_n^m < \ldots; \quad t_n^m \uparrow \infty \text{ as } n \uparrow \infty \tag{4.5.1}$$

be a sequence of partitions of $[0, \infty)$ such that for all $T < \infty$,

$$\delta_m(T) = (\sup_{\{n \,:\, t_n^m \leq T\}} (t_{n+1}^m - t_n^m)) \to 0 \text{ as } m \uparrow \infty. \tag{4.5.2}$$

Let

$$Z_t^m = \sum_{n=0}^{\infty} Y_{t_n^m \wedge t}(X_{t_{n+1}^m \wedge t} - X_{t_n^m \wedge t}) \tag{4.5.3}$$

and $Z = \int Y^- dX$. Note that for each t, m, the sum in (4.5.3) is a finite sum since $t_n^m \wedge t = t$ from some n onwards. Then

$$Z^m \xrightarrow{ucp} Z \tag{4.5.4}$$

or in other words

$$\sum_{n=0}^{\infty} Y_{t_n^m \wedge t}(X_{t_{n+1}^m \wedge t} - X_{t_n^m \wedge t}) \xrightarrow{ucp} \int_0^t Y^- dX.$$

Proof Let Y^m be defined by

$$Y_t^m = \sum_{n=0}^{\infty} Y_{t_n^m \wedge t} 1_{(t_n^m,\, t_{n+1}^m]}(t).$$

We will first prove

$$\int Y^m dX = Z^m. \tag{4.5.5}$$

For this, let $V_t = \sup_{s \leq t} |Y|$. Then V is an r.c.l.l. adapted process, and hence V^- is locally bounded.

Let us fix m and let $\phi^k(x) = \max(\min(x, k), -k)$, so that $|\phi^k(x)| \leq k$ for all $x \in \mathbb{R}$. Let

$$U_t^k = \sum_{n=0}^{k} \phi^k(Y_{t_n^m \wedge t}) 1_{(t_n^m, t_{n+1}^m]}(t).$$

Then $U^k \in \mathbb{S}$ and

$$\int_0^t U^k dX = \sum_{n=0}^{k} \phi^k(Y_{t_n^m \wedge t})(X_{t_{n+1}^m \wedge t} - X_{t_n^m \wedge t}).$$

Note that U^k converges pointwise (as k increases to ∞) to Y^m. Further, $|U_t^k| \le V_t^- = V(t-)$ for all t and hence by the Dominated Convergence Theorem 4.29, $\int_0^t U^k dX$ converges to $\int Y^m dX$. On the other hand,

$$\sum_{n=0}^{k} \phi^k(Y_{t_n^m \wedge t})(X_{t_{n+1}^m \wedge t} - X_{t_n^m \wedge t}) \rightarrow \sum_{n=0}^{\infty} Y_{t_n^m \wedge t}(X_{t_{n+1}^m \wedge t} - X_{t_n^m \wedge t}) \text{ pointwise.}$$

This proves (4.5.5).

Now, Y^m converges pointwise to Y^- and are dominated by the locally bounded process V^-. Hence again by Theorem 4.29,

$$\int Y^m dX \xrightarrow{ucp} \int Y^- dX$$

which is same as (4.5.4). □

We will next show that the preceding result is true when the sequence of deterministic partitions is replaced by a sequence of random partitions via stopping times. For this, we need the following lemma.

Lemma 4.56 *Let X be a stochastic integrator, Z be an r.c.l.l. adapted process and τ be stopping time. Let*

$$h = Z_\tau 1_{(\tau, \infty)}. \tag{4.5.6}$$

Then h is locally bounded predictable and

$$\int_0^t h dX = Z_{\tau \wedge t}(X_t - X_{\tau \wedge t}). \tag{4.5.7}$$

Proof We have seen in Proposition 4.1 that h is predictable. Since

$$\sup_{0 \le s \le t} |h_s| \le \sup_{0 \le s \le t} |Z_s^-|$$

and Z^- is locally bounded, it follows that h is locally bounded and thus $h \in \mathbb{L}(X)$.

If Z is a bounded r.c.l.l. adapted process and τ takes finitely many values, then easy to see that h belongs to $\mathbb{S}$ and that (4.5.7) is true. Now if τ is a bounded stopping time, then for $m \ge 1$, τ^m defined by

$$\tau^m = 2^{-m}([2^m \tau] + 1) \tag{4.5.8}$$

are stopping times, each taking finitely many values and $\tau^m \downarrow \tau$. One can then verify (4.5.7) by approximating τ by τ^m defined via (4.5.8) and then using the fact that $h^m = Z_{\tau^m} 1_{(\tau^m, \infty)}$ converges boundedly pointwise to h, validity of (4.5.7) for τ^m implies the same for τ. For a general τ, we approximate it by $\tau_n = \tau \wedge n$. Thus, it follows that (4.5.6)–(4.5.7) are true for bounded r.c.l.l. adapted processes Z. For a general Z, let

$$Z^n = \max(\min(Z, n), -n).$$

Noting that $h^n = Z_\tau^n 1_{(\tau,\infty)}$ converges to h and $|h^n| \leq |h|$, the validity of (4.5.7) for Z^n implies the same for Z by Theorem 4.29. $\qquad\square$

Corollary 4.57 *Let Z and τ be as in the previous lemma and σ be another stopping time with $\tau \leq \sigma$. Let*

$$g = Z_\tau 1_{(\tau,\sigma]}. \tag{4.5.9}$$

Then

$$\int_0^t g dX = Z_{\tau \wedge t}(X_{t \wedge \sigma} - X_{t \wedge \tau}) = Z_\tau(X_{t \wedge \sigma} - X_{t \wedge \tau}). \tag{4.5.10}$$

The first equality follows from the observation that $g = h 1_{[0,\sigma]}$ where h is as in Lemma 4.56 and hence $\int_0^t g dX = (\int_0^\cdot h dX)_{t \wedge \sigma}$ (using Lemma 4.36). The second equality can be directly verified.

Exercise 4.58 Express $1_{(\tau,\sigma]}$ as $1_{(\tau,\infty)} - 1_{(\sigma,\infty)}$ and thereby deduce the above Corollary from Lemma 4.56.

Corollary 4.59 *Let X be a stochastic integrator and $\tau \leq \sigma$ be stopping times. Let ξ be a $\mathcal{F}_\tau$ measurable random variable. Let*

$$f = \xi 1_{(\tau,\sigma]}.$$

Then $f \in \mathbb{L}(X)$ and

$$\int_0^t f dX = \xi(X_{t \wedge \sigma} - X_{t \wedge \tau}).$$

Proof This follows from the Corollary 4.57 by taking $Z = \xi 1_{[\tau,\infty)}$ and noting that as shown in Lemma 2.41, Z is adapted. $\qquad\square$

Definition 4.60 For $\delta > 0$, a δ-*partition* for an r.c.l.l. adapted process Z is a sequence of stopping times $\{\tau_n; : n \geq 0\}$ such that $0 = \tau_0 < \tau_1 < \ldots < \tau_n < \ldots; \tau_n \uparrow \infty$ and

$$|Z_t - Z_{\tau_n}| \leq \delta \quad \text{for } \tau_n \leq t < \tau_{n+1}, \ n \geq 0. \tag{4.5.11}$$

Remark 4.61 Given r.c.l.l. adapted processes $Z^i, 1 \leq i \leq k$ and $\delta > 0$, we can get a sequence of partitions $\{\tau_n : n \geq 0\}$ such that $\{\tau_n : n \geq 0\}$ is a δ partition for each of $Z^1, Z^2, \ldots Z^k$. Indeed, let $\{\tau_n : n \geq 0\}$ be defined inductively via $\tau_0 = 0$ and

$$\tau_{n+1} = \inf\{t > \tau_n : \max(\max_{1 \leq i \leq k} |Z_t^i - Z_{\tau_n}^i|, \max_{1 \leq i \leq k} |Z_{t-}^i - Z_{\tau_n}^i|) \geq \delta\}. \tag{4.5.12}$$

Invoking Theorem 2.46, we can see that $\{\tau_n : n \geq 0\}$ are stopping times and that $\lim_{n \uparrow \infty} \tau_n = \infty$.

Let $\delta_m \downarrow 0$ and for each m, let $\{\tau_n^m : n \geq 0\}$ be a δ_m-partition for Z. We implicitly assume that $\delta_m > 0$. Let

$$Z_t^m = \sum_{n=0}^{\infty} Z_{t \wedge \tau_n^m} 1_{(\tau_n^m, \tau_{n+1}^m]}(t).$$

Then it follows that $|Z_t^m - Z_t| \leq \delta_m$ and hence $Z^m \xrightarrow{ucp} Z^-$. For $k \geq 1$, let

$$Z_t^{m,k} = \sum_{n=0}^{k} Z_{t \wedge \tau_n^m} 1_{(\tau_n^m, \tau_{n+1}^m]}(t).$$

Now, using Corollary 4.57 and linearity of stochastic integrals, we have

$$\int_0^t Z^{m,k} dX = \sum_{n=0}^{k} Z_{t \wedge \tau_n^m} (X_{\tau_{n+1}^m \wedge t} - X_{\tau_n^m \wedge t}).$$

Let $V_t = \sup_{s \leq t} |Z_s|$. Then V is r.c.l.l. and thus, as noted earlier, V^- is locally bounded. Easy to see that $|Z^{m,k}| \leq V^-$ and $Z^{m,k}$ converges pointwise to Z^m. Hence by Theorem 4.29, $\int_0^t Z^{m,k} dX \xrightarrow{ucp} \int_0^t Z^m dX$. Thus

$$\int_0^t Z^m dX = \sum_{n=0}^{\infty} Z_{t \wedge \tau_n^m} (X_{\tau_{n+1}^m \wedge t} - X_{\tau_n^m \wedge t}). \qquad (4.5.13)$$

Since Z^m converges pointwise to Z^- and $|Z^n| \leq V^-$ with V^- locally bounded, invoking Theorem 4.29 it follows that

$$\int Z^m dX \xrightarrow{ucp} \int Z^- dX. \qquad (4.5.14)$$

We have thus proved another version of Theorem 4.55.

Theorem 4.62 *Let X be a stochastic integrator. Let Z be an r.c.l.l. adapted process. Let $\delta_m \downarrow 0$ and for $m \geq 1$ let $\{\tau_n^m : n \geq 0\}$ be a δ_m-partition for Z. Then*

$$\sum_{n=0}^{\infty} Z_{t \wedge \tau_n^m} (X_{\tau_{n+1}^m \wedge t} - X_{\tau_n^m \wedge t}) \xrightarrow{ucp} \int_0^t Z^- dX. \qquad (4.5.15)$$

Remark **4.63** When $\sum_m (\delta_m)^2 < \infty$, say $\delta_m = 2^{-m}$, then the convergence in (4.5.15) is stronger: it is uniform convergence on $[0, T]$ almost surely for each $T < \infty$. We will prove this in Chap. 6.

4.6 Quadratic Variation of Stochastic Integrators

We now show that stochastic integrators also, like Brownian motion, have finite quadratic variation.

Theorem 4.64 *Let X be a stochastic integrator. Then there exists an adapted increasing process A, written as $[X, X]$, such that*

$$X_t^2 = X_0^2 + 2 \int_0^t X^- dX + [X, X]_t, \quad \forall t. \tag{4.6.1}$$

Further, let $\delta_m \downarrow 0$ and for $m \geq 1$ let $\{\tau_n^m : n \geq 0\}$ be a δ_m-partition for X. Then one has

$$\sum_{n=0}^{\infty} (X_{\tau_{n+1}^m \wedge t} - X_{\tau_n^m \wedge t})^2 \xrightarrow{ucp} [X, X]_t. \tag{4.6.2}$$

Proof For $a, b \in \mathbb{R}$,

$$b^2 - a^2 = 2a(b - a) + (b - a)^2.$$

Using this with $b = X_{\tau_{n+1}^m \wedge t}$ and $a = X_{\tau_n^m \wedge t}$ and summing with respect to n, we get

$$X_t^2 = X_0^2 + 2V_t^m + Q_t^m$$

where

$$V_t^m = \sum_{n=0}^{\infty} X_{\tau_n^m \wedge t} (X_{\tau_{n+1}^m \wedge t} - X_{\tau_n^m \wedge t})$$

and

$$Q_t^m = \sum_{n=0}^{\infty} (X_{\tau_{n+1}^m \wedge t} - X_{\tau_n^m \wedge t})^2.$$

Note that after some n that may depend upon $\omega \in \Omega$, $\tau_n^m > t$ and hence $X_{\tau_{n+1}^m \wedge t} = X_t$. In view of this, the two sums above have only finitely many nonzero terms.

By Theorem 4.62, $V^m \xrightarrow{ucp} \int X^- dX$. Hence, writing

$$A_t = X_t^2 - X_0^2 - 2 \int_0^t X^- dX, \tag{4.6.3}$$

we conclude

$$Q_t^m \xrightarrow{ucp} A_t.$$

This proves (4.6.1). Remains to show that A_t is an increasing process. Fix $\omega \in \Omega$, $s \leq t$, and note that if $\tau_j^m \leq s < \tau_{j+1}^m$, then $|X_s - X_{\tau_j^m}| \leq \delta_m$ and

$$Q_s^m = \sum_{n=0}^{j-1} (X_{\tau_{n+1}^m} - X_{\tau_n^m})^2 + (X_s - X_{\tau_j^m})^2$$

and

$$Q_t^m = \sum_{n=0}^{j-1} (X_{\tau_{n+1}^m} - X_{\tau_n^m})^2 + \sum_{n=j}^{\infty} (X_{\tau_{n+1}^m \wedge t} - X_{\tau_n^m \wedge t})^2.$$

Thus

$$Q_s^m \leq Q_t^m + \delta_m^2. \tag{4.6.4}$$

Since $Q^m \xrightarrow{ucp} A$, it follows that A is an increasing process. $\qquad\square$

Remark **4.65** From the identity (4.6.1), it follows that the process $[X, X]$ does not depend upon the choice of partitions.

Definition 4.66 For a stochastic integrator X, the process $[X, X]$ obtained in the previous theorem is called the quadratic variation of X.

From the definition of quadratic variation as in (4.6.2), it follows that for a stochastic integrator X and a stopping time σ

$$[X^{[\sigma]}, X^{[\sigma]}]_t = [X, X]_{t \wedge \sigma} \ \forall t. \tag{4.6.5}$$

For stochastic integrators X, Y, let us define cross-quadratic variation between X, Y via the polarization identity

$$[X, Y]_t = \frac{1}{4}([X + Y, X + Y]_t - [X - Y, X - Y]_t) \tag{4.6.6}$$

By definition $[X, Y] \in \mathbb{V}$ since it is defined as difference of two increasing processes. Also, it is easy to see that $[X, Y] = [Y, X]$.

By applying Theorem 4.64 to $X + Y$ and $X - Y$ and using that the mapping $(f, X) \mapsto \int f dX$ is bilinear one can deduce the following result.

Theorem 4.67 (Integration by Parts Formula) *Let X, Y be stochastic integrators. Then*

$$X_t Y_t = X_0 Y_0 + \int_0^t Y^- dX + \int_0^t X^- dY + [X, Y]_t, \ \forall t. \tag{4.6.7}$$

Let $\delta_m \downarrow 0$ and for $m \geq 1$ let $\{\tau_n^m : n \geq 0\}$ be a δ_m-partition for X and Y. Then one has

$$\sum_{n=0}^{\infty} (X_{\tau_{n+1}^m \wedge t} - X_{\tau_n^m \wedge t})(Y_{\tau_{n+1}^m \wedge t} - Y_{\tau_n^m \wedge t}) \xrightarrow{ucp} [X, Y]_t. \tag{4.6.8}$$

Corollary 4.68 *If X, Y are stochastic integrators, then so is $Z = XY$.*

This follows from the integration by parts formula and Theorem 4.33.

Remark **4.69** Like (4.6.5), one also has for any stopping time σ and stochastic integrators X, Y:

$$[X^{[\sigma]}, Y]_t = [X, Y^{[\sigma]}]_t = [X^{[\sigma]}, Y^{[\sigma]}]_t = [X, Y]_{t \wedge \sigma} \quad \forall t. \qquad (4.6.9)$$

This follows easily from (4.6.8).

Exercise 4.70 For stochastic integrators X, Y and stopping times σ and τ show that

$$[X^{[\sigma]}, Y^{[\tau]}] = [X, Y]^{[\sigma \wedge \tau]}.$$

Exercise 4.71 Let $X^1, X^2, \ldots, X^m$ be stochastic integrators and let p be a polynomial in m variables. Then $Z = p(X^1, X^2, \ldots, X^m)$ is also a stochastic integrator.

Corollary 4.72 *Let X, Y be stochastic integrators. Then*

(i) $\Delta[X, X]_t = ((\Delta X)_t)^2$, *for $t > 0$.*
(ii) $\sum_{0 < s \le t} ((\Delta X)_s)^2 \le [X, X]_t < \infty$.
(iii) $\Delta[X, Y]_t = (\Delta X)_t (\Delta Y)_t$ *for $t > 0$.*
(iv) *If X (or Y) is a continuous process, then $[X, Y]$ is also a continuous process.*

Proof For (i), using (4.6.1) and (4.3.7), we get for every $t > 0$

$$X_t^2 - X_{t-}^2 = 2X_{t-}(X_t - X_{t-}) + [X, X]_t - [X, X]_{t-}.$$

Using $b^2 - a^2 - 2a(b - a) = (b - a)^2$, we get

$$(X_t - X_{t-})^2 = [X, X]_t - [X, X]_{t-}$$

which is same as (i). (ii) follows from (i) as $[X, X]_t$ is an increasing process. (iii) follows from (i) via the polarization identity (4.6.6). And lastly, (iv) is an easy consequence of (iii). $\qquad \square$

Remark **4.73** Let us note that by definition, $[X, X] \in \mathbb{V}_0^+$ and $[X, Y] \in \mathbb{V}_0$. In particular, $[X, X]_0 = 0$ and $[X, Y]_0 = 0$.

The next result shows that for a continuous process $A \in \mathbb{V}$, $[X, A] = 0$ for all stochastic integrators X.

Theorem 4.74 *Let X be a stochastic integrator and $A \in \mathbb{V}$, i.e. an r.c.l.l. process with finite variation paths. Then*

$$[X, A]_t = \sum_{0 < s \le t} (\Delta X)_s (\Delta A)_s. \qquad (4.6.10)$$

In particular, if X, A have no common jumps, then $[X, A] = 0$. This is clearly the case if one of the two processes is continuous.

Proof Since $A \in \dot{\mathbb{V}}$, for $\{\tau_n^m\}$ as in Theorem 4.67 above one has

$$\sum_{n=0}^{\infty} X_{\tau_n^m \wedge t}(A_{\tau_{n+1}^m \wedge t} - A_{\tau_n^m \wedge t}) \xrightarrow{ucp} \int_0^t X^- dA$$

and

$$\sum_{n=0}^{\infty} X_{\tau_{n+1}^m \wedge t}(A_{\tau_{n+1}^m \wedge t} - A_{\tau_n^m \wedge t}) \xrightarrow{ucp} \int_{0+}^t X dA.$$

Using (4.6.8) we conclude

$$[X, A]_t = \int_{0+}^t X dA - \int_0^t X^- dA$$

and hence (4.6.10) follows.   □

For stochastic integrators X, Y, let

$$j(X, Y)_t = \sum_{0 < s \leq t} (\Delta X)_s (\Delta Y)_s.$$

The sum above is absolutely convergent in view of part (ii) Corollary 4.72. Clearly, $j(X, X)$ is an increasing process. Also we have seen that

$$j(X, X)_t \leq [X, X]_t. \tag{4.6.11}$$

We can directly verify that $j(X, Y)$ satisfies the polarization identity

$$j(X, Y)_t = \frac{1}{4}(j(X + Y, X + Y)_t - j(X - Y, X - Y)_t). \tag{4.6.12}$$

The identity (4.6.7) characterizes $[X, Y]_t$, and it shows that $(X, Y) \mapsto [X, Y]_t$ is a bilinear form. The relation (4.6.8) also yields the parallelogram identity for $[X, Y]$:

Lemma 4.75 *Let X, Y be stochastic integrators. Then we have*

$$[X + Y, X + Y]_t + [X - Y, X - Y]_t = 2([X, X]_t + [Y, Y]_t), \quad \forall t \geq 0. \tag{4.6.13}$$

Proof Let $\delta_m \downarrow 0$ and for $m \geq 1$, let $\{\tau_n^m : n \geq 0\}$ be a δ_m-partition for X and Y. Take $a_n^m = (X_{\tau_{n+1}^m \wedge t} - X_{\tau_n^m \wedge t})$, $b_n^m = (Y_{\tau_{n+1}^m \wedge t} - Y_{\tau_n^m \wedge t})$. Use the identity $(a + b)^2 + (a - b)^2 = 2(a^2 + b^2)$ with $a = a_n^m$ and $b = b_n^m$; sum over n and take limit over m. We will get the required identity by (4.6.2).   □

Exercise 4.76 Deduce (4.6.13) from the integration by parts formula (4.6.7).

Here is an analogue of the inequality $|a^2 - b^2| \leq \sqrt{2(a-b)^2(a^2+b^2)}$.

Lemma 4.77 *Let X, Y be stochastic integrators. Then we have*

$$|[X, X]_t - [Y, Y]_t| \leq \sqrt{2[X - Y, X - Y]_t([X, X]_t + [Y, Y]_t)}. \tag{4.6.14}$$

Proof Let τ_n^m, a_n^m, b_n^m be as in the proof of Lemma 4.75 above. Then we have

$$\left|\sum_n (a_n^m)^2 - \sum_n (b_n^m)^2\right| \leq \sum_n |(a_n^m)^2 - (b_n^m)^2|$$

$$\leq \sum_n \sqrt{2(a_n^m - b_n^m)^2((a_n^m)^2 + (b_n^m)^2)}$$

$$\leq \sqrt{2\sum_n (a_n^m - b_n^m)^2} \sqrt{\sum_n ((a_n^m)^2 + (b_n^m)^2)}$$

and taking limit over m, using (4.6.2), we get the required result (4.6.14).   □

Also, using that

$$[aX + bY, aX + bY]_t \geq 0 \quad \forall a, b \in \mathbb{R}$$

we can deduce that

$$|[X, Y]_t| \leq \sqrt{[X, X]_t [Y, Y]_t}.$$

Indeed, one has to do this carefully (in view of the null sets lurking around). We can prove a little bit more.

Theorem 4.78 *Let X, Y be stochastic integrators. Then for any $s \leq t$*

$$\text{VAR}_{(s,t]}([X, Y]) \leq \sqrt{([X, X]_t - [X, X]_s).([Y, Y]_t - [Y, Y]_s)}, \tag{4.6.15}$$

$$\text{VAR}_{[s,t]}([X, Y]) \leq \sqrt{([X, X]_t - [X, X]_{s-}).([Y, Y]_t - [Y, Y]_{s-})}, \tag{4.6.16}$$

and

$$\text{VAR}_{[0,t]}([X, Y]) \leq \sqrt{[X, X]_t [Y, Y]_t}, \tag{4.6.17}$$

$$\text{VAR}_{[0,t]}(j(X, Y)) \leq \sqrt{[X, X]_t [Y, Y]_t}. \tag{4.6.18}$$

Proof Let

$$\Omega_{a,b,s,r} = \{\omega \in \Omega : [aX + bY, aX + bY]_r(\omega) \geq [aX + bY, aX + bY]_s(\omega)\}$$

and

$$\Omega_0 = \cup\{\Omega_{a,b,s,r} : s, r, a, b \in \mathbb{Q}, r \geq s\}.$$

Then it follows that $P(\Omega_0) = 1$ (since for any process Z, $[Z, Z]$ is an increasing process). For $\omega \in \Omega_0$, for $0 \leq s \leq r, s, r, a, b \in \mathbb{Q}$

$$(a^2([X, X]_r - [X, X]_s) + b^2([Y, Y]_r - [Y, Y]_s) + 2ab([X, Y]_r - [X, Y]_s))(\omega) \geq 0.$$

Since the quadratic form above remains positive, we conclude

$$|([X, Y]_r(\omega) - [X, Y]_s(\omega))|$$
$$\leq \sqrt{([X, X]_r(\omega) - [X, X]_s(\omega))([Y, Y]_r(\omega) - [Y, Y]_s(\omega))}. \qquad (4.6.19)$$

Since all the processes occurring in (4.6.19) are r.c.l.l., it follows that (4.6.19) is true for all $s \leq r, s, r \in [0, \infty)$.

Now given $s < t$ and $s = t_0 < t_1 < \ldots < t_m = t$, we have

$$\sum_{j=0}^{m-1} |[X, Y]_{t_{j+1}} - [X, Y]_{t_j}|$$
$$\leq \sum_{j=0}^{m-1} \sqrt{([X, X]_{t_{j+1}} - [X, X]_{t_j})([Y, Y]_{t_{j+1}} - [Y, Y]_{t_j})} \qquad (4.6.20)$$
$$\leq \sqrt{([X, X]_t - [X, X]_s)([Y, Y]_t - [Y, Y]_s)}$$

where the last step follows from Cauchy–Schwarz inequality and the fact that $[X, X], [Y, Y]$ are increasing processes. Now taking supremum over partitions of $[s, t]$ in (4.6.20) we get (4.6.15). For (4.6.16), recalling definition of $\text{VAR}_{[a,b]}(G)$ we have

$$\text{VAR}_{[s,t]}([X, Y]) = \text{VAR}_{(s,t]}([X, Y]) + |(\Delta[X, Y])_s|$$
$$\leq \sqrt{([X, X]_t - [X, X]_s)([Y, Y]_t - [Y, Y]_s)} + |(\Delta X)_s(\Delta Y)_s|$$
$$\leq \sqrt{([X, X]_t - [X, X]_s + (\Delta X)_s^2)([Y, Y]_t - [Y, Y]_s + (\Delta Y)_s^2)}$$
$$\leq \sqrt{([X, X]_t - [X, X]_{s-})([Y, Y]_t - [Y, Y]_{s-})}.$$

Now (4.6.17) follows from (4.6.16) taking $s = 0$. As for (4.6.18), note that

$$\text{VAR}_{[0,t]}(j(X, Y)) = \sum_{0 < s \leq t} |(\Delta X)_s(\Delta Y)_s|$$
$$\leq \sqrt{\sum_{0 < s \leq t} (\Delta X)_s^2 \sum_{0 < s \leq t} (\Delta Y)_s^2}$$
$$\leq \sqrt{[X, X]_t [Y, Y]_t}. \qquad \square$$

Corollary 4.79 *For stochastic integrators X, Y, one has*

$$|[X, Y]_t| \le \sqrt{[X, X]_t[Y, Y]_t} \qquad (4.6.21)$$

and

$$\sqrt{[X + Y, X + Y]_t} \le \sqrt{[X, X]_t} + \sqrt{[Y, Y]_t} \qquad (4.6.22)$$

Proof Taking $s = 0$ and using $|[X, Y]_t| \le \text{VAR}_{[0,t]}([X, Y])$ (4.6.21) follows from (4.6.17). In turn using (4.6.21) we note that

$$
\begin{aligned}
[X + Y, X + Y]_t &= [X, X]_t + [Y, Y]_t + 2[X, Y]_t \\
&\le [X, X]_t + [Y, Y]_t + 2\sqrt{[X, X]_t[Y, Y]_t} \\
&= (\sqrt{[X, X]_t} + \sqrt{[Y, Y]_t})^2 .
\end{aligned}
$$

Thus (4.6.22) follows.   □

The next inequality is a version of the Kunita–Watanabe inequality that was proven in the context of square integrable martingales.

Theorem 4.80 *Let X, Y be stochastic integrators and let f, g be predictable processes. Then for all $T < \infty$*

$$\int_0^T |f_s g_s| d|[X, Y]|_s \le \left(\int_0^T |f_s|^2 d[X, X]_s\right)^{\frac{1}{2}} \left(\int_0^T |g_s|^2 d[Y, Y]_s\right)^{\frac{1}{2}}. \qquad (4.6.23)$$

Proof Let us write $A_t = |[X, Y]|_t = \text{VAR}_{[0,t]}([X, Y])$. Note $A_0 = 0$ by definition of quadratic variation. We first observe that (4.6.23) holds for simple predictable processes $f, g \in \mathbb{S}$. As seen in the proof of Theorem 4.33, we can assume that f, g are given by (4.3.11) and (4.3.12). Using (4.6.15), it follows that for $0 \le s < t$

$$|A_t - A_s| \le ([X, X]_t - [X, X]_s)^{\frac{1}{2}}([Y, Y]_t - [Y, Y]_s)^{\frac{1}{2}}$$

and hence

$$
\begin{aligned}
&\int_0^T |f_s g_s| d|[X, Y]|_s \\
&= \sum_{j=0}^k |c_{j+1} d_{j+1}|(A_{r_{j+1}} - A_{r_j}) \\
&\le \sum_{j=0}^k |c_{j+1} d_{j+1}|([X, X]_{r_{j+1}} - [X, X]_{r_j})^{\frac{1}{2}}([Y, Y]_{r_{j+1}} - [Y, Y]_{r_j})^{\frac{1}{2}}
\end{aligned}
$$

$$\leq (\sum_{j=0}^{k} c_{j+1}^2 ([X, X]_{r_{j+1}} - [X, X]_{r_j}))^{\frac{1}{2}} (\sum_{j=0}^{k} d_{j+1}^2 ([Y, Y]_{r_{j+1}} - [Y, Y]_{r_j}))^{\frac{1}{2}}$$

$$= (\int_0^T |f_s|^2 d[X, X]_s)^{\frac{1}{2}} \cdot (\int_0^T |g_s|^2 d[Y, Y]_s)^{\frac{1}{2}}.$$

This proves (4.6.23) for $f, g \in \mathbb{S}$. Now using functional version of monotone class theorem, Theorem 2.66, one can deduce that (4.6.23) continues to hold for all bounded predictable processes f, g. Finally, for general f, g, the inequality follows by approximating f, g by $f^n = f 1_{\{|f| \leq n\}}$ and $g^n = g 1_{\{|g| \leq n\}}$, respectively, and using monotone convergence theorem (recall that integrals appearing in this result are Lebesgue–Stieltjes integrals with respect to increasing processes.) $\qquad \square$

Remark **4.81** *Equivalent Probability Measures continued:* Let X be a stochastic integrator on $(\Omega, \mathcal{F}, P)$. Let Q be a probability measure equivalent to P. We have seen in Remark 4.26 that X is also a stochastic integrator on $(\Omega, \mathcal{F}, Q)$ and the class $\mathbb{L}(X)$ under the two measures is the same and for $f \in \mathbb{L}(X)$, the stochastic integral $\int f dX$ on the two spaces is identical.

It follows (directly from definition or from (4.6.1)) that the quadratic variation of X is the same when X is considered on $(\Omega, \mathcal{F}, P)$ or $(\Omega, \mathcal{F}, Q)$.

4.7 Quadratic Variation of Stochastic Integrals

In this section, we will relate the quadratic variation of $Y = \int f dX$ with the quadratic variation of X. We will show that for stochastic integrators X, Z and $f \in \mathbb{L}(X)$, $g \in \mathbb{L}(Z)$

$$[\int f dX, \int g dZ] = \int f g d[X, Z]. \tag{4.7.1}$$

We begin with a simple result.

Lemma 4.82 *Let X be a stochastic integrator, $f \in \mathbb{L}(X)$, $0 \leq u < \infty$, b be a $\mathcal{F}_u$ measurable bounded random variable. Then*

$$h = b 1_{(u,\infty)} f$$

is predictable, $h \in \mathbb{L}(X)$ and

$$\int_0^t b 1_{(u,\infty)} f dX = b \int_0^t 1_{(u,\infty)} f dX = b (\int_0^t f dX - \int_0^{u \wedge t} f dX). \tag{4.7.2}$$

Proof When $f \in \mathbb{S}$, validity of (4.7.2) can be verified directly as then h is also simple predictable. Then, the class of f such that (4.7.2) is true can be seen to be closed under bp-convergence and hence by Theorem 2.66, (4.7.2) is valid for all bounded predictable processes. Finally, since b is bounded, say by c, $|h| \leq c|f|$ and hence

$h \in \mathbb{L}(X)$. Now (4.7.2) can be shown to be true for all $f \in \mathbb{L}(X)$ by approximating f by $f^n = f1_{\{|f| \le n\}}$ and using Dominated Convergence Theorem—Theorem 4.29.

$\square$

Lemma 4.83 *Let X, Z be a stochastic integrators, $0 \le u < \infty$, b be a $\mathcal{F}_u$ measurable bounded random variable. Let $g = b1_{(u,\infty)}$ and $Y = \int g dX$. Then*

$$[Y, Z]_t = \int_0^t g_s d[X, Z]_s. \tag{4.7.3}$$

As a consequence,

$$Y_t Z_t = \int_0^t Z_{s-} dY_s + \int_0^t Y_{s-} dZ_s + \int_0^t g_s d[X, Z]_s. \tag{4.7.4}$$

Proof Let $\{t_n^m : n \ge 1\}$, $m \ge 1$ be a sequence of partitions satisfying (4.5.1) and (4.5.2) such that for each m, $t_n^m = u$ for some n. For $m \ge 1$, let

$$A_t^m = \sum_{n=0}^{\infty} (X_{t_{n+1}^m \wedge t} - X_{t_n^m \wedge t})(Z_{t_{n+1}^m \wedge t} - Z_{t_n^m \wedge t})$$

and

$$B_t^m = \sum_{n=0}^{\infty} (Y_{t_{n+1}^m \wedge t} - Y_{t_n^m \wedge t})(Z_{t_{n+1}^m \wedge t} - Z_{t_n^m \wedge t}).$$

Noting that $Y_t = b(X_t - X_{u \wedge t})$, it follows that if $s < t \le u$, then $(Y_t - Y_s) = 0$ and if $u \le s < t$ then $(Y_t - Y_s) = g (X_t - X_s)$ and as a consequence,

$$B_t^m = g (A_t^m - A_{u \wedge t}^m).$$

Using (4.6.8), it now follows that

$$[Y, Z]_t = g([X, Z]_t - [X, Z]_{u \wedge t}).$$

Of course this is same as (4.7.3). Now (4.7.4) follows from the integration by parts formula and (4.7.3).

$\square$

Theorem 4.84 *Let X, Y be stochastic integrators and let f, h be bounded predictable processes. Then*

$$[\int f dX, \int h dY]_t = \int_0^t f h d[X, Y]. \tag{4.7.5}$$

Proof Fix a stochastic integrator Z and let $\mathbb{K}$ be the class of bounded predictable processes f such that with

$$W_t = \int f dX \tag{4.7.6}$$

we have

$$W_t Z_t = W_0 Z_0 + \int_0^t W^- dZ + \int_0^t Z^- dW + \int_0^t f_s d[X, Z]_s. \tag{4.7.7}$$

Easy to see that $\mathbb{K}$ is a linear space and that it is closed under bounded pointwise convergence of sequences. It trivially contains $f = a1_{\{0\}}$ where a is bounded $\mathcal{F}_0$ measurable and we have seen in Lemma 4.83 that $\mathbb{K}$ contains $g = b1_{(u,\infty)}$, where $0 \leq u < \infty$ and b is $\mathcal{F}_u$ measurable bounded random variable. Since $\mathbb{S}$ is contained in the linear span of such processes, it follows that $\mathbb{S} \subseteq \mathbb{K}$.

Now Theorem 2.66 implies that (4.7.7) holds for all bounded predictable processes where W is given by (4.7.6). Comparing (4.7.7) with (4.6.7), we conclude that for any stochastic integrator Z

$$\left[\int f dX, Z \right] = \int f d[X, Z]. \tag{4.7.8}$$

For $Z = \int h dY$, we can use (4.7.8) to conclude

$$[Z, X] = \int h d[Y, X]$$

and using symmetry of the cross-quadratic variation $[X, Y]$, we conclude

$$[X, Z] = \int h d[X, Y]. \tag{4.7.9}$$

The two Eqs. (4.7.8)–(4.7.9) together give

$$\begin{aligned}
\left[\int f dX, \int h dY \right] &= \int f d[X, \int h dY] \\
&= \int f h d[X, Y].
\end{aligned} \tag{4.7.10}$$

$\square$

We would like to show that (4.7.5) is true for all $f \in \mathbb{L}(X)$ and $h \in \mathbb{L}(Y)$.

Theorem 4.85 *Let X, Y be stochastic integrators and let $f \in \mathbb{L}(X)$, $g \in \mathbb{L}(Y)$ and let $U = \int f dX$, $V = \int g dY$. Then*

$$[U, V]_t = \int_0^t f g d[X, Y]. \tag{4.7.11}$$

Proof Let us approximate f, g by $f^n = f1_{\{|f| \leq n\}}$ and $g^n = g1_{\{|g| \leq n\}}$ and let $U^n = \int f^n dX$, $V^n = \int g^n dY$. Since $f \in \mathbb{L}(X)$ and $g \in \mathbb{L}(Y)$, by definition $U^n \xrightarrow{ucp} U$

and $V^n \xrightarrow{ucp} V$. Now using Theorem 2.72 (also see the Remark following the result) we can get a subsequence $\{n^k\}$ and a r.c.l.l. adapted increasing process H such that

$$|U_t^{n^k}| + |V_t^{n^k}| \leq H_t \quad \forall k \geq 1. \tag{4.7.12}$$

Using (4.7.5) for f^n, g^n (since they are bounded) we get invoking Theorem 4.33

$$U_t^n V_t^n = U_0^n V_0^n + \int_0^t U_{s-}^n dV_s^n + \int_0^t V_{s-}^n dU_s^n + [U^n, V^n]_t$$

$$= U_0^n V_0^n + \int_0^t U_{s-}^n g_s^n dY_s + \int_0^t V_{s-}^n f_s^n dX_s + \int_0^t f_s^n g_s^n d[X, Y]_s. \tag{4.7.13}$$

Taking $Y = X$ and $g = f$ in (4.7.13) we get

$$(U_t^n)^2 = (U_0^n)^2 + 2 \int_0^t U_{s-}^n f_s^n dX_s + \int_0^t (f_s^n)^2 d[X, X]_s. \tag{4.7.14}$$

In (4.7.14), we would like to take limit as $n \to \infty$. Since $(f_s^n)^2 = f_s^2 1_{\{|f| \leq n\}}$ increases to f_s^2, using monotone convergence theorem, we get

$$\int_0^t (f_s^n)^2 d[X, X]_s \to \int_0^t f_s^2 d[X, X]_s. \tag{4.7.15}$$

For the stochastic integral term, taking limit along the subsequence $\{n^k\}$ (chosen so that (4.7.12) holds) and using $H^- f \in \mathbb{L}(X)$ (see Corollary 4.44) and dominated convergence theorem (Theorem 4.29), we get

$$\int_0^t U_{s-}^{n^k} f_s^{n^k} dX_s \xrightarrow{ucp} \int_0^t U_{s-} f_s dX_s. \tag{4.7.16}$$

Thus putting together (4.7.14)–(4.7.16) along with $U^n \xrightarrow{ucp} U$ we conclude

$$(U_t)^2 = (U_0)^2 + 2 \int_0^t U_{s-} f_s dX_s + \int_0^t (f_s)^2 d[X, X]_s. \tag{4.7.17}$$

This implies

$$[U, U]_t = \int_0^t (f_s)^2 d[X, X]_s. \tag{4.7.18}$$

More importantly, this implies

$$\int_0^t (f_s(\omega))^2 d[X, X]_s(\omega) < \infty \quad \forall t < \infty \quad a.s. \tag{4.7.19}$$

Likewise, we also have

$$\int_0^t (g_s(\omega))^2 d[Y, Y]_s(\omega) < \infty \quad \forall t < \infty \quad a.s. \qquad (4.7.20)$$

Now invoking Theorem 4.80 along with (4.7.19)–(4.7.20), we get

$$\int_0^t |f_s(\omega)g_s(\omega)| d|[X, Y]|_s(\omega) < \infty \quad \forall t < \infty \quad a.s. \qquad (4.7.21)$$

and then using dominated convergence theorem (for signed measures) we conclude

$$\int_0^t f_s^n(\omega)g_s^n(\omega)d[X, Y]_s(\omega) \to \int_0^t f_s(\omega)g_s(\omega)d[X, Y]_s(\omega) \quad \forall t < \infty \quad a.s. \qquad (4.7.22)$$

In view of (4.7.22), taking limit in (4.7.13) along the subsequence $\{n^k\}$ and using argument similar to the one leading to (4.7.17), we conclude

$$U_t V_t = U_0 V_0 + \int_0^t U_{s-}g_s dY_s + \int_0^t V_{s-}f_s dX_s + \int_0^t f_s g_s d[X, Y]_s$$

which in turn implies

$$U_t V_t = U_0 V_0 + \int_0^t U_{s-}dV_s + \int_0^t V_{s-}dU_s + \int_0^t f_s g_s d[X, Y]_s$$

and hence that

$$[U, V]_t = \int_0^t f_s g_s d[X, Y]_s. \qquad \square$$

The earlier proof contains a proof of the following theorem. Of course, this can also be deduced by taking $f = h$ and $X = Y$.

Theorem 4.86 *Let X be stochastic integrator and let $f \in \mathbb{L}(X)$ and let $U = \int f dX$. Then*

$$[U, U]_t = \int_0^t f^2 d[X, X] \qquad (4.7.23)$$

Remark 4.87 In particular, it follows that for a stochastic integrator X, if $f \in \mathbb{L}(X)$ then

$$\int_0^t f_s^2 d[X, X]_s < \infty \quad a.s. \quad \forall t < \infty.$$

4.8 Ito's Formula

Ito's formula is a change of variable formula for stochastic integral. Let us look at the familiar change of variable formula in usual calculus. Let G be a continuously differentiable function on $[0, \infty)$ with derivative $G' = g$ and f be a continuously differentiable function on $\mathbb{R}$. Then

$$f(G(t)) = f(G(0)) + \int_0^t f'(G(s))g(s)ds. \tag{4.8.1}$$

This can be proven by observing that $\frac{d}{dt} f(G(t)) = f'(G(t))g(t)$ and using the fundamental theorem of integral calculus. What can we say when $G(t)$ is not continuously differentiable? Let us recast the change of variable formula as

$$f(G(t)) = f(G(0)) + \int_{0+}^t f'(G(s))dG(s). \tag{4.8.2}$$

Now this is true as long as G is a continuous function with finite variation. Fix $t > 0$ and let $|G(s)| \leq K$ for $0 \leq s \leq t$. For $\varepsilon > 0$ and $\delta > 0$ let

$$h(\varepsilon) = \sup\{|f'(x_1) - f'(x_2)| \ : \ -K \leq x_1, x_2 \leq K, \ |x_1 - x_2| \leq \varepsilon\},$$

$$a(\delta) = \sup\{|G(t_1) - G(t_2)| \ : \ 0 \leq t_1, t_2 \leq t, \ |t_1 - t_2| \leq \delta\},$$

so that $h(a(\frac{t}{n})) \to 0$ as $n \to \infty$ in view of uniform continuity of f on $[-K, K]$ and G on $[0, t]$.

Let us write $t_i^n = \frac{it}{n}$. Now using the mean value theorem, we get

$$\begin{aligned}
f(G(t_{i+1}^n)) - f(G(t_i^n)) &= f'(\theta_i^n)(G(t_{i+1}^n) - G(t_i^n)) \\
&= [f'(G(t_i^n)) + \{f'(\theta_i^n) - f'(G(t_i^n))\}](G(t_{i+1}^n) - G(t_i^n))
\end{aligned} \tag{4.8.3}$$

where $\theta_i^n = G(t_i^n) + u_i^n(G(t_{i+1}^n) - G(t_i^n))$, for some $u_i^n, 0 \leq u_i^n \leq 1$. Now, it is easily seen that

$$f(G(t)) - f(G(0)) = \sum_{i=0}^{n-1} [f(G(t_{i+1}^n)) - f(G(t_i^n))] \tag{4.8.4}$$

and using dominated convergence theorem for Lebesgue–Stieltjes integrals, we have

$$\lim_{n \to \infty} \sum_{i=0}^{n-1} [f'(G(t_i^n))(G(t_{i+1}^n) - G(t_i^n))] = \int_{0+}^t f'(G(s))dG(s). \tag{4.8.5}$$

Since $|G(t_{i+1}^n) - G(t_i^n)| \leq a(tn^{-1})$, it follows that

$$|(f'(\theta_i^n) - f'(G(t_i^n)))| \leq h(a(tn^{-1})) \tag{4.8.6}$$

since $\theta_i^n = G(t_i^n) + u_i^n(G(t_{i+1}^n) - G(t_i^n))$, with $u_i^n, 0 \leq u_i^n \leq 1$. Hence,

$$\left| \sum_{i=0}^{n-1} [f'(\theta_i^n) - f'(G(t_i^n))](G(t_{i+1}^n) - G(t_i^n)) \right|$$

$$\leq h(a(tn^{-1})) \sum_{i=0}^{n-1} |G(t_{i+1}^n) - G(t_i^n)| \qquad (4.8.7)$$

$$\leq h(a(tn^{-1})) \mathrm{VAR}_{(0,t]}(G)$$

$$\to 0.$$

Now putting together (4.8.3)–(4.8.7) we conclude that (4.8.2) is true.

From the proof we see that unless the *integrator* has finite variation on $[0, T]$, the sum of error terms may not go to zero. For a stochastic integrator, we have seen that the quadratic variation is finite. This means we should keep track of first two terms in Taylor expansion and take their limits (and prove that remainder goes to zero). Note that (4.8.3) is essentially using Taylor expansion up to one term with remainder, but we had assumed that f is only once continuously differentiable.

The following lemma is a crucial step in the proof of the Ito's formula. First part is proven earlier, stated here for comparison and ease of reference.

Lemma 4.88 *Let X, Y be stochastic integrators and let Z be an r.c.l.l. adapted process. Let $\delta_m \downarrow 0$ and for $m \geq 1$ let $\{\tau_n^m : n \geq 0\}$ be a δ_m-partition for X, Y and Z. Then*

$$\sum_{n=0}^{\infty} Z_{\tau_n^m \wedge t}(X_{\tau_{n+1}^m \wedge t} - X_{\tau_n^m \wedge t}) \xrightarrow{ucp} \int_0^t Z^- dX \qquad (4.8.8)$$

and

$$\sum_{n=0}^{\infty} Z_{\tau_n^m \wedge t}(X_{\tau_{n+1}^m \wedge t} - X_{\tau_n^m \wedge t})(Y_{\tau_{n+1}^m \wedge t} - Y_{\tau_n^m \wedge t}) \xrightarrow{ucp} \int_0^t Z^- d[X, Y]. \qquad (4.8.9)$$

Remark **4.89** Observe that if Z is continuous, then $\int_0^t Z^- dX = \int_{0+}^t Z dX$.

Proof The first part (4.8.8) has been proved in Theorem 4.62. The second part for the special case $Z = 1$ is proven in Theorem 4.67. For (4.8.9), note that

$$A_t^n = B_t^n - C_t^n - D_t^n$$

where

$$A_t^n = \sum_{n=0}^{\infty} Z_{\tau_n^m \wedge t}(X_{\tau_{n+1}^m \wedge t} - X_{\tau_n^m \wedge t})(Y_{\tau_{n+1}^m \wedge t} - Y_{\tau_n^m \wedge t}),$$

$$B_t^n = \sum_{n=0}^{\infty} Z_{\tau_n^m \wedge t}(X_{\tau_{n+1}^m \wedge t} Y_{\tau_{n+1}^m \wedge t} - X_{\tau_n^m \wedge t} Y_{\tau_n^m \wedge t}),$$

$$C_t^n = \sum_{n=0}^{\infty} Z_{\tau_n^m \wedge t} X_{\tau_n^m \wedge t}(Y_{\tau_{n+1}^m \wedge t} - Y_{\tau_n^m \wedge t}),$$

$$D_t^n = \sum_{n=0}^{\infty} Z_{\tau_n^m \wedge t} Y_{\tau_n^m \wedge t}(X_{\tau_{n+1}^m \wedge t} - X_{\tau_n^m \wedge t}).$$

Recall that XY is a stochastic integrator as seen in Corollary 4.68. Now using (4.8.8), we have

$$B_t^n \xrightarrow{ucp} \int_0^t Z^- d(XY)$$

$$C_t^n \xrightarrow{ucp} \int_0^t Z^- X^- dY = \int_0^t Z^- dS \quad \text{where } S = \int X^- dY$$

$$D_t^n \xrightarrow{ucp} \int_0^t Z^- Y^- dX = \int_0^t Z^- dR \quad \text{where } R = \int Y^- dX.$$

Here we have used Theorem 4.33. Using bilinearity of integral, it follows that

$$A_t^n \xrightarrow{ucp} \int_0^t Z^- dV$$

where $V_t = X_t Y_t - X_0 Y_0 - \int_0^t X^- dY - \int_0^t Y^- dX$. As seen in Theorem 4.67, $V_t = [X, Y]_t$. This completes the proof. □

Remark **4.90** For each $m \geq 1$, let $\{\tau_n^m : m \geq 1\}$ be a partition of $[0, \infty)$ via stopping times

$$0 = \tau_0^m < \tau_1^m < \tau_2^m \ldots ; \tau_n^m \uparrow \infty, \quad m \geq 1$$

such that for all n, m

$$(\tau_{n+1}^m - \tau_n^m) \leq 2^{-m},$$

then (4.8.9) holds for all r.c.l.l. adapted processes Z and stochastic integrators X, Y.

We will first prove a single variable change of variable formula for a continuous stochastic integrator and then go on to the multivariate version.

Now let us fix a twice continuously differentiable function f. The standard version of Taylor's theorem gives

$$f(b) = f(a) + f'(a)(b - a) + \int_a^b f''(s)(b - s)ds. \tag{4.8.10}$$

However, we need the expansion up to two terms with an estimate on the remainder. Let us write

$$f(b) = f(a) + f'(a)(b - a) + \frac{1}{2}f''(a)(b - a)^2 + R_f(a, b) \qquad (4.8.11)$$

where

$$R_f(a, b) = \int_a^b [f''(s) - f''(a)](b - s)ds.$$

Since f'' is assumed to be continuous, for any $K < \infty$, f'' is uniformly continuous and bounded on $[-K, K]$ so that

$$\lim_{\delta \to 0} \Lambda_f(K, \delta) = 0$$

where

$$\Lambda_f(K, \delta) = \sup\{|R_f(a, b)(b - a)^{-2}| : a, b \in [-K, K], 0 < |b - a| < \delta\}.$$

Here is the univariate version of the Ito's formula for continuous stochastic integrators.

Theorem 4.91 (Ito's formula) *Let f be a twice continuously differentiable function on $\mathbb{R}$ and X be a continuous stochastic integrator. Then*

$$f(X_t) = f(X_0) + \int_0^t f'(X_u)1_{\{u>0\}}dX_u + \frac{1}{2}\int_0^t f''(X_u)d[X, X]_u.$$

Corollary 4.92 *Equivalently, we can write the formula as*

$$f(X_t) = f(X_0) + \int_{0+}^t f'(X_u)dX_u + \frac{1}{2}\int_0^t f''(X_u)d[X, X]_u.$$

Proof Fix t. Let $t_i^n = \frac{ti}{n}$ for $i \geq 0, n \geq 1$. Let

$$U_i^n = f(X_{t_{i+1}^n}) - f(X_{t_i^n}), \qquad (4.8.12)$$

$$V_i^n = f'(X_{t_i^n})(X_{t_{i+1}^n} - X_{t_i^n}), \qquad (4.8.13)$$

$$W_i^n = \frac{1}{2}f''(X_{t_i^n})(X_{t_{i+1}^n} - X_{t_i^n})^2, \qquad (4.8.14)$$

$$R_i^n = R_f(X_{t_i^n}, X_{t_{i+1}^n}). \qquad (4.8.15)$$

Then one has

$$U_i^n = V_i^n + W_i^n + R_i^n, \qquad (4.8.16)$$

$$\sum_{i=0}^{n-1} U_i^n = f(X_t) - f(X_0).$$ (4.8.17)

Further using (4.8.8) and (4.8.9) (see Remark 4.89), we get

$$\sum_{i=0}^{n-1} V_i^n \to \int_0^t f'(X_u) 1_{\{u>0\}} dX_u \text{ in probability,}$$ (4.8.18)

$$\sum_{i=0}^{n-1} W_i^n \to \frac{1}{2} \int_0^t f''(X_u) d[X,X]_u \text{ in probability.}$$ (4.8.19)

It suffices to prove that,

$$\sum_{i=0}^{n-1} R_i^n \to 0 \text{ in probability.}$$ (4.8.20)

Observe that
$$|R_i^n(\omega)| \le |R_f(X_{t_i^n}(\omega), X_{t_{i+1}^n}(\omega))|.$$

For each ω, $u \mapsto X_u(\omega)$ is uniformly continuous on $[0, t]$ and hence

$$\delta_n(\omega) = [\sup_i |X_{t_{i+1}^n}(\omega) - X_{t_i^n}(\omega)|] \to 0.$$

Let $K_t(\omega) = \sup_{0 \le u \le t} |X_u(\omega)|$. Now

$$|R_i^n(\omega)| \le \Lambda_f(K_t(\omega), \delta_n(\omega))(X_{t_{i+1}^n} - X_{t_i^n})^2$$

and hence
$$\sum_{i=0}^{n-1} |R_i^n(\omega)| \le \Lambda_f(K_t(\omega), \delta_n(\omega)) \sum_{i=0}^{n-1} (X_{t_{i+1}^n} - X_{t_i^n})^2.$$

Since
$$\sum_{i=0}^{n-1} (X_{t_{i+1}^n} - X_{t_i^n})^2 \to [X,X]_t \text{ in probability}$$

and $\delta_n(\omega) \to 0$, it follows that (4.8.20) is valid completing the proof.   □

Applying the Ito's formula with $f(x) = x^m$, $m \ge 2$, we get

$$X_t^m = X_0^m + \int_0^t m X_u^{m-1} 1_{\{u>0\}} dX_u + \frac{m(m-1)}{2} \int_0^t X_u^{m-2} 1_{\{u>0\}} d[X,X]u$$

and taking $f(x) = \exp(x)$ we get

$$\exp(X_t) = \exp(X_0) + \int_{0+}^t \exp(X_u)dX_u + \frac{1}{2}\int_0^t \exp(X_u)d[X, X]_u.$$

We now turn to the multidimensional version of the Ito's formula. Its proof given below is in the same spirit as the one given above in the one-dimensional case and is based on the Taylor series expansion of a function. This idea is classical, and the proof given here is a simplification of the proof presented in Metivier [50]. A similar proof was also given in Kallianpur and Karandikar [32].

We will first prove the required version of Taylor's theorem. Here, $|\cdot|$ denotes the Euclidean norm on $\mathbb{R}^d$, U will denote a fixed open convex subset of $\mathbb{R}^d$, and $C^{1,2}([0, \infty) \times U)$ denotes the class of functions $f : [0, \infty) \times U \mapsto \mathbb{R}$ that are once continuously differentiable in $t \in [0, \infty)$ and twice continuously differentiable in $x \in U$. Also, for $f \in C^{1,2}([0, \infty) \times \mathbb{U})$, f_0 denotes the partial derivative of f in the t variable, f_j denotes the partial derivative of f w.r.t. jth coordinate of $x = (x^1, \ldots, x^d)$, and f_{jk} denotes the partial derivative of f_j w.r.t. kth coordinate of $x = (x^1, \ldots, x^d)$.

Lemma 4.93 *Let $f \in C^{1,2}([0, \infty) \times U)$. Define $h : [0, \infty) \times U \times U \to \mathbb{R}$ as follows. For $t \in [0, \infty)$, $y = (y^1, \ldots, y^d)$, $x = (x^1, \ldots, x^d) \in U$, let*

$$
\begin{aligned}
h(t, y, x) =& f(t, y) - f(t, x) - \sum_{j=1}^d (y^j - x^j)f_j(t, x) \\
& - \frac{1}{2}\sum_{j,k=1}^d (y^j - x^j)(y^k - x^k)f_{jk}(t, x).
\end{aligned}
\tag{4.8.21}
$$

Then there exist continuous functions $g_{jk} : [0, \infty) \times U \times U \to \mathbb{R}$ such that for $t \in [0, \infty)$, $y = (y^1, \ldots, y^d)$, $x = (x^1, \ldots, x^d) \in U$,

$$h(t, y, x) = \sum_{j,k=1}^d g_{jk}(t, y, x)(y^j - x^j)(y^k - x^k). \tag{4.8.22}$$

Further, the following holds. Define for $T < \infty$, $K \subseteq U$ and $\delta > 0$,

$$\Gamma(T, K, \delta) = \sup\{\frac{|h(t, y, x)|}{|y - x|^2} : 0 \le t \le T, x \in K, y \in K, 0 < |x - y| \le \delta\}$$

and

$$\Lambda(T, K) = \sup\{\frac{|h(t, y, x)|}{|y - x|^2} : 0 \le t \le T, x \in K, y \in K, x \ne y\}.$$

Then for $T < \infty$ and a compact subset $K \subseteq U$, we have

$$\Lambda(T, K) < \infty$$

and

$$\lim_{\delta \downarrow 0} \Gamma(T, K, \delta) = 0.$$

Proof Fix $t \in [0, \infty)$. For $0 \le s \le 1$, define

$$g(s, y, x) = f(t, x + s(y - x)) - f(t, x) - s \sum_{j=1}^{d} (y^j - x^j) f_j(t, x)$$

$$- \frac{s^2}{2} \sum_{j,k=1}^{d} (y^j - x^j)(y^k - x^k) f_{jk}(t, x)$$

where $x = (x^1, \ldots, x^d)$, $y = (y^1, \ldots, y^d) \in \mathbb{R}^d$. Then $g(0, y, x)=0$ and $g(1, y, x) = h(t, y, x)$. Writing $\frac{d}{ds}g = g'$ and $\frac{d^2}{ds^2}g = g''$, we can check that $g'(0, y, x) = 0$ and

$$g''(s, y, x) = \sum_{j,k=1}^{d} (f_{jk}(t, x + s(y - x)) - f_{jk}(t, x))(y^j - x^j)(y^k - x^k).$$

Noting that $g(0, y, x) = g'(0, y, x) = 0$, by Taylor's theorem (see remainder form given in Eq. (4.8.11)) we have

$$h(t, y, x) = g(1, y, x)$$

$$= \int_0^1 (1 - s)g''(s, y, x)ds. \tag{4.8.23}$$

Thus (4.8.22) is satisfied where $\{g_{jk}\}$ are defined by

$$g_{jk}(t, y, x) = \int_0^1 (1 - s)(f_{jk}(t, x + s(y - x)) - f_{jk}(t, x))ds. \tag{4.8.24}$$

The desired estimates on h follow from (4.8.22) and (4.8.24). $\square$

Theorem 4.94 (Ito's Formula for Continuous Stochastic Integrators) *Let $U \subseteq \mathbb{R}^d$ be a convex open set and let $f \in C^{1,2}([0, \infty) \times U)$. Let $X_t = (X_t^1, \ldots, X_t^d)$ be an U-valued continuous process where each X^j is a stochastic integrator. Then*

$$f(t, X_t) = f(0, X_0) + \int_0^t f_0(s, X_s)ds + \sum_{j=1}^{d} \int_{0+}^t f_j(s, X_s)dX_s^j$$

$$+ \frac{1}{2} \sum_{j=1}^{d} \sum_{k=1}^{d} \int_0^t f_{jk}(s, X_s)d[X^j, X^k]_s. \tag{4.8.25}$$

Proof Suffices to prove the equality (4.8.25) for a fixed t a.s. since both sides are continuous processes. Once again let $t_i^n = \frac{ti}{n}$ and

$$V_i^n = \sum_{j=1}^{d} f_j(t_i^n, X_{t_i^n})(X_{t_{i+1}^n}^j - X_{t_i^n}^j)$$

$$W_i^n = \frac{1}{2} \sum_{j,k=1}^{d} f_{jk}(t_i^n, X_{t_i^n})(X_{t_{i+1}^n}^j - X_{t_i^n}^j)(X_{t_{i+1}^n}^k - X_{t_i^n}^k)$$

$$R_i^n = h(t_i^n, X_{t_i^n}, X_{t_{i+1}^n})$$

From the definition of h—(4.8.21)—it follows that

$$f(t_i^n, X_{t_{i+1}^n}) - f(t_i^n, X_{t_i^n}) = V_i^n + W_i^n + R_i^n \tag{4.8.26}$$

Now

$$
\begin{aligned}
f(t, X_t) - f(0, X_0) &= \sum_{i=0}^{n-1} (f(t_{i+1}^n, X_{t_{i+1}^n}) - f(t_i^n, X_{t_i^n})) \\
&= \sum_{i=0}^{n-1} (f(t_{i+1}^n, X_{t_{i+1}^n}) - f(t_i^n, X_{t_{i+1}^n})) \\
&\quad + \sum_{i=0}^{n-1} (f(t_i^n, X_{t_{i+1}^n}) - f(t_i^n, X_{t_i^n})) \\
&= \sum_{i=0}^{n-1} (U_i^n + V_i^n + W_i^n + R_i^n)
\end{aligned}
$$

in view of (4.8.26), where

$$
\begin{aligned}
U_i^n &= f(t_{i+1}^n, X_{t_{i+1}^n}) - f(t_i^n, X_{t_{i+1}^n}) \\
&= \int_{t_i^n}^{t_{i+1}^n} f_0(u, X_{t_{i+1}^n})\,du
\end{aligned}
\tag{4.8.27}
$$

and hence, by the dominated convergence theorem for Lebesgue integrals, we have

$$\sum_{i=0}^{n-1} U_i^n \to \int_0^t f_0(s, X_s)\,ds. \tag{4.8.28}$$

Using (4.8.8) and (4.8.9), we get (see Remark 4.89)

$$\sum_{i=0}^{n-1} V_i^n \to \sum_{j=1}^{d} \int_{0+}^{t} f_j(s, X_s)\,dX_s^j \quad \text{in probability} \tag{4.8.29}$$

and

$$\sum_{i=0}^{n-1} W_i^n \to \frac{1}{2} \sum_{j,k=1}^{d} \int_0^t f_{jk}(s, X_s) d[X^j, X^k]_s \text{ in probability.} \qquad (4.8.30)$$

In view of these observations, the result, namely (4.8.25), would follow once we show that

$$\sum_{i=0}^{n-1} R_i^n \to 0 \text{ in probability.} \qquad (4.8.31)$$

Now let $K^{t,\omega} = \{X_s(\omega) : 0 \leq s \leq t\}$ and $\delta_n(\omega) = \sup_i |(X_{t_{i+1}^n}(\omega) - X_{t_i^n}(\omega))|$. Then $K^{t,\omega}$ is compact, $\delta_n(\omega) \to 0$ and hence by Lemma 4.93 for every $\omega \in \Omega$

$$\Gamma(K^{t,\omega}, \delta_n(\omega)) \to 0. \qquad (4.8.32)$$

Since

$$|R_i^n| \leq \Gamma(K^{t,\omega}, \delta_n(\omega)) |X_{t_{i+1}^n}(\omega) - X_{t_i^n}(\omega)|^2$$

we have

$$\sum_{i=0}^{n-1} |R_i^n| \leq \Gamma(K^{t,\omega}, \delta_n(\omega)) \sum_{i=0}^{n-1} |X_{t_{i+1}^n}(\omega) - X_{t_i^n}(\omega)|^2.$$

The first factor above converges to 0 pointwise as seen in (4.8.32), and the second factor converges to $\sum_{j=1}^{d} [X^j, X^j]_t$ in probability, and we conclude that (4.8.31) is true completing the proof as noted earlier. $\qquad \square$

Theorem 4.95 (Ito's Formula for r.c.l.l. Stochastic Integrators) *Let U be a convex open subset of $\mathbb{R}^d$. Let $f \in C^{1,2}([0, \infty) \times U)$. Let $X^1, \ldots, X^d$ be stochastic integrators and $X_t := (X_t^1, \ldots, X_t^d)$. Further suppose both X and X^- are U-valued. Then*

$$f(t, X_t) = f(0, X_0) + \int_0^t f_0(s, X_{s-})ds + \sum_{j=1}^{d} \int_{0+}^t f_j(s, X_{s-})dX_s^j$$

$$+ \frac{1}{2} \sum_{j=1}^{d} \sum_{k=1}^{d} \int_0^t f_{jk}(s, X_{s-})d[X^j, X^k]_s$$

$$+ \sum_{0 < s \leq t} \{f(s, X_s) - f(s, X_{s-}) - \sum_{j=1}^{d} f_j(s, X_{s-}) \Delta X_s^j \qquad (4.8.33)$$

$$- \sum_{j=1}^{d} \sum_{k=1}^{d} \frac{1}{2} f_{jk}(s, X_{s-})(\Delta X_s^j)(\Delta X_s^k)\}.$$

Proof Let us begin by examining the last term. Firstly, what appears to be a sum of uncountably many terms is really a sum of countable number of terms since the summand is zero whenever $X_s = X_{s-}$. Also, the last term equals

$$D_t = \sum_{0 < s \leq t} h(s, X_s, X_{s-}) \tag{4.8.34}$$

where h is defined by (4.8.21). By Lemma 4.93 for $0 < s \leq t$

$$|h(s, X_s(\omega), X_{s-}(\omega))| \leq \Lambda(K^{t,\omega})|\Delta X_s(\omega)|^2$$

where

$$K^{t,\omega} = \{X_s(\omega) : 0 \leq s \leq T\} \cup \{X_{s-}(\omega) : 0 \leq s \leq T\} \tag{4.8.35}$$

Here $K^{t,\omega}$ is compact (see Exercise 2.1) and thus by Lemma 4.93, $\Lambda(K^{t,\omega}) < \infty$. As a consequence, we have

$$\sum_{0 < s \leq t} |h(s, X_s(\omega), X_{s-}(\omega))| \leq \Lambda(K^{t,\omega}) \sum_{0 < s \leq t} |\Delta X_s(\omega)|^2$$

and invoking Corollary 4.72 we conclude that series in (4.8.34) converges absolutely, a.s. The rest of the argument is on the lines of the proof in the case of continuous stochastic integrators except that this time the remainder term after expansion up to two terms does not go to zero yielding an additional term. Also, the proof given below requires use of partitions via stopping times.

For each $n \geq 1$, define a sequence $\{\tau_i^n : i \geq 1\}$ of stopping times inductively as follows: $\tau_0^n = 0$ and for $i \geq 0$,

$$\tau_{i+1}^n = \inf\{t > \tau_i^n : \max\{|X_t - X_{\tau_i^n}|, |X_{t-} - X_{\tau_i^n}|, |t - \tau_i^n|\} \geq 2^{-n}\} \tag{4.8.36}$$

Let us note that each τ_i^n is a stopping time (see Theorem 2.46),

$$0 = \tau_0^n < \tau_1^n < \ldots < \tau_m^n < \ldots,$$

$$\forall n \geq 1, \quad \tau_m^n \uparrow \infty \text{ as } m \to \infty$$

and

$$(\tau_{i+1}^n - \tau_i^n) \leq 2^{-n}.$$

Thus, $\{\tau_m^n : m \geq 0\}, n \geq 1$ satisfies the conditions of Lemma 4.88.

Fix $t > 0$. On the lines of the proof in the continuous case, let

$$U_i^n = f(\tau_{i+1}^n \wedge t, X_{\tau_{i+1}^n \wedge t}) - f(\tau_i^n \wedge t, X_{\tau_{i+1}^n \wedge t})$$

$$V_i^n = \sum_{j=1}^{d} f_j(\tau_i^n \wedge t, X_{\tau_i^n \wedge t})(X_{\tau_{i+1}^n \wedge t}^j - X_{\tau_i^n \wedge t}^j)$$

$$W_i^n = \frac{1}{2} \sum_{j,k=1}^{d} f_{jk}(\tau_i^n \wedge t, X_{\tau_i^n \wedge t})(X_{\tau_{i+1}^n \wedge t}^j - X_{\tau_i^n \wedge t}^j)(X_{\tau_{i+1}^n \wedge t}^k - X_{\tau_i^n \wedge t}^k)$$

$$R_i^n = h(\tau_i^n \wedge t, X_{\tau_{i+1}^n \wedge t}, X_{\tau_i^n \wedge t})$$

Then

$$f(\tau_{i+1}^n \wedge t, X_{\tau_{i+1}^n \wedge t}) - f(\tau_i^n \wedge t, X_{\tau_i^n \wedge t}) = U_i^n + V_i^n + W_i^n + R_i^n$$

and hence

$$f(t, X_t) - f(0, X_0) = \sum_{i=0}^{\infty} (U_i^n + V_i^n + W_i^n + R_i^n). \qquad (4.8.37)$$

Now,

$$\sum_{i=0}^{\infty} U_i^n = \sum_{i=0}^{\infty} \int_{\tau_i^n \wedge t}^{\tau_{i+1}^n \wedge t} f_0(s, X_{\tau_{i+1}^n \wedge t}) ds$$

and hence for all ω

$$\sum_{i=0}^{\infty} U_i^n(\omega) \to \int_0^t f_0(s, X_s(\omega)) ds$$

Since for every ω, $X_s(\omega) = X_{s-}(\omega)$ for all but countably many s, we can conclude

$$\sum_{i=0}^{\infty} U_i^n \to \int_0^t f_0(s, X_{s-}) ds \text{ in probability.} \qquad (4.8.38)$$

By Lemma 4.88 and the remark following it,

$$\sum_{i=0}^{\infty} V_i^n \to \sum_{j=1}^{d} \int_{0+}^{t} f_j(s, X_{s-}) dX_s^j \qquad (4.8.39)$$

and

$$\sum_{i=0}^{\infty} W_i^n \to \frac{1}{2} \sum_{j=1}^{d} \sum_{k=1}^{d} \int_0^t f_{jk}(s, X_{s-}) d[X^j, X^k] \qquad (4.8.40)$$

in probability. In view of (4.8.37)–(4.8.40), to complete the proof of the result it suffices to prove

$$\sum_{i=0}^{\infty} R_i^n = \sum_{i=0}^{\infty} h(\tau_i^n \wedge t, X_{\tau_{i+1}^n \wedge t}, X_{\tau_i^n \wedge t}) \to \sum_{0<s\le t} h(s, X_s, X_{s-}). \qquad (4.8.41)$$

Let us partition indices i into three sets: $h(\tau_i^n \wedge t, X_{\tau_{i+1}^n \wedge t}, X_{\tau_i^n \wedge t})$ is zero, is small and is large as follows:

$$H^n(\omega) = \{i \ge 0 : \tau_{i+1}^n(\omega) \wedge t = \tau_i^n(\omega) \wedge t\}$$
$$E^n(\omega) = \{i \notin H^n : |\Delta X_{\tau_{i+1}^n \wedge t}(\omega)| \le 2 \cdot 2^{-n}\}$$
$$F^n(\omega) = \{i \notin H^n : |\Delta X_{\tau_{i+1}^n \wedge t}(\omega)| > 2 \cdot 2^{-n}\}.$$

For $i \in H^n(\omega)$, $h(\tau_i^n(\omega) \wedge t, X_{\tau_{i+1}^n \wedge t}(\omega), X_{\tau_i^n \wedge t}(\omega)) = 0$ and thus writing

$$B^n(\omega) = \sum_{i \in E^n(\omega)} h(\tau_i^n(\omega) \wedge t, X_{\tau_{i+1}^n \wedge t}(\omega), X_{\tau_i^n \wedge t}(\omega)),$$

$$C^n(\omega) = \sum_{i \in F^n(\omega)} h(\tau_i^n(\omega) \wedge t, X_{\tau_{i+1}^n \wedge t}(\omega), X_{\tau_i^n \wedge t}(\omega))$$

we observe that

$$\sum_{i=0}^{\infty} R_i^n = B^n + C^n.$$

Note that for any j if $u, v \in (\tau_j^n, \tau_{j+1}^n)$, then $|X_u - X_v| \le 2.2^{-n}$ as X_u, X_v are within 2^{-n} distance from $X_{\tau_j^n}$. As a result, for any $v \in (\tau_j^n, \tau_{j+1}^n)$, $|X_{v-} - X_v| \le 2.2^{-n}$. Thus, if $|\Delta X_s(\omega)| > 2 \cdot 2^{-n}$ for $s \in (0, t]$, then s must equal $\tau_j^n(\omega) \wedge t$ for some j with $i = j - 1 \in F^n$, i.e.

$$\text{if } s \in (0, t] \text{ and } |\Delta X_s(\omega)| > 2 \cdot 2^{-n} \text{ then } s = \tau_{i+1}^n \wedge t \text{ for } i \in F^n(\omega). \qquad (4.8.42)$$

Hence for $i \in E^n(\omega)$,

$$|X_{\tau_{i+1}^n \wedge t}(\omega) - X_{\tau_i^n \wedge t}(\omega)| \le |(X^-)_{\tau_{i+1}^n \wedge t}(\omega) - X_{\tau_i^n \wedge t}(\omega)| + |\Delta X_{\tau_{i+1}^n \wedge t}(\omega)|$$
$$\le 3 \cdot 2^{-n}$$

and hence

$$|B^n(\omega)| \le \sum_{i \in E^n(\omega)} |h(\tau_i^n(\omega) \wedge t, X_{\tau_{i+1}^n \wedge t}(\omega), X_{\tau_i^n \wedge t}(\omega))|$$
$$\le \Gamma(K^{t,\omega}, 3 \cdot 2^{-n}) \sum_i |X_{\tau_{i+1}^n \wedge t}(\omega) - X_{\tau_i^n \wedge t}(\omega)|^2$$

where $K^{t,\omega}$ defined by (4.8.35) is compact. Since $\sum_i |X_{\tau^n_{i+1}\wedge t} - X_{\tau^n_i \wedge t}|^2$ converges to $\sum_j [X^j, X^j]_t$ in probability and $\Gamma(K^{t,\omega}, 3 \cdot 2^{-n}) \to 0$ for all ω, invoking Lemma 4.93 it follows that

$$B^n \to 0 \text{ in probability.}$$

Thus to complete the proof of (4.8.41), it would suffice to show that

$$C^n \to \sum_{0 < s \leq t} h(s, X_s, X_{s-}) \text{ in probability.} \tag{4.8.43}$$

Let $G(\omega) = \{s \in (0, t] : |\Delta X_s(\omega)| > 0\}$. Since X is an r.c.l.l. process, $G(\omega)$ is a countable set for every ω. Fix ω and for $s \in G(\omega)$, define

$$a^n_s(\omega) := \sum_{i \in F^n(\omega)} h(\tau^n_i(\omega) \wedge t, X_{\tau^n_{i+1}(\omega) \wedge t}, X_{\tau^n_i \wedge t}(\omega)) 1_{\{(\tau^n_{i+1}(\omega) \wedge t) = s\}}.$$

Then

$$C^n(\omega) = \sum_{i \in F^n(\omega)} h(\tau^n_i \wedge t(\omega), X_{\tau^n_{i+1} \wedge t}(\omega), X_{\tau^n_i \wedge t}(\omega)) = \sum_{s \in G(\omega)} a^n_s(\omega).$$

If $|\Delta X_s(\omega)| > 2 \cdot 2^{-n}$, then $a^n_s(\omega) = h(\tau^n_i(\omega) \wedge t, X_s(\omega), X_{\tau^n_i \wedge t}(\omega))$ with $s = \tau^n_{i+1}(\omega) \wedge t$ (as seen in (4.8.42)) and hence

$$a^n_s(\omega) \to h(s, X_s(\omega), X_{s-}(\omega)) \quad \text{for all} \quad \omega. \tag{4.8.44}$$

For $i \in F^n(\omega)$,

$$|X_{\tau^n_{i+1} \wedge t}(\omega) - X_{\tau^n_i \wedge t}(\omega)| \leq |(X^-)_{\tau^n_{i+1} \wedge t}(\omega) - X_{\tau^n_i \wedge t}(\omega)| + |\Delta X_{\tau^n_{i+1} \wedge t}(\omega)|$$
$$\leq 2^{-n} + |\Delta X_{\tau^n_{i+1} \wedge t}(\omega)|$$
$$\leq 2|\Delta X_{\tau^n_{i+1} \wedge t}(\omega)|$$

Thus if $|a^n_s(\omega)| \neq 0$, then $s = \tau^n_{i+1}(\omega) \wedge t$ for some $i \in F^n(\omega)$ and then

$$|a^n_s(\omega)| \leq \Lambda(K^{t,\omega}) |X_{\tau^n_{i+1} \wedge t}(\omega) - X_{\tau^n_i \wedge t}(\omega)|^2$$
$$\leq 4\Lambda(K^{t,\omega}) |\Delta X_{\tau^n_{i+1} \wedge t}(\omega)|^2 \tag{4.8.45}$$
$$= 4\Lambda(K^{t,\omega}) |\Delta X_s(\omega)|^2.$$

Let $C_s(\omega) = 4\Lambda(K^{t,\omega}) |\Delta X_s(\omega)|^2$. Then

$$\sum_s C_s(\omega) = 4\Lambda(K^{t,\omega}) \sum_s |\Delta X_s(\omega)|^2$$

and hence by Corollary 4.72

$$\sum_s C_s(\omega) < \infty \quad a.s.$$

Using Weierstrass's M-test (series version of the dominated convergence theorem) along (4.8.44) and (4.8.45), we get

$$C^n(\omega) = \sum_{s \in G(\omega)} a_s^n(\omega) \to \sum_{s \in G(\omega)} h(s, X_s(\omega), X_{s-}(\omega)) \quad a.s. \qquad (4.8.46)$$

We have proved that the convergence in (4.8.43) holds almost surely and hence in probability. This completes the proof. $\qquad \square$

Corollary 4.96 *Let f, X be as in Theorem 4.95. Then $Z_t = f(t, X_t)$ is a stochastic integrator.*

Proof In the Ito's formula (4.8.33) that expresses $f(t, X_t)$, it is clear that the terms involving integral are stochastic integrators. We had seen that the last term is

$$D_t = \sum_{0 < s \le t} h(s, X_s, X_{s-})$$

where h is defined by (4.8.21). By Lemma 4.93

$$|h(s, X_s(\omega), X_{s-}(\omega))| \le \Lambda(K^{t,\omega})|\Delta X_s(\omega)|^2$$

for $0 < s \le t$, where

$$K^{t,\omega} = \{X_s(\omega) : 0 \le s \le t\} \cup \{X_{s-}(\omega) : 0 \le s \le t\}.$$

Hence

$$\mathrm{VAR}_{[0,T]}(D)(\omega) \le \Lambda(K^{T,\omega})[X, X]_T.$$

Thus D is a process with finite variation and hence is a stochastic integrator, completing the proof. $\qquad \square$

We have seen earlier in Corollary 4.72 that $\Delta[X, Y]_t = (\Delta X)_t(\Delta Y)_t$. Thus carefully examining the right-hand side of the Ito's formula (4.8.33), we see that we are adding and subtracting a term

$$\sum_{j=1}^d \sum_{k=1}^d \frac{1}{2} f_{jk}(s, X_{s-})(\Delta X_s^j)(\Delta X_s^k).$$

Let us introduce for now in an ad hoc manner $[X, Y]^{(c)}$ for stochastic integrators X, Y:

$$[X, Y]_t^{(c)} = [X, Y]_t - \sum_{0 < s \le t} (\Delta X)_s (\Delta Y)_s. \tag{4.8.47}$$

Later we will show that this is the cross-quadratic variation of continuous martingale parts of X and Y (see Theorem 8.83). For now we observe the following.

Lemma 4.97 *Let X, Y be stochastic integrators and h be a locally bounded predictable process. Then $[X, Y]^{(c)}$ is a continuous process and further*

$$\int h d[X, Y]^{(c)} = \int h d[X, Y] - \sum_{0 < s \le t} h_s (\Delta X)_s (\Delta Y)_s. \tag{4.8.48}$$

Proof Continuity of $[X, Y]^{(c)}$ follows from its definition (4.8.47) and part (iii) of Corollary 4.72. The identity (4.8.48) for simple functions $h \in \mathbb{S}$ follows by direct verification and hence follows for all bounded predictable processes by monotone class theorem. $\square$

Using Lemma 4.97, we can recast the Ito's formula in an alternate form.

Theorem 4.98 (Ito's formula for r.c.l.l. stochastic integrators)
Let U be a convex open subset of $\mathbb{R}^d$. Let $f \in C^{1,2}([0, \infty) \times U)$. Let $X^1, \ldots, X^d$ be stochastic integrators $X_t := (X_t^1, \ldots, X_t^d)$ is U-valued. Further, suppose X^- is also U-valued. Then

$$f(t, X_t) = f(0, X_0) + \int_0^t f_0(s, X_{s-}) ds + \sum_{j=1}^d \int_{0+}^t f_j(s, X_{s-}) dX_s^j$$

$$+ \frac{1}{2} \sum_{j=1}^d \sum_{k=1}^d \int_{0+}^t f_{jk}(s, X_{s-}) d[X^j, X^k]^{(c)} \tag{4.8.49}$$

$$+ \sum_{0 < s \le t} \{ f(s, X_s) - f(s, X_{s-}) - \sum_{j=1}^d f_j(s, X_{s-}) \Delta X_s^j \}.$$

Exercise 4.99 Let S be a $(0, \infty)$-valued continuous stochastic integrator. Show that $R_t = (S_t)^{-1}$ is also a stochastic integrator and

$$R_t = R_0 - \int_{0+}^t (S_s)^{-2} dS_s + \int_{0+}^t (S_s)^{-3} d[S, S]_s.$$

Exercise 4.100 Let S be a $(0, \infty)$-valued r.c.l.l. stochastic integrator with $S_{t-} > 0$ for $t > 0$. Show that $R_t = (S_t)^{-1}$ is also a stochastic integrator and

$$R_t = R_0 - \int_{0+}^t (S_{s-})^{-2} dS_s + \int_{0+}^t (S_{s-})^{-3} d[S, S]_s^{(c)} + \sum_{0 < s \le t} u(S_s, S_{s-})$$

where $u(y, x) = \frac{1}{y} - \frac{1}{x} + (y - x)\frac{1}{x^2}$.

Exercise 4.101 Let X be a continuous stochastic integrator and let $S_t = \exp(X_t - \frac{1}{2}[X, X]_t)$.

(i) Show that S satisfies

$$S_t = \exp(X_0) + \int_{0+}^{t} S_u dX_u. \tag{4.8.50}$$

HINT: Apply Ito's formula for $f(X, [X, X])$ for suitable function f.

(ii) Let Z satisfy

$$Z_t = \exp(X_0) + \int_{0+}^{t} Z_u dX_u. \tag{4.8.51}$$

Show that $Z = S$.

HINT: Let $Y_t = Z_t \exp(-X_t + \frac{1}{2}[X, X]_t)$ and applying Ito's formula, conclude that $Y_t = Y_0 = 1$.

(iii) Show that

$$X_t = X_0 + \int_{0+}^{t} S_u^{-1} dS_u.$$

(iv) Let $R_t = (S_t)^{-1}$. Show that R satisfies

$$R_t = (S_0)^{-1} - \int_{0+}^{t} (S_u)^{-1} dX_u + \int_{0+}^{t} (S_u)^{-1} d[X, X]_u \tag{4.8.52}$$

which in turn is same as (see also Exercise 4.99)

$$R_t = (S_0)^{-1} - \int_{0+}^{t} (S_u)^{-2} dS_u + \int_{0+}^{t} (S_u)^{-3} d[S, S]_u. \tag{4.8.53}$$

HINT: Use part (i) above and $R_t = \exp(Y_t - \frac{1}{2}[Y, Y]_t)$ where $Y_t = -X_t + [X, X]_t$.

Exercise 4.102 Let X, Y be a $(0, \infty)$-valued continuous stochastic integrators and let U, V be solutions to

$$U_t = \exp(X_0) + \int_{0+}^{t} U_u dX_u \tag{4.8.54}$$

and

$$V_t = \exp(Y_0) + \int_{0+}^{t} V_u dY_u. \tag{4.8.55}$$

Let $W_t = U_t V_t$. Show that W is the unique solution to

$$W_t = \exp(X_0 Y_0) + \int_{0+}^t W_u dX_u + \int_{0+}^t W_u dY_u + \int_{0+}^t W_u d[X, Y]_u. \quad (4.8.56)$$

4.9 The Emery Topology

On the class of stochastic integrators, we define a natural metric $\mathbf{d}_{em}$ as follows. Let $\mathbb{S}_1$ be the class of simple predictable processes f such that $|f| \leq 1$. For stochastic integrators X, Y, let

$$\mathbf{d}_{em}(X, Y) = \sup\{\mathbf{d}_{ucp}(\int f dX, \int f dY) : f \in \mathbb{S}_1\} \quad (4.9.1)$$

Easy to see that $\mathbf{d}_{em}$ is a metric bounded by 1 on the class of stochastic integrators. This metric was defined by Emery [16] and the induced topology is called the Emery topology. If X^n converges to X in $\mathbf{d}_{em}$ metric, we will write it as $X^n \xrightarrow{em} X$. Taking $f = 1_{[0,T]}$ in (4.9.1) and then taking limit as $T \uparrow \infty$, it follows that

$$\mathbf{d}_{ucp}(X, Y) \leq \mathbf{d}_{em}(X, Y) \quad (4.9.2)$$

and thus convergence in Emery topology implies convergence in $\mathbf{d}_{ucp}$ metric. If A is an increasing process, then it is easy to see that $\mathbf{d}_{em}(A, 0) \leq \mathbf{d}_{ucp}(A, 0)$ and hence we have

$$\mathbf{d}_{em}(A, 0) = \mathbf{d}_{ucp}(A, 0). \quad (4.9.3)$$

Let us define a metric $\mathbf{d}_{var}$ on $\mathbb{V}$ as follows: for $B, C \in \mathbb{V}$

$$\mathbf{d}_{var}(B, C) = \mathbf{d}_{ucp}(\text{VAR}(B - C), 0). \quad (4.9.4)$$

Lemma 4.103 *Let $B, C \in \mathbb{V}$. Then*

$$\mathbf{d}_{em}(B, C) \leq \mathbf{d}_{var}(B, C).$$

Proof Note that for any predictable f with $|f| \leq 1$

$$|\int_0^T f dB - \int_0^T f dC| \leq \text{VAR}_{[0,T]}(B - C).$$

The result follows from this observation.   □

As a consequence of (2.5.2)–(2.5.3), it can be seen that $\mathbf{d}_{em}(X^n, X) \to 0$ if and only if for all $T > 0, \delta > 0$

$$\lim_{n\to\infty} [\sup_{f:\in \mathbb{S}_1} \mathsf{P}(\sup_{0\le t\le T} |\int_0^t f\,dX^n - \int_0^t f\,dX| > \delta)] = 0 \qquad (4.9.5)$$

and likewise that $\{X^n\}$ is Cauchy in $\mathbf{d}_{em}$ is and only if for all $T > 0, \delta > 0$

$$\lim_{n,k\to\infty} [\sup_{f:\in \mathbb{S}_1} \mathsf{P}(\sup_{0\le t\le T} |\int_0^t f\,dX^n - \int_0^t f\,dX^k| > \delta)] = 0. \qquad (4.9.6)$$

Theorem 4.104 *For a bounded predictable process f with $|f| \le 1$ one has*

$$\mathbf{d}_{ucp}(\int f\,dX, \int f\,dY) \le \mathbf{d}_{em}(X, Y). \qquad (4.9.7)$$

Proof Let $\mathbb{K}_1$ be the class of bounded predictable processes for which (4.9.7) is true. Then using the dominated convergence theorem it follows that $\mathbb{K}_1$ is closed under bp-convergence. Using the definition of $\mathbf{d}_{ucp}$ we can conclude that $\mathbb{K}_1$ contains $\mathbb{S}_1$.

Let $\mathbb{K}$ denote the class of bounded predictable processes g such that the process $\tilde{g}$ defined by $\tilde{g} = \max(\min(g, 1), -1) \in \mathbb{K}_1$. Then it is easy to see that $\mathbb{K}$ is closed under bp-convergence and contains S. Thus $\mathbb{K}$ is the class of all bounded predictable processes g and hence $\mathbb{K}_1$ contains all predictable processes bounded by 1. $\qquad\square$

Corollary 4.105 *If X_n, X are stochastic integrators such that $X^n \xrightarrow{em} X$ then for all $T > 0, \delta > 0, \eta > 0$, there exists n_0 such that for $n \ge n_0$, we have*

$$\sup_{f:\in \mathbb{K}_1} \mathsf{P}(\sup_{0\le t\le T} |\int_0^t f\,dX^n - \int_0^t f\,dX| > \delta) < \eta \qquad (4.9.8)$$

where $\mathbb{K}_1$ is the class of predictable processes bounded by 1.

Linearity of the stochastic integral and the definition of the metric $\mathbf{d}_{em}$ yields the inequality

$$\mathbf{d}_{em}(U + V, X + Y) \le \mathbf{d}_{em}(U, X) + \mathbf{d}_{em}(V, Y)$$

which in turn implies that if $X^n \xrightarrow{em} X$ and $Y^n \xrightarrow{em} Y$ then $(X^n + Y^n) \xrightarrow{em} (X + Y)$.

We will now prove an important property of the metric $\mathbf{d}_{em}$.

Theorem 4.106 *The space of stochastic integrators is complete under the metric $\mathbf{d}_{em}$.*

Proof Let $\{X^n\}$ be a Cauchy sequence in $\mathbf{d}_{em}$ metric. Then by (4.9.2) it is also Cauchy in $\mathbf{d}_{ucp}$ metric and so by Theorem 2.71 there exists an r.c.l.l. adapted process X such that $X^n \xrightarrow{ucp} X$. We will show that X is a stochastic integrator and $X^n \xrightarrow{em} X$.

Let $a_n = \sup_{k\ge 1} \mathbf{d}_{em}(X^n, X^{n+k})$. Then $a_n \to 0$ since X^n is Cauchy in $\mathbf{d}_{em}$. For $f \in \mathbb{B}(\tilde{\Omega}, \mathcal{P})$, consider $Y^n(f) = \int f\,dX^n$. Then (in view of (4.9.7)) for bounded predictable f with $|f| \le c$, we have

$$\mathbf{d}_{ucp}(Y^n(f), Y^{n+k}(f)) \le c\, a_n, \quad \forall k \ge 1 \tag{4.9.9}$$

and hence for bounded predictable processes f, $\{Y^n(f)\}$ is Cauchy in $\mathbf{d}_{ucp}$ and hence by Theorem 2.71, $Y^n(f)$ converges to say $Y(f) \in \mathbb{R}^0(\Omega, (\mathcal{F}_.), \mathsf{P})$ and

$$\mathbf{d}_{ucp}(Y^n(f), Y(f)) \le c\, a_n, \quad \text{for all predictable } f, \ |f| \le c. \tag{4.9.10}$$

For simple predictable f, we can directly verify that $Y(f) = J_X(f)$. We will show (using the standard $\frac{\varepsilon}{3}$) argument that $Y(f^m) \xrightarrow{ucp} Y(f)$ if $f^m \xrightarrow{bp} f$ and $\{f^m\}$ are uniformly bounded. $Y(f)$ would then be the required extension in the definition of stochastic integrator proving that X is a stochastic integrator. Let us fix $f^m \xrightarrow{bp} f$ where $\{f^m\}$ are uniformly bounded. Dividing by the uniform upper bound if necessary, we can assume that $|f^m| \le 1$. We wish to show that $\mathbf{d}_{ucp}(Y(f^m), Y(f)) \to 0$ as $m \to \infty$.

Given $\varepsilon > 0$, first choose and fix n^* such that $a_{n^*} < \frac{\varepsilon}{3}$. Then

$$\begin{aligned}
\mathbf{d}_{ucp}(Y(f^m), Y(f)) &\le \mathbf{d}_{ucp}(Y(f^m), Y^{n^*}(f^m)) + \mathbf{d}_{ucp}(Y^{n^*}(f^m), Y^{n^*}(f)) \\
&\quad + \mathbf{d}_{ucp}(Y^{n^*}(f), Y(f)) \\
&\le a_{n^*} + \mathbf{d}_{ucp}(Y^{n^*}(f^m), Y^{n^*}(f)) + a_{n^*} \\
&\le \frac{\varepsilon}{3} + \mathbf{d}_{ucp}(Y^{n^*}(f^m), Y^{n^*}(f)) + \frac{\varepsilon}{3}.
\end{aligned} \tag{4.9.11}$$

Since $f^m \xrightarrow{bp} f$ and $\{f^m\}$ is bounded by 1, $\int f^m dX^{n^*} \xrightarrow{ucp} \int f dX^{n^*}$ and hence we can choose m^* (depends upon n^* which has been chosen and fixed earlier) such that for $m \ge m^*$ one has

$$\mathbf{d}_{ucp}(Y^{n^*}(f^m), Y^{n^*}(f)) \le \frac{\varepsilon}{3}$$

and hence using (4.9.11) we get for $m \ge m^*$,

$$\mathbf{d}_{ucp}(Y(f^m), Y(f)) \le \varepsilon.$$

Thus, $Y(f)$ is the required extension of $\{\int f dX : f \in \mathbb{S}\}$ proving that X is a stochastic integrator and

$$Y(f) = \int f dX.$$

Recalling that $Y^n(f) = \int f dX^n$, Eq. (4.9.10) implies

$$\mathbf{d}_{em}(X^n, X) \le a_n$$

with $a_n \to 0$. This completes the proof of completeness of $\mathbf{d}_{em}$. $\qquad\square$

We will now strengthen the conclusion in Theorem 4.50 by showing that the convergence is actually in the Emery topology.

Theorem 4.107 *Suppose* $Y^n, Y \in \mathbb{R}^0(\Omega, (\mathcal{F}_\cdot), \mathsf{P})$, $Y^n \xrightarrow{ucp} Y$ *and* X *is a stochastic integrator. Then*

$$\int (Y^n)^- dX \xrightarrow{em} \int Y^- dX.$$

Proof As noted in Theorem 4.33, $\int (Y^n)^- dX$ and $\int Y^- dX$ are stochastic integrators. Further, we need to show that for $T < \infty, \delta > 0$

$$\lim_{n \to \infty} [\sup_{f \in \mathbb{S}_1} \mathsf{P}(\sup_{0 \le t \le T} |\int_0^t (Y^n)^- f dX - \int_0^t Y^- f dX| > \delta)] = 0. \qquad (4.9.12)$$

We will prove this by contradiction. Suppose (4.9.12) is not true. Then there exists an ε and a subsequence $\{n^k\}$ such that

$$\sup_{f \in \mathbb{S}_1} \mathsf{P}(\sup_{0 \le t \le T} |\int_0^t (Y^{n^k})^- f dX - \int_0^t Y^- f dX| > \delta) > \varepsilon \;\; \forall k \ge 1. \qquad (4.9.13)$$

For each k get $f^k \in \mathbb{S}_1$ such that

$$\mathsf{P}(\sup_{0 \le t \le T} |\int_0^t (Y^{n^k})^- f^k dX - \int_0^t Y^- f^k dX| > \delta) > \varepsilon. \qquad (4.9.14)$$

Now let $g^k = (Y^{n^k} - Y)^- f^k$. Since $Y^n \xrightarrow{ucp} Y$ and f^k are uniformly bounded by 1, it follows that $g^k \xrightarrow{ucp} 0$ and hence by Theorem 4.50,

$$\mathsf{P}(\sup_{0 \le t \le T} |\int_0^t (Y^{n^k})^- f^k dX - \int_0^t Y^- f^k dX| > \delta) \to 0.$$

This contradicts (4.9.14). This proves (4.9.12). $\qquad\qquad\square$

We will show that indeed $Y^n \xrightarrow{ucp} Y$ and $X^n \xrightarrow{em} X$ implies that $\int Y^n dX^n \xrightarrow{em} \int Y dX$. We will first prove a lemma and then go on to this result.

Lemma 4.108 *Suppose* $U^n \in \mathbb{R}^0(\Omega, (\mathcal{F}_\cdot), \mathsf{P})$ *be such that*

$$\sup_{n \ge 1} \sup_{0 \le s \le t} |U_s^n| \le H_t$$

where $H \in \mathbb{R}^0(\Omega, (\mathcal{F}_\cdot), \mathsf{P})$ *is an increasing process. Let* X^n, X *be stochastic integrators such that* $X^n \xrightarrow{em} X$. *Let* $Z^n = \int (U^n)^- dX^n$ *and* $W^n = \int (U^n)^- dX$. *Then* $(Z^n - W^n) \xrightarrow{em} 0$.

Proof Note that in view of Theorem 4.33 for bounded predictable f

$$\int_0^t f \, dZ^n - \int_0^t f \, dW^n = \int (U^n)^- f \, dX^n - \int (U^n)^- f \, dX.$$

So we need to show (see (4.9.5)) for $T < \infty$, $\delta > 0$ and $\varepsilon > 0$, $\exists n_0$ such that for $n \geq n_0$ we have

$$\sup_{f : \in \mathbb{S}_1} \mathsf{P}(\sup_{0 \leq t \leq T} |\int_0^t (U^n)^- f \, dX^n - \int_0^t (U^n)^- f \, dX| > \delta) < \varepsilon. \tag{4.9.15}$$

Recall, $\mathbb{S}_1$ is the class of predictable processes that are bounded by 1.
   First, get $\lambda < \infty$ such that

$$\mathsf{P}(H_T \geq \lambda) \leq \frac{\varepsilon}{2}$$

and let a stopping time σ be defined by

$$\sigma = \inf\{t > 0 : \ H_t \geq \lambda \text{ or } H_{t-} \geq \lambda\} \wedge (T+1).$$

Then we have

$$\mathsf{P}(\sigma < T) \leq \mathsf{P}(H_T \geq \lambda) \leq \frac{\varepsilon}{2}$$

and for $f \in \mathbb{S}_1$ writing $h^n = (U^n)^- f \frac{1}{\lambda} 1_{[0,\sigma]}$ we see that

$$\mathsf{P}(\sup_{0 \leq t \leq T} |\int_0^t (U^n)^- f \, dX^n - \int_0^t (U^n)^- f \, dX| > \delta) \tag{4.9.16}$$

$$\leq \mathsf{P}(\sup_{0 \leq t \leq \sigma \wedge T} |\int_0^t (U^n)^- f \, dX^n - \int_0^t (U^n)^- f \, dX| > \delta) + \mathsf{P}(\sigma < T)$$

$$\leq \mathsf{P}(\sup_{0 \leq t \leq \sigma \wedge T} |\int_0^t h^n \, dX^n - \int_0^t h^n \, dX| > \frac{\delta}{\lambda}) + \frac{\varepsilon}{2}.$$

   Finally, since $X^n \xrightarrow{em} X$, invoking Corollary 4.105 get n_0 such that for $n \geq n_0$ one has

$$\sup_{g \in \mathbb{K}_1} \mathsf{P}(\sup_{0 \leq t \leq T} |\int_0^t g \, dX^n - \int_0^t g \, dX| > \frac{\delta}{\lambda}) < \frac{\varepsilon}{2}$$

where $\mathbb{K}_1$ is the class of predictable processes bounded by 1. Since $h^n \in \mathbb{K}_1$, using (4.9.16) it follows that

$$\mathsf{P}(\sup_{0 \leq t \leq T} |\int_0^t (U^n)^- f \, dX^n - \int_0^t (U^n)^- f \, dX| > \delta) < \varepsilon.$$

Note that the choice of λ and n_0 did not depend upon $f \in \mathbb{K}_1$ and hence (4.9.15) holds completing the proof. $\qquad \Box$

Here is our main result that connects convergence in the Emery topology and stochastic integration.

Theorem 4.109 *Let* $Y^n, Y \in \mathbb{R}^0(\Omega, (\mathcal{F}.), \mathbb{P})$ *be such that* $Y^n \xrightarrow{ucp} Y$ *and let* X^n, X *be stochastic integrators such that* $X^n \xrightarrow{em} X$. *Let* $Z^n = \int (Y^n)^- dX^n$ *and* $Z = \int Y^- dX$. *Then* $Z^n \xrightarrow{em} Z$.

Proof We have seen in Theorem 4.107 that

$$\int (Y^n)^- dX - \int Y^- dX \xrightarrow{em} 0.$$

Thus suffices to prove that

$$\int (Y^n)^- dX^n - \int (Y^n)^- dX \xrightarrow{em} 0. \tag{4.9.17}$$

Let $b_n = \mathbf{d}_{em}(\int (Y^n)^- dX^n, \int (Y^n)^- dX)$. To prove that $b_n \to 0$ suffices to prove the following: (see proof of Theorem 4.50) For any subsequence $\{n_k : k \geq 1\}$, there exists a further subsequence $\{m_j : j \geq 1\}$ of $\{n_k : k \geq 1\}$ (i.e. there exists a subsequence $\{k_j : j \geq 1\}$ such that $m_j = n_{k_j}$) such that

$$b_{m_j} = \mathbf{d}_{em}(\int (Y^{m_j})^- dX^{m_j}, \int (Y^{m_j})^- dX) \to 0.$$

So now, given a subsequence $\{n_k : k \geq 1\}$, using $\mathbf{d}_{ucp}(Y^{n_k}, Y) \to 0$, let us choose $m_j = n_{k_j}$ with $k_{j+1} > k_j$ and $\mathbf{d}_{ucp}(Y^{m_j}, Y) \leq 2^{-j}$ for each $j \geq 1$. Then as seen earlier, this would imply

$$\sum_{j=1}^{\infty} [\sup_{t \leq T} |Y_t^{m_j} - Y_t|] < \infty, \quad \forall T < \infty.$$

Thus defining

$$H_t = \sup_{0 \leq s \leq t} [\sum_{j=1}^{\infty} |Y_s^{m_j} - Y_s| + |Y_s|]$$

we have that H is an r.c.l.l. process and $|Y^{m_j}| \leq H$. Then by Lemma 4.108, it follows that

$$\int (Y^{m_j})^- dX^{m_j} - \int (Y^{m_j})^- dX \xrightarrow{em} 0.$$

This proves $b_{m_j} \to 0$ and thus (4.9.17), completing the proof of the theorem. $\qquad \Box$

Essentially the same arguments also proves the following:

Proposition 4.110 *Let* $Y^n, Y \in \mathbb{R}^0(\Omega, (\mathcal{F}_.), \mathsf{P})$ *be such that* $Y^n \xrightarrow{ucp} Y$ *and let* $X^n, X \in \mathbb{V}$ *be such that* $\mathbf{d}_{var}(X^n, X) \to 0$. *Let* $Z^n = \int (Y^n) dX^n$ *and* $Z = \int Y dX$. *Then*

$$\mathbf{d}_{var}(Z^n, Z) \to 0.$$

In particular,

$$\mathbf{d}_{em}(Z^n, Z) \to 0.$$

We will now show that $X \mapsto [X, X]$ and $(X, Y) \mapsto [X, Y]$ are continuous mappings in the Emery topology.

Theorem 4.111 *Suppose* X^n, X, Y^n, Y *are stochastic integrators such that* $X^n \xrightarrow{em} X$, $Y^n \xrightarrow{em} Y$. *Then*

$$\mathbf{d}_{var}([X^n, Y^n], [X, Y]) \to 0 \tag{4.9.18}$$

and as a consequence

$$[X^n, Y^n] \xrightarrow{em} [X, Y]. \tag{4.9.19}$$

Proof Let $U^n = X^n - X$. Then $\mathbf{d}_{em}(U^n, 0) \to 0$ implies that $\mathbf{d}_{ucp}(U^n, 0) \to 0$ and hence as noted earlier in Proposition 4.52 $(U^n)^2 \xrightarrow{ucp} 0$. Also, by Theorem 4.109, $\int (U^n)^- dU^n \xrightarrow{ucp} 0$. Since

$$[U^n, U^n] = (U^n)^2 - 2 \int (U^n)^- dU^n,$$

it follows that $[U^n, U^n] \xrightarrow{ucp} 0$ and so

$$\mathbf{d}_{var}([X^n - X, X^n - X], 0) \to 0. \tag{4.9.20}$$

Now, (4.6.17) gives

$$\mathrm{VAR}_{[0,T]}([U^n, Z]) \le \sqrt{[U^n, U^n]_T [Z, Z]_T}$$

and hence $[U^n, U^n] \xrightarrow{ucp} 0$ implies that for all stochastic integrators Z

$$\mathbf{d}_{var}([U^n, Z], 0) \to 0.$$

Since $U^n = X^n - X$, one has $[U^n, Z] = [X^n, Z] - [X, Z]$ and so

$$\mathbf{d}_{var}([X^n, Z] - [X, Z], 0) \to 0. \tag{4.9.21}$$

Noting that

$$[X^n, X^n] = [X^n - X, X^n - X] + 2[X^n, X] - [X, X]$$

we have

$$\begin{aligned}
\mathbf{d}_{var}&([X^n, X^n], [X, X]) \\
&= \mathbf{d}_{var}([X^n, X^n] - [X, X], 0) \\
&= \mathbf{d}_{var}([X^n - X, X^n - X] + [X^n, 2X] - [X, 2X], 0) \\
&\le \mathbf{d}_{var}([X^n - X, X^n - X], 0) + \mathbf{d}_{var}([X^n, 2X] - [X, 2X], 0).
\end{aligned}$$

Thus, using (4.9.20) and (4.9.21) it follows that

$$\mathbf{d}_{var}([X^n, X^n], [X, X]) \to 0. \tag{4.9.22}$$

Now the required relation (4.9.18) follows from (4.9.22) by polarization identity (4.6.6). □

The following result is proven on similar lines.

Theorem 4.112 *Suppose X^n, X, Y^n, Y are stochastic integrators such that $X^n \xrightarrow{em} X$, $Y^n \xrightarrow{em} Y$. Then*

$$\mathbf{d}_{var}(j(X^n, Y^n), j(X, Y)) \to 0. \tag{4.9.23}$$

Proof As seen in (4.6.11)

$$j(X^n - X, X^n - X)_t \le [X^n - X, X^n - X]_t$$

and hence by Theorem 4.111,

$$\mathbf{d}_{var}(j(X^n - X, X^n - X), 0) \to 0.$$

Now proceeding as in the proof of Theorem 4.111, we can first prove that for all stochastic integrators Z,

$$\mathbf{d}_{var}(j(X^n, Z), j(X, Z)) \to 0.$$

Once again, noting that

$$j(X^n, X^n) = j(X^n - X, X^n - X) + 2j(X^n, X) - j(X, X)$$

we conclude that

$$\mathbf{d}_{var}(j(X^n, X^n), j(X, X)) \to 0.$$

The required result, namely (4.9.23), follows from this by polarization identity (4.6.12).   □

Next we will prove that $(X_t) \mapsto (f(t, X_t))$ is a continuous mapping in the Emery topology for smooth functions f. Here is a lemma that would be needed in the proof.

Lemma 4.113 *Let X^n, Y^n be stochastic integrators such that $X^n \xrightarrow{em} X$, $Y^n \xrightarrow{em} Y$. Suppose Z^n is a sequence of r.c.l.l. processes such that $Z^n \xrightarrow{ucp} Z$. Let*

$$A_t^n = \sum_{0 < s \le t} Z_s^n (\Delta X^n)_s (\Delta Y^n)_s$$

$$A_t = \sum_{0 < s \le t} Z_s (\Delta X)_s (\Delta Y)_s.$$

Then $\mathbf{d}_{var}(A^n, A) \to 0$.

Proof Note that writing $B^n = j(X^n, Y^n)$ and $B = j(X, Y)$, we have

$$A^n = \int Z^n dB^n, \quad A = \int Z dB.$$

We have seen in (4.9.23) that $\mathbf{d}_{var}(B^n, B) \to 0$. The conclusion now follows from Proposition 4.110.   □

Theorem 4.114 *Let $X^{(n)} = (X^{1,n}, X^{2,n}, \ldots, X^{d,n})$ be a sequence of $\mathbb{R}^d$-valued r.c.l.l. processes such that $X^{j,n}$ are stochastic integrators for $1 \le j \le d$, $n \ge 1$. Let*

$$X^{j,n} \xrightarrow{em} X^j, \ 1 \le j \le d.$$

Let $f \in C^{1,2}([0, \infty) \times \mathbb{R}^d)$. Let

$$Z_t^n = f(t, X_t^{(n)}), \quad Z_t = f(t, X_t)$$

where $X = (X^1, X^2, \ldots, X^d)$, Then $Z^n \xrightarrow{em} Z$.

Proof By Ito's formula, (4.8.33) (see also (4.8.34)), we can write

$$Z_t^n = Z_0^n + A_t^n + Y_t^n + B_t^n + V_t^n, \quad Z_t = Z_0 + A_t + Y_t + B_t + V_t$$

where

$$A_t^n = \int_0^t f_0(s, X_s^{(n)}) ds$$

$$A^t = \int_0^t f_0(s, X_s) ds$$

$$Y_t^n = \sum_{j=1}^{d} \int_{0+}^{t} f_j(s, X_{s-}^{(n)}) dX_s^{j,n}$$

$$Y_t = \sum_{j=1}^{d} \int_{0+}^{t} f_j(s, X_{s-}) dX_s^{j}$$

$$B_t^n = \frac{1}{2} \sum_{j=1}^{d} \sum_{k=1}^{d} \int_{0+}^{t} f_{jk}(s, X_{s-}^{(n)}) d[X^{j,n}, X^{k,n}]_s$$

$$B_t = \frac{1}{2} \sum_{j=1}^{d} \sum_{k=1}^{d} \int_{0}^{t} f_{jk}(s, X_{s-}) d[X^j, X^k]_s$$

$$V^n = \sum_{0 < s \le t} h(s, X_s^{(n)}, X_{s-}^{(n)})$$

$$V = \sum_{0 < s \le t} h(s, X_s, X_{s-}).$$

Since $X^n \xrightarrow{em} X$, it follows that X_0^n converges to X_0 in probability and hence Z_0^n converges to Z_0 in probability.

By continuity of f_0, f_j, f_{jk} (the partial derivatives of $f(t, x)$ w.r.t. t, x_j and x_j, x_k, respectively), and Proposition 4.52 we have

$$f_0(\cdot, X_\cdot^{(n)}) \xrightarrow{ucp} f_0(\cdot, X_\cdot);$$

$$f_j(\cdot, X_\cdot^{(n)}) \xrightarrow{ucp} f_j(\cdot, X_\cdot), \ 1 \le j \le d;$$

$$f_{jk}(\cdot, X_\cdot^{(n)}) \xrightarrow{ucp} f_{jk}(\cdot, X_\cdot), \ 1 \le j, k \le d.$$

Also, $X^{(n)} \xrightarrow{em} X$ implies that for all j, k, $1 \le j, k \le d$, $[X^{j,n}, X^{k,n}] \xrightarrow{em} [X^j, X^k]$ as seen in (4.9.19), Theorem 4.111. Thus, by Theorem 4.109, it follows that

$$A^n \xrightarrow{em} A, \ Y^n \xrightarrow{em} Y, \ B^n \xrightarrow{em} B. \tag{4.9.24}$$

If X^n, X were continuous processes so that V^n, V are identically equal to zero then the result follows from (4.9.24). For the r.c.l.l. case, recall that h can be expressed as

$$h(t, y, x) = \sum_{j,k=1}^{d} g_{jk}(t, y, x)(y^j - x^j)(y^k - x^k)$$

where g_{jk} are defined by (4.8.24). Thus,

$$V^n = \sum_{j,k=1}^{d} \sum_{0<s\leq t} g_{jk}(s, X_s^{(n)}, X_{s-}^{(n)})(\Delta X^{j,n})_s (\Delta X^{k,n})_s,$$

$$V = \sum_{j,k=1}^{d} \sum_{0<s\leq t} g_{jk}(s, X_s, X_{s-})(\Delta X^j)_s (\Delta X^k)_s.$$

Now Lemma 4.113 implies that $\mathbf{d}_{var}(V^n, V) \to 0$ and as a consequence, $V^n \xrightarrow{em} V$. Now combining this with (4.9.24) we finally get $Z^n \xrightarrow{em} Z$. $\square$

Essentially the same proof (invoking Proposition 4.52) would yield the slightly stronger result. Recall that for $g \in C^{1,2}([0, \infty) \times \mathbb{U})$, g_0 denotes the partial derivative of g in the t variable, g_j denotes the partial derivative of g w.r.t. jth coordinate of $x = (x^1, \ldots, x^d)$, and g_{jk} denotes the partial derivative of g_j w.r.t. kth coordinate of $x = (x^1, \ldots, x^d)$.

Theorem 4.115 *Let $X^{(n)} = (X^{1,n}, X^{2,n}, \ldots, X^{d,n})$ be a sequence of $\mathbb{R}^d$-valued r.c.l.l. processes such that $X^{j,n}$ are stochastic integrators for $1 \leq j \leq d$, $n \geq 1$ such that*

$$X^{j,n} \xrightarrow{em} X^j, \ 1 \leq j \leq d.$$

Let $X = (X^1, X^2, \ldots, X^d)$. Let $f^n, f \in C^{1,2}([0, \infty) \times \mathbb{R}^d)$ be functions such that $f^n, f_0^n, f_j^n, f_{jk}^n$ converge to f, f_0, f_j, f_{jk} (respectively) uniformly on compact subsets of $[0, \infty) \times \mathbb{R}^d$. Let

$$Z_t^n = f^n(t, X_t^{(n)}), \ Z_t = f(t, X_t).$$

Then $Z^n \xrightarrow{em} Z$.

4.10   Extension Theorem

We have defined stochastic integrator as an r.c.l.l. process X for which J_X defined by (4.2.1), (4.2.2) for $f \in \mathbb{S}$ admits an extension to $\mathbb{B}(\widetilde{\Omega}, \mathcal{P})$ satisfying (4.2.3). Indeed, just assuming that J_X satisfies

$$f^n \xrightarrow{bp} 0 \text{ implies } J_X(f^n) \xrightarrow{ucp} 0 \tag{4.10.1}$$

it can be shown that J_X admits a required extension. The next exercise gives steps to construct such an extension. The steps are like the usual proof of Caratheodory extension theorem from measure theory. This exercise can be skipped on the first reading.

Exercise 4.116 Let X be an r.c.l.l. adapted process. Let $J_X(f)$ be defined by (4.2.1), (4.2.2) for $f \in \mathbb{S}$. Suppose X satisfies

$$f^n \in \mathbb{S}, \quad f^n \xrightarrow{bp} 0 \text{ implies } J_X(f^n) \xrightarrow{ucp} 0. \tag{4.10.2}$$

For $f \in \mathbb{S}$, let

$$\Gamma_X(f) = \sup\{\mathbf{d}_{ucp}(J_X(g), 0) \ : \ |g| \le |f|\} \tag{4.10.3}$$

and for $\xi \in \mathbb{B}(\widetilde{\Omega}, \mathcal{P})$, let

$$\Gamma_X^*(\xi) = \inf\{\sum_{n=1}^{\infty} \Gamma_X(g^n) \ : \ g^n \in \mathbb{S}, \ |\xi| \le \sum_{n=1}^{\infty} |g^n|\}. \tag{4.10.4}$$

Let

$$\mathbb{A} = \{\xi \in \mathbb{B}(\widetilde{\Omega}, \mathcal{P}) \ : \ \exists f^m \in \mathbb{S} \text{ s.t. } \Gamma_X^*(\xi - f^m) \to 0\}$$

(i) Let $f^n, f \in \mathbb{S}$. Show that

$$f^n \xrightarrow{bp} f \text{ implies } J_X(f^n) \xrightarrow{ucp} J_X(f). \tag{4.10.5}$$

(ii) Let $f, f^1, f^2 \in \mathbb{S}$ and $c \in \mathbb{R}$. Show that

$$\Gamma_X(f) = \Gamma_X(|f|). \tag{4.10.6}$$
$$\Gamma_X(cf) = |c|\Gamma_X(|f|). \tag{4.10.7}$$
$$|f^1| \le |f^2| \Rightarrow \Gamma_X(f^1) \le \Gamma_X(f^2). \tag{4.10.8}$$
$$\Gamma_X(f^1 + f^2) \le \Gamma_X(f^1) + \Gamma_X(f^2). \tag{4.10.9}$$

(iii) Show that for $\xi_1, \xi_2 \in \mathbb{B}(\widetilde{\Omega}, \mathcal{P})$

$$\Gamma_X^*(\xi_1 + \xi_2) \le \Gamma_X^*(\xi_1) + \Gamma_X^*(\xi_2).$$

(iv) Show that $\mathbb{A}$ is a vector space.

(v) Let $f^n \in \mathbb{S}$ for $n \ge 1$ be such that $f^n \xrightarrow{bp} 0$. Show that

$$\Gamma_X(f^n) \to 0. \tag{4.10.10}$$

(vi) Let $h^n, h \in \mathbb{S}$ be such that $h^n \xrightarrow{bp} h$. Show that

$$\Gamma_X(h^n) \to \Gamma_X(h). \tag{4.10.11}$$

(vii) Let $f^n \in \mathbb{S}$ be such that $f^n \le f^{n+1} \le K$ for all n, where K is a constant. Show that $\forall \epsilon > 0 \ \exists n^*$ such that for $m, n \ge n^*$, we have

$$\Gamma_X(f^m - f^n) < \epsilon. \tag{4.10.12}$$

[Hint: Prove by contradiction. If not true, get $\epsilon > 0$ and subsequences $\{n^k\}$, $\{m^k\}$ both increasing to ∞ such that

$$\Gamma_x(f^{n^k} - f^{m^k}) \geq \epsilon \quad \forall k \geq 1.$$

Observe that g^k defined by $g^k = f^{n^k} - f^{m^k}$ satisfy $g^k \xrightarrow{bp} 0$.]

(viii) For $\xi^1, \xi^2 \in \mathbb{B}(\widetilde{\Omega}, \mathcal{P})$

$$|\xi^1| \leq |\xi^2| \Rightarrow \Gamma_x^*(\xi^1) \leq \Gamma_x^*(\xi^2). \tag{4.10.13}$$

(ix) Let $\xi^j, \xi \in \mathbb{B}(\widetilde{\Omega}, \mathcal{P})$ be such that $\xi = \sum_{j=1}^{\infty} |\xi^j|$. Then

$$\Gamma_x^*\left(\sum_{j=1}^{\infty} \xi^j\right) \leq \sum_{j=1}^{\infty} \Gamma_x^*(\xi^j). \tag{4.10.14}$$

(x) Let $g \in \mathbb{S}$ and $f^n \in \mathbb{S}$, $n \geq 1$, satisfy

$$|g| \leq \sum_{n=1}^{\infty} |f^n|. \tag{4.10.15}$$

Show that

$$\Gamma_x(|g|) \leq \sum_{k=1}^{\infty} \Gamma_x(|f^k|). \tag{4.10.16}$$

[Hint: Let $g^n = \min(|g|, \sum_{k=1}^{n} |f^k|)$ and note that $g^n \xrightarrow{bp} |g|$.]

(xi) For $f \in \mathbb{S}$ show that
$$\Gamma_x^*(f) = \Gamma_x(f). \tag{4.10.17}$$

(xii) For $f \in \mathbb{S}$ show that

$$\mathbf{d}_{ucp}(J_X(f), 0) \leq \Gamma_x^*(f). \tag{4.10.18}$$

(xiii) For $\xi \in \mathbb{A}$, define
$$J_X(\xi) = \lim_{m \to \infty} J_X(f^m)$$

in $\mathbf{d}_{ucp}$ where $f^m \in \mathbb{S}$ are such that $\Gamma_x^*(\xi - f^m) \to 0$. Show that J_X is well defined on $\mathbb{A}$ and that

$$\mathbf{d}_{ucp}((J_X(\xi), 0) \leq \Gamma_x^*(\xi) \quad \forall \xi \in \mathbb{A}. \tag{4.10.19}$$

(xiv) Let $h^m = 1_{[0,m]}$. If $\xi \in \mathbb{B}(\widetilde{\Omega}, \mathcal{P})$ is such that $|\xi| \leq K$, then for each $m \geq 1$ show that
$$\Gamma_x^*(\xi) \leq \Gamma_x(K|h^m|) + 2^{-m+1}. \tag{4.10.20}$$

(xv) Let $\xi \in \mathbb{B}(\widetilde{\Omega}, \mathcal{P})$ and let $a_j \in \mathbb{R}$ be such that $a_j \to 0$. Show that

$$\lim_{j \to \infty} \Gamma_x^*(a_j \xi) = 0. \tag{4.10.21}$$

(xvi) Let $\xi, \xi^n \in \mathbb{B}(\widetilde{\Omega}, \mathcal{P})$ for $n \geq 1$ be such that $\xi^n \to \xi$ uniformly. Show that

$$\lim_{n \to \infty} \Gamma_x^*(\xi^n - \xi) = 0. \tag{4.10.22}$$

(xvii) Let $\xi \in \mathbb{B}(\widetilde{\Omega}, \mathcal{P})$ and $\xi^n \in \mathbb{A}$ be such that $\lim_{m \to \infty} \Gamma_x^*(\xi^m - \xi) = 0$. Show that $\xi \in \mathbb{A}$.

(xviii) Let $\xi^n \in \mathbb{A}$ be such that $\xi^n \leq \xi^{n+1}$ $\forall n \geq 1$ and let

$$\xi = \lim_{n \to \infty} \xi^n \in \mathbb{B}(\widetilde{\Omega}, \mathcal{P}). \tag{4.10.23}$$

(a) Given $\epsilon' > 0$, show that there exists a sequence $\{g^m\}$ in $\mathbb{S}$ such that $g^n \leq g^{n+1}$ $\forall n \geq 1$ and $\Gamma_x^*(g^n - \xi^n) \leq \epsilon'$.

(b) Given ϵ show that $\exists n^*$ such that for $n, m \geq n^*$ we have

$$\Gamma_x^*(\xi^m - \xi^n) < \epsilon.$$

[Hint: Use (a) above along with (vii).]

(c) For $k \geq 1$ show that $\exists n^k$ such that $n^k > n^{k-1}$ (here $n^0 = 1$) and

$$\Gamma_x^*(\xi^{n^k} - \xi^{n^{k-1}}) < 2^{-k}.$$

(d) Show that

$$\Gamma_x^*(\xi - \xi^{n^k}) \leq \sum_{m=k}^{\infty} \Gamma_x^*(\xi^{n^{m+1}} - \xi^{n^m}) \leq 2^{-k+1}. \tag{4.10.24}$$

(e) Show that $\xi \in \mathbb{A}$.

(f) Show that $\Gamma_x^*(\xi - \xi^n) \to 0$.
[Hint: Use (d) above to conclude that every subsequence of $a_n = \Gamma_x^*(\xi - \xi^n)$ has a further subsequence converging to 0.]

(xix) Show that $\mathbb{A} = \mathbb{B}(\widetilde{\Omega}, \mathcal{P})$.
[Hint: Use Monotone Class Theorem 2.63].

(xx) Let $\eta^n \in \mathbb{B}(\widetilde{\Omega}, \mathcal{P})$ be a sequence such that $\eta^n \geq \eta^{n+1} \geq 0$ for all n with $\lim_n \eta^n = 0$. Show that
$$\Gamma^*(\eta^n) \to 0.$$

[Hint: Let $\xi^n = \eta^1 - \eta^n$ and use (f) above].

(xxi) Let $f^n, f \in \mathbb{B}(\widetilde{\Omega}, \mathcal{P})$ be such that $f^n \xrightarrow{bp} f$. Show that

$$\Gamma^*(f^n - f) \to 0. \tag{4.10.25}$$

[Hint: Let $g^n = \sup_{m \geq n} |f^m - f|$. Then use $g^n \downarrow 0$ and $|f^m - f| \leq g^m$.]

(xxii) Let $f^n, f \in \mathbb{B}(\widetilde{\Omega}, \mathcal{P})$ be such that $f^n \xrightarrow{bp} f$. Show that $\mathbb{J}_X(f^n) \xrightarrow{ucp} \mathbb{J}_X(f)$.

(xxiii) Show that X is a stochastic integrator.

Chapter 5
Semimartingales

The reader would have noticed that in the development of stochastic integration in the previous chapter, we have not talked about either martingales or semimartingales.

A semimartingale is any process which can be written as a sum of a local martingale and a process with finite variation paths.

The main theme of this chapter is to show that the class of stochastic integrators is the same as the class of semimartingales, thereby showing that stochastic integral is defined for all semimartingales and the Ito's formula holds for them. This is the Dellacherie–Meyer–Mokobodzky–Bichteler Theorem.

Traditionally, the starting point for integration with respect to square integrable martingales is the Doob–Meyer decomposition theorem. We follow a different path, proving that for a square integrable martingale M, the quadratic variation $[M, M]$ can be defined directly and then $X_t = M_t^2 - [M, M]_t$ is itself a martingale. This along with Doob's maximal inequality would show that square integrable martingales (and locally square integrable martingales) are integrators. We would then go on to show that a local martingale (and hence any semimartingale) is an integrator.

Next we show that every stochastic integrator is a semimartingale, thus proving a weak version of the Dellacherie–Meyer–Mokobodzky–Bichteler Theorem. Subsequently, we prove the full version of this result.

5.1 Notations and Terminology

We begin with some definitions.

Definition 5.1 An r.c.l.l. adapted process M is said to be a square integrable martingale if M is a martingale and

$$\mathsf{E}[M_t^2] < \infty \quad \forall t < \infty.$$

© Springer Nature Singapore Pte Ltd. 2018
R. L. Karandikar and B. V. Rao, *Introduction to Stochastic Calculus*,
Indian Statistical Institute Series, https://doi.org/10.1007/978-981-10-8318-1_5

Exercise 5.2 Let M be a square integrable martingale. Show that

(i) $E[\sup_{0 \le t \le T} M_t^2] < \infty$ for each $T < \infty$.
(ii) $\sup_{0 \le t \le T} E[M_t^2] < \infty$ for each $T < \infty$.

Definition 5.3 An r.c.l.l. adapted process L is said to be a local martingale if there exist stopping times $\tau^n \uparrow \infty$ such that for each n, the process M^n defined by $M^n = L^{[\tau^n]}$; i.e. $M_t^n = L_{t \wedge \tau^n}$ is a martingale. Such a sequence $\{\tau^n\}$ is called a localizing sequence.

Exercise 5.4 Let W be the Brownian motion and let $U = \exp(W_1^2)$. Let

$$L_t = \begin{cases} W_t & \text{if } t < 1 \\ W_1 + U(W_t - W_1) & \text{if } t \ge 1. \end{cases}$$

Let $\tau^n = 1$ if $U \ge n$ and $\tau^n = n$ if $U < n$. Let $M^n = L^{[\tau^n]}$ be the stopped process. Show that

(i) For each n, τ^n is a stopping time for the filtration $(\mathcal{F}_\cdot^W)$.
(ii) $\tau^n \uparrow \infty$.
(iii) M^n is a martingale (w.r.t. the filtration $(\mathcal{F}_\cdot^W)$).
(iv) $E[|L_t|] = \infty$ for $t > 1$

This gives an example of a local martingale that is not a martingale.

Here is a simple condition under which a local martingale is a martingale.

Lemma 5.5 *Let an r.c.l.l. process L be a local martingale such that*

$$E[\sup_{0 \le s \le t} |L_s|] < \infty \ \forall t < \infty. \tag{5.1.1}$$

Then L is a martingale.

Proof Let $\tau^n \uparrow \infty$ such that for each n, the process M^n defined by $M^n = L^{[\tau^n]}$ is a martingale. Then M_t^n converges to L_t pointwise and (5.1.1) implies that $\sup_{0 \le s \le t} |L_s|$ serves as a dominating function. Thus, for each t, M_t^n converges to L_t in $\mathbb{L}^1(\mathsf{P})$. Now the required result follows using Theorem 2.23. $\square$

Lemma 5.6 *Let an r.c.l.l. process L be a local martingale such that for some $p > 1$*

$$\sup_{\tau \in \mathbb{T}_b} E[|L_\tau|^p] < \infty \tag{5.1.2}$$

where $\mathbb{T}_b$ denotes the class of all bounded stopping times. Then L is a uniformly integrable martingale.

Proof Fix $\sigma \in \mathbb{T}_b$ and let τ_n be stopping times increasing to ∞ such that $L^{[\tau_n]}$ are martingales. Then

$$\mathsf{E}[L_{\sigma \wedge \tau_n}] = \mathsf{E}[L_0] \quad \forall n. \tag{5.1.3}$$

Since $\{L_{\sigma \wedge \tau_n} : n \geq 1\}$ is $\mathbb{L}^p$ bounded and converges to L_σ as $n \to \infty$, it follows that (see Exercise 1.30) $\mathsf{E}[L_\sigma] = \mathsf{E}[L_0]$. Using Theorem 2.57 it follows that L is a martingale. Since $\{L_t : t \geq 0\}$ is $\mathbb{L}^p$ bounded, we conclude that L is uniformly integrable. $\qquad \square$

Here is another observation on positive local martingales.

Lemma 5.7 *Let L be an r.c.l.l. local martingale such that $\mathsf{P}(L_t \geq 0 \ \forall t) = 1$. Then*

(i) For $\sigma \in \mathbb{T}_b$

$$\mathsf{E}[L_\sigma] \leq \mathsf{E}[L_0]. \tag{5.1.4}$$

(ii) L is a supermartingale.
(iii) If $\mathsf{E}[L_t] = \mathsf{E}[L_0]$ for all $t \geq 0$ then L is a martingale.

Proof Let τ_n be as in the proof of Lemma 5.6 above. Since $L_t \geq 0$, (5.1.4) follows by using Fatou's lemma in the relation (5.1.3). As a consequence, L_t is integrable for each t. Also, for $s \leq t$ we have

$$\mathsf{E}[L_{t \wedge \tau_n} \mid \mathcal{F}_s] = L_{s \wedge \tau_n}.$$

It follows that

$$\mathsf{E}[\inf_{m \geq n} L_{t \wedge \tau_m} \mid \mathcal{F}_s] \leq L_{s \wedge \tau_n}. \tag{5.1.5}$$

Since $L_{t \wedge \tau_m}$ converges to L_t as $m \to \infty$, writing $\xi_n = \inf_{m \geq n} L_{t \wedge \tau_m}$, it follows that ξ_n increases to L_t and thus

$$\mathsf{E}[\inf_{m \geq n} L_{t \wedge \tau_m} \mid \mathcal{F}_s] \uparrow \mathsf{E}[L_t \mid \mathcal{F}_s].$$

Taking limit in (5.1.5), we conclude

$$\mathsf{E}[L_t \mid \mathcal{F}_s] \leq L_s. \tag{5.1.6}$$

This proves *(ii)*. If $\mathsf{E}[L_t] = \mathsf{E}[L_s]$, then (5.1.6) implies $\mathsf{E}[L_t \mid \mathcal{F}_s] = L_s$ and thus *(iii)* follows. $\qquad \square$

Definition 5.8 An r.c.l.l. adapted process N is said to be a locally square integrable martingale if there exist stopping times $\tau^n \uparrow \infty$ such that for each n, the process M^n defined by $M^n = N^{[\tau^n]}$; i.e. $M_t^n = N_{t \wedge \tau^n}$ is a square integrable martingale.

Exercise 5.9 Show that the process L constructed in Exercise 5.4 is a locally square integrable martingale.

Definition 5.10 An r.c.l.l. adapted process X is said to be a semimartingale if X can be written as $X = M + A$ where M is an r.c.l.l. local martingale and A is an r.c.l.l. process whose paths have finite variation on $[0, T]$ for all $T < \infty$, i.e. $\text{VAR}_{[0,T]}(A) < \infty$ for all $T < \infty$.

Let us denote by $\mathbb{M}$ the class of all r.c.l.l. martingales M with $M_0 = 0$, $\mathbb{M}^2$ the class of all r.c.l.l. square integrable martingales with $M_0 = 0$. We will also denote by $\mathbb{M}_{\text{loc}}$ the class of r.c.l.l. local martingales with $M_0 = 0$ and $\mathbb{M}^2_{\text{loc}}$ the class of r.c.l.l. locally square integrable martingales with $M_0 = 0$. Thus, $M \in \mathbb{M}^2_{\text{loc}}$ ($\mathbb{M}_{\text{loc}}$) if there exist stopping times τ^n increasing to infinity such that the stopped processes $M^{[\tau^n]}$ belong to $\mathbb{M}^2$ (respectively belong to $\mathbb{M}$). The sequence τ^n is called a localizing sequence. Let $\mathbb{M}_c$ be the class of all continuous martingales M with $M_0 = 0$, $\mathbb{M}_{c,\text{loc}}$ be the class of all continuous local martingales M with $M_0 = 0$ and $\mathbb{M}^2_c$ be the class of square integrable continuous martingales M with $M_0 = 0$.

Exercise 5.11 Show that $\mathbb{M}_{c,\text{loc}} \subseteq \mathbb{M}^2_{\text{loc}}$.

Thus X is a semimartingale if we can write $X = M + A$ where $M \in \mathbb{M}_{\text{loc}}$ and $A \in \mathbb{V}$. We will first show that semimartingales are stochastic integrators. Recall that all semimartingales and stochastic integrators are by definition r.c.l.l. processes. We begin by showing that square integrable r.c.l.l. martingales are stochastic integrators. Usually this step is done involving Doob–Meyer decomposition theorem. We bypass the same by a study of quadratic variation as a functional on the path space

5.2   The Quadratic Variation Map

Let $\mathbb{D}([0, \infty), \mathbb{R})$ denote the space of r.c.l.l. functions on $[0, \infty)$. Recall our convention that for $\gamma \in \mathbb{D}([0, \infty), \mathbb{R})$, $\gamma(t-)$ denotes the left limit at t (for $t > 0$) and $\gamma(0-) = 0$ and $\Delta\gamma(t) = \gamma(t) - \gamma(t-)$. Note that by definition, $\Delta\gamma(0) = \gamma(0)$.

Exercise 5.12 Suppose $f_n \in \mathbb{D}([0, \infty), \mathbb{R})$ are such that f_n converges to a function f uniformly on compact subsets of $[0, \infty)$; i.e. $\sup_{0 \leq t \leq T} |f_n(t) - f(t)|$ converges to zero for all $T < \infty$. Show that

(i) $f \in \mathbb{D}([0, \infty), \mathbb{R})$.
(ii)   Let $s \in [0, \infty)$. If $s_n \in [0, s)$ converges to s then $f_n(s_n)$ converges to $f(s-)$.
(iii) $f_n(s-)$ converges to $f(s-)$ for all $s \in [0, \infty)$.
(iv) $(\Delta f_n)(s)$ converges to $(\Delta f)(s)$ for all $s \in [0, \infty)$.

We will now define *quadratic variation* $\Psi(\gamma)$ of a function $\gamma \in \mathbb{D}([0, \infty), \mathbb{R})$.

For each $n \geq 1$; let $\{t^n_i(\gamma) : i \geq 1\}$ be defined inductively as follows: $t^n_0(\gamma) = 0$ and having defined $t^n_i(\gamma)$, let

$$t^n_{i+1}(\gamma) = \inf\{t > t^n_i(\gamma) : |\gamma(t) - \gamma(t^n_i(\gamma))| \geq 2^{-n} \text{ or } |\gamma(t-) - \gamma(t^n_i(\gamma))| \geq 2^{-n}\}.$$

If $\lim_i t_i^n(\gamma) = t^* < \infty$, then the function γ cannot have a left limit at t^*. Hence for each $\gamma \in \mathbb{D}([0, \infty), \mathbb{R})$, $t_i^n(\gamma) \uparrow \infty$ as $i \uparrow \infty$ for each n. Let

$$\Psi_n(\gamma)(t) = \sum_{i=0}^{\infty} (\gamma(t_{i+1}^n(\gamma) \wedge t) - \gamma(t_i^n(\gamma) \wedge t))^2. \tag{5.2.1}$$

Since $t_i^n(\gamma)$ increases to infinity, for each γ and n fixed, the infinite sum appearing above is essentially a finite sum and hence $\Psi_n(\gamma)$ is itself an r.c.l.l. function. Recall that the space $\mathbb{D} = \mathbb{D}([0, \infty), \mathbb{R})$ is equipped with the topology of uniform convergence on compact subsets (abbreviated as *ucc*). Let $\widetilde{\mathbb{D}}$ denote the set of $\gamma \in \mathbb{D}$ such that $\Psi_n(\gamma)$ converges in the ucc topology and

$$\Psi(\gamma) = \begin{cases} \lim_n \Psi_n(\gamma) & \text{if } \gamma \in \widetilde{\mathbb{D}} \\ 0 & \text{if } \gamma \notin \widetilde{\mathbb{D}}. \end{cases} \tag{5.2.2}$$

Here are some basic properties of the quadratic variation map Ψ.

Lemma 5.13 *For $\gamma \in \widetilde{\mathbb{D}}$*

(i) $\Psi(\gamma)$ *is an increasing r.c.l.l. function.*
(ii) $\Delta\Psi(\gamma)(t) = (\Delta\gamma(t))^2$ *for all $t \in (0, \infty)$.*
(iii) $\sum_{s \le t}(\Delta\gamma(s))^2 < \infty$ *for all $t \in (0, \infty)$.*
(iv) *Let $\Psi_c(\gamma)(t) = \Psi(\gamma)(t) - \sum_{0 < s \le t}(\Delta\gamma(s))^2$. Then $\Psi_c(\gamma)$ is a continuous function.*

Proof For *(i)*, note that for $s \le t$, if $t_j^n \le s < t_{j+1}^n$, then $|(\gamma(s) - \gamma(t_j^n))| \le 2^{-n}$, and

$$\Psi_n(\gamma)(s) = \sum_{i=0}^{j-1}(\gamma(t_{i+1}^n(\gamma)) - \gamma(t_i^n(\gamma)))^2 + (\gamma(s) - \gamma(t_j^n))^2$$

$$\Psi_n(\gamma)(t) = \sum_{i=0}^{j-1}(\gamma(t_{i+1}^n(\gamma)) - \gamma(t_i^n(\gamma)))^2$$

$$+ \sum_{i=j}^{\infty}(\gamma(t_{i+1}^n(\gamma) \wedge t) - \gamma(t_i^n(\gamma) \wedge t))^2$$

and hence

$$\Psi_n(\gamma)(s) \le \Psi_n(\gamma)(t) + 2^{-2n}. \tag{5.2.3}$$

Compare with (4.6.4). Thus (5.2.3) is valid for all $n \ge 1$ and $s \le t$. Hence it follows that the limiting function $\Psi(\gamma)$ is an increasing function. Convergence of the r.c.l.l. function $\Psi_n(\gamma)$ to $\Psi(\gamma)$ in ucc topology implies that $\Psi(\gamma)$ is an r.c.l.l. function.

For *(ii)*, it is easy to see that the set of points of discontinuity of $\Psi_n(\gamma)$ are contained in the set of points of discontinuity of γ for each n. Uniform convergence of $\Psi_n(\gamma)(t)$ to $\Psi(\gamma)(t)$ for $t \in [0, T]$ for every $T < \infty$ implies that the same is true for $\Psi(\gamma)$; i.e. for $t > 0$, $\Delta\Psi(\gamma)(t) \neq 0$ implies that $\Delta\gamma(t) \neq 0$.

On the other hand, let $t > 0$ be a discontinuity point for γ. Let us note that by the definition of $t_j^n(\gamma)$,

$$|\gamma(u) - \gamma(v)| \leq 2.2^{-n} \ \forall u, v \in [t_j^n(\gamma), t_{j+1}^n(\gamma)). \tag{5.2.4}$$

Thus for n such that $2.2^{-n} < \Delta\gamma(t)$, t must be equal to $t_k^n(\gamma)$ for some $k \geq 1$ since (5.2.4) implies $\Delta\gamma(v) \leq 2.2^{-n}$ for any $v \in \cup_j(t_j^n(\gamma), t_{j+1}^n(\gamma))$. Let $s_n = t_{k-1}^n(\gamma)$ where k (depending on n) is such that $t = t_k^n(\gamma)$. Note that $s_n < t$. Let $s^* = \liminf_n s_n$. We will prove that

$$\lim_n \gamma(s_n) = \gamma(t-), \quad \lim_n \Psi_n(\gamma)(s_n) = \Psi(\gamma)(t-). \tag{5.2.5}$$

If $s^* = t$, then $s_n < t$ for all $n \geq 1$ implies $s_n \to t$. Thus $\lim_n \gamma(s_n) = \gamma(t-)$. Second part of (5.2.5) follows from uniform convergence of $\Psi_n(\gamma)$ to $\Psi(\gamma)$ on $[0, t]$ (see Exercise 5.12).

If $s^* < t$, using (5.2.4) it follows that $|\gamma(u) - \gamma(v)| = 0$ for $u, v \in (s^*, t)$ and hence the function $\gamma(u)$ is constant on the interval (s^*, t) and implying that $s_n \to s^*$. Also, $\gamma(s^*) = \gamma(t-)$ and $\Psi(\gamma)(s^*) = \Psi(\gamma)(t-)$. So if γ is continuous at s^*, once again uniform convergence of $\Psi_n(\gamma)$ to $\Psi(\gamma)$ on $[0, t]$ shows that (5.2.5) is valid in this case too.

Remains to consider the case $s^* < t$ and $\Delta\gamma(s^*) = \delta > 0$. In this case, for n such that $2.2^{-n} < \delta$, $s_n = s^*$ and uniform convergence of $\Psi_n(\gamma)$ to $\Psi(\gamma)$ on $[0, t]$ shows that (5.2.5) is true in this case as well.

We have (for large n)

$$\Psi_n(\gamma)(t) = \Psi_n(\gamma)(s_n) + (\gamma(s_n) - \gamma(t))^2 \tag{5.2.6}$$

and hence (5.2.5) yields

$$\Psi(\gamma)(t) = \Psi(\gamma)(t-) + [\Delta\gamma(t)]^2$$

completing the proof of (ii).

(iii) follows from (i) and (ii) since for an increasing function that is non-negative at zero, the sum of jumps up to t is almost equal to its value at t:

$$\sum_{0 < s \leq t} (\Delta\gamma(s))^2 \leq \Psi(\gamma)(t).$$

The last part (iv) follows from (ii) and (iii). $\square$

Remark **5.14** Ψ is the *quadratic variation* map. It may depend upon the choice of the partitions. If, instead of 2^{-n}, we had used any other sequence $\{\varepsilon_n\}$, it would yield another mapping $\tilde{\Psi}$ which will have similar properties. Our proof in the next section will show that if $\sum_n \varepsilon_n < \infty$, then for a square integrable local martingale (M_t),

$$\Psi(M.(\omega)) = \tilde{\Psi}(M.(\omega)) \ a.s. \ P.$$

We note two more properties of the quadratic variation map Ψ. Recall that the total variation $\mathrm{VAR}_{(0,T]}(\gamma)$ of γ on the interval $(0, T]$ is defined by

$$\mathrm{VAR}_{(0,T]}(\gamma) = \sup\{\sum_{j=0}^{m-1} |\gamma(s_{j+1}) - \gamma(s_j)| : 0 \le s_1 \le s_2 \le \ldots s_m = T, m \ge 1\}.$$

If $\mathrm{VAR}_{(0,T]}(\gamma) < \infty$, γ is said to have finite variation on $[0, T]$ and then on $[0, T]$ it can be written as difference of two increasing functions.

Lemma 5.15 *The quadratic variation map Ψ satisfies the following.*

(i) *For $\gamma \in \tilde{\mathbb{D}}$ and $0 < s < \infty$ fixed, let $\gamma^s \in \mathbb{D}$ be defined by: $\gamma^s(t) = \gamma(t \wedge s)$. Then $\gamma^s \in \tilde{\mathbb{D}}$.*

(ii) *For $\gamma \in \mathbb{D}$ and $s_k \uparrow \infty$, γ^k be defined via $\gamma^k(t) = \gamma(t \wedge s_k)$. If $\gamma^k \in \tilde{\mathbb{D}}$ for all k, then $\gamma \in \tilde{\mathbb{D}}$ and*

$$\Psi(\gamma)(t \wedge s_k) = \Psi(\gamma^k)(t), \ \forall t < \infty, \ \forall k \ge 1. \tag{5.2.7}$$

(iii) *Suppose γ is continuous, and $\mathrm{VAR}_{(0,T]}(\gamma) < \infty$. Then $\Psi(\gamma)(t) = 0, \ \forall t \in [0, T]$.*

Proof (i) is immediate. For (ii), it can be checked from the definition that

$$\Psi_n(\gamma)(t \wedge s_k) = \Psi_n(\gamma^k)(t), \ \forall t. \tag{5.2.8}$$

Since $\gamma^k \in \tilde{\mathbb{D}}$, it follows that $\Psi_n(\gamma)(t)$ converges uniformly on $[0, s_k]$ for every k and hence using (5.2.8) we conclude that $\gamma \in \tilde{\mathbb{D}}$ and that (5.2.7) holds.

For (iii), note that γ being a continuous function,

$$|\gamma(t_{i+1}^n(\gamma) \wedge t) - \gamma(t_i^n(\gamma) \wedge t)| \le 2^{-n}$$

for all i, n and hence we have

$$\Psi_n(\gamma)(t) = \sum_{i=0}^{\infty} (\gamma(t_{i+1}^n(\gamma) \wedge t) - \gamma(t_i^n(\gamma) \wedge t))^2$$

$$\le 2^{-n} \times \sum_{i=0}^{\infty} |\gamma(t_{i+1}^n(\gamma) \wedge t) - \gamma(t_i^n(\gamma) \wedge t)|$$

$$\le 2^{-n} \times \mathrm{VAR}_{[0,T]}(\gamma).$$

This shows that $\Psi(\gamma)(t) = 0$ for $t \in [0, T]$. $\qquad \square$

5.3   Quadratic Variation of a Square Integrable Martingale

The next lemma connects the quadratic variation map Ψ and r.c.l.l. martingales.

Lemma 5.16 *Let $(N_t, \mathcal{F}_t)$ be an r.c.l.l. martingale such that $E(N_t^2) < \infty$ for all $t > 0$. Suppose there is a constant $C < \infty$ such that with*

$$\tau = \inf\{t > 0 : |N_t| \geq C \text{ or } |N_{t-}| \geq C\}$$

one has

$$N_t = N_{t \wedge \tau}.$$

Let

$$A_t(\omega) = \Psi(N.(\omega))(t).$$

Then (A_t) is an $(\mathcal{F}_t)$ adapted r.c.l.l. increasing process such that $X_t := N_t^2 - A_t$ is also a martingale.

Proof Let $\Psi_n(\gamma)$ and $t_i^n(\gamma)$ be as in the previous section.

$$\begin{aligned}
A_t^n(\omega) &= \Psi_n(N.(\omega))(t) \\
\sigma_i^n(\omega) &= t_i^n(N.(\omega)) \\
Y_t^n(\omega) &= N_t^2(\omega) - N_0^2(\omega) - A_t^n(\omega)
\end{aligned} \tag{5.3.1}$$

It is easy to see that for each n, $\{\sigma_i^n : i \geq 1\}$ are stopping times (see Theorem 2.46) and that

$$A_t^n = \sum_{i=0}^{\infty}(N_{\sigma_{i+1}^n \wedge t} - N_{\sigma_i^n \wedge t})^2.$$

Further, for each n, $\sigma_i^n(\omega)$ increases to ∞ as $i \uparrow \infty$.

We will first prove that for each n, (Y_t^n) is an $(\mathcal{F}_t)$-martingale. Using the identity $b^2 - a^2 - (b-a)^2 = 2a(b-a)$, we can write

$$\begin{aligned}
Y_t^n &= N_t^2 - N_0^2 - \sum_{i=0}^{\infty}(N_{\sigma_{i+1}^n \wedge t} - N_{\sigma_i^n \wedge t})^2 \\
&= \sum_{i=0}^{\infty}(N_{\sigma_{i+1}^n \wedge t}^2 - N_{\sigma_i^n \wedge t}^2) - \sum_{i=0}^{\infty}(N_{\sigma_{i+1}^n \wedge t} - N_{\sigma_i^n \wedge t})^2 \\
&= 2\sum_{i=0}^{\infty} N_{\sigma_i^n \wedge t}(N_{\sigma_{i+1}^n \wedge t} - N_{\sigma_i^n \wedge t})
\end{aligned}$$

Let us define

$$X_t^{n,i} = N_{\sigma_i^n \wedge t}(N_{\sigma_{i+1}^n \wedge t} - N_{\sigma_i^n \wedge t}).$$

Then

$$Y_t^n = 2 \sum_{i=0}^{\infty} X_t^{n,i}. \tag{5.3.2}$$

Noting that for $s < \tau$, $|N_s| \leq C$ and for $s \geq \tau$, $N_s = N_\sigma$, it follows that

$$|N_{\sigma_{i+1}^n \wedge t} - N_{\sigma_i^n \wedge t}| > 0 \text{ implies that } |N_{\sigma_i^n \wedge t}| \leq C.$$

Thus, writing $\Gamma_C(x) = \max\{\min\{x, C\}, -C\}$ (x truncated at C), we have

$$X_t^{n,i} = \Gamma_C(N_{\sigma_i^n \wedge t})(N_{\sigma_{i+1}^n \wedge t} - N_{\sigma_i^n \wedge t}) \tag{5.3.3}$$

and hence, $X_t^{n,i}$ is a martingale. Using the fact that $E(X_t^{n,i} | \mathcal{F}_{t \wedge \sigma_i^n}) = 0$ and that $X_t^{n,i}$ is $\mathcal{F}_{t \wedge \sigma_{i+1}^n}$ measurable, it follows that for $i \neq j$,

$$E[X_t^{n,i} X_t^{n,j}] = 0. \tag{5.3.4}$$

Also, using (5.3.3) and the fact that N is a martingale we have

$$\begin{aligned} E(X_t^{n,i})^2 &\leq C^2 E(N_{\sigma_{i+1}^n \wedge t} - N_{\sigma_i^n \wedge t})^2 \\ &= C^2 E(N_{\sigma_{i+1}^n \wedge t}^2 - N_{\sigma_i^n \wedge t}^2). \end{aligned} \tag{5.3.5}$$

Using (5.3.4) and (5.3.5), it follows that for $s \leq r$,

$$E(\sum_{i=s}^{r} X_t^{n,i})^2 \leq C^2 E(N_{\sigma_{r+1}^n \wedge t}^2 - N_{\sigma_s^n \wedge t}^2). \tag{5.3.6}$$

Since σ_i^n increases to ∞ as i tends to infinity, $E(N_{\sigma_s^n \wedge t}^2)$ and $E(N_{\sigma_{r+1}^n \wedge t}^2)$ both tend to $E[N_t^2]$ as r, s tend to ∞ and hence $\sum_{i=1}^{r} X_t^{n,i}$ converges in $\mathbb{L}^2(P)$. In view of (5.3.2), one has

$$2 \sum_{i=0}^{r} X_t^{n,i} \to Y_t^n \text{ in } \mathbb{L}^2(P) \text{ as } r \to \infty$$

and hence (Y_t^n) is an $(\mathcal{F}_t)$-martingale for each $n \geq 1$.

For $n \geq 1$, define a process N^n by

$$N_t^n = N_{\sigma_i^n} \text{ if } \sigma_i^n \leq t < \sigma_{i+1}^n.$$

Observe that by the choice of $\{\sigma_i^n : i \geq 1\}$, one has

$$|N_t - N_t^n| \leq 2^{-n} \text{ for all } t. \tag{5.3.7}$$

For now let us fix n. For each $\omega \in \Omega$, let us define

$$H(\omega) = \{\sigma_i^n(\omega) : i \geq 1\} \cup \{\sigma_i^{n+1}(\omega) : i \geq 1\} \tag{5.3.8}$$

It may be noted that for ω such that $t \mapsto N_t(\omega)$ is continuous, each $\sigma_j^n(\omega)$ is necessarily equal to $\sigma_i^{n+1}(\omega)$ for some i, but this need not be the case when $t \mapsto N_t(\omega)$ has jumps. Let $\theta_0(\omega) = 0$ and for $j \geq 0$, let

$$\theta_{j+1}(\omega) = \inf\{s > \theta_j(\omega) : s \in H(\omega)\}.$$

It can be verified that

$$\{\theta_i(\omega) : i \geq 1\} = \{\sigma_i^n(\omega) : i \geq 1\} \cup \{\sigma_i^{n+1}(\omega) : i \geq 1\}. \tag{5.3.9}$$

To see that each θ_i is a stopping time, fix $i \geq 1, t < \infty$. Let

$$A_{kj} = \{(\sigma_k^n \wedge t) \neq (\sigma_j^{n+1} \wedge t)\}.$$

Since $\sigma_k^n, \sigma_j^{n+1}$ are stopping times, $A_{kj} \in \mathcal{F}_t$ for all k, j. It is not difficult to see that

$$\{\theta_i \leq t\} = \cup_{k=0}^{i}(\{\sigma_{i-k}^n \leq t\} \cap B_k)$$

where $B_0 = \Omega$ and for $1 \leq k \leq i$,

$$B_k = \cup_{0 < j_1 < j_2 < \ldots < j_k}((\cap_{l=0}^{i-k} \cap_{m=1}^{k} A_{lj_m}) \cap \{\sigma_{j_k}^{n+1} \leq t\})$$

and hence θ_i is a stopping time.

Using (5.3.9) and using the fact that $N_t^n = N_{t \wedge \sigma_j^n}$ for $\sigma_j^n \leq t < \sigma_{j+1}^n$, one can write Y^n and Y^{n+1} as

$$Y_t^n = \sum_{j=0}^{\infty} 2N_{t \wedge \theta_j}^n (N_{t \wedge \theta_{j+1}} - N_{t \wedge \theta_j}),$$

$$Y_t^{n+1} = \sum_{j=0}^{\infty} 2N_{t \wedge \theta_j}^{n+1} (N_{t \wedge \theta_{j+1}} - N_{t \wedge \theta_j}).$$

Hence

$$Y_t^{n+1} - Y_t^n = 2\sum_{j=0}^{\infty} Z_t^{n,j} \tag{5.3.10}$$

where

$$Z_t^{n,j} = (N_{t \wedge \theta_j}^{n+1} - N_{t \wedge \theta_j}^n)(N_{t \wedge \theta_{j+1}} - N_{t \wedge \theta_j}).$$

Also, using (5.3.7) one has

$$|N_t^{n+1} - N_t^n| \leq |N_t^{n+1} - N_t| + |N_t - N_t^n| \leq 2^{-(n+1)} + 2^{-n} \leq 2.2^{-n} \quad (5.3.11)$$

and hence (using that (N_s) is a martingale), one has

$$\mathsf{E}[(Z_t^{n,j})^2] \leq \frac{4}{2^{2n}} \mathsf{E}[(N_{t\wedge\theta_{j+1}} - N_{t\wedge\theta_j})^2] = \frac{4}{2^{2n}} \mathsf{E}[(N_{t\wedge\theta_{j+1}})^2 - (N_{t\wedge\theta_j})^2]. \quad (5.3.12)$$

It is easy to see that $\mathsf{E}(Z_t^{n,j} | \mathcal{F}_{t\wedge\theta_j}) = 0$ and $Z_t^{n,j}$ is $\mathcal{F}_{t\wedge\theta_{j+1}}$ measurable. It then follows that for $i \neq j$

$$\mathsf{E}[Z_t^{n,j}, Z_t^{n,i}] = 0$$

and hence (using (5.3.12))

$$\mathsf{E}(Y_t^{n+1} - Y_t^n)^2 = 4\mathsf{E}[(\sum_{j=0}^{\infty} Z_t^{n,j})^2]$$

$$= 4\mathsf{E}[\sum_{j=0}^{\infty} (Z_t^{n,j})^2] \qquad (5.3.13)$$

$$\leq \frac{16}{2^{2n}} \sum_{j=0}^{\infty} \mathsf{E}[(N_{t\wedge\theta_{j+1}})^2 - (N_{t\wedge\theta_j})^2]$$

$$\leq \frac{16}{2^{2n}} \mathsf{E}[(N_t)^2].$$

Thus, recalling that Y_t^{n+1}, Y_t^n are martingales, it follows that $Y_t^{n+1} - Y_t^n$ is also a martingale and thus invoking Doob's maximal inequality, one has (using (5.3.13))

$$\mathsf{E}[\sup_{s \leq T} |Y_s^{n+1} - Y_s^n|^2] \leq 4\mathsf{E}(Y_T^{n+1} - Y_T^n)^2$$

$$\leq \frac{64}{2^{2n}} \mathsf{E}[N_T^2]. \qquad (5.3.14)$$

Thus, for each $n \geq 1$,

$$\| [\sup_{s \leq T} |Y_s^{n+1} - Y_s^n|] \|_2 \leq \frac{8}{2^n} \|N_T\|_2. \qquad (5.3.15)$$

It follows that

$$\xi = \sum_{n=1}^{\infty} \sup_{s \leq T} |Y_s^{n+1} - Y_s^n| < \infty \ a.s.$$

as $\|\xi\|_2 < \infty$ by (5.3.15). Hence (Y_s^n) converges uniformly in $s \in [0, T]$ for every T a.s. to an r.c.l.l. process say (Y_s). As a result, (A_s^n) also converges uniformly in $s \in [0, T]$ for every $T < \infty$ a.s. to say $(\tilde{A}_s)$ with $Y_t = N_t^2 - N_0^2 - \tilde{A}_t$. Further, (5.3.15) also implies that for each s, convergence of Y_s^n to Y_s is also in $\mathbb{L}^2$ and thus

(Y_t) is a martingale. Since A_s^n converges uniformly in $s \in [0, T]$ for all $T < \infty$ a.s., it follows that

$$P(\omega : N_{\cdot}(\omega) \in \widetilde{\mathbb{D}}) = 1$$

and $\tilde{A}_t = A_t$. We have already proven that $Y_t = N_t^2 - N_0^2 - A_t$ is a martingale. This completes the proof.   □

Exercise 5.17 Construct an example of a martingale N that is unbounded, but satisfies the conditions of the Lemma 5.16.

Exercise 5.18 Use completeness of the underlying σ-field to show that the set $\{\omega : N_{\cdot}(\omega) \in \widetilde{\mathbb{D}}\}$ appearing above is measurable.

We are now in a position to prove an analogue of the Doob–Meyer decomposition theorem for the square of an r.c.l.l. locally square integrable martingale. We will use the notation $[N, N]^{\psi}$ for the process $A = \Psi(N_{\cdot}(\omega))$ of the previous result and call it quadratic variation of N. We will later show that square integrable martingales and locally square integrable martingales are stochastic integrators. Then it would follow that the quadratic variation defined for a stochastic integrator X via (4.6.2) agrees with the definition given below for a square integrable martingale and a locally square integrable martingale.

Theorem 5.19 *Let $(M_t, \mathcal{F}_t)$ be an r.c.l.l. locally square integrable martingale. Let*

$$[M, M]_t^{\psi}(\omega) = \Psi(M_{\cdot}(\omega))(t). \tag{5.3.16}$$

Then

  (i) *$[M, M]^{\psi}$ is an $(\mathcal{F}_t)$ adapted r.c.l.l. increasing process such that $X_t = M_t^2 - [M, M]_t^{\psi}$ is a local martingale.*
 (ii)
$$P(\Delta[M, M]_t^{\psi} = (\Delta M_t)^2, \ \forall t > 0) = 1.$$

(iii) *If (B_t) is an r.c.l.l. adapted increasing process such that $B_0 = 0$ and*

$$P(\Delta B_t = (\Delta M_t)^2, \ \forall t > 0) = 1$$

   and $V_t = M_t^2 - B_t$ is a local martingale, then $P(B_t = [M, M]_t^{\psi}, \ \forall t) = 1$.
 (iv) *If M is a martingale and $E(M_t^2) < \infty$ for all t, then $E([M, M]_t^{\psi}) < \infty$ for all t and $X_t = M_t^2 - [M, M]_t^{\psi}$ is a martingale.*
  (v) *If $E([M, M]_t^{\psi}) < \infty$ for all t and $M_0 = 0$, then $E(M_t^2) < \infty$ for all t, (M_t) is a martingale and $X_t = M_t^2 - [M, M]_t^{\psi}$ is a martingale.*

Proof Let θ_n be stopping times increasing to ∞ such that for each n, $\{M_t^n = M_{t \wedge \theta_n} : t \geq 0\}$ is a martingale with $E[(M_{t \wedge \theta_n})^2] < \infty$ for all t, n. For $k \geq 1$, let τ_k be the stopping time defined by

$$\tau_k = \inf\{t > 0 : |M_t| \geq k \text{ or } |M_{t-}| \geq k\} \wedge \theta_k \wedge k.$$

Then τ_k increases to ∞ and let $M_t^k = M_{t \wedge \tau_k}$. Then for each k, M^k is a martingale satisfying conditions of Lemma 5.16 with $C = k$ and $\tau = \tau_k$. Hence $X_t^k = (M_t^k)^2 - [M^k, M^k]_t^{\psi}$ is a martingale, where $[M^k, M^k]_t^{\psi} = \Psi(M_{\cdot}^k(\omega))_t$. Also,

$$P(\{\omega : M_{\cdot}^k(\omega) \in \widetilde{\mathbb{D}}\}) = 1, \quad \forall k \geq 1. \tag{5.3.17}$$

Since $M_t^k = M_{t \wedge \tau_k}$ it follows from Lemma 5.15 that

$$P(\{\omega : M_{\cdot}(\omega) \in \widetilde{\mathbb{D}}\}) = 1 \tag{5.3.18}$$

and

$$P(\{\omega : [M^k, M^k]_t^{\psi}(\omega) = [M, M]_{t \wedge \tau_k(\omega)}^{\psi}(\omega) \ \forall t\}) = 1.$$

It follows that $X_{t \wedge \tau_k} = X_t^k$ a.s. and since X^k is a martingale for all k, it follows that X_t is a local martingale. This completes the proof of part (i).

Part (ii) follows from Lemma 5.13.

For (iii), note that from part (ii) and the hypothesis on B it follows that

$$U_t = [M, M]_t^{\psi} - B_t$$

is a continuous process. Recalling $X_t = M_t^2 - [M, M]_t^{\psi}$ and $V_t = M_t^2 - B_t$ are local martingales, it follows that $U_t = V_t - X_t$ is also a local martingale with $U_0 = 0$. Being continuous, U is locally square integrable. By part (i) above, $W_t = U_t^2 - [U, U]_t^{\psi}$ is a local martingale. On the other hand, U_t being a difference of two increasing functions has finite variation, i.e. $\text{VAR}_{(0,T]}(U) < \infty$ for all $T < \infty$. Continuity of U and part (iii) of Lemma 5.15 gives

$$[U, U]_t^{\psi} = 0 \ \forall t.$$

Hence $W_t = U_t^2$ is a local martingale. Now if σ_k are stopping times increasing to ∞ such that $W_{t \wedge \sigma_k}$ is a martingale for $k \geq 1$, then we have

$$E[W_{t \wedge \sigma_k}] = E[U_{t \wedge \sigma_k}^2] = E[U_0^2] = 0.$$

and hence $U_{t \wedge \sigma_k}^2 = 0$ for each k. This yields $U_t = 0$ a.s. for every t. This completes the proof of (iii).

For (iv), we have proven in (i) that $X_t = M_t^2 - [M, M]_t^{\psi}$ is a local martingale. Let σ_k be stopping times increasing to ∞ such that $X_t^k = X_{t \wedge \sigma_k}$ are martingales. Hence, $E[X_t^k] = 0$, or

$$E([M, M]_{t \wedge \sigma_k}^{\psi}) = E(M_{t \wedge \sigma_k}^2) - E(M_0^2). \tag{5.3.19}$$

Hence

$$E([M, M]_{t \wedge \sigma_k}^{\psi}) \leq E(M_{t \wedge \sigma_k}^2) \leq E(M_t^2). \tag{5.3.20}$$

Now Fatou's lemma (or monotone convergence theorem) gives

$$E([M, M]_t^{\psi}) \leq E(M_t^2) < \infty. \tag{5.3.21}$$

Since $M_{t \wedge \sigma_k}^2$ converges to M_t^2 in $\mathbb{L}^1(P)$ and $[M, M]_{t \wedge \sigma_k}^{\psi}$ converges to $[M, M]_t^{\psi}$ in $\mathbb{L}^1(P)$, it follows that X_t^k converges to X_t in $L^1(P)$ and hence (X_t) is a martingale.

For (v) let σ_k be as in part (iv). Using $M_0 = 0$, that $[M, M]_t^{\psi}$ is increasing and (5.3.19) we conclude

$$\begin{aligned} E[M_{t \wedge \sigma_k}^2] &= E([M, M]_{t \wedge \sigma_k}^{\psi}) \\ &\leq E([M, M]_t^{\psi}) \end{aligned}$$

Now using Fatou's lemma, one gets

$$E[M_t^2] \leq E([M, M]_t^{\psi}) < \infty.$$

Now we can invoke part (iv) to complete the proof. □

Corollary 5.20 *For an r.c.l.l. martingale M with $M_0 = 0$ and $E[M_T^2] < \infty$, one has*

$$E[[M, M]_T^{\psi}] \leq E[\sup_{0 \leq s \leq T} |M_s|^2] \leq 4E[[M, M]_T^{\psi}] \tag{5.3.22}$$

Proof Let $X_t = M_t^2 - [M, M]_t^{\psi}$. As noted above X is a martingale. Since $X_0 = 0$, it follows that $E[X_T] = 0$ and thus

$$E[[M, M]_T^{\psi}] = E[M_T^2]. \tag{5.3.23}$$

The inequality (5.3.22) now follows from Doob's maximal inequality, Theorem 2.26. □

Corollary 5.21 *For an r.c.l.l. locally square integrable martingale M, for any stopping time σ, one has*

$$E[[M, M]_\sigma^{\psi}] \leq E[\sup_{0 \leq s \leq \sigma} |M_s|^2] \leq 4E[[M, M]_\sigma^{\psi}] \tag{5.3.24}$$

Proof If $\{\tau_n\}$ is a localizing sequence, then using (5.3.22) for the square integrable martingale $M_t^n = M_{t \wedge \tau_n}$ we get

$$\mathsf{E}[[M,M]^{\psi}_{\sigma\wedge\tau_n}] \le \mathsf{E}[\sup_{0\le s\le\sigma\wedge\tau_n}|M_s|^2] \le 4\mathsf{E}[[M,M]^{\psi}_{\sigma\wedge\tau_n}].$$

Now the required result follows using monotone convergence theorem.   □

Corollary 5.22 *If $M \in \mathbb{M}^2_{loc}$ with $\mathsf{E}[[M,M]^{\psi}_T] < \infty$ for all $T < \infty$ then M is a square integrable martingale.*

Proof Using (5.3.24), we conclude that $\mathsf{E}[\sup_{0\le s\le T}|M_s|^2] < \infty$.   □

Corollary 5.23 *If $M \in \mathbb{M}^2_{loc}$ then $[M,M]^{\psi}$ is locally square integrable.*

Theorem 5.24 *Let M be a continuous local martingale with $M_0 = 0$. If $M \in \mathbb{V}$ then $M_t = 0$ for all t.*

Proof Invoking (iii) in Lemma 5.15, we conclude that $[M,M]^{\psi}_t = 0$ for all t and thus the conclusion follows from Corollary 5.21.   □

Remark **5.25** The pathwise formula for quadratic variation of a continuous local martingale M was proven in Karandikar [34], but the proof required the theory of stochastic integration. A proof involving only Doob's inequality as presented above for the case of continuous local martingales was the main theme of Karandikar [35]. The formula for r.c.l.l. case was given in Karandikar [38] but the proof required again the theory of stochastic integration. The treatment given above is adapted from Karandikar–Rao [42].

Exercise 5.26 If P is Weiner measure on $C([0,\infty),\mathbb{R}^d)$ and Q is a probability measure absolutely continuous w.r.t. P such that the coordinate process is a local martingale (in the sense that each component is a local martingale), then P = Q.
HINT: Use Levy's characterization of Brownian motion.

For locally square integrable r.c.l.l. martingales M, N, we define cross-quadratic variation $[M,N]^{\psi}$ by the polarization identity as in the case of stochastic integrators (see (4.6.6))

$$[M,N]^{\psi}_t = \frac{1}{4}([M+N,M+N]^{\psi}_t - [M-N,M-N]^{\psi}_t). \tag{5.3.25}$$

It is easy to see that $M_t N_t - [M,N]^{\psi}_t$ is a local martingale. It can be checked that $[M,N]^{\psi}$ is the only process B in $\mathbb{V}_0$ such that $M_t N_t - B_t$ is a local martingale and $\mathsf{P}((\Delta B)_t = (\Delta M)_t(\Delta N)_t \; \forall t) = 1$. Also, $[M,N]^{\psi}$ is locally square integrable.

5.4 Square Integrable Martingales Are Stochastic Integrators

The main aim of this section is to show that square integrable martingales are stochastic integrators.

The treatment is essentially classical, as in Kunita–Watanabe [46], but with an exception. The role of $\langle M, M \rangle$—the predictable quadratic variation in the Kunita–Watanabe treatment—is here played by the quadratic variation $[M, M]$.

Recall that $\mathbb{M}^2$ denotes the class of r.c.l.l. martingales M such that $\mathsf{E}[M_t^2] < \infty$ for all $t < \infty$ with $M_0 = 0$.

Lemma 5.27 *Let $M, N \in \mathbb{M}^2$ and $f, g \in \mathbb{S}$. Let $X = J_M(f)$ and $Y = J_N(g)$. Let $Z_t = X_t Y_t - \int_0^t f_s g_s d[M, N]_s^\psi$. Then X, Y, Z are martingales.*

Proof The proof is almost the same as proof of Lemma 3.10, and it uses $M_t N_t - [M, N]_t^\psi$ is a martingale along with Theorem 2.59, Corollary 2.60 and Theorem 2.61.
$\qquad\qquad\square$

Corollary 5.28 *Let $M \in \mathbb{M}^2$ and $f \in \mathbb{S}$. Then $Y_t = \int_0^t f dM$ and $Z_t = (Y_t)^2 - \int_0^t f_s^2 d[M, M]_s^\psi$ are martingales and*

$$\mathsf{E}[\sup_{0 \le t \le T} |\int_0^t f dM|^2] \le 4\mathsf{E}[\int_0^T f_s^2 d[M, M]_s^\psi]. \qquad (5.4.1)$$

Proof Lemma 5.27 gives Y, Z are martingales. The estimate (5.4.1) now follows from Doob's inequality.
$\qquad\qquad\square$

Theorem 5.29 *Let $M \in \mathbb{M}^2$. Then M is a stochastic integrator. Further, for $f \in \mathbb{B}(\widetilde{\Omega}, \mathcal{P})$, the processes $Y_t = \int_0^t f dM$ and $Z_t = Y_t^2 - \int_0^t f_s^2 d[M, M]_s^\psi$ are martingales, $[Y, Y]_t^\psi = \int_0^t f_s^2 d[M, M]_s^\psi$ and*

$$\mathsf{E}[\sup_{0 \le t \le T} |\int_0^t f dM|^2] \le 4\mathsf{E}[\int_0^T f_s^2 d[M, M]_s^\psi], \quad \forall T < \infty. \qquad (5.4.2)$$

Proof Fix $T < \infty$. Suffices to prove the result for the case when $M_t = M_{t \wedge T}$. The rest follows by localization. See Theorem 4.49. Recall that $\widetilde{\Omega} = [0, \infty) \times \Omega$ and $\mathcal{P}$ is the predictable σ-field on $\widetilde{\Omega}$. Let μ be the measure on $(\widetilde{\Omega}, \mathcal{P})$ defined for $A \in \mathcal{P}$

$$\mu(A) = \int [\int_0^T 1_A(\omega, s) d[M, M]_s^\psi(\omega)] d\mathsf{P}(\omega). \qquad (5.4.3)$$

Note that

$$\mu(\widetilde{\Omega}) = \mathsf{E}[[M, M]_T^\psi] = \mathsf{E}[|M_T|^2] < \infty$$

and for $f \in \mathbb{B}(\widetilde{\Omega}, \mathcal{P})$ the norm on $\mathbb{L}^2(\widetilde{\Omega}, \mathcal{P}, \mu)$ is given by

$$\|f\|_{2,\mu} = \sqrt{\mathsf{E}[\int_0^T f_s^2 d[M, M]_s^\psi]}. \qquad (5.4.4)$$

Clearly, $\mathbb{B}(\widetilde{\Omega}, \mathcal{P}) \subseteq \mathbb{L}^2(\widetilde{\Omega}, \mathcal{P}, \mu)$. Since $\sigma(\mathbb{S}) = \mathcal{P}$, it follows from Theorem 2.67 that $\mathbb{S}$ is dense in $\mathbb{L}^2(\widetilde{\Omega}, \mathcal{P}, \mu)$. Thus, given $f \in \mathbb{B}(\widetilde{\Omega}, \mathcal{P})$, we can get $f^n \in \mathbb{S}$ such that

$$\|f - f^n\|_{2,\mu} \le 2^{-n-1}.$$
(5.4.5)

Letting $Y_t^n = \int_0^t f^n dM$ for $t \le T$ and $Y_t^n = Y_T^n$ for $t > T$ one has for $m \ge n$

$$E[\sup_{0 \le t \le T} |Y_t^n - Y_t^m|^2] \le 4\|f^m - f^n\|_{2,\mu}^2 \le 4.4^{-n}.$$

It then follows that (as in the proof of Lemma 3.12) Y_t^n converges uniformly in t to Y_t (a.s.), where Y is an r.c.l.l. adapted process with $Y_t = Y_{t \wedge T}$. For any $g \in \mathbb{S}$, using the estimate (5.4.1) for $f^n - g$, we get

$$E[\sup_{0 \le t \le T} |Y_t^n - \int_0^t g dM|^2] \le 4\|f^n - g\|_{2,\mu}^2$$

and taking limit as n tends to infinity in the inequality above we get

$$E[\sup_{0 \le t \le T} |Y_t - \int_0^t g dM|^2] \le 4\|f - g\|_{2,\mu}^2.$$
(5.4.6)

Let us denote Y as $J_M(f)$. Equation (5.4.6) implies that for $f \in \mathbb{S}$, $J_M(f) = \int_0^t f dM$. Also, (5.4.6) implies that the process Y does not depend upon the choice of the particular sequence $\{f^n\}$ in (5.4.5). Further, taking $h \in \mathbb{B}(\widetilde{\Omega}, \mathcal{P})$, a sequence $h^m \in \mathbb{S}$ approximating h, using (5.4.6) for h^m and taking limit as $m \to \infty$

$$E[\sup_{0 \le t \le T} |(J_M(f))_t - (J_M(h))_t|^2] \le 4\|f - h\|_{2,\mu}^2.$$
(5.4.7)

The estimate (5.4.7) implies that if $f_n \xrightarrow{bp} f$, then $J_M(f_n)$ converges to $J_M(f)$ in ucp topology and thus M is a stochastic integrator.

The estimate (5.4.2) follows from (5.4.7) by taking $h = 0$.

Remains to show that $Y_t = \int_0^t f dM$ and $Z_t = Y_t^2 - \int_0^t f_s^2 d[M, M]_s^\psi$ are martingales. We have seen in Corollary 5.28 that Y^n and $Z_t^n = (Y_t^n)^2 - \int_0^T (f^n)_s^2 d[M, M]_s^\psi$ are martingales. Here Y_t^n converges to Y_t in $\mathbb{L}^2(P)$, and hence in $\mathbb{L}^1(P)$, and so Y is a martingale. Further, $(Y_t^n)^2 \to (Y_t)^2$ in $\mathbb{L}^1(P)$ and moreover $\|f^n - f\|_{2,\mu} \to 0$ implies

$$E[\int_0^t |(f_s^n)^2 - f_s^2| d[M, M]_s^\psi] \to 0$$

and thus

$$E|\int_0^t ((f_s^n)^2 - f_s^2) d[M, M]_s^\psi| \to 0.$$

So Z_t^n converges to Z_t in $\mathbb{L}^1(P)$ and thus Z is also a martingale. Since

$$\Delta(\int f_s^2 d[M, M]_s^\psi) = f^2(\Delta M)^2 = (\Delta Y)^2$$

where we have used Theorem 4.32 for the last equality. Using part (iii) of Theorem 5.19, it now follows that $[Y, Y]_t^\psi = \int_0^t f_s^2 d[M, M]_s^\psi$.   □

Let us now introduce a class of integrands f such that $\int f dM$ is a locally square integrable martingale.

Definition 5.30 For $M \in \mathbb{M}_{loc}^2$ let $\mathbb{L}_m^2(M)$ denote the class of predictable processes f such that there exist stopping times $\sigma_k \uparrow \infty$ with

$$\mathsf{E}[\int_0^{\sigma_k} f_s^2 d[M, M]_s^\psi] < \infty \text{ for } k \geq 1. \tag{5.4.8}$$

We then have the following.

Theorem 5.31 *Let $M \in \mathbb{M}_{loc}^2$ i.e. M be a locally square integrable r.c.l.l. martingale with $M_0 = 0$. Then M is a stochastic integrator,*

$$\mathbb{L}_m^2(M) \subseteq \mathbb{L}(M) \tag{5.4.9}$$

and for $f \in \mathbb{L}_m^2(M)$, the process $Y_t = \int_0^t f dM$ is a locally square integrable martingale and $U_t = Y_t^2 - \int_0^t f_s^2 d[M, M]_s^\psi$ is a local martingale, $[Y, Y]_t^\psi = \int_0^t f_s^2 d [M, M]_s^\psi$. Further, for any stopping time σ,

$$\mathsf{E}[\sup_{0 \leq t \leq \sigma} |\int_0^t f dM|^2] \leq 4\mathsf{E}[\int_0^\sigma f_s^2 d[M, M]_s^\psi]. \tag{5.4.10}$$

Proof Let $\{\theta_k\}$ be stopping times such that $M^k = M^{[\theta_k]} \in \mathbb{M}^2$. Then M^k is a stochastic integrator by Theorem 5.29 and thus so is M by Theorem 4.49.

Now given $f \in \mathbb{L}_m^2(M)$, let $\sigma_k \uparrow \infty$ be stopping times such that (5.4.8) is true and let $\tau_k = \sigma_k \wedge \theta_k \wedge k$.

For $n \geq 1$, let g_n be a bounded predictable process with $|g_n| \leq |f|$ such that g_n converges to 0 pointwise. Let $Z^n = \int g_n dM$. To prove $f \in \mathbb{L}(M)$, we need to show that $\mathbf{d}_{ucp}(Z^n, 0)$ converges to 0. In view of Lemma 2.75, suffices to show that for each $k \geq 1$,

$$Y^{n,k} = (Z^n)^{[\tau_k]} \xrightarrow{ucp} 0 \quad \text{as } n \to \infty. \tag{5.4.11}$$

Note that $Y^{n,k} = \int g_n 1_{[0,\tau_k]} dM^k$. Also, $Y^{n,k}$ is a square integrable martingale since g_n is bounded and M^k is a square integrable martingale. Moreover,

$$\mathsf{E}([Y^{n,k}, Y^{n,k}]_T) = \mathsf{E}(\int_0^T (g_n)^2 1_{[0,\tau_k]} d[M^k, M^k])$$

$$\leq \mathsf{E}(\int_0^{\theta_k} (g_n)^2 d[M, M]).$$

Since g_n converge pointwise to 0 and $|g_n| \leq |f|$ and f satisfies (5.4.8), it follows that for each k fixed, $\mathsf{E}([Y^{n,k}, Y^{n,k}]_T) \to 0$ as $n \to \infty$. Invoking (5.3.22), we thus conclude

$$\lim_{n\to\infty} \mathsf{E}[\sup_{0\leq t\leq T} |Y^{n,k}_t|^2] = 0.$$

Thus, (5.4.11) holds completing the proof that $f \in \mathbb{L}(M)$.

The proof that Y is a square integrable martingale and that U is a martingale is similar to the proof of part (iv) in Theorem 5.19. The estimate (5.4.10) follows invoking (5.3.24) as $Y \in \mathbb{M}^2_{\mathrm{loc}}$. $\Box$

Remark **5.32** Now that we have shown that locally square integrable r.c.l.l. martingales M are stochastic integrators, the quadratic variation of M defined via the mapping Ψ is consistent with the definition of $[M, M]$ given in Theorem 4.6. As a consequence various identities and inequalities that were proven for the quadratic variation of a stochastic integrator in Sect. 4.6 also apply to quadratic variation of locally square integrable martingales. Thus from now on we will drop the superfix Ψ in $[M, M]^{\Psi}$, $[M, N]^{\Psi}$.

Remark **5.33** When M is a continuous martingale with $M_0 = 0$, it follows that M is locally square integrable (since it is locally bounded). Further, $[M, M]_t$ is continuous and hence for any predictable f such that for each $t > 0$

$$D_t = \int_0^t f_s^2 d[M, M]_s < \infty \ \ a.s., \qquad (5.4.12)$$

D itself is continuous. Thus D is locally bounded and hence $f \in \mathbb{L}^2_m(M)$. It is easy to see that if $f \in \mathbb{L}^2_m(M)$ then f satisfies (5.4.12).

The estimate (5.4.10) has the following implication.

Theorem 5.34 *Let M^n, $M \in \mathbb{M}^2_{\mathrm{loc}}$ be such that for a sequence $\{\sigma_j\}$ of stopping times increasing to ∞, one has for each $j \geq 1$,*

$$\mathsf{E}[[M^n - M, M^n - M]_{\sigma_j}] \to 0. \qquad (5.4.13)$$

Then M^n converges to M in Emery topology.

Proof Given predictable f bounded by 1, using (5.4.10) one has

$$\mathsf{P}(\sup_{0\leq t\leq T} |\int_0^t f dM^n - \int_0^t f dM| > \delta)$$

$$\leq \mathsf{P}(\sup_{0\leq t\leq \sigma_j} |\int_0^t f dM^n - \int_0^t f dM| > \delta) + \mathsf{P}(\sigma_j < T)$$

$$\leq \frac{1}{\delta^2} \mathsf{E}[\int_0^{\sigma_j} |f_s|^2 d[M^n - M, M^n - M]_s] + \mathsf{P}(\sigma_j < T)$$

$$\leq \frac{1}{\delta^2} \mathsf{E}[[M^n - M, M^n - M]_{\sigma_j}] + \mathsf{P}(\sigma_j < T).$$

Taking limit first as $n \to \infty$ and then as $j \to \infty$, we get

$$\lim_{n \to \infty} \sup_{f \in \mathbb{S}_1} \mathsf{P}(\sup_{0 \le t \le T} | \int_0^t f dM^n - \int_0^t f dM | > \delta) = 0.$$

In view of the observation (4.9.5), this proves convergence of M^n to M in Emery topology.   □

Exercise 5.35 Let $M^n, M \in \mathbb{M}_{loc}^2$ be such that for a sequence $\{\sigma_j\}$ of stopping times increasing to ∞, one has for each $j \ge 1$,

$$\mathsf{E}[\sup_{t \le \sigma_j} |M_t^n - M_t|^2] \to 0. \tag{5.4.14}$$

Show that

(i) M^n converges to M in Emery topology.
(ii) $[M^n - M, M^n - M] \xrightarrow{ucp} 0$.
(iii) $[M^n, M^n] \xrightarrow{ucp} [M, M]$.
  HINT: Use Theorem (4.111) for (ii) and (iii) above.

5.5 Semimartingales Are Stochastic Integrators

In the previous section, we have shown that locally square integrable martingales are stochastic integrators. In this section, we propose to show that all martingales are integrators and hence by localization it would follow that local martingales are integrators as well.

Earlier we have shown that processes whose paths have finite variation on $[0, T]$ for every T are stochastic integrators. It would then follow that all semimartingales are stochastic integrators. Here is the continuous analogue of the Burkholder's inequality, Theorem 1.44.

Lemma 5.36 *Let Z be a r.c.l.l. martingale with $Z_0 = 0$ and $f \in \mathbb{S}_1$, namely a simple predictable process bounded by 1. Then for all $\lambda > 0$, $T < \infty$ we have*

$$\mathsf{P}(\sup_{0 \le t \le T} | \int_0^t f dZ | > \lambda) \le \frac{20}{\lambda} \mathsf{E}[|Z_T|]. \tag{5.5.1}$$

Proof Let $f \in \mathbb{S}_1$ be given by

$$f(s) = a_0 1_{\{0\}}(s) + \sum_{j=0}^{m-1} a_{j+1} 1_{(s_j, s_{j+1}]}(s) \tag{5.5.2}$$

where $0 = s_0 < s_1 < s_2 < \ldots < s_m < \infty$, a_j is $\mathcal{F}_{s_{j-1}}$ measurable random variable, $|a_j| \le 1, 1 \le j \le m$ and a_0 is $\mathcal{F}_0$ measurable and $|a_0| \le 1$. Without loss of generality, we assume that $s_{m-1} < T = s_m$. Then

$$\int_0^t f dZ = \sum_{k=1}^m a_k (Z_{s_k \wedge t} - Z_{s_{k-1} \wedge t}). \tag{5.5.3}$$

Let us define a discrete process $M_k = Z_{s_k}$ and discrete filtration $\mathcal{G}_k = \mathcal{F}_{s_k}$ for $0 \le k \le m$. Then M is a martingale with respect to the filtration $\{\mathcal{G}_k : 0 \le k \le m\}$. Let us also define $U_k = a_k$, $1 \le k \le m$ and $U_0 = 0$. Then U is predictable (with respect to $\{\mathcal{G}_k : 0 \le k \le m\}$) and is bounded by 1, and hence using Theorem 1.44 we conclude that for $\alpha > 0$,

$$P(\max_{1 \le n \le m} |\sum_{k=1}^n U_k (M_k - M_{k-1})| \ge \alpha) \le \frac{9}{\alpha} E[|M_m|] = \frac{9}{\alpha} E[|Z_T|]. \tag{5.5.4}$$

Note that for $s_{k-1} \le t \le s_k$, defining $V_t^k = \int_0^t f dZ - \int_0^{s_{k-1}} f dZ$, we have $V_t^k = a_k (Z_t - Z_{s_{k-1}})$ and hence

$$\sup_{s_{k-1} \le t \le s_k} |V_t^k| \le 2 \sup_{0 \le t \le T} |Z_t| \tag{5.5.5}$$

Also, note that $\int_0^{s_k} f dZ = \sum_{j=1}^k U_j (M_j - M_{j-1})$ and hence

$$\sup_{0 \le t \le T} |\int_0^t f dZ| \le \max_{1 \le k \le m} |\int_0^{s_k} f dZ| + \max_{1 \le k \le m} \sup_{s_{k-1} \le t \le s_k} |V_t^k|$$
$$\le \max_{1 \le k \le m} |\sum_{j=1}^k U_j (M_j - M_{j-1})| + 2 \sup_{0 \le t \le T} |Z_t|. \tag{5.5.6}$$

Thus using (5.5.4) and (5.5.6) along with Theorem 2.26 we get

$$P(\sup_{0 \le t \le T} |\int_0^t f dZ| > \lambda) \le P(\max_{1 \le k \le m} |\sum_{j=1}^k U_j (M_j - M_{j-1})| > \frac{3}{4}\lambda)$$

$$+ P(2 \sup_{0 \le t \le T} |Z_t| > \frac{1}{4}\lambda) \tag{5.5.7}$$

$$\le \frac{36}{3\lambda} E[|Z_T|] + \frac{8}{\lambda} E[|Z_T|]$$

$$= \frac{20}{\lambda} E[|Z_T|].$$

$\square$

Lemma 5.37 *Let M be a square integrable r.c.l.l. martingale with $M_0 = 0$, and let g be a bounded predictable process, $|g| \leq C$. Then*

$$P(\sup_{0 \leq t \leq T} |\int_0^t g \, dM| > \lambda) \leq C \frac{20}{\lambda} E[|M_T|]. \qquad (5.5.8)$$

Proof When g is simple predictable process, the inequality (5.5.8) follows from Lemma 5.36. Since M is a stochastic integrator, the class of predictable processes g bounded by C for which (5.5.8) is true is *bp*- closed and hence it includes all such processes.   □

Theorem 5.38 *Let M be a uniformly integrable r.c.l.l. martingale with $M_0 = 0$. Then M is a stochastic integrator and (5.5.8) continues to be true for all bounded predictable process g.*

Proof In view of Theorem 4.34, in order to show that M is a stochastic integrator for the filtration $(\mathcal{F}.)$, suffices to show that it is a stochastic integrator w.r.t. $(\mathcal{F}.^+)$. Recall that M being r.c.l.l., remains a uniformly integrable martingale w.r.t. $(\mathcal{F}.^+)$.

Since M is a uniformly integrable martingale, by Theorem 2.25, $\xi = \lim_{t \to \infty} M_t$ exists *a.e* and in $\mathbb{L}^1(P)$ and further $M_t = E[\xi \mid \mathcal{F}_t^+]$. For $k \geq 1$, let M^k be the r.c.l.l. $(\mathcal{F}.^+)$-martingale given by

$$M_t^k = E[\xi 1_{\{|\xi| \leq k\}} \mid \mathcal{F}_t^+] - E[\xi 1_{\{|\xi| \leq k\}} \mid \mathcal{F}_0^+].$$

Note that for any $T < \infty$, and $k, j \geq n$

$$E[|M_T^k - M_T^j|] \leq 2E[|\xi| 1_{\{|\xi| > n\}}] \qquad (5.5.9)$$

and

$$E[|M_T^k - M_T|] \leq 2E[|\xi| 1_{\{|\xi| > n\}}]. \qquad (5.5.10)$$

Doob's maximal inequality—Theorem 2.26 now implies that M^k converges to M in $\mathbf{d}_{ucp}$ metric.

Let $a_n = E[|\xi| 1_{\{|\xi| > n\}}]$. Since ξ is integrable, it follows that $a_n \to 0$ as $n \to \infty$.

Since M^k is a bounded $(\mathcal{F}.^+)$-martingale, it is a square integrable $(\mathcal{F}.^+)$-martingale and hence a $(\mathcal{F}.^+)$-stochastic integrator. We will first prove that M^k is Cauchy in $\mathbf{d}_{em}$ metric.

Note that for any $f \in \mathbb{S}_1$, using (5.5.8) we have for $k, j \geq n$,

$$P(\sup_{0 \leq t \leq T} |\int_0^t f \, dM^k - \int_0^t f \, dM^j| > \lambda) \leq \frac{20}{\lambda} E[|M_T^k - M_T^j|] = \frac{40}{\lambda} a_n$$

and hence, using (4.9.6) it follows that $\{M^k : k \geq 1\}$ is Cauchy in $\mathbf{d}_{em}$ metric.

Since the class of stochastic integrators is complete in $\mathbf{d}_{em}$ metric as seen in Theorem 4.106, and M^k converges to M in $\mathbf{d}_{ucp}$, it would follow that indeed M is also a $(\mathcal{F}.^+)$-stochastic integrator and M^k converges to M in $\mathbf{d}_{em}$.

Since (5.5.8) holds for M^k and for any bounded predictable g, $\int g \, dM^k$ converges to $\int g \, dM$ in $\mathbf{d}_{ucp}$, it follows that (5.5.8) continues to be true for uniformly integrable martingales M. $\quad\square$

On the same lines, one can also prove the following.

Theorem 5.39 *Let* N, N^k, M^k *for* $k \geq 1$ *be r.c.l.l. martingales.*

(i) *If for all* $T < \infty$

$$E[\,|N_T^k - N_T|\,] \to 0 \text{ as } k \to \infty,$$

then N^k *converges to* N *in the Emery topology.*

(ii) *If for all* $T < \infty$

$$E[\,|M_T^k - M_T^n|\,] \to 0 \text{ as } k, n \to \infty,$$

then M^k *is Cauchy in the* $\mathbf{d}_{em}$ *metric for the Emery topology.*

Proof Noting that

$$\sup_{f: \in \mathbb{S}_1} P(\sup_{0 \leq t \leq T} | \int_0^t f \, dN^n - \int_0^t f \, dN | > \delta) \leq \frac{20}{\delta} E[|N_T^n - N_T|] \quad (5.5.11)$$

and

$$\sup_{f: \in \mathbb{S}_1} P(\sup_{0 \leq t \leq T} | \int_0^t f \, dM^n - \int_0^t f \, dM^k | > \delta) \leq \frac{20}{\delta} E[|M_T^n - M_T^k|] \quad (5.5.12)$$

the conclusions follow from (4.9.5) and (4.9.6) $\quad\square$

Here is the final result of this section.

Theorem 5.40 *Let an r.c.l.l. process* X *be a semimartingale; i.e.,* X *can be decomposed as* $X = M + A$ *where* M *is an r.c.l.l. local martingale and* A *is an r.c.l.l. process with finite variation paths. Then* X *is a stochastic integrator.*

Proof We have shown in Theorem 5.38 that uniformly integrable r.c.l.l. martingales are stochastic integrators and hence by localization, all r.c.l.l. local martingales are stochastic integrators. Thus M is a stochastic integrator.

Earlier, in Theorem 4.23 we had observed that r.c.l.l. processes A with finite variation paths are stochastic integrators and thus $X = M + A$ is a stochastic inte-breakgrator. $\quad\square$

5.6 Stochastic Integrators Are Semimartingales

The aim of this section is to prove the converse to Theorem 5.40. These two results taken together constitute one version of The Dellacherie–Meyer–Mokobodzky–Bichteler Theorem.

Let Z be a stochastic integrator. Let

$$B_t = Z_0 + \sum_{0 < s \leq t} (\Delta Z)_s 1_{\{|(\Delta Z)_s| > 1\}}. \tag{5.6.1}$$

Since paths of Z are r.c.l.l., for every ω, there are only finitely many jumps of $Z(\omega)$ of size greater than 1 in $[0, t]$ and thus B is a well-defined r.c.l.l. adapted process whose paths are of finite variation and hence B itself is a stochastic integrator. Thus $Y = Z - B$ is a stochastic integrator and now jumps of Y are of magnitude at most 1. Now defining stopping times τ_n for $n \geq 1$ via

$$\tau^n = \inf\{t > 0 \; : \; |Y_t| \geq n \text{ or } |Y_{t-}| \geq n\} \tag{5.6.2}$$

and $Y^n = Y^{[\tau^n]}$ (i.e. $Y_t^n = Y_{t \wedge \tau^n}$), it follows that for each n, Y^n is a stochastic integrator by Lemma 4.47. Further Y^n is bounded by $n + 1$, since its jumps are bounded by 1.

We will show that bounded stochastic integrators X can be decomposed as $X = M + A$ where M is a r.c.l.l. square integrable martingale and A is an r.c.l.l. process with finite variation paths. We will also show that this decomposition is unique under a certain condition on A. This would help in piecing together $\{M^n\}$, $\{A^n\}$ obtained in the decomposition $Y^n = M^n + A^n$ of Y^n to get a decomposition of Y into an r.c.l.l. locally square integrable martingale and an r.c.l.l. process with finite variation paths.

The proof of these steps is split into several lemmas.

Lemma 5.41 *Let $M^n \in \mathbb{M}^2$ be a sequence of r.c.l.l. square integrable martingales such that $M_0^n = 0$. Suppose $\exists\, T < \infty$ such that $M_t^n = M_{t \wedge T}^n$ for all n and*

$$E[[M^n - M^k, M^n - M^k]_T] \to 0 \text{ as } \min(k, n) \to \infty. \tag{5.6.3}$$

Then there exists an r.c.l.l. square integrable martingale $M \in \mathbb{M}^2$ such that

$$\lim_{n \to \infty} E[[M^n - M, M^n - M]_T] = 0, \tag{5.6.4}$$

$d_{em}(M^n, M) \to 0$ *and*

$$\lim_{n \to \infty} E[\sup_{0 \leq t \leq T} |M_t^n - M_t|^2] = 0. \tag{5.6.5}$$

Proof The relation (5.3.22) and the hypothesis (5.6.3) imply that

$$E[\sup_{0 \leq t \leq T} |M_s^n - M_s^k|^2] \to 0 \text{ as } \min(k, n) \to \infty. \tag{5.6.6}$$

Hence the sequence of processes $\{M^n\}$ is Cauchy in *ucp* metric and thus in view of Theorem 2.71 converges to an r.c.l.l. adapted process M and (5.6.5) is satisfied. Further, (5.6.6) also implies that M_s^n converges to M_s in $\mathbb{L}^2(P)$ for each s and hence using Theorem 2.23 it follows that M is a martingale, indeed a square integrable martingale. As a consequence $(M_T^n - M_T) \to 0$ in $\mathbb{L}^2(P)$ and using (5.3.23) it follows that (5.6.4) is true. The convergence of M^n to M in Emery topology follows from Theorem 5.34.   □

The following localized version of this result follows easily.

Corollary 5.42 *Let $M^n \in \mathbb{M}_{loc}^2$ be such that for a sequence of bounded stopping times $\sigma_i \uparrow \infty$, we have*

$$\mathsf{E}[[M^n - M^k, M^n - M^k]_{\sigma_i}] \to 0 \text{ as } \min(k, n) \to \infty \qquad (5.6.7)$$

for all $i \geq 1$, then there exists $M \in \mathbb{M}_{loc}^2$ such that

$$\lim_{n \to \infty} \mathsf{E}[\sup_{0 \leq t \leq \sigma_i} |M_t^n - M_t|^2] = 0. \qquad (5.6.8)$$

If M^n in the Lemma above are continuous, it follows that M is also continuous. This gives us the following.

Corollary 5.43 *Let $M^n \in \mathbb{M}^2$ be a sequence of continuous square integrable martingales such that $M_0^n = 0$. Suppose $\exists \, T < \infty$ such that $M_t^n = M_{t \wedge T}^n$ and*

$$\mathsf{E}[[M^n - M^k, M^n - M^k]_T] \to 0 \text{ as } \min(k, n) \to \infty. \qquad (5.6.9)$$

Then there exists a continuous square integrable martingale $M \in \mathbb{M}^2$ such that

$$\lim_{n \to \infty} \mathsf{E}[[M^n - M, M^n - M]_T] = 0 \qquad (5.6.10)$$

$d_{em}(M^n, M) \to 0$ *and*

$$\lim_{n \to \infty} \mathsf{E}[\sup_{0 \leq t \leq T} |M_t^n - M_t|^2] = 0. \qquad (5.6.11)$$

Theorem 5.44 *Let X be a stochastic integrator such that*

(i) $X_t = X_{t \wedge T}$ for all t,
(ii) $\mathsf{E}[[X, X]_T] < \infty$.

Then X admits a decomposition $X = M + A$ where M is an r.c.l.l. square integrable martingale and A is a stochastic integrator satisfying

$$\mathsf{E}[[N, A]_T] = 0 \qquad (5.6.12)$$

for all r.c.l.l. square integrable martingales N.

Proof The proof is very similar to the proof of existence of the projection operator on a Hilbert space onto a closed subspace of the Hilbert space. Let

$$\alpha = \inf\{\mathsf{E}[[X - M, X - M]_T] : M \in \mathbb{M}^2\}.$$

Since $\mathsf{E}[[X, X]_T] < \infty$, it follows that $\alpha < \infty$. For $k \geq 1$, let $\tilde{M}^k \in \mathbb{M}^2$ be such that

$$\mathsf{E}[[X - \tilde{M}^k, X - \tilde{M}^k]_T] \leq \alpha + \frac{1}{k}.$$

Define $M_t^k = \tilde{M}_{t \wedge T}^k, t \geq 0, k \geq 1$. Then

$$\mathsf{E}[[X - M^k, X - M^k]_T] = \mathsf{E}[[X - \tilde{M}^k, X - \tilde{M}^k]_T] \leq \alpha + \frac{1}{k}.$$

Applying the parallelogram identity (4.6.13) to $Y^k = \frac{1}{2}(X - M^k)$, $Y^n = \frac{1}{2}(X - M^n)$ we get

$$[Y^k - Y^n, Y^k - Y^n]_T = 2[Y^k, Y^k]_T + 2[Y^n, Y^n]_T - [Y^k + Y^n, Y^k + Y^n]_T$$
$$(5.6.13)$$

Note that $Y^k + Y^n = X - \frac{1}{2}(M^n + M^k)$ and since $\frac{1}{2}(M^n + M^k) \in \mathbb{M}^2$, we have

$$\mathsf{E}[[Y^k + Y^n, Y^k + Y^n]_T] \geq \alpha$$

and hence

$$\mathsf{E}[[Y^k - Y^n, Y^k - Y^n]_T] \leq 2(\frac{1}{4}(\alpha + \frac{1}{k})) + 2(\frac{1}{4}(\alpha + \frac{1}{n})) - \alpha \qquad (5.6.14)$$

Since $Y^k - Y^n = \frac{1}{2}(M^n - M^k)$, (5.6.13)–(5.6.14) yields

$$\frac{1}{4}\mathsf{E}[[M^n - M^k, M^n - M^k]_T] \leq \frac{1}{2}(\frac{1}{k} + \frac{1}{n})$$

Thus by Lemma 5.41, it follows that there exists $M \in \mathbb{M}^2$ such that

$$\lim_{n \to \infty} \mathsf{E}[[M^n - M, M^n - M]_T] = 0. \qquad (5.6.15)$$

We now show that α is attained for this M. Let us define $Y = \frac{1}{2}(X - M)$. Then we have $Y^n - Y = \frac{1}{2}(M - M^n)$ and hence (5.6.15) yields

$$\mathsf{E}[[Y^n - Y, Y^n - Y]_T] \to 0. \qquad (5.6.16)$$

Using (4.6.14) we have

$$|\mathsf{E}[[Y^n, Y^n]_T] - \mathsf{E}[[Y, Y]_T]|$$
$$\leq \mathsf{E}[|[Y^n, Y^n]_T - [Y, Y]_T|]$$
$$\leq \mathsf{E}[\sqrt{2([Y^n - Y, Y^n - Y]_T).([Y^n, Y^n]_T + [Y, Y]_T)]} \qquad (5.6.17)$$
$$\leq \sqrt{2(\mathsf{E}[[Y^n - Y, Y^n - Y]_T])(\mathsf{E}[[Y^n, Y^n]_T + [Y, Y]_T])}.$$

Since $\mathsf{E}[[Y^n, Y^n]_T] \leq \frac{1}{4}(\alpha + \frac{1}{n})$ and $\mathsf{E}[[Y, Y]_T] \leq 2(\mathsf{E}[[X, X]_T + [M, M]_T]) < \infty$, (5.6.16) and (5.6.17) together yield $\mathsf{E}[[Y^n, Y^n]_T] \to \mathsf{E}[[Y, Y]_T]$ as $n \to \infty$ and hence $\mathsf{E}[[Y, Y]_T] \leq \frac{1}{4}\alpha$. On the other hand, since $Y = \frac{1}{2}(X - M)$ where $M \in \mathbb{M}^2$, we have $\mathsf{E}[[Y, Y]_T] \geq \frac{1}{4}\alpha$ and hence $\mathsf{E}[[Y, Y]_T] = \frac{1}{4}\alpha$. Since $Y = \frac{1}{2}(X - M)$ we conclude $\mathsf{E}[[X - M, X - M]_T] = \alpha$.

By definition of α we have, for any $N \in \mathbb{M}^2$, for all $u \in \mathbb{R}$

$$\mathsf{E}[[X - M - uN, X - M - uN]_T] \geq \alpha = \mathsf{E}[[X - M, X - M]_T]$$

since $M + uN \in \mathbb{M}^2$. We thus have

$$u^2\mathsf{E}[[N, N]_T] - 2u\mathsf{E}[[N, X - M]_T] \geq 0 \text{ for all } u \in \mathbb{R}. \qquad (5.6.18)$$

This implies $\mathsf{E}[[N, X - M]_T] = 0$. Now the result follows by setting $A = X - M$. A is a stochastic integrator because X is so by hypothesis and M has been proven to be so. $\qquad \square$

Recall that $\mathbb{M}_c^2$ denotes the class of continuous square integrable martingales. A small modification of the proof above yields the following.

Theorem 5.45 *Let X be a stochastic integrator such that*

(i) $X_t = X_{t \wedge T}$ for all t,
(ii) $\mathsf{E}[[X, X]_T] < \infty$.

Then X admits a decomposition $X = N + Y$ where N is a continuous square integrable martingale and Y is a stochastic integrator satisfying

$$\mathsf{E}[[Y, U]_T] = 0 \quad \forall U \in \mathbb{M}_c^2. \qquad (5.6.19)$$

Proof This time we define

$$\alpha' = \inf\{\mathsf{E}[[X - M, X - M]_T] : M \in \mathbb{M}_c^2\}$$

and proceed as in the proof of Theorem 5.44. $\qquad \square$

The next lemma shows that (5.6.12) implies an apparently stronger conclusion that $[N, A]$ is a martingale for $N \in \mathbb{M}^2$.

Lemma 5.46 *Let A be a stochastic integrator such that*

(i) $A_t = A_{t \wedge T}$, $\forall t < \infty$,

(ii) $\mathsf{E}[[A, A]_T] < \infty$,

(iii) $\mathsf{E}[[N, A]_T] = 0$ for all $N \in \mathbb{M}^2$.

Then $[N, A]$ is a martingale for all $N \in \mathbb{M}^2$.

Proof Fix $N \in \mathbb{M}^2$. For any stopping time σ, $N^{[\sigma]}$ is also a square integrable martingale and (4.6.9) gives $[N^{[\sigma]}, A]_T = [N, A]_T^{[\sigma]} = [N, A]_{T \wedge \sigma}$ and thus we conclude $\mathsf{E}[[N, A]_{T \wedge \sigma}] = 0$. Theorem 2.57 now implies that $[N, A]$ is a martingale. $\square$

Remark **5.47** If U is a continuous square integrable martingale and σ is a stopping time, $U^{[\sigma]}$ is also a continuous square integrable martingale. Hence, arguments as in the proof of the previous result yield that if a stochastic integrator Y satisfies (5.6.19), then

$$[Y, U] \text{ is a martingale } \forall U \in \mathbb{M}_c^2.$$

The next result would tell us that essentially, the integrator A obtained above has finite variation paths (under some additional conditions).

Lemma 5.48 *Let A be a stochastic integrator such that*

(i) $A_t = A_{t \wedge T}$, $\forall t < \infty$,

(ii) $\mathsf{E}[B] < \infty$ *where* $B = \sup_{t \leq T} |A_t| + [A, A]_T$,

(iii) $[N, A]$ *is a martingale for all* $N \in \mathbb{M}^2(\mathcal{F}_\cdot^+)$.

Then A is a process with finite variation paths: $\mathrm{VAR}_{[0,T]}(A) < \infty$ *a.s.*

Proof For a partition $\tilde{\pi} = \{0 = s_0 < s_1 < \ldots < s_m = T\}$ of $[0, T]$ let us denote

$$V^{\tilde{\pi}} = \sum_{j=1}^m |A_{s_j} - A_{s_{j-1}}|.$$

We will show that for all $\varepsilon > 0$, $\exists K < \infty$ such that

$$\sup_\pi \mathsf{P}(V^\pi \geq K) < \varepsilon \qquad (5.6.20)$$

where the supremum above is taken over all partitions of $[0, T]$. Taking a sequence π^n of successively finer partitions such that $\delta(\pi^n) \to 0$ (e.g. $\pi^n = \{kT2^{-n} : 0 \leq k \leq 2^n\}$), it follows that $V^{\pi^n} \uparrow \mathrm{VAR}_{[0,T]}(A)$ and thus (5.6.20) would imply

$$\mathsf{P}(\mathrm{VAR}_{[0,T]}(A) \geq K) < \varepsilon$$

and hence that $\mathsf{P}(\mathrm{VAR}_{[0,T]}(A) < \infty) = 1$. This would complete the proof.

Fix $\varepsilon > 0$. Since A is a stochastic integrator for the filtration $(\mathcal{F}_\cdot)$, it is also a stochastic integrator for the filtration $(\mathcal{F}_\cdot^+)$ in view of Theorem 4.44. Thus we can get (see Remark 4.25) $J_1 < \infty$ such that

$$\sup_{f \in \mathbb{S}_1(\mathcal{F}_\cdot^+)} P(|\int_0^T f dA| \geq J_1) \leq \frac{\varepsilon}{6}. \tag{5.6.21}$$

Since $\mathsf{E}[B] < \infty$, we can get $J_2 < \infty$ such that

$$P(B \geq J_2) \leq \frac{\varepsilon}{6}. \tag{5.6.22}$$

Let $J = \max\{J_1, J_2, \mathsf{E}[B]\}$ and n be such that $\frac{24}{n} < \varepsilon$. Let $K = (n+1)J$. We will show that (5.6.20) holds for this choice of K. Note that the choice has been made independent of a partition.

Now fix a partition $\pi = \{0 = t_0 < t_1 < \ldots < t_m = T\}$. Recall that for $x \in \mathbb{R}$, $\mathsf{sgn}(x) = 1$ for $x \geq 0$ and $\mathsf{sgn}(x) = -1$ for $x < 0$, so that $|x| = \mathsf{sgn}(x)x$. For $1 \leq j \leq m$, let us consider the $(\mathcal{F}_\cdot^+)$-martingale

$$\tilde{Z}_t^j = \mathsf{E}[\mathsf{sgn}(A_{t_j} - A_{t_{j-1}}) \mid \mathcal{F}_t^+].$$

Since the filtration $(\mathcal{F}_\cdot^+)$ is right continuous, the martingale $\tilde{Z}_t^j$ admits an r.c.l.l. version Z_t^j. Then $Z_{t_j}^j = \mathsf{sgn}(A_{t_j} - A_{t_{j-1}})$ and hence

$$Z_{t_j}^j(A_{t_j} - A_{t_{j-1}}) = |(A_{t_j} - A_{t_{j-1}})|.$$

Writing $C_t^j = \int_0^t 1_{(t_{j-1}, t_j]} dA_s = (A_{t \wedge t_j} - A_{t \wedge t_{j-1}})$, we get by integration by parts formula (4.6.7)

$$|(A_{t_j} - A_{t_{j-1}})| = Z_{t_j}^j C_{t_j}^j$$
$$= \int_0^{t_j} Z_{s-}^j dC_s^j + \int_0^{t_j} C_{s-}^j dZ_s^j + [Z^j, C^j]_{t_j}$$
$$= \int_{t_{j-1}}^{t_j} Z_{s-}^j dA_s + \int_{t_{j-1}}^{t_j} C_{s-}^j dZ_s^j + [Z^j, A]_{t_j} - [Z^j, A]_{t_{j-1}}$$

and for $t_{j-1} \leq t \leq t_j$

$$Z_t^j C_t^j = \int_{t_{j-1}}^t Z_{s-}^j dA_s + \int_{t_{j-1}}^t C_{s-}^j dZ_s^j + [Z^j, A]_t - [Z^j, A]_{t_{j-1}}.$$

Let us define

$$Z_t = \sum_{j=1}^m Z_t^j 1_{(t_{j-1}, t_j]}(t),$$

$$C_t = \sum_{j=1}^m C_t^j 1_{(t_{j-1}, t_j]}(t),$$

$$M_t = \sum_{j=1}^{m} (Z_{t \wedge t_j}^j - Z_{t \wedge t_{j-1}}^j).$$

It follows that $|C_s| \le 2B$, $|Z_s| \le 1$, M is a bounded $(\mathcal{F}_\cdot^+)$-martingale and

$$[M, A]_t = \sum_{j=1}^{m} ([Z^j, A]_{t \wedge t_j} - [Z^j, A]_{t \wedge t_{j-1}}).$$

Thus, defining

$$Y_t = \int_0^t Z_{s-} dA_s + \int_0^t C_{s-} dM_s + [M, A]_t$$

we get, for $t_{k-1} \le t < t_k$, $1 \le k \le m$

$$Y_t = \sum_{j=1}^{k-1} |(A_{t_j} - A_{t_{j-1}})| + Z_t^k C_t^k$$

and thus $Y_t \ge -2B$. Also note that

$$Y_T = \sum_{j=1}^{m} |(A_{t_j} - A_{t_{j-1}})| = V^\pi.$$

Let $U_t = \int_0^t C_{s-} dM_s + [M, A]_t$, it follows that U is a $(\mathcal{F}_\cdot^+)$ local martingale since by assumption on A, $[M, A]$ is itself a $(\mathcal{F}_\cdot^+)$-martingale, and thus $\int_0^t C_{s-} dM_s$ is a $(\mathcal{F}_\cdot^+)$ local martingale by Theorem 5.31. Further

$$Y_t = \int_0^t Z_{s-} dA_s + U_t.$$

Now defining

$$\tau = \inf\{t \ge 0 : U_t < -3J\} \wedge T$$

it follows that τ is a stopping time (see Lemma 2.48) and

$$\{\tau < T\} \subseteq \{U_\tau \le -3J\}.$$

Since $Y_\tau \ge -2B$, we note that

$$(\{\tau < T\} \cap \{B < J\}) \subseteq (\{U_\tau \le -3J\} \cap \{Y_\tau \ge -2J\}$$

$$\subseteq \{\int_0^\tau Z_{s-} dA_s \ge J\}.$$

Hence using (5.6.21) and (5.6.22) we conclude

$$P(\tau < T) \leq P(\int_0^T Z_{s-} dA_s \geq J) + P(B \geq J) \leq \frac{\varepsilon}{3}. \tag{5.6.23}$$

Now $(\Delta U)_t = C_{t-}(\Delta M)_t + (\Delta M)_t (\Delta A)_t$. Since M is bounded by 1, $(\Delta M)_t \leq 2$. Also, $|C_t| \leq 2B$ and $|(\Delta A)_t| \leq 2B$. Thus $|(\Delta U)_t| \leq 8B$. Let σ_n be $(\mathcal{F}_\cdot^+)$ stopping times increasing to ∞ such that $U^{[\sigma_n]}$ is a $(\mathcal{F}_\cdot^+)$-martingale. By definition of τ, it follows that $U_{t \wedge \sigma_n \wedge \tau} \geq -(3J + 8B)$. Since $U^{[\sigma_n]}$ is a $(\mathcal{F}_\cdot^+)$-martingale, we have $E[U_{T \wedge \sigma_n \wedge \tau}] = 0$. Recall that B is integrable and hence using Fatou's lemma we conclude

$$E[U_{T \wedge \tau}] \leq 0. \tag{5.6.24}$$

Since $U_{T \wedge \tau} \geq -(3J + 8B)$, it follows that $E[(U_{T \wedge \tau})^-] \leq E[(3J + 8B)]$ (here, $(U_{T \wedge \tau})^-$ denotes the negative part of $U_{T \wedge \tau}$). Since $J \geq E[B]$, using (5.6.24) we conclude

$$E[(U_{T \wedge \tau})^+] \leq 11J. \tag{5.6.25}$$

Since $V^\pi = Y_T = \int_0^T Z_{s-} dA_s + U_T$ and recalling that $K = (n+1)J$

$$\begin{aligned}
P(V^\pi \geq K) &\leq P(\int_0^T Z_{s-} dA_s \geq J) + P(U_T \geq nJ) \\
&\leq \frac{\varepsilon}{6} + P(\tau < T) + P(U_{T \wedge \tau} \geq nJ) \\
&\leq \frac{\varepsilon}{6} + \frac{\varepsilon}{3} + \frac{1}{nJ} E[(U_{T \wedge \tau})^+] \\
&\leq \frac{\varepsilon}{2} + \frac{11J}{nJ} \\
&< \frac{\varepsilon}{2} + \frac{\varepsilon}{2} = \varepsilon
\end{aligned}$$

since by our choice $\frac{24}{n} < \varepsilon$. This proves (5.6.20) and completes the proof as noted earlier. □

We now put together the results obtained earlier in this chapter to get the following key step in the main theorem of the section. We have to avoid assuming right continuity of the filtration in the main theorem. However, it was required in the previous result and we avoid the same by an interesting argument. Here is a lemma that is useful here and in later chapter.

Lemma 5.49 *Let $(\Omega, \mathcal{F}, P)$ be a complete probability space and let $\mathcal{H} \subseteq \mathcal{G}$ be sub-σ-fields of $\mathcal{F}$ such that $\mathcal{H}$ contains all the P null sets in $\mathcal{F}$. Let Z be an integrable $\mathcal{G}$ measurable random variable such that for all $\mathcal{G}$ measurable bounded random variables U one has*

$$E[ZU] = E[ZE[U \mid \mathcal{H}]]. \tag{5.6.26}$$

Then Z is $\mathcal{H}$ measurable.

Proof Noting that

$$E[ZE[U \mid \mathcal{H}]] = E[E[Z \mid \mathcal{H}]E[U \mid \mathcal{H}]] = E[E[Z \mid \mathcal{H}]U]$$

using (5.6.26) it follows that for all $\mathcal{G}$ measurable bounded random variables U

$$E[(Z - E[Z \mid \mathcal{H}])U] = 0. \qquad (5.6.27)$$

Taking $U = \mathrm{sgn}(Z - E[Z \mid \mathcal{H}])$ (here $\mathrm{sgn}(x) = 1$ for $x \geq 0$ and $\mathrm{sgn}(x) = -1$ for $x < 0$), we conclude from (5.6.27) that

$$E[\,|(Z - E[Z \mid \mathcal{H}])|\,] = 0.$$

Since $\mathcal{H}$ is assumed to contain all P null sets in $\mathcal{F}$, it follows that Z is $\mathcal{H}$ measurable. $\qquad\square$

Theorem 5.50 *Let X be a stochastic integrator such that*

(i) $X_t = X_{t \wedge T}$ *for all t,*
(ii) $E[\sup_{s \leq T} |X_s|] < \infty$,
(iii) $E[[X, X]_T] < \infty$.

Then X admits a decomposition

$$X = M + A, \ M \in \mathbb{M}^2, \ A \in \mathbb{V}, \qquad (5.6.28)$$

such that
$$[N, A] \text{ is a martingale for all } N \in \mathbb{M}^2 \qquad (5.6.29)$$

and

$$E[[X, X]_T] = E[[M, M]_T] + E[[A, A]_T] \qquad (5.6.30)$$

and further, the decomposition (5.6.28) is unique under the requirement (5.6.29).

Proof As seen in Theorem 4.34, X being a stochastic integrator for the filtration $(\mathcal{F}_\cdot)$ implies that X is also a stochastic integrator for the filtration $(\mathcal{F}_\cdot^+)$. Also, (4.6.2) implies that $[X, X]$ does not depend upon the underlying filtration, so the assumptions of the theorem continue to be true when we take the underlying filtration to be $(\mathcal{F}_\cdot^+)$. Now Theorem 5.44 yields a decomposition $X = \tilde{M} + \tilde{A}$ with $\tilde{M} \in \mathbb{M}^2(\mathcal{F}_\cdot^+)$ and $E[[N, \tilde{A}]_T] = 0$ for all $N \in M^2(\mathcal{F}_\cdot^+)$. Let $M_t = \tilde{M}_{t \wedge T}$, $A_t = \tilde{A}_{t \wedge T}$. Then $M \in \mathbb{M}^2(\mathcal{F}_\cdot^+)$ and $E[[N, A]_T] = 0$ for all $N \in M^2(\mathcal{F}_\cdot^+)$ since $[N, A]_t = [N, \tilde{A}]_{t \wedge T}$ (by (4.6.9)). As a consequence,

$$E[[X, X]_T] = E[[M, M]_T] + E[[A, A]_T].$$

Thus $E[[A, A]_T] < \infty$ and then by Lemma 5.46, we have $[N, A]$ is a $(\mathcal{F}_{\cdot}^+)$-martingale for all $N \in \mathbb{M}^2(\mathcal{F}_{\cdot}^+)$. Since the underlying filtration $(\mathcal{F}_{\cdot}^+)$ is right continuous, Lemma 5.48 implies that $A \in \mathbb{V}$; namely, paths of A have finite variation. By construction, M, A are $(\mathcal{F}_{\cdot}^+)$ adapted. Since for $s < t$, $\mathcal{F}_s^+ \subseteq \mathcal{F}_t$, it follows that $A_t^- = A_{t-}$ is $\mathcal{F}_t$ measurable. We will show that

$$(\Delta A)_t \text{ is } \mathcal{F}_t \text{ measurable } \forall t > 0. \tag{5.6.31}$$

Since X_t is $\mathcal{F}_t$ measurable, it would follow that $M_t = X_t - A_t$ is also $\mathcal{F}_t$ measurable. This will also imply $M \in \mathbb{M}^2 = \mathbb{M}^2(\mathcal{F}_{\cdot})$ completing the proof.

Fix $t > 0$. Let U be a bounded $\mathcal{F}_t^+$ measurable random variable, and let $V = U - E[U \mid \mathcal{F}_t]$. Let

$$N_s = V 1_{[t,\infty)}(s)$$

i.e. $N_s = 0$ for $s < t$ and $N_s = V$ for $s \geq t$. It is easy to see that $N \in \mathbb{M}^2(\mathcal{F}_{\cdot}^+)$ and that $[N, A]_s = V(\Delta A)_t 1_{[t,\infty)}(s)$ (using (4.6.10)). Thus $[N, A]$ is a $(\mathcal{F}_{\cdot}^+)$-martingale, and in particular $E[[N, A]_t] = 0$. Hence we conclude that for all bounded $\mathcal{F}_t^+$ measurable U

$$E[(\Delta A)_t U] = E[(\Delta A)_t E[U \mid \mathcal{F}_t]]. \tag{5.6.32}$$

Invoking Lemma 5.49 we conclude that $(\Delta A)_t$ is $\mathcal{F}_t$ measurable and hence that A_t is $\mathcal{F}_t$ measurable. As noted earlier, this implies $M \in \mathbb{M}^2$. Only remains to prove uniqueness of decomposition satisfying (5.6.29). Let $X = Z + B$ be another decomposition with $Z \in \mathbb{M}^2$, $B \in \mathbb{V}$ and $[N, B]$ being a martingale for all $N \in \mathbb{M}^2$.

Now $X = M + A = Z + B$, $B - A = M - Z \in \mathbb{M}^2$ and $[N, B - A]$ is a martingale for all $N \in \mathbb{M}^2$. Let $N = M - Z = B - A$. By definition, $M_0 = Z_0 = 0$ and so $N_0 = 0$. Now $[N, B - A]$ is a martingale implies $[B - A, B - A] = [M - Z, M - Z]$ is a martingale, and as a consequence, we have $E[[M - Z, M - Z]_T] = E[[M - Z, M - Z]_0] = 0$ (see Remark 4.73). Now invoking (5.3.22), we conclude (since $M - Z \in \mathbb{M}^2$ and $M_0 = Z_0 = 0$)

$$E[\sup_{s \leq T}|M_s - Z_s|] = 0.$$

Thus $M = Z$ and as a consequence $A = B$. This completes the proof. $\qquad\square$

Corollary 5.51 *The processes* M, A *in* (5.6.28) *satisfy*

$$M_t = M_{t \wedge T}, \quad A_t = A_{t \wedge T} \quad \forall t \geq 0 \tag{5.6.33}$$

Proof Note that $X_t = X_{t \wedge T}$. Let $R_t = M_{t \wedge T}$ and $B_t = A_{t \wedge T}$. Then $X = R + B$ is also a decomposition that satisfies (5.6.28) since R is also a square integrable martingale, B is a process with finite variation paths and if R is any square integrable martingale, then

$$[N, B]_t = [N, A]_{t \wedge T}$$

and hence $[N, B]$ is also a martingale. Now uniqueness part of Theorem 5.50 implies (5.6.33). □

Corollary 5.52 *Suppose* X, Y *are stochastic integrators satisfying conditions of Theorem 5.50, and let* $X = M + A$ *and* $Y = N + B$ *be decompositions with* $M, N \in \mathbb{M}^2$, $A, B \in \mathbb{V}$ *and* $[U, A], [U, B]$ *being martingales for all* $U \in \mathbb{M}^2$. *If for a stopping time* σ, $X^{[\sigma]} = Y^{[\sigma]}$, *then* $M^{[\sigma]} = N^{[\sigma]}$ *and* $A^{[\sigma]} = B^{[\sigma]}$.

Proof Follows by observing that $X^{[\sigma]}$ is also a stochastic integrator and $X^{[\sigma]} = M^{[\sigma]} + A^{[\sigma]}$ and $X^{[\sigma]} = N^{[\sigma]} + B^{[\sigma]}$ are two decompositions, both satisfying (5.6.29). The conclusion follows from the uniqueness part of Theorem 5.50. □

We now introduce two important definitions.

Definition 5.53 An adapted process B is said to be locally integrable if there exist stopping times τ_n increasing to ∞ and random variables D_n such that $\mathsf{E}[D_n] < \infty$ and

$$\mathsf{P}(\omega : \sup_{0 \le t \le \tau_n(\omega)} |B_t(\omega)| \le D_n(\omega)) = 1 \quad \forall n \ge 1.$$

The condition above is meaningful even if $\sup_{0 \le t \le \tau_n(\omega)} |B_t(\omega)|$ is not measurable. It is to be interpreted as—there exists a set $\Omega_0 \in \mathcal{F}$ with $\mathsf{P}(\Omega_0) = 1$ such that the above inequality holds for $\omega \in \Omega_0$.

Definition 5.54 An adapted process B is said to be locally square integrable if there exist stopping times τ_n increasing to ∞ and random variables D_n such that $\mathsf{E}[D_n^2] < \infty$ and

$$\mathsf{P}(\omega : \sup_{0 \le t \le \tau_n(\omega)} |B_t(\omega)| \le D_n(\omega)) = 1 \quad \forall n \ge 1.$$

Clearly, if B is locally bounded process then it is locally square integrable. It is easy to see that a continuous adapted processes Y is locally bounded if Y_0 is bounded, is locally integrable if $\mathsf{E}[|Y_0|] < \infty$ and locally square integrable if $\mathsf{E}[|Y_0|^2] < \infty$. Indeed, the same is true for an r.c.l.l. adapted process if its jumps are bounded.

We have seen that for an r.c.l.l. adapted process Z, the process Z^- is locally bounded and hence it follows that Z is locally integrable if and only if the process ΔZ is locally integrable and likewise, Z is locally square integrable if and only if the process ΔZ is locally square integrable.

Theorem 5.55 *Let* X *be locally square integrable stochastic integrator. Then* X *admits a decomposition* $X = M + A$ *where* $M \in \mathbb{M}^2_{loc}$ *(M is a locally square integrable martingale) and* $A \in \mathbb{V}$ *(A is a process with finite variation paths) satisfying*

$$[A, N] \in \mathbb{M}^2_{loc} \quad \forall N \in \mathbb{M}^2_{loc}. \tag{5.6.34}$$

Further, such a decomposition is unique.

Proof Since X is a locally square integrable process, it follows that ΔX is locally square integrable. Since $\Delta[X, X] = (\Delta X)^2$, it follows that $[X, X]$ is locally integrable and thus so is $D_t = \sup_{s \leq t} |X_s|^2 + [X, X]_t$. Let $\sigma^n \uparrow \infty$ be stopping times such that $\mathsf{E}[D_{\sigma^n}] < \infty$, and let $\tau^n = n \wedge \sigma^n$. Let $X^n = X^{[\tau^n]}$. Then X^n satisfies conditions of Theorem 5.50 (with $T = n$), and thus we can get decomposition $X^n = M^n + A^n$ such that $M^n \in \mathbb{M}^2$ and $A^n \in \mathbb{V}$ and

$$[U, A^n] \text{ is a martingale for all } U \in \mathbb{M}^2. \tag{5.6.35}$$

Using Corollary 5.52, we can see that

$$\mathsf{P}(M_t^n = M_t^k \ \forall t \leq \tau^n \wedge \tau^k) = 1, \quad \forall n, k.$$

Thus we can define r.c.l.l. processes M, A such that $M^{[\tau^n]} = M^n$ and $A^{[\tau^n]} = A^n$ for all n. This decomposition satisfies the asserted properties.

Uniqueness follows as in Theorem 5.50 and the observation that if $Y \in \mathbb{M}^2_{loc}$, $Y_0 = 0$ and $[Y, Y]_t = 0$ for all t then $Y = 0$ (i.e. $\mathsf{P}(Y_t = 0 \ \forall t) = 1$.)   □

Remark 5.56 The process A with finite variation r.c.l.l. paths appearing in the above theorem was called a *Natural* process by Meyer, and it appeared in the Doob Meyer decomposition of supermartingales. Later it was shown that such a process is indeed a predictable process. A is also known as the *compensator* of X. We will come back to this in Chap. 8 later.

Corollary 5.57 *Let X be a locally square integrable stochastic integrator and A be its compensator and $M = X - A \in \mathbb{M}^2_{loc}$. Then for any stopping time σ such that $\mathsf{E}[[X, X]_\sigma] < \infty$,*

$$\mathsf{E}[[A, A]_\sigma] \leq \mathsf{E}[[X, X]_\sigma] \tag{5.6.36}$$

and

$$\mathsf{E}[[M, M]_\sigma] \leq \mathsf{E}[[X, X]_\sigma] \tag{5.6.37}$$

Proof If $\tau_n \uparrow \infty$ are as in the proof of Theorem 5.55, then by Theorem 5.50 we have

$$\mathsf{E}[[X, X]_{\sigma \wedge \tau_n}] = \mathsf{E}[[M, M]_{\sigma \wedge \tau_n}] + \mathsf{E}[[A, A]_{\sigma \wedge \tau_n}]$$

and the required inequalities follow by taking limit as $n \to \infty$ and using monotone convergence theorem.   □

Arguments similar to the ones leading to Theorem 5.55 yield the following (we use Theorem 5.45 and Remark 5.47). We also use the fact that every continuous process is locally bounded and hence locally square integrable.

Theorem 5.58 *Let X be locally square integrable stochastic integrator. Then X admits a decomposition $X = N + Y$ with $N \in \mathbb{M}_{c,loc}$; i.e. N is a continuous locally square integrable martingale with $N_0 = 0$, and Y is a locally square integrable stochastic integrator satisfying*

$$[Y, U] \text{ is a local martingale } \forall U \in \mathbb{M}_c^2. \tag{5.6.38}$$

Further, such a decomposition is unique. Indeed,

$$[Y, U] = 0 \ \forall U \in \mathbb{M}_c^2. \tag{5.6.39}$$

Proof The only new part is to show that (5.6.38) yields (5.6.39). For this note that on the one hand $[Y, U]$ is continuous as $\Delta[Y, U] = (\Delta Y)(\Delta U)$ and U is continuous. On the other hand by definition $[Y, U]$ has finite variation paths. Thus $[Y, U]$ is a continuous local martingale with finite variation paths. Hence by Theorem 5.24, $[Y, U] = 0$. $\qquad\square$

As an immediate consequence of Theorem 5.55, here is a version of the Della-cherie–Meyer–Mokobodzky–Bichteler Theorem. See Theorem 5.89 for the final version.

Theorem 5.59 *Let X be an r.c.l.l. adapted process. Then X is a stochastic integrator if and only if X is a semimartingale.*

Proof We have already proved (in Theorem 5.40) that if X is a semimartingale then X is a stochastic integrator.

For the other part, let X be a stochastic integrator. Let us define

$$B_t = X_0 + \sum_{0 < s \le t} (\Delta X)_s 1_{\{|(\Delta X)_s| > 1\}}. \tag{5.6.40}$$

Then as noted at the beginning of the section, B is an adapted r.c.l.l. process with finite variation paths and is thus a stochastic integrator. Hence $Z = X - B$ is also a stochastic integrator. By definition,

$$(\Delta Z) = (\Delta X) 1_{\{|\Delta X| \le 1\}}$$

and hence jumps of Z are bounded by 1. Hence Z is locally square integrable. Hence by Theorem 5.55, Z admits a decomposition $Z = M + A$ where $M \in \mathbb{M}_{\text{loc}}^2$ and A is a process with finite variation paths. Thus $X = M + (B + A)$ and thus X is a semimartingale. $\qquad\square$

The result proven above contains a proof of the following fact, which we record here for later reference.

Corollary 5.60 *Every semimartingale X can be written as $X = M + A$ where $M \in \mathbb{M}_{loc}^2$ and $A \in \mathbb{V}$; i.e. M is a locally square integrable r.c.l.l. martingale with $M_0 = 0$ and A is an r.c.l.l. process with finite variation paths.*

Exercise 5.61 Let X be a semimartingale for the filtration $(\mathcal{F}_.)$. Let $(\mathcal{H}_.)$ be a filtration such that $\mathcal{H}_t \subseteq \mathcal{F}_t$ for all t. Suppose X is $(\mathcal{H}_.)$ adapted. Show that X admits a decomposition $X = N + B$, where N, B are $(\mathcal{H}_.)$ adapted, N a local martingale and B a process with finite variation paths.

Exercise 5.62 Let $G(t)$ be a deterministic function. Suppose, G considered as a stochastic process on some probability space and some filtration is a semimartingale. Show that G is a function with finite variation on $[0, T]$ for every $T < \infty$.
HINT: Use Exercise 5.61 with $\mathcal{H}_t = \{B \in \mathcal{F} : P(B) = 0 \text{ or } P(B) = 1\}$ for all t.

Every local martingale X is a semimartingale by definition. In that case, the process A appearing in the corollary above is also a local martingale. Thus we have

Corollary 5.63 *Every r.c.l.l. local martingale N can be written as $N = M + L$ where M is a locally square integrable r.c.l.l. martingale with $M_0 = 0$ and L is an r.c.l.l. process with finite variation paths that is also a local martingale, i.e. $M \in \mathbb{M}^2_{loc}$ and $L \in \mathbb{V} \cap \mathbb{M}_{loc}$.*

Using the technique of separating large jumps from a semimartingale to get a locally bounded semimartingale used in proof of Theorem 5.59, we can get the following extension of Theorem 5.58.

Theorem 5.64 *Let X be a stochastic integrator. Then X admits a decomposition $X = N + S$ with $N \in \mathbb{M}_{c,loc}$ (N is a continuous locally square integrable martingale with $N_0 = 0$) and S is a stochastic integrator satisfying*

$$[S, U] = 0 \quad \forall U \in \mathbb{M}^2_c. \tag{5.6.41}$$

Further, such a decomposition $X = N + S$ is unique.

Proof Let B be defined by (5.6.40), and let $Z = X - B$. Then Z is locally bounded, and thus invoking Theorem 5.58, we can decompose Z as $Z = N + Y$ with $N \in \mathbb{M}_{c,loc}$ and Y satisfying $[Y, U] = 0$ for all $U \in \mathbb{M}^2_c$. Let $S = Y + B$. It follows that $X = N + S$. Since $[B, U] = 0$ for all $U \in \mathbb{M}^2_c$, $[S, U] = 0$ for all $U \in \mathbb{M}^2_c$. To see that such a decomposition is unique, if $X = M + R$ is another such decomposition with $M \in \mathbb{M}^2_c$ and $[R, U] = 0$ for all $U \in \mathbb{M}^2_c$, then $V = N - M = R - S \in \mathbb{M}_{c,loc}$ with $[V, V] = 0$ and hence $V = 0$ by Theorem 5.24. $\qquad \square$

Definition 5.65 Let X be a semimartingale. The continuous local martingale N such that $X = N + S$ and $[S, U] = 0$ for all $U \in \mathbb{M}_{c,loc}$ is said to be the continuous local martingale part of X and is denoted by $X^{(c)}$.

For a semimartingale X with $X = X^{(c)} + Z$, we can see that $[X, X] = [X^{(c)}, X^{(c)}] + [Z, Z]$. We will later show in Theorem 8.83 that $[X^{(c)}, X^{(c)}] = [X, X]^{(c)}$—the continuous part of the quadratic variation of X.

The next result shows that if X is a continuous process that is a semimartingale, then it can be uniquely decomposed as a sum of a continuous local martingale and a continuous process with finite variation paths.

Theorem 5.66 *Let X be a continuous process, and further let X be a semimartingale. Then X can be uniquely decomposed as*

$$X = M + A$$

where M and A are continuous processes, $M_0 = 0$, M a local martingale and A a process with finite variation paths.

Proof Without loss of generality, we assume $X_0 = 0$. Now X being continuous, it follows that X is locally square integrable and thus X admits a decomposition $X = M + A$ with $M \in \mathbb{M}_{\text{loc}}^2$ and $A \in \mathbb{V}$ with A satisfying (5.6.34). On the one hand, continuity of X implies $(\Delta M)_t = -(\Delta A)_t$ for all $t > 0$ and since $A \in \mathbb{V}$, we have

$$[A, M]_t = \sum_{0 < s \leq t} (\Delta A)_s (\Delta M)_s.$$

Thus,

$$[A, M]_t = -\sum_{0 < s \leq t} (\Delta A)_s^2 = -[A, A]_t.$$

Since A satisfies (5.6.34), it follows that $[A, M] = -[A, A]$ is a local martingale. If σ_n is a localizing sequence, it follows that $\mathsf{E}[[A, A]_{t \wedge \sigma_n}] = 0$ for all n. Since $[A, A]_s \geq 0$ for all s, it follows that $[A, A]_{t \wedge \sigma_n} = 0$ *a.s.* for all t, n. This implies

$$[A, A]_t = \sum_{0 < s \leq t} (\Delta A)_s^2 = 0 \quad a.s. \; \forall t$$

and hence A is a continuous process and hence so is $M = X - A$. Uniqueness follows from Theorem 5.24. $\square$

Exercise 5.67 Let X be a continuous semimartingale, and let $X = M + A$ be a decomposition as in Theorem 5.66 with $M \in \mathbb{M}_{c,\text{loc}}$ with $M_0 = 0$ and $A \in \mathbb{V}$. Let $X = N + B$ be any decomposition with $N \in \mathbb{M}_{\text{loc}}$ and $B \in \mathbb{V}$. Then $[N, N] - [M, M]$ is an increasing process, i.e.

$$[N, N] = [M, M] + C, \quad \text{for some } C \in \mathbb{V}^+. \tag{5.6.42}$$

HINT: Observe that $[M, A - B] = 0$ since M is continuous. Write $N = M + (A - B)$ and take $C = [A - B, A - B]$.

Definition 5.68 A local martingale M is said to be *purely discontinuous* if

$$[M, N] = 0 \quad \forall N \in \mathbb{M}_{c,\text{loc}}. \tag{5.6.43}$$

Let $\mathbb{M}_d$, $\mathbb{M}_d^2$, $\mathbb{M}_{d,\text{loc}}$, $\mathbb{M}_{d,\text{loc}}^2$ denote the class of purely discontinuous martingales, purely discontinuous square integrable martingales, purely discontinuous local martingales, and purely discontinuous locally square integrable martingales, respectively.

Exercise 5.69 Let $M \in \mathbb{M}_{\text{loc}} \cap \mathbb{V}$. Then show that $M \in \mathbb{M}_{d,\text{loc}}$.
HINT: Use Theorem 4.74.

We will now show that every local martingale can be (uniquely) decomposed as a sum of a continuous local martingale and a purely discontinuous local martingale.

Theorem 5.70 *Let X be an r.c.l.l. local martingale. Then X admits a decomposition*

$$X = M + N, \quad M \in \mathbb{M}_{c,loc}, \quad N \in \mathbb{M}_{d,loc}. \tag{5.6.44}$$

Proof Let

$$B_t = X_0 + \sum_{0 < s \le t} (\Delta X)_s 1_{\{|(\Delta X)_s| > 1\}}$$

and let $Y = X - B$. Then B is an adapted r.c.l.l. process with finite variation paths, and Y is a stochastic integrator with jumps bounded by 1 and hence is locally square integrable. Then invoking Theorem 5.58, we get a decomposition

$$Y = M + A$$

where $M \in \mathbb{M}_{c,loc}$ and A is a locally square integrable stochastic integrator such that $[A, S]_t = 0$ for all $S \in \mathbb{M}_c^2$. Since $B \in \mathbb{V}$, it follows that $[B, S] = 0$ for all $S \in \mathbb{M}_c^2$. Defining $N = A + B = X - M$, we get that $[N, S] = 0$ for all $S \in \mathbb{M}_c^2$ and since X, M are local martingales, it follows that so is N. $\qquad\square$

Remark 5.71 Here, if $X \in \mathbb{M}_{loc}^2$ then $M \in \mathbb{M}_{c,loc}^2$ and $N \in \mathbb{M}_{d,loc}^2$. We will later show in Theorem 8.80 that in this case $[M, M]$ is the continuous part of $[X, X]$ and $[N, N]$ is the sum of squares of jumps of X.

5.7 The Class $\mathbb{L}(X)$

In the previous section, we have given a characterization of stochastic integrators. Like stochastic integrators, the class $\mathbb{L}(X)$ of integrands for the integral $\int f dX$ was defined (see Definition 4.17) in an ad hoc fashion. Here we give a concrete description of $\mathbb{L}(X)$.

Theorem 5.72 *Let X be a stochastic integrator. Then a predictable process f belongs to $\mathbb{L}(X)$ if and only if X admits a decomposition $X = M + A$, where $M \in \mathbb{M}_{loc}^2$ (M is a locally square integrable martingale with $M_0 = 0$) and $A \in \mathbb{V}$ (A is a process with finite variation paths) such that*

$$\int_0^t |f_s| d|A|_s < \infty \quad \forall t < \infty \ a.s. \tag{5.7.1}$$

and there exist stopping times $\sigma_k \uparrow \infty$ such that

$$\mathsf{E}[\int_0^{\sigma_k} |f_s|^2 d[M, M]_s] < \infty \quad \forall t < \infty. \tag{5.7.2}$$

Proof If f satisfies (5.7.1), then as seen in Remark 4.24 $f \in \mathbb{L}(A)$ and if f satisfies (5.7.2), then $f \in \mathbb{L}_m^2(M)$ and hence $f \in \mathbb{L}(M)$ in view of Theorem 5.31. Thus if X admits a decomposition $X = M + A$ such that (5.7.1)–(5.7.2) holds then $f \in \mathbb{L}(M) \cap \mathbb{L}(A) \subseteq \mathbb{L}(M + A)$.

Conversely, let $f \in \mathbb{L}(X)$. Then it is easy to see that $h = (1 + |f|) \in \mathbb{L}(X)$. So let $Y = \int (1 + |f|) dX$. Then as noted in Theorem 4.33, Y is a stochastic integrator. Hence by Corollary 5.60, Y admits a decomposition $Y = N + B$ and with $N \in \mathbb{M}_{loc}^2$ and $B \in \mathbb{V}$. Let

$$M = \int (1 + |f|)^{-1} dN,$$

$$A = \int (1 + |f|)^{-1} dB.$$

The two integrals are defined as $(1 + |f|)^{-1}$ is bounded. Further, this also yields, $M \in \mathbb{M}_{loc}^2$ and $A \in \mathbb{V}$. Clearly

$$M + A = \int (1 + |f|)^{-1} dY = \int (1 + |f|)^{-1}(1 + |f|) dX = X.$$

Since $g = f \cdot (1 + |f|)^{-1}$ is bounded, $g \in \mathbb{L}(Y)$ and hence $f = g \cdot (1 + |f|) \in \mathbb{L}(X)$ (see Theorem 4.33). $\qquad\square$

Exercise 5.73 Let M be a continuous martingale with $M_0 = 0$.

(i) Let $M = N + B$ be any decomposition with $N \in \mathbb{M}_{loc}^2$ and $B \in \mathbb{V}$, and let $f \in \mathbb{L}_m^2(N)$. Show that $f \in \mathbb{L}_m^2(M)$.
   HINT: Use (5.6.42).
(ii) Show that $\mathbb{L}(M) = \mathbb{L}_m^2(M)$.
(iii) Show that

$$\mathbb{L}(M) = \{f : \text{ predictable such that } \int_0^t f_s^2 d[M, M]_s < \infty \text{ a.s.}\}. \quad (5.7.3)$$

HINT: Use Remark 5.33.

Remark **5.74** It follows that for a Brownian motion B,

$$\mathbb{L}(B) = \{f : \text{ predictable such that } \int_0^t f_s^2 ds < \infty \text{ a.s.}\}.$$

Definition 5.75 For a process A with finite variation paths, let $\mathbb{L}_l^1(A)$ be the class of predictable processes f satisfying (5.7.1).

Thus, $\mathbb{L}_l^1(A) \subseteq \mathbb{L}(A)$. Theorem 5.72 can be recast as: for a stochastic integrator X, $f \in \mathbb{L}(X)$ if and only if X admits a decomposition $X = M + A$, where M is a locally square integrable martingale with $M_0 = 0$ and A is a process with finite variation paths such that

$$f \in \mathbb{L}_m^2(M), \quad \text{and} \quad f \in \mathbb{L}_l^1(A).$$

As a consequence, for a semimartingale X

$$\mathbb{L}(X) = \bigcup_{\{M,A\,:\,X=M+A,\,M\in\mathbb{M}_{\text{loc}}^2,\,A\in\mathbb{V}\}} \mathbb{L}_m^2(M) \cap \mathbb{L}_l^1(A).$$

Exercise 5.76 Let X be a continuous semimartingale, and let $X = M + A$ be the unique decomposition with M being continuous local martingale and $A \in \mathbb{V}$. Then show that

$$\mathbb{L}(X) = \mathbb{L}_m^2(M) \cap \mathbb{L}_l^1(A).$$

HINT: To show, $\mathbb{L}(X) \subseteq \mathbb{L}_m^2(M) \cap \mathbb{L}_l^1(A)$ use Exercise 5.67.

The following exercise gives an example of a process $A \in \mathbb{V}$ such that $\mathbb{L}(A)$ is strictly larger than $\mathbb{L}_l^1(A)$.

Exercise 5.77 Let $\{\xi^{k,m} : 1 \le k \le 2^{m-1}, m \ge 1\}$ be a family of independent identically distributed random variables with

$$P(\xi^{1,1} = 1) = P(\xi^{1,1} = -1) = 0.5$$

and let $a^{k,m} = \frac{2k-1}{2^m}$. Let

$$\mathcal{F}_t = \sigma\{\xi^{k,m} : a^{k,m} \le t\},$$

$$A_t^n = \sum_{m=1}^{n} \sum_{k=1}^{2^{m-1}} \frac{1}{2^{2m}} \xi^{k,m} 1_{[a^{k,m},\infty)}(t),$$

$$A_t = \sum_{m=1}^{\infty} \sum_{k=1}^{2^{m-1}} \frac{1}{2^{2m}} \xi^{k,m} 1_{[a^{k,m},\infty)}(t)$$

and $f : [0, \infty) \mapsto [0, \infty)$ be defined by

$$f(a^{k,m}) = 2^m$$

with $f(t) = 0$ otherwise. Show that

(i) For each n, $(A_t^n, \mathcal{F}_t)$ is a martingale.
(ii) For each t, A_t^n converges to A_t in $\mathbb{L}^2(P)$.
(iii) $(A_t, \mathcal{F}_t)$ is a martingale.

(iv) $A \in \mathbb{V}$.

(v) Let $B_t = |A|_t$. Show that $\int_0^t f(s)dB_s = \infty$.

(vi) Show that $[A, A]_t = \sum_{m=1}^{\infty} \sum_{k=1}^{2^{m-1}} \frac{1}{2^{4m}} 1_{[a^{n,m}, \infty)}(t)$.

(vii) Show that $\int_0^1 f^2(s)d[A, A]_s < \infty$.

Thus $\int_0^t f dA$ is defined as a stochastic integral but not defined as a Riemann–Stieltjes integral.

5.8 The Dellacherie–Meyer–Mokobodzky–Bichteler Theorem

In Theorem 5.59 we have shown that an r.c.l.l. adapted process is a stochastic integrator if and only if it is a semimartingale. Even if we demand seemingly weaker requirements on an adapted r.c.l.l. process X, it implies that it is a semimartingale.

Indeed, if we demand that $f^n \to f$ uniformly implies that $J_X(f^n)_t \to J_X(f)_t$ in probability for every t, then it follows that X is a semimartingale. Thus demanding continuity of the mapping J_X with the strongest form of convergence on the domain $\mathbb{S}$ and the weakest form of convergence on the range leads to the same conclusion.

Definition 5.78 Let X be an r.c.l.l. adapted process. Then X is said to be a weak stochastic integrator if

$$f^n \in \mathbb{S}, \ f^n \to 0 \text{ uniformly} \ \Rightarrow J_X(f^n)_t \to 0 \text{ in probability for each } t < \infty.$$

The Dellacherie–Meyer–Mokobodzky–Bichteler Theorem (first proven by Dellacherie with contributions from Meyer, Mokobodzky and then independently proven by Bichteler) states that X is a weak stochastic integrator if and only if it is a semimartingale. If X is a semimartingale then it is a stochastic integrator. Clearly, if X is a stochastic integrator, then it is a weak stochastic integrator. To complete the circle, we will now show that if X is a weak stochastic integrator, then it is a semimartingale.

Towards this goal, we introduce some notation. Let $\mathbb{S}^+$ be the class of stochastic processes f of the form

$$f_s(\omega) = \sum_{j=0}^{m} a_{j+1}(\omega) 1_{(s_j, s_{j+1}]}(s) \tag{5.8.1}$$

where $0 = s_0 < s_1 < s_2 < \ldots < s_{m+1} < \infty$, a_{j+1} is bounded $\mathcal{F}_{s_j}^+$ measurable random variable, $0 \le j \le m$, $m \ge 1$ and let $\mathbb{C}$ be the class of stochastic processes f of the form

$$f_s(\omega) = \sum_{j=0}^{n} b_{j+1}(\omega) 1_{(\sigma_j(\omega), \sigma_{j+1}(\omega)]}(s) \tag{5.8.2}$$

where $0 = \sigma_0 \leq \sigma_1 \leq \sigma_2 \leq \ldots, \leq \sigma_{n+1} < \infty$ are $(\mathcal{F}_\cdot^+)$ bounded stopping times and b_{j+1} is bounded $\mathcal{F}_{\sigma_j}^+$ measurable random variable, $0 \leq j \leq n$, $n \geq 1$.

For $f \in \mathbb{C}$, let us define

$$I_X(f)_t(\omega) = \sum_{j=0}^{n} b_{j+1}(\omega)(X_{\sigma_{j+1} \wedge t}(\omega) - X_{\sigma_j \wedge t}(\omega)). \qquad (5.8.3)$$

So for $f \in \mathbb{S}$, $J_X(f) = I_X(f)$. Indeed, if X is a stochastic integrator, then as noted earlier, $I_X(f) = \int f \, dX$, $\forall f \in \mathbb{C}$.

We start with a few observations.

Lemma 5.79 *Let X be an r.c.l.l. adapted process, $f \in \mathbb{C}$ and τ be a stopping time. Then we have, for all $t < \infty$*

$$I_X(f 1_{[0,\tau)})_t = I_X(f)_{t \wedge \tau}. \qquad (5.8.4)$$

Proof Let f be given by (5.8.2), and let $g = f 1_{[0,\tau]}$. Then we have

$$g_s = \sum_{j=0}^{n} b_{j+1} 1_{(\sigma_j \wedge \tau, \sigma_{j+1} \wedge \tau]}(s)$$

and writing $d_{j+1} = b_{j+1} 1_{\{\sigma_j \leq \tau\}}$, it follows that d_{j+1} is $\mathcal{F}_{\sigma_j \wedge \tau}^+$ measurable and we have

$$g_s = \sum_{j=0}^{n} d_{j+1} 1_{(\sigma_j \wedge \tau, \sigma_{j+1} \wedge \tau]}(s).$$

Since $I_X(f)$ and $I_X(g)$ are defined pathwise, we can verify that the relation (5.8.4) is true. $\square$

Lemma 5.80 *Let X be an r.c.l.l. adapted process. Then X is a weak stochastic integrator if and only if X satisfies the following condition for each $t < \infty$:*

$$\forall \varepsilon > 0 \, \exists K_\varepsilon < \infty \text{ s.t. } [\sup_{f \in \mathbb{S}, |f| \leq 1} P(|J_X(f)_t| > K_\varepsilon)] \leq \varepsilon. \qquad (5.8.5)$$

Proof If X satisfies (5.8.5), then given $f^n \in \mathbb{S}$, $a_n = \sup_{t,\omega} |f_t^n(\omega)| \to 0$, we need to show $J_X(f^n)_t \to 0$ in probability. So given $\varepsilon > 0$, get K_ε as in (5.8.5). Given $\eta > 0$ let n_0 be such that for $n \geq n_0$, we have $a_n K_\varepsilon < \eta$. Then $g_n = \frac{1}{a_n} f^n \in \mathbb{S}$ and $|g_n| \leq 1$. Hence

$$P(|J_X(f^n)_t| \geq \eta) \leq P(|J_X(g^n)_t| > K_\varepsilon) \leq \varepsilon.$$

Thus X is a weak stochastic integrator. Conversely let X be a weak stochastic integrator. If for some $\varepsilon, t < \infty$, no such $K_\varepsilon < \infty$ exists, then for each n, we will get f^n such that $|f^n| \leq 1$ and

$$P(|J_X(f^n)_t| > n) \geq \varepsilon. \tag{5.8.6}$$

Then $g^n = \frac{1}{n} f^n$ converges uniformly to zero, but in view of (5.8.6),

$$P(|J_X(g^n)_t| > 1) \geq \varepsilon \ \forall n \geq 1.$$

This contradicts the assumption that X is a weak stochastic integrator. $\qquad \square$

Lemma 5.81 *Let X be an r.c.l.l. adapted process. Then X satisfies (5.8.5) if and only if for each $t < \infty$ the following condition holds:*

$$\forall \varepsilon > 0 \ \exists K_\varepsilon < \infty \ s.t. \ [\sup_{f \in \mathbb{C}, \ |f| \leq 1} P(|I_X(f)_t| > K_\varepsilon)] \leq \varepsilon. \tag{5.8.7}$$

Proof Since $f \in \mathbb{S}$ implies $f 1_{(0,\infty)} \in \mathbb{C}$, it is easy to see that (5.8.7) implies (5.8.5).

So now suppose K_ε is such that (5.8.5) holds. We show that (5.8.7) holds. First let $f \in \mathbb{S}^+$ be given by

$$f_s(\omega) = \sum_{j=0}^{m-1} a_{j+1}(\omega) 1_{(s_j, s_{j+1}]}(s)$$

where $0 = s_0 < s_1 < s_2 < \ldots < s_m < \infty$, a_{j+1} is bounded $\mathcal{F}_{s_j}^+$ measurable random variable, $0 \leq j \leq (m-1)$. Let $0 < \delta_k < \frac{1}{k}$ be such that $s_j + \delta_k < s_{j+1}, 0 \leq j \leq m$ and let

$$g_s^k(\omega) = \sum_{j=0}^{m-1} a_{j+1}(\omega) 1_{(s_j + \delta_k, s_{j+1}]}(s).$$

Since $\mathcal{F}_{s_j}^+ \subseteq \mathcal{F}_{s_j + \delta_k}$, it follows that $g^k \in \mathbb{S}$. Noting that g^k converges to f and using the explicit formula for I_X, J_X along with the fact that paths of X are r.c.l.l. we see that (the number of terms in the sum remain fixed!)

$$J_X(g^k)_t \to I_X(f)_t \quad pointwise.$$

Hence we conclude that

$$P(|I_X(f)_t| > K_\varepsilon) \leq \varepsilon.$$

In other words, (5.8.5) implies

$$\forall \varepsilon > 0 \ \exists K_\varepsilon < \infty \ s.t. \ [\sup_{f \in \mathbb{S}^+, \ |f| \leq 1} P(|I_X(f)_t| > K_\varepsilon)] \leq \varepsilon. \tag{5.8.8}$$

Let us note that if $f \in \mathbb{C}$ is given by (5.8.2) with σ_j being simple stopping times, namely taking finitely many values, then $f \in \mathbb{S}^+$. To see this, let us order all the values taken by the stopping times $\sigma_j, \ j = 0, 1, \ldots, n$ and let this ordered list be

$s_0 < s_1 < s_2 < s_3 < \ldots < s_k$. Recall that by construction, f is l.c.r.l. adapted and thus

$$b_j = \lim_{u \downarrow s_j} f(u)$$

exists and is $\mathcal{F}_{s_j}^+$ measurable. And then

$$f_s = \sum_{j=0}^{k-1} b_j 1_{(s_j, s_{j+1}]}(s)$$

proving that $f \in \mathbb{S}^+$.

Now returning to the proof of (5.8.7), let $f \in \mathbb{C}$ be given by

$$f_s = \sum_{j=0}^{n-1} b_{j+1} 1_{(\sigma_j, \sigma_{j+1}]}(s)$$

with $0 = \sigma_0 \leq \sigma_1 \leq \sigma_2 \leq \ldots, \leq \sigma_n < \infty$ are $(\mathcal{F}_\cdot^+)$ bounded stopping times and b_j is bounded $\mathcal{F}_{\sigma_j}^+$ measurable random variable, $0 \leq j \leq (n-1)$. Let

$$\tau_j^m = \frac{[2^m \sigma_j] + 1}{2^m}, \quad m \geq 1$$

be simple stopping times decreasing to σ_j, $0 \leq j \leq n$, and let

$$g_s^m = \sum_{j=0}^{n-1} b_{j+1} 1_{(\tau_j^m, \tau_{j+1}^m]}(s).$$

Then g^m converges to f, and from the explicit expression for $I_X(g^m)$ and $I_X(f)$, it follows that $I_X(g^m)$ converges to $I_X(f)$. As noted above $g^m \in \mathbb{S}^+$ and hence

$$P(|I_X(g^m)_t| > K_\varepsilon) \leq \varepsilon$$

and then $I_X(g^m)_t \to I_X(f)_t$ implies

$$P(|I_X(f)_t| > K_\varepsilon) \leq \varepsilon.$$

Since this holds for every $f \in \mathbb{C}$, (5.8.7) follows. □

The preceding two lemmas lead to the following interesting result. The convergence for each t in the definition of weak stochastic integrator leads to the apparently stronger result—namely convergence in $\mathbf{d}_{ucp}$.

Theorem 5.82 *Let X be a weak stochastic integrator. Then $f^n \in \mathbb{C}$, $f^n \to 0$ uniformly implies*

$$I_X(f^n) \xrightarrow{ucp} 0. \tag{5.8.9}$$

As a consequence, X satisfies (4.2.20).

Proof Fix $f^n \in \mathbb{C}$ such that $f^n \to 0$ uniformly. Proceeding as in the proof of Lemma 5.80, one can show using (5.8.7) that for each t,

$$I_X(f^n)_t \to 0 \quad \text{in probability.} \tag{5.8.10}$$

Fix $\eta > 0$, $T < \infty$. For $n \geq 1$, let

$$\tau^n = \inf\{t \geq 0 : |I_X(f^n)_t| > 2\eta\}.$$

τ^n is a stopping time with respect to the filtration $(\mathcal{F}^+_\cdot)$. Let $g^n = f^n 1_{[0,\tau^n]}$. Note that $g^n \in \mathbb{C}$. In view of Lemma 5.79 we have

$$I_X(g^n)_T = I_X(f^n)_{T \wedge \tau^n}.$$

Also, from definition of τ^n, we have

$$\{ \sup_{0 \leq t \leq T} |I_X(f^n)_t| > 2\eta \} \subseteq \{|I_X(g^n)_T| > \eta\}. \tag{5.8.11}$$

Clearly, g^n converges to 0 uniformly and hence as noted in (5.8.10),

$$I_X(g^n)_T \to 0 \quad \text{in probability.}$$

In view of (5.8.11), this proves (5.8.9). As seen in Remark 4.25, (5.8.9) implies (4.2.20). $\qquad\qquad\square$

Here is one last observation in this theme.

Theorem 5.83 *Let X be a weak stochastic integrator, and let $h^n \in \mathbb{C}$. Then*

$$h^n \xrightarrow{ucp} 0 \Rightarrow I_X(h^n) \xrightarrow{ucp} 0. \tag{5.8.12}$$

Proof Fix $T < \infty$ and let $n_1 = 1$. For each $k \geq 2$, get n_k such that $n_k > n_{k-1}$ and

$$n \geq n_k \Rightarrow P(\sup_{0 \leq t \leq T} |h^n_t| > \tfrac{1}{k}) \leq \tfrac{1}{k}.$$

For $n_k \leq n < n_{k+1}$, let
$$\sigma_n = \inf\{t : |h^n_t| > \tfrac{1}{k}\}.$$

Since $h^n \in \mathbb{C}$, (a left continuous step function) it can be seen that σ_n is a stopping time and that

$$f^n = h^n 1_{[0,\sigma_n]}$$

satisfies

$$|f_t^n| \le \tfrac{1}{k} \quad \text{for } n_k \le n < n_{k+1}$$

and

$$\{ \sup_{0 \le t \le T} |h_t^n| > \tfrac{1}{k} \} = \{\sigma_n \le T\}.$$

So

$$P(\sigma_n \le T) \le \tfrac{1}{k} \quad \text{for } n_k \le n < n_{k+1}. \tag{5.8.13}$$

Thus

$$f^n \text{ converges to } 0 \text{ uniformly} \tag{5.8.14}$$

and

$$P(I_X(f^n)_t = I_X(h^n)_t \ \forall t \in [0, T]) \ge P(\sigma_n > T). \tag{5.8.15}$$

In view of Theorem 5.82, (5.8.14) implies $I_X(f^n) \xrightarrow{ucp} 0$ and then (5.8.13) and (5.8.15) yield the desired conclusion, namely

$$I_X(h^n) \xrightarrow{ucp} 0.$$

$\square$

In view of this result, for a weak stochastic integrator, we can extend I_X continuously to the closure $\bar{\mathbb{C}}$ of $\mathbb{C}$ in the $\mathbf{d}_{ucp}$ metric.

For $g \in \bar{\mathbb{C}}$, let $I_X(g)$ be defined as limit of $I_X(g^n)$ where $g^n \xrightarrow{ucp} g$. It is easy to see that I_X is well defined. If X is a stochastic integrator, I_X agrees with $\int g \, dX$.

We identify $\bar{\mathbb{C}}$ in the next result.

Lemma 5.84 *The closure $\bar{\mathbb{C}}$ of $\mathbb{C}$ in the $\mathbf{d}_{ucp}$ metric is given by*

$$\bar{\mathbb{C}} = \{Z^- : Z \in \mathbb{R}^0\}.$$

Proof Let $f \in \bar{\mathbb{C}}$ and let $f^n \in \mathbb{C}$, $f^n \xrightarrow{ucp} f$. Then $V_t^n = \lim_{u \downarrow t} f_u^n$ are r.c.l.l. adapted processes, and V^n can be seen to be Cauchy in $\mathbf{d}_{ucp}$ and hence converge to V. Further, using Theorem 2.72, it follows that a subsequence of V^n converges to V uniformly on $[0, T]$ for every $T < \infty$, almost surely. Thus $V \in \mathbb{R}^0$. Now it follows that $f = V^-$. $\square$

Theorem 5.85 *Let X be a weak stochastic integrator and $Z \in \mathbb{R}^0$. Let $\{\tau_n^m : n \ge 1\}$ be a sequence of partitions of $[0, \infty)$ via stopping times:*

$$0 = \tau_0^m < \tau_1^m < \tau_2^m \ldots ; \tau_n^m \uparrow \infty, \quad m \ge 1$$

such that for some sequence $\delta_m \downarrow 0$ (of real numbers)

$$|Z_{t-} - Z_{\tau_n^m}| \le \delta_m \quad for \; \tau_n^m < t \le \tau_{n+1}^m, \quad n \ge 0, \quad m \ge 1. \tag{5.8.16}$$

For $m \ge 1$, *let*

$$Z_t^m = \sum_{n=0}^{\infty} Z_{t \wedge \tau_n^m} 1_{(\tau_n^m, \tau_{n+1}^m]}(t).$$

Then

$$I_X(Z^m) = \sum_{n=0}^{\infty} Z_{t \wedge \tau_n^m}(X_{\tau_{n+1}^m \wedge t} - X_{\tau_n^m \wedge t}) \tag{5.8.17}$$

and

$$I_X(Z^m) \xrightarrow{ucp} I_X(Z^-). \tag{5.8.18}$$

Thus, if X *is also a stochastic integrator,* $I_X(Z^-) = \int Z^- dX$.

Proof Let us note that given $\delta_m \downarrow 0$, there exist stopping times $\{\tau_n^m : n \ge 1\} \, m \ge 1$, satisfying (5.8.16), (say given in (4.5.12)).
    For $k \ge 1$, let

$$Z_t^{m,k} = \sum_{n=0}^{k} Z_{t \wedge \tau_n^m} 1_{(\tau_n^m, \tau_{n+1}^m]}(t).$$

Then $Z^{m,k} \in \mathbb{C}$ and

$$I_X(Z^{m,k}) = \sum_{n=0}^{k} Z_{t \wedge \tau_n^m}(X_{\tau_{n+1}^m \wedge t} - X_{\tau_n^m \wedge t}).$$

Now $P(\sup_{0 \le t \le T} |Z_t^{m,k} - Z_t^m| > 0) \le P(\tau_k^m < T)$ and hence $Z^m \in \bar{\mathbb{C}}$ and

$$I_X(Z^{m,k}) \xrightarrow{ucp} I_X(Z^m).$$

This proves (5.8.17). Now Z^m converges to Z^- uniformly and hence $Z \in \bar{\mathbb{C}}$ and

$$I_X(Z^m) \xrightarrow{ucp} I_X(Z^-).$$

If X is also a stochastic integrator, using Theorem 4.62 it follows that for $Z \in \mathbb{R}^0$, $I_X(Z^-)$ agrees with $\int Z^- dX$. $\square$

Remark **5.86** Note that while dealing with weak stochastic integrators, we considered the filtration $\mathcal{F}_\cdot^+$ for defining the class $\mathbb{C}$, but the definition of weak stochastic integrator did not require the underlying filtration to be right continuous.

Thus for $Z \in \mathbb{R}^0$ and a weak stochastic integrator X, we define $I_X(Z^-)$ to be the stochastic integral $\int Z^- dX$. When a weak stochastic integrator is also a stochastic integrator, this does not lead to any ambiguity as noted in the previous theorem.

Remark 5.87 A careful look at results in Sect. 4.6 shows that Theorems 4.64, 4.67 continue to be true if the underlying processes are weak stochastic integrators instead of stochastic integrators. Instead of invoking Theorem 4.62, we can invoke Theorem 5.85. Thus weak stochastic integrators X, Y admit quadratic variations $[X, X]$, $[Y, Y]$ and cross-quadratic variation $[X, Y]$. Various results on quadratic variation obtained in Sect. 4.6 continue to be true for weak stochastic integrators.

Moreover, Theorems 5.44, 5.45 and Lemma 5.46 are true for weak stochastic integrators as well since the proof only relies on quadratic variation. Likewise, Lemma 5.48 is true for weak stochastic integrators since apart from quadratic variation, it relies on (4.2.20), a property that holds for weak stochastic integrators as noted earlier in Theorem 5.82.

As a consequence, Theorems 5.50, 5.55 and 5.58 are true for weak stochastic integrators.

This discussion leads to the following result, whose proof is same as that of Theorem 5.59.

Theorem 5.88 *Let X be a weak stochastic integrator. Then X is a semimartingale.*

Here is the full version of the Dellacherie–Meyer–Mokobodzky–Bichteler Theorem.

Theorem 5.89 *Let X be an r.c.l.l. $(\mathcal{F}_\centerdot)$ adapted process. Let J_X be defined by (4.2.1)–(4.2.2). Then the following are equivalent.*

(i) *X is a weak stochastic integrator; i.e. if $f^n \in \mathbb{S}$, $f^n \to 0$ uniformly, then $J_X(f^n)_t \to 0$ in probability $\forall t < \infty$.*

(ii) *If $f^n \in \mathbb{S}$, $f^n \to 0$ uniformly, then $J_X(f^n) \xrightarrow{ucp} 0$.*

(iii) *If $f^n \in \mathbb{S}$, $f^n \xrightarrow{ucp} 0$, then $J_X(f^n) \xrightarrow{ucp} 0$.*

(iv) *If $f^n \in \mathbb{S}$, $f^n \xrightarrow{bp} 0$, then $J_X(f^n) \xrightarrow{ucp} 0$.*

(v) *X is a stochastic integrator; i.e. the mapping J_X from $\mathbb{S}$ to $\mathbb{R}^0(\Omega, (\mathcal{F}_\centerdot), \mathrm{P})$ has an extension $J_X \colon \mathbb{B}(\tilde{\Omega}, \mathcal{P}) \mapsto \mathbb{R}^0(\Omega, (\mathcal{F}_\centerdot), \mathrm{P})$ satisfying*

$$f^n \xrightarrow{bp} f \text{ implies } J_X(f^n) \xrightarrow{ucp} J_X(f).$$

(vi) *X is a semimartingale; i.e. X admits a decomposition $X = M + A$ where M is a local martingale and A is a process with finite variation paths.*

(vii) *X admits a decomposition $X = N + B$ where N is a locally square integrable martingale and B is a process with finite variation paths.*

Proof Equivalence of (v), (vi), (vii) has been proven in Theorem 5.59 and Corollary 5.60. Clearly,

$$(v) \Rightarrow (iv) \Rightarrow (iii) \Rightarrow (ii) \Rightarrow (i).$$

Theorems 5.82 and 5.83 tell us that

$$(i) \Rightarrow (ii) \Rightarrow (iii).$$

And we have observed in Theorem 5.88 that (iii) implies (vi). This completes the proof. $\square$

5.9 Enlargement of Filtration

The main result of the section is about enlargement of the underlying filtration $(\mathcal{F}_\cdot)$ by adding a set $A \in \mathcal{F}$ to each $\mathcal{F}_t$. The surprising result is that a semimartingale for the original filtration remains a semimartingale for the enlarged filtration. In the traditional approach, this was a deep result as it required decomposition of the semimartingale into local martingale w.r.t. the enlarged filtration and a finite variation process.

Let $A \in \mathcal{F}$ be fixed, and let us define a filtration $(\mathcal{G}_\cdot)$ by

$$\mathcal{G}_t = \{(B \cap A) \cup (C \cap A^c) : B, C \in \mathcal{F}_t\}. \tag{5.9.1}$$

It is easy to see that $\mathcal{G}_t$ is a σ-field for all $t \geq 0$ and $(\mathcal{G}_\cdot)$ is a filtration. Using the description (5.9.1) of sets in $\mathcal{G}_t$, it can be seen that if ξ is a $\mathcal{G}_t$ measurable bounded random variable, then $\exists \, \mathcal{F}_t$ measurable bounded random variables η, η' such that

$$\xi = \eta 1_A + \eta' 1_{A^c}. \tag{5.9.2}$$

Let $\mathbb{S}(\mathcal{G}_\cdot)$ denote the class of simple predictable process for the filtration $(\mathcal{G}_\cdot)$. Using (5.9.2) it is easy to verify that for $f \in \mathbb{S}(\mathcal{G}_\cdot) \, \exists \, g, h \in \mathbb{S}(\mathcal{F}_\cdot)$ such that

$$f(t, \omega) = g(t, \omega) 1_A(\omega) + h(t, \omega) 1_{A^c}(\omega) \quad \forall (t, \omega) \in \widetilde{\Omega}. \tag{5.9.3}$$

With this we can now describe the connection between predictable processes for the two filtrations.

Theorem 5.90 *Let f be an $(\mathcal{G}_\cdot)$ predictable process. We can choose $(\mathcal{F}_\cdot)$ predictable processes g, h such that (5.9.3) holds. Further, if f is bounded by a constant c, then g, h can also be chosen to be bounded by the same constant c.*

Proof Let $\mathbb{K}$ be the set of $(\mathcal{G}_{.})$ predictable process f for which the conclusion is true. We have seen that $\mathbb{K}$ contains $\mathbb{S}(\mathcal{G}_{.})$. We next show that $\mathbb{K}$ is closed under pointwise convergence. This will show in particular that $\mathbb{K}$ is closed under bp-convergence and hence that the conclusion is true for all bounded predictable processes. If $f^n \in \mathbb{K}$ with

$$f^n = g^n 1_A + h^n 1_{A^c}$$

and $f^n \to f$ then we can take

$$g(t, \omega) = \limsup_{n \to \infty} g^n(t, \omega) 1_{\{\limsup_{n \to \infty} |g^n|(t,\omega) < \infty\}} \tag{5.9.4}$$

$$h(t, \omega) = \limsup_{n \to \infty} h^n(t, \omega) 1_{\{\limsup_{n \to \infty} |h^n|(t,\omega) < \infty\}} \tag{5.9.5}$$

and then it follows that $f = g 1_A + h 1_{A^c}$ and thus $f \in \mathbb{K}$. As noted above, this proves $\mathbb{K}$ contains all bounded predictable processes and in turn all predictable processes since $f^n = f 1_{\{|f| \le n\}}$ converges pointwise to f.

If f is bounded by c, we can replace g by $\tilde{g} = g 1_{\{|g| \le c\}}$ and h by $\tilde{h} = h 1_{\{|h| \le c\}}$. $\square$

Theorem 5.91 *Let X be a semimartingale for the filtration $(\mathcal{F}_{.})$. Let $A \in \mathcal{F}$ and $(\mathcal{G}_{.})$ be defined by (5.9.1).*

Then X is also a semimartingale for the filtration $(\mathcal{G}_{.})$.

Proof Let us denote the mapping defined by (4.2.1) and (4.2.2) for the filtration $(\mathcal{G}_{.})$ by H_X, and let J_X be the mapping for the filtration $(\mathcal{F}_{.})$. Since these mappings are defined pathwise, it is easy to verify that if $f \in \mathbb{S}(\mathcal{G}_{.})$ and $g, h \in \mathbb{S}(\mathcal{F}_{.})$ are as in (5.9.3), then

$$H_X(f) = J_X(g) 1_A + J_X(h) 1_{A^c} \tag{5.9.6}$$

We will prove that X is a weak stochastic integrator for the filtration $(\mathcal{G}_{.})$. For this, let $f^n \in \mathbb{B}(\tilde{\Omega}, \mathcal{P}(\mathcal{G}_{.}))$ decrease to 0 uniformly. Let $a_n \downarrow 0$ be such that $|f^n| \le a_n$. Then for each n invoking Theorem 5.90 we choose $g^n, h^n \in \mathbb{S}(\mathcal{F}_{.})$ with $|g^n| \le a_n$, $|h^n| \le a_n$ such that

$$f^n = g^n 1_A + h^n 1_{A^c}.$$

As noted above this gives, for $n \ge 1$

$$H_X(f^n) = J_X(g^n) 1_A + J_X(h^n) 1_{A^c}. \tag{5.9.7}$$

Since X is a semimartingale for the filtration $(\mathcal{F}_{.})$, it is a stochastic integrator. Thus, $J_X(g^n) \xrightarrow{ucp} 0$ and $J_X(h^n) \xrightarrow{ucp} 0$ and then (5.9.7) implies $H_X(f^n) \xrightarrow{ucp} 0$. Hence X is a weak stochastic integrator for the filtration $(\mathcal{F}_{.})$. Invoking Theorem 5.89, we conclude that X is a semimartingale for the filtration $(\mathcal{G}_{.})$. $\square$

Remark **5.92** As noted in Theorem 4.13, when f is $(\mathcal{F}.)$ predictable bounded process, $H_X(f) = J_X(f)$ and thus the stochastic integral $\int f dX$ is unambiguously defined.

Theorem 5.93 *Let X be a stochastic integrator on $(\Omega, \mathcal{F}, \mathsf{P})$ with a filtration $(\mathcal{F}.)$. Let Q be a probability measure on $(\Omega, \mathcal{F})$ that is absolutely continuous w.r.t. P. Then X is a stochastic integrator on $(\Omega, \mathcal{F}, \mathsf{Q})$, and the stochastic integral under P is a version of the integral under Q.*

Proof Let $\mathcal{H}$ be the completion of $\mathcal{F}$ under Q, and for $t \geq 0$ let $\mathcal{H}_t$ be the σ-field obtained by adding all Q null sets to $\mathcal{F}_t$. Let ξ denote the Radon–Nikodym derivative of Q with respect to P, and let $\Omega_0 = \{\omega : \xi(\omega) > 0\}$. Let $\mathcal{G}$ be the filtration defined by (5.9.1) with $A = \Omega_0$. It can be checked that a process f is $(\mathcal{H}.)$ predictable if and only if $f 1_{\Omega_0}$ is $(\mathcal{G}.)$ predictable. Let

$$H_X(f) = J_X(f 1_{\Omega_0}) \quad f \in \mathbb{B}(\tilde{\Omega}, \mathcal{P}(\mathcal{H}.)).$$

It is easy to see that H_X is the required extension of the integral of simple predictable processes.   □

Remark **5.94** Suppose we start with a filtration $(\tilde{\mathcal{F}}.)$ that may not satisfy the condition that each $\tilde{\mathcal{F}}_t$ contains all null sets. Suppose X is a $(\tilde{\mathcal{F}}.)$ adapted process that satisfies (4.2.3) for this filtration. Let $\mathcal{F}_t$ be the smallest σ-field containing $\tilde{\mathcal{F}}_t$ and all the null sets. It is easy to see that X continues to satisfy (4.2.3) w.r.t. the filtration $(\mathcal{F}.)$ and is thus a stochastic integrator.

Exercise 5.95 Let X be a semimartingale for a filtration $(\mathcal{F}.)$ on $(\Omega, \mathcal{F}, \mathsf{P})$. Let $\{A_m : m \geq 1\}$ be a partition of Ω with $A_n \in \mathcal{F}$ for all $n \geq 1$. For $t \geq 0$, let

$$\mathcal{G}_t = \sigma(\mathcal{F}_t \cup \{A_m : m \geq 1\}).$$

Show that

(i) For every $(\mathcal{G}.)$ predictable process f, there exists $(\mathcal{F}.)$ predictable processes f^m such that

$$f = \sum_{m=1}^{\infty} 1_{A_m} f^m.$$

(ii) Suppose for each $m \geq 1$, $\{Y^{m,n} : n \geq 1\}$ are r.c.l.l. processes such that $Y^{m,n} \xrightarrow{ucp} Y^m$ as $n \to \infty$. Let

$$Z^n = \sum_{m=1}^{\infty} 1_{A_m} Y^{m,n}$$

and

$$Z = \sum_{m=1}^{\infty} 1_{A_m} Y^m.$$

Then prove that $Z^n \xrightarrow{ucp} Z$ as $n \to \infty$.

(iii) Show that X is a stochastic integrator for the filtration $(\mathcal{G}_.)$.

Chapter 6
Pathwise Formula for the Stochastic Integral

In the previous chapter, we had obtained a pathwise formula for the quadratic variation of a martingale. We will show in this chapter that the same formula yields the quadratic variation of a semimartingale. We will also obtain a pathwise formula for the stochastic integral.

6.1 Preliminaries

For a simple predictable process f, the stochastic integral $\int f \, dX$ has been defined explicitly, path by path. In other words, the path $t \mapsto (\int_0^t f \, dX)(\omega)$ is a function of the paths $\{f_s(\omega) : 0 \le s \le t\}$ and $\{X_s(\omega) : 0 \le s \le t\}$ of the integrand f and integrator X. For a general (bounded) predictable f the integral has been defined as limit in probability of suitable approximations and it is not clear if we can obtain a *pathwise* version. In statistical inference for stochastic processes the estimate; in stochastic filtering theory the filter; and in stochastic control theory the control in most situations involves stochastic integral, where the integrand and integrator are functionals of the observation path and to be meaningful, the integral should also be a functional of the observation.

How much does the integral depend upon the underlying filtration or the underlying probability measure? Can we get one fixed version of the integral when we have not one but a family of probability measures $\{P_\alpha\}$ such that X is a semimartingale under each P_α.

If we have one probability measure Q such that each P_α is absolutely continuous w.r.t. Q and the underlying process X is a semimartingale under Q then the answer to the question above is yes—simply take the integral defined under Q and that will agree with the integral under P_α for each α by Remark 4.26.

© Springer Nature Singapore Pte Ltd. 2018
R. L. Karandikar and B. V. Rao, *Introduction to Stochastic Calculus*,
Indian Statistical Institute Series, https://doi.org/10.1007/978-981-10-8318-1_6

When the family $\{P_\alpha\}$ is countable family, such a Q can always be constructed. However, such a Q may not exist in general. One concrete instance where such a situation arises and has been considered in the literature is the following. In the context of Markov Processes, one considers a family of measures $\{P_x : x \in E\}$, where P_x represents the distribution of the Markov Process conditioned on $X_0 = x$. See [18]. In this context, Cinlar et al. [9] showed the following: For processes S such that S is semimartingale under P_x for every x and for f in a suitable class of predictable processes, there exists a process Z such that Z is a version of $\int f\, dS$ under P_x for every x.

In Bichteler [3], Karandikar [33, 34, 38] it was shown that for an r.c.l.l. adapted process Z and a semimartingale X, suitably constructed Riemann sums converge almost surely to the stochastic integral $\int Z^-\, dX$. This result was recast in [41] to obtain a universal mapping $\Phi : \mathbb{D}([0,\infty),\mathbb{R}) \times \mathbb{D}([0,\infty),\mathbb{R}) \mapsto \mathbb{D}([0,\infty),\mathbb{R})$ such that if X is a semimartingale and Z is an r.c.l.l. adapted process, then $\Phi(Z, X)$ is a version of the stochastic integral $\int Z^-\, dX$.

As in the previous chapter, we fix a filtration $(\mathcal{F}_\cdot)$ on a complete probability space $(\Omega, \mathcal{F}, P)$ and we assume that $\mathcal{F}_0$ contains all P-null sets in $\mathcal{F}$.

First we will prove a simple result which enables us to go from $\mathbb{L}^2$ estimates to almost sure convergence.

Lemma 6.1 *Let V^m be a sequence of r.c.l.l. process and τ_k an increasing sequence of stopping times, increasing to ∞ such that for all $k \geq 1$,*

$$\sum_{m=1}^{\infty} \|\sup_{t \leq \tau_k} |V_t^m|\|_2 < \infty. \tag{6.1.1}$$

Then we have

$$\sup_{t \leq T} |V_t^m| \to 0 \quad \forall T < \infty, \quad a.s.$$

Proof The condition (6.1.1) implies

$$\|\sum_{m=1}^{\infty} \sup_{t \leq \tau_k} |V_t^m|\|_2 < \infty$$

and hence for each k,

$$\sum_{m=1}^{\infty} \sup_{t \leq \tau_k} |V_t^m| < \infty \quad a.s.$$

Since τ_k increase to ∞, the required result follows. $\square$

6.2 Pathwise Formula for the Stochastic Integral

Recall that $\mathbb{D}([0, \infty), \mathbb{R})$ denotes the space of r.c.l.l. functions on $[0, \infty)$ and that for $\gamma \in \mathbb{D}([0, \infty), \mathbb{R})$, $\gamma(t-)$ denotes the left limit at t (for $t > 0$) and $\gamma(0-) = 0$.

Fix $\gamma \in \mathbb{D}([0, \infty), \mathbb{R})$. For each $n \geq 1$; let $\{t_i^n(\gamma) : i \geq 1\}$ be defined inductively as follows : $t_0^n(\gamma) = 0$ and having defined $t_i^n(\gamma)$, let

$$t_{i+1}^n(\gamma) = \inf\{t > t_i^n(\gamma) : |\gamma(t) - \gamma(t_i^n(\gamma))| \geq 2^{-n} \text{ or } |\gamma(t-) - \gamma(t_i^n(\gamma))| \geq 2^{-n}\}.$$
(6.2.1)

Note that for each $\gamma \in \mathbb{D}([0, \infty), \mathbb{R})$ and for $n \geq 1$, $t_i^n(\gamma) \uparrow \infty$ as $i \uparrow \infty$ (if $\lim_i t_i^n(\gamma) = t^* < \infty$, then the function γ cannot have a left limit at t^*). For $\gamma, \gamma_1 \in \mathbb{D}([0, \infty), \mathbb{R})$ let

$$\Phi_n(\gamma, \gamma_1)(t) = \sum_{i=0}^{\infty} \gamma(t_i^n(\gamma) \wedge t)(\gamma_1(t_{i+1}^n(\gamma) \wedge t) - \gamma_1(t_i^n(\gamma) \wedge t)).$$
(6.2.2)

Since $t_i^n(\gamma)$ increases to infinity, for each γ and t fixed, the infinite sum appearing above is essentially a finite sum and hence $\Phi_n(\gamma, \gamma_1)$ is itself an r.c.l.l. function. We now define a mapping $\Phi : \mathbb{D}([0, \infty), \mathbb{R}) \times \mathbb{D}([0, \infty), \mathbb{R}) \mapsto \mathbb{D}([0, \infty), \mathbb{R})$ as follows: Let $\mathbb{D}^* \subseteq \mathbb{D}([0, \infty), \mathbb{R}) \times \mathbb{D}([0, \infty), \mathbb{R})$ be defined by

$$\mathbb{D}^* = \{(\gamma, \gamma_1) : \Phi_n(\gamma, \gamma_1) \text{ converges in ucc topology}\}$$

and for $\gamma, \gamma_1 \in \mathbb{D}([0, \infty), \mathbb{R})$

$$\Phi(\gamma, \gamma_1) = \begin{cases} \lim_n \Phi_n(\gamma, \gamma_1) & \text{if } (\gamma, \gamma_1) \in \mathbb{D}^* \\ 0 & \text{otherwise.} \end{cases}$$
(6.2.3)

Note that the mapping Φ has been defined without any reference to a probability measure or a process. Here is the main result on pathwise integration formula.

Theorem 6.2 *Let X be a semimartingale on a probability space $(\Omega, \mathcal{F}, P)$ with filtration $(\mathcal{F}.)$ and let U be an r.c.l.l. adapted process. Let*

$$Z.(\omega) = \Phi(U.(\omega), X.(\omega))$$
(6.2.4)

Then

$$Z = \int U^- dX.$$
(6.2.5)

Proof For each fixed n, define $\{\sigma_i^n : i \geq 0\}$ inductively with $\sigma_0^n = 0$ and

$$\sigma_{i+1}^n = \inf\{t > \sigma_i^n : |U_t - U_{\sigma_i^n}| \geq 2^{-n} \text{ or } |U_{t-} - U_{\sigma_i^n}| \geq 2^{-n}\}.$$

For all n, i, σ_i^n is a stopping time. Let us note that $\sigma_i^n(\omega) = t_i^n(U.(\omega))$. Let

$$Z_\cdot^n(\omega) = \Phi_n(U_\cdot(\omega), X_\cdot(\omega)).$$

Then we can see that

$$Z_t^n = \sum_{j=0}^{\infty} U_{t \wedge \sigma_j^n}(X_{t \wedge \sigma_{j+1}^n} - X_{t \wedge \sigma_j^n})$$

and thus $Z^n = \int U^n dX$ where

$$U_t^n = \sum_{j=0}^{\infty} U_{t \wedge \sigma_j^n} 1_{(\sigma_j^n, \sigma_{j+1}^n]}(t).$$

By definition of $\{\sigma_i^n\}$, $\{U^n\}$, we have

$$|U_t^n - U_{t-}| \le 2^{-n} \tag{6.2.6}$$

and hence $U^n \to U^-$ in ucp. Then by Theorem 4.50, $Z^n \to Z = \int U^- dX$ in the ucp metric.

The crux of the argument is to show that the convergence is indeed almost sure-

$$\sup_{t \le T} |\int_0^t U^n dX - \int_0^t U^- dX| \to 0 \quad \forall T < \infty \ a.s. \tag{6.2.7}$$

Once this is shown, it would follow that $(U_\cdot(\omega), X_\cdot(\omega)) \in \mathbb{D}^*$ $a.s.$ and then by definition of Φ and Z we conclude that $Z_n = \Phi(U^n, X)$ converges to $\Phi(U, X)$ in ucc topology almost surely. Since $Z^n \to Z$ in ucp, we have $Z = \Phi(U, X)$ completing the proof.

Remains to prove (6.2.7). For this, first using Corollary 5.60, let us decompose X as $X = M + A$, $M \in \mathbb{M}_{loc}^2$ and $A \in \mathbb{V}$. Now using the fact that the dA integral is just the Lebesgue–Stieltjes integral and the estimate (6.2.6) we get

$$|\int_0^t U^n dA - \int_0^t U^- dA| \le \int_0^t |U_s^n - U_{s-}| d|A|_t$$

$$\le 2^{-n} |A|_t.$$

and hence

$$\sup_{t \le T} |\int_0^t U^n dA - \int_0^t U^- dA| \le 2^{-n} |A|_T \tag{6.2.8}$$

$$\to 0.$$

Thus (6.2.7) would follow in view of linearity of the integral once we show

$$\sup_{t \le T} |\int_0^t U^n dM - \int_0^t U^- dM| \to 0 \; \forall T < \infty \; a.s. \qquad (6.2.9)$$

Let τ_k be stopping times increasing to ∞ such that $\tau_k \le k$ and $M^{[\tau^k]}$ is a square integrable martingale so that

$$E[[M,M]_{\tau_k}] = E[[M^{[\tau^k]}, M^{[\tau^k]}]_k] < \infty. \qquad (6.2.10)$$

Thus using the estimate (5.4.10) on the growth of the stochastic integral with respect to a local martingale (Theorem 5.31), we get

$$E[\sup_{t \le \tau_k} |\int_0^t U^n dM - \int_0^t U^- dM|^2] \le 4E[\int_0^{\tau_k} |U_s^n - U_{s-}|^2 d[M,M]_s] \qquad (6.2.11)$$

$$\le 4(2^{-2n})E[[M,M]_{\tau_k}].$$

Thus, writing $\xi_t^n = \int_0^t U^n dM - \int_0^t U^- dM$ and $\alpha_k = \sqrt{E[[M,M]_{\tau_k}]}$, we have

$$\|\sup_{t \le \tau_k} |\xi_t^n|\|_2 \le 2^{-n+1} \alpha_k. \qquad (6.2.12)$$

Since $\alpha_k < \infty$ as seen in (6.2.10), Lemma 6.1 implies that (6.2.9) is true completing the proof.   □

Remark **6.3** This result implies that the integral $\int U^- dX$ for an r.c.l.l. adapted process U does not depend upon the underlying filtration or the probability measure or on the decomposition of the semimartingale X into a (local) martingale and a process with finite variation paths. An ω-path $t \mapsto \int_0^t U^- dX(\omega)$ of the integral depends only on the ω-paths $t \mapsto U_t(\omega)$ of the integrand and $t \mapsto X_t(\omega)$ of the integrator. The same however cannot be said in general about $\int f dX$ if f is given to be a predictable process.

Remark **6.4** In Karandikar [38, 41] the same result was obtained with $t_i^n(\gamma)$ defined via

$$t_{i+1}^n(\gamma) = \inf\{t \ge t_i^n(\gamma) : |\gamma(t) - \gamma(t_i^n(\gamma))| \ge 2^{-n}\}$$

instead of (6.2.1). The result is of course true, but requires the underlying σ-fields to be right continuous to prove that the resulting σ_j^n are stopping times.

6.3   Pathwise Formula for Quadratic Variation

In Sect. 5.2 we had obtained a pathwise formula for the quadratic variation process of a locally square integrable martingale—namely (5.3.16). We now observe that the same formula gives quadratic variation of local martingales as well—indeed of semimartingales. We will need to use notations from Sect. 5.2 as well as Sect. 6.2.

Theorem 6.5 *Let Ψ be the mapping defined in Sect. 5.2 by (5.2.2). For a semimartingale X, let*

$$[X, X]_t^{\psi}(\omega) = [\Psi(X.(\omega))](t). \tag{6.3.1}$$

Then $[X, X]^{\psi}$ is a version of the quadratic variation $[X, X]$ i.e.

$$\mathsf{P}([X, X]_t^{\psi} = [X, X]_t \ \forall t) = 1.$$

Proof For $\gamma \in \mathbb{D}([0, \infty), \mathbb{R})$ and for $n \geq 1$, let $\{t_i^n(\gamma) : i \geq 1\}$ be defined inductively by (6.2.1). Recall the Definition 5.2.1 of Ψ_n and (6.2.2) of Φ_n. Using the identity $(b - a)^2 = b^2 - a^2 - 2a(b - a)$ with $b = \gamma(t_{i+1}^n(\gamma) \wedge t)$ and $a = \gamma(t_i^n(\gamma) \wedge t)$ and summing over $i \in \{0, 1, 2 \ldots\}$, we get the identity

$$\Psi_n(\gamma) = (\gamma(t))^2 - (\gamma(0))^2 - 2\Phi_n(\gamma, \gamma). \tag{6.3.2}$$

Let $\Psi, \widetilde{\mathbb{D}}$ be as defined in Sect. 5.2 and $\Phi, \mathbb{D}^*$ be as defined in Sect. 6.2. Let

$$\widehat{\mathbb{D}} = \{\gamma \in \mathbb{D} : (\gamma, \gamma) \in \mathbb{D}^*\}.$$

Then using (6.3.2) along with the definition (6.2.3) of Φ, it follows that

$$\widehat{\mathbb{D}} \subseteq \widetilde{\mathbb{D}}$$

and

$$\Psi(\gamma) = (\gamma(t))^2 - (\gamma(0))^2 - 2\Phi(\gamma, \gamma) \quad \gamma \in \widehat{\mathbb{D}}. \tag{6.3.3}$$

As noted in Sect. 6.2, $(X.(\omega), X.(\omega)) \in \mathbb{D}^*$ almost surely and

$$\int X^- dX = \Phi(X, X) \tag{6.3.4}$$

From (6.3.1), (6.3.3) and (6.3.4) and it follows that

$$[X, X]_t^{\psi} = X_t^2 - X_0^2 - 2 \int_0^t X^- dX. \tag{6.3.5}$$

This along with (4.6.1) implies $[X, X]_t^{\psi} = [X, X]_t$ completing the proof. $\square$

Chapter 7
Continuous Semimartingales

In this chapter, we will consider continuous semimartingales and show that stochastic differential equations driven by these can be analysed essentially using the same techniques as in the case of SDE driven by Brownian motion. This can be done using *random time change*. The use of random time change in study of solutions to stochastic differential equations was introduced in [33, 34].

We introduce random time change and we then obtain a growth estimate on $\int f\, dX$ where X is a continuous semimartingale and f is a predictable process. Then we observe that if a semimartingale satisfies a condition (7.2.2), then the growth estimate on $\int f\, dX$ is very similar to the growth estimate on $\int f\, d\beta$, where β is a Brownian motion. We also note that by changing time via a suitable random time, any semimartingale can be transformed to a semimartingale satisfying (7.2.2). Thus, without loss of generality we can assume that the driving semimartingale satisfies (7.2.2) and then use techniques used for Brownian motion case. We thus show that stochastic differential equation driven by continuous semimartingales admits a solution when the coefficients are Lipschitz functions. We also show that in this case, one can get a pathwise formula for the solution, like the formula for the integral obtained in the previous chapter.

7.1 Random Time Change

Change of variable plays an important role in calculations involving integrals of functions of a real variable. As an example, let G be a continuous increasing function with $G[0] = 0$. Let us look at the formula

$$f(G(t)) = f(0) + \int_0^t f'(G(s))dG(s). \tag{7.1.1}$$

© Springer Nature Singapore Pte Ltd. 2018

R. L. Karandikar and B. V. Rao, *Introduction to Stochastic Calculus*,
Indian Statistical Institute Series, https://doi.org/10.1007/978-981-10-8318-1_7

which was derived in Sect. 4.8. We had seen that when G is absolutely continuous, this formula follows by the chain rule for derivatives. Let $a : [0, \infty) \mapsto [0, \infty)$ be a continuous strictly increasing one-one onto function. Let us write $\tilde{G}(s) = G(a(s))$. It can be seen that (7.1.1) can equivalently be written as

$$f(\tilde{G}(t)) = f(0) + \int_0^t f'(\tilde{G}(s))d\tilde{G}(s). \qquad (7.1.2)$$

Exercise 7.1 Show that (7.1.1) holds if and only if (7.1.2) is true.

So to prove (7.1.1), suffices to prove (7.1.2) for a suitable choice of $a(t)$. Let

$$a(s) = \inf\{t \geq 0 : (t + G(t)) \geq s\}.$$

For this choice of a it can be seen that $\tilde{G}$ is a continuous increasing function and that for $0 \leq u \leq v < \infty$, $\tilde{G}(v) - \tilde{G}(u) \leq v - u$ so that $\tilde{G}$ is absolutely continuous and thus (7.1.2) follows from chain rule.

When working with continuous semimartingales, the same idea yields interesting results—of course, the time change $t \mapsto a(t)$ has to be replaced by $t \mapsto \phi_t$, where ϕ_t is a stopping time.

Definition 7.2 A $(\mathcal{F}_.)$-random time change $\phi = (\phi_t)$ is a family of $(\mathcal{F}_.)$ stopping times $\{\phi_t : 0 \leq t < \infty\}$ such that for all $\omega \in \Omega$, $t \mapsto \phi_t(\omega)$ is a continuous strictly increasing function from $[0, \infty)$ onto $[0, \infty)$.

Example **7.3** Let A be a $(\mathcal{F}_.)$ adapted continuous increasing process with $A_0 = 0$. Then

$$\phi_s = \inf\{t \geq 0 : (t + A_t) \geq s\}$$

can be seen to be a $(\mathcal{F}_.)$-random time change.

Example **7.4** Let B be a $(\mathcal{F}_.)$ adapted continuous increasing process with $B_0 = 0$ such that B is strictly increasing and $\lim_{t \to \infty} B_t = \infty$ a.s.. Then

$$\phi_s = \inf\{t \geq 0 : B_t \geq s\}$$

can be seen to be a $(\mathcal{F}_.)$-random time change.

Recall Definition 2.37 of the stopped σ-field. Given a $(\mathcal{F}_.)$-random time change $\phi = (\phi_t)$, we define a new filtration $(\mathcal{G}_.) = (\mathcal{G}_t)$ as follows:

$$\mathcal{G}_t = \mathcal{F}_{\phi_t}, \ 0 \leq t < \infty. \qquad (7.1.3)$$

Clearly, for $s \leq t$, we have $\phi_s \leq \phi_t$ and hence $\mathcal{G}_s \subseteq \mathcal{G}_t$ and so $\{\mathcal{G}_s\}$ is a filtration. Further, $\mathcal{G}_0 = \mathcal{F}_0$. We will denote the filtration $(\mathcal{G}_.)$ defined by (7.1.3) as $(\phi\mathcal{F}_.)$. Given a process f, we define the process $g = \phi[f]$ via

$$g_s = f_{\phi_s} \quad 0 \le s < \infty. \tag{7.1.4}$$

The map $f \to g$ is linear. We also define $\psi = \{\psi_t : 0 \le t < \infty\}$ via

$$\psi_t = \inf\{s \ge 0 : \phi_s \ge t\} \tag{7.1.5}$$

and denote ψ by $[\phi]^{-1}$. Here ψ is the reverse time change.

Exercise 7.5 Show that if $\phi_u(\omega) = v$ then $\psi_v(\omega) = u$.

Given a $(\mathcal{F}.)$ stopping time τ, we define $\sigma = \phi[\tau]$ by

$$\sigma = [\phi]_\tau^{-1} = \psi_\tau \tag{7.1.6}$$

that is given ω consider the map $t \mapsto \phi_t(\omega)$ and consider its inverse and evaluate it at $\tau(\omega)$—in other words, $\sigma(\omega)$ equals $\psi_{\tau(\omega)}(\omega)$. Note the appearance of $[\phi]^{-1}$ in the definition above. It is not difficult to see that

$$\phi_\sigma = \tau.$$

Recall definition (4.4.12) of $X^{[\tau]}$, the process X stopped at τ. For $Y = \phi[X]$, note that

$$(\phi[X^{[\tau]}])_s = X^{[\tau]}_{\phi_s} = X_{\phi_s \wedge \tau} = X_{\phi_s \wedge \phi_\sigma} = X_{\phi_{s \wedge \sigma}} = Y_{s \wedge \sigma} = Y^{[\sigma]}_s$$

and thus we have

$$Y^{[\sigma]} = \phi X^{[\tau]}. \tag{7.1.7}$$

We will now prove few relations about random time change and its interplay with notions discussed in the earlier chapters such as stopping times, predictable processes, local martingales, semimartingales and stochastic integrals.

Theorem 7.6 $\phi = (\phi_t)$ be a $(\mathcal{F}.)$- random time change. Let $\psi = [\phi]^{-1}$ be defined via (7.1.5). Then we have

(i) $\psi = (\psi_s)$ is a $(\mathcal{G}.)$- random time change.
(ii) Let τ be a $(\mathcal{F}.)$ stopping time. Then $\sigma = \phi[\tau] = \psi_\tau$ is a $(\mathcal{G}.)$ stopping time. Further, if τ, α are $(\mathcal{F}.)$ stopping times, then

$$\phi[\tau \wedge \alpha] = \phi[\tau] \wedge \phi[\alpha].$$

(iii) Let X be a $(\mathcal{F}.)$ adapted r.c.l.l. process. Then $Y = \phi[X]$ is a $(\mathcal{G}.)$ adapted r.c.l.l. process.
(iv) Let f be a $(\mathcal{F}.)$ bounded predictable process. Then $g = \phi[f]$ is a $(\mathcal{G}.)$ bounded predictable process. If f is a $(\mathcal{F}.)$ locally bounded predictable process then $g = \phi[f]$ is a $(\mathcal{G}.)$ locally bounded predictable process.
(v) Let A be a $(\mathcal{F}.)$ adapted r.c.l.l. process with finite variation paths. Then $B = \phi[A]$ is a $(\mathcal{G}.)$ adapted r.c.l.l. process with finite variation paths.

(vi) Let M be a $(\mathcal{F}_\cdot)$-local martingale. Then $N = \phi[M]$ is a $(\mathcal{G}_\cdot)$-local martingale.

(vii) Let X be a $(\mathcal{F}_\cdot)$-semimartingale. Then $Y = \phi[X]$ is a $(\mathcal{G}_\cdot)$-semimartingale.

(viii) Let $Z = \int f dX$. Then $\phi[Z] = \int g dY$ (where f, g, X, Y are as in *(iv)* and *(vii)*).

(ix) $[Y, Y] = \phi[[X, X]]$, where X, Y are as in *(vii)*.

(x) Let X^n, X be $(\mathcal{F}_\cdot)$ adapted r.c.l.l. processes such that $X^n \xrightarrow{ucp} X$. Then $Y^n = \phi[X^n] \xrightarrow{ucp} Y = \phi[X]$.

Proof Note that by Corollary 2.52, ϕ is $(\mathcal{G}_\cdot)$ adapted. For any $a, t \in [0, \infty)$, note that

$$\{\psi_a \leq t\} = \{a \leq \phi_t\}.$$

Since ϕ_t is $\mathcal{F}_{\phi_t} = \mathcal{G}_t$ measurable, it follows that $\{\psi_a \leq t\} \in \mathcal{G}_t$ and hence ψ_a is a $(\mathcal{G}_\cdot)$ stopping time. Since $s \mapsto \phi_s$ is continuous strictly increasing function from $[0, \infty)$ onto itself, same is true of $s \mapsto \psi_s$, and hence, $\psi = (\psi_s)$ is a random time change. This proves *(i)*.

Now, for $s \in [0, \infty)$, using Corollary 2.40, we have

$$\{\sigma \leq s\} = \{\tau \leq \phi_s\} \in \mathcal{F}_{\phi_s} = \mathcal{G}_s.$$

Thus σ is a $(\mathcal{G}_\cdot)$ stopping time. The last part of *(ii)* follows since ϕ is an increasing function.

For *(iii)* since X is r.c.l.l., $(\mathcal{F}_\cdot)$ adapted and ϕ_s is a $(\mathcal{F}_\cdot)$ stopping time, using Lemma 2.38, we conclude that $Y_s = X_{\phi_s}$ is $\mathcal{G}_s = \mathcal{F}_{\phi_s}$ measurable. Thus Y is $(\mathcal{G}_\cdot)$ adapted and is clearly r.c.l.l. When X is continuous, so is Y.

For *(iv)* the class of bounded processes f such that $g = \phi[f]$ is $(\mathcal{G}_\cdot)$ predictable is bp-closed and by part *(iii)*, it contains bounded continuous $(\mathcal{F}_\cdot)$ adapted processes and thus also contains bounded $(\mathcal{F}_\cdot)$ predictable processes. Now if f is $(\mathcal{F}_\cdot)$ predictable and locally bounded, let τ^n be sequence of stopping times, $\tau^n \uparrow \infty$ such that $f_n = f^{[\tau^n]}$ is bounded predictable. Then as shown above, $\phi[f_n]$ is also bounded predictable. Let $\sigma^n = \phi[\tau^n]$. As seen in (7.1.7),

$$g^{[\sigma^n]} = \phi[f^{[\tau^n]}] = \phi[f_n]$$

and thus $g^{[\sigma^n]}$ is predictable. Now $\tau^n \uparrow \infty$ implies $\sigma^n \uparrow \infty$ and thus g is locally bounded $(\mathcal{G}_\cdot)$ predictable process. This proves *(iv)*.

For *(v)*, we have already noted that B is $(\mathcal{G}_\cdot)$ adapted. And clearly,

$$\mathrm{VAR}_{[0,s]}(B(\omega)) = \mathrm{VAR}_{[0,\phi_s(\omega)]}(A(\omega))$$

and hence paths of B have finite variation.

For *(vi)*, in order to prove that N is a $(\mathcal{G}_\cdot)$-local martingale, we will obtain a sequence σ_n of $(\mathcal{G}_\cdot)$ stopping times increasing to ∞ such that for all $(\mathcal{G}_\cdot)$ stopping times β,

$$\mathsf{E}[N_{\sigma_n \wedge \beta}] = \mathsf{E}[N_0].$$

This will prove $N^{[\sigma_n]}$ is a martingale and hence that N is a local martingale. Since M is a local martingale, let $\tilde{\tau}_n$ be stopping times such $M^{[\tilde{\tau}_n]}$ is a martingale for each n where $\tilde{\tau}_n \uparrow \infty$. Let

$$\tau_n = \tilde{\tau}_n \wedge n \wedge \phi_n.$$

Then $\tau_n \leq \tilde{\tau}_n$ and $\tau_n \uparrow \infty$. Then $M^{[\tau_n]}$ is a martingale for each n and hence for all stopping times η one has

$$E[M_{\tau_n \wedge \eta}] = E[M_0]. \qquad (7.1.8)$$

Now let $\sigma_n = \phi[\tau_n] = \psi_{\tau_n}$. Since $\tau_n \leq \phi_n$, it follows that $\sigma_n \leq \psi_{\phi_n} = n$. Now for any $(\mathcal{G}_.)$ stopping time β, we will show

$$E[N_{\sigma_n \wedge \beta}] = E[N_0]. \qquad (7.1.9)$$

Let $\eta = \psi[\beta] = \phi_\beta$. Then by part (ii) above, η is a $(\mathcal{F}_.)$ stopping time. Note that

$$N_{\sigma_n \wedge \beta} = M_{\tau_n \wedge \eta}. \qquad (7.1.10)$$

Further, $M_0 = N_0$ and thus (7.1.8) and (7.1.10) together imply (7.1.9) proving (vi).

Part (vii) follows from (v) and (vi) by decomposing the semimartingale X into a local martingale M and a process with finite variation paths A: $X = M + A$. Then $Y = \phi[X] = \phi[M] + \phi[A] = N + B$.

We can verify the validity of $(viii)$ when f is a simple predictable process and then easy to see that the class of processes for which $(viii)$ is true is bp-closed and thus contains all bounded predictable processes. We can then get the general case (of f being locally bounded) by localization.

For (ix), note that

$$[X, X] = X_t^2 - X_0^2 - 2 \int_0^t X^- dX.$$

Changing time in this equation, and using $(viii)$, we get

$$\begin{aligned} \phi[[X, X]]_t &= X_{\phi_t}^2 - X_0^2 - 2 \int_0^{\phi_t} X^- dX \\ &= Y_t^2 - Y_0^2 - 2 \int_0^t Y^- dY \\ &= [Y, Y]_t. \end{aligned}$$

For the last part, note that for $T < \infty$, $T_0 < \infty$, $\delta > 0$ one has (using $Y_s = X_{\phi_s}$)

$$P(\sup_{0 \leq t \leq T} |Y_s^n - Y_s| \geq \delta) \leq P(\sup_{0 \leq t \leq T_0} |X_s^n - X_s| \geq \delta) + P(\phi_T \geq T_0)$$

Now given $T < \infty, \delta > 0$ and $\epsilon > 0$, first get T_0 such that $P(\phi_T \geq T_0) < \frac{\epsilon}{2}$ and then for this T_0, using $X^n \xrightarrow{ucp} X$ get n_0 such that for $n \geq n_0$ one has $P(\sup_{0 \leq t \leq T_0} |X_s^n - X_s| \geq \delta) < \frac{\epsilon}{2}$. Now, for $n \geq n_0$ we have

$$P(\sup_{0 \leq t \leq T} |Y_s^n - Y_s| \geq \delta) < \epsilon.$$

Remark 2.70 now completes the proof.   □

Exercise 7.7 Let X be a $(\mathcal{F}_.)$-semimartingale, $f \in \mathbb{L}(X)$ and let $\phi = (\phi_t)$ be a $(\mathcal{F}_.)$- random time change. Let $g = \phi[f]$ and $Y = \phi[X]$. Show that $g \in \mathbb{L}(Y)$.

Exercise 7.8 Let X^1, X^2 be $(\mathcal{F}_.)$-semimartingales and let $\phi = (\phi_t)$ be a $(\mathcal{F}_.)$-random time change. Let $Y^i = \phi[X^i]$. Show that

$$[Y^1, Y^2] = \phi[[X^1, X^2]].$$

Remark 7.9 It should be noted that if M is a martingale then $N = \phi[M]$ may not be a $(\mathcal{G}_.)$-martingale. In fact, N_t may not be integrable as seen in the next exercise.

Exercise 7.10 Let W be a Brownian motion and ξ be a $(0, \infty)$-valued random variable independent of W such that $E[\xi] = \infty$. Let $\mathcal{F}_t = \sigma(\xi, W_s : 0 \leq s \leq t)$. Let $\phi_t = t\xi$. Show that

(i) ϕ_t is a stopping time for each t.
(ii) $\phi = (\phi_t)$ is a random time change.
(iii) $E[|Z_t|] = \infty$ for all $t > 0$ where $Z = \phi[W]$.
(iv) Z is a local martingale but not a martingale.

7.2 Growth Estimate

Let X be a continuous semimartingale and let $X = M + A$ be the decomposition of X with M being a continuous local martingale, A being a process with finite variation paths. We will call this as the canonical decomposition. Recall that the quadratic variation $[M, M]$ is itself a continuous process and $|A|_t = \text{VAR}_{[0,t]}(A)$ is also a continuous process. For a locally bounded predictable process f, for any stopping time σ such that the right-hand side in (7.2.1) below is finite one has

$$E[\sup_{0 \leq s \leq \sigma} |\int_{0+}^{s} f dX|^2]$$
$$\leq 8E[\int_{0+}^{\sigma} |f_s|^2 d[M, M]_s] + 2E[(\int_{0+}^{\sigma} |f_s| d|A|_s)^2]. \tag{7.2.1}$$

To see this, we note $\int f dX = \int f dM + \int f dA$ and for the dM integral we use Theorem 5.31 and for the dA integral we use $|\int f dA| \leq \int |f| d|A|$.

For process $A, B \in \mathbb{V}^+$ (increasing adapted processes), we define $A \ll B$ if $C_t = B_t - A_t$ is an increasing process. The following observation will be used repeatedly in the rest of this chapter: if $A \ll B$, then for all f

$$|\int_{0+}^{t} f_s dA_s| \leq \int_{0+}^{t} |f_s| d B_s.$$

We introduce a notion of a amenable semimartingale and obtain a growth estimate on integrals w.r.t. a amenable semimartingale which is similar to the one for Brownian motion.

Definition 7.11 A continuous semimartingale Y is said to be a amenable semimartingale if the canonical decomposition $Y = N + B$ satisfies, for $0 \leq s \leq t < \infty$

$$[N, N]_t - [N, N]_s \leq (t - s), \quad |B|_t - |B|_s \leq (t - s). \tag{7.2.2}$$

Remark **7.12** The condition (7.2.2) can be equivalently stated as

$$s - [N, N]_s \leq t - [N, N]_t, \quad s - |B|_s \leq t - |B|_t \text{ for } 0 \leq s \leq t < \infty \tag{7.2.3}$$

or, writing $I_t = t$, it is same as

$$[N, N] \ll I, \quad |B| \ll I. \tag{7.2.4}$$

Theorem 7.13 *Suppose Y is a continuous amenable semimartingale. Then for any locally bounded predictable f, and a stopping time σ, one has*

$$\mathrm{E}[\sup_{0 \leq s \leq \sigma \wedge T} |\int_{0+}^{s} f dX|^2] \leq 2(4 + T)\mathrm{E}[\int_{0+}^{\sigma \wedge T} |f_s|^2 ds]. \tag{7.2.5}$$

Proof The condition (7.2.2) implies that $t - [N, N]_t$ and $t - |B|_t$ are increasing processes. This observation along with (7.2.1) yields

$$\mathrm{E}[\sup_{0 \leq s \leq \sigma \wedge T} |\int_{0+}^{s} f dX|^2] \leq 8\mathrm{E}[\int_{0+}^{\sigma \wedge T} |f_s|^2 ds] + 2(\mathrm{E}[\int_{0+}^{\sigma \wedge T} |f_s| ds])^2.$$

Now the required estimate follows by the Cauchy–Schwarz inequality:

$$(\mathrm{E}[\int_{0+}^{\sigma \wedge T} |f_s| ds])^2 \leq T\mathrm{E}[\int_{0+}^{\sigma \wedge T} |f_s|^2 ds].$$

$\square$

Remark **7.14** We see that for a amenable semimartingale X, the stochastic integral $\int f dX$ satisfies a growth estimate similar to the one when X is a Brownian motion. Thus, results such as existence, uniqueness, approximation of solution to an SDE driven by a Brownian motion continue to hold for a amenable semimartingale X. We will come back to this later in this chapter.

Remark **7.15** In the definition of amenable semimartingale, instead of (7.2.2) we could have required that for some constant $K < \infty$

$$[N, N]_t - [N, N]_s \le K(t - s), \quad |B|_t - |B|_s \le K(t - s). \tag{7.2.6}$$

The only difference is that a constant K^2 would appear in the estimate (7.2.5)

$$E[\sup_{0 \le s \le \sigma \wedge T} |\int_{0+}^{s} f dX|^2] \le 2(4 + T) K^2 E[\int_{0+}^{\sigma \wedge T} |f_s|^2 ds]. \tag{7.2.7}$$

A simple but important observation is that given a continuous semimartingale X one can get a random time change $\phi = (\phi_{\cdot})$ such that the semimartingale $Y = \phi[X]$ satisfies (7.2.2). Indeed, given finitely many semimartingales, we can choose one random time change that does it as we see in the next result.

Theorem 7.16 *Let $X^1, X^2, \ldots, X^m$ be continuous semimartingales with respect to the filtration $(\mathcal{F}_{\cdot})$. Then there exists a random time change $\phi = (\phi_t)$ (with respect to the filtration $(\mathcal{F}_{\cdot})$) such that for $1 \le j \le m$, $Y^j = \phi[X^j]$ is a amenable semimartingale.*

Proof For $1 \le j \le m$, let $X^j = M^j + A^j$ be the canonical decomposition of the semimartingale X^j with M^j being a continuous local martingale, $M_0^j = 0$ and A^j are continuous processes with finite variation paths. Define an increasing process V by

$$V_t = t + \sum_{j=1}^{m} ([M^j, M^j]_t + |A^j|_t).$$

Then V is strictly increasing adapted process with $V_0 = 0$. Now defining

$$\phi_t = \inf\{s \ge 0 : V_s \ge t\}$$

it follows that $\phi = (\phi_t)$ is a random time change. As noted earlier, $Y^j = \phi[X^j]$ is a semimartingale with canonical decomposition $Y^j = N^j + B^j$ where $N^j = \phi[M^j]$, $B^j = \phi[A^j]$. Further, observing that for $1 \le j \le m, 0 \le s \le t < \infty$,

$$[N^j, N^j]_t - [N^j, N^j]_s = [M^j, M^j]_{\phi_t} - [M^j, M^j]_{\phi_s} \le V_{\phi_t} - V_{\phi_s} = t - s$$

and

$$|B^j|_t - |B^j|_s = |A^j|_{\phi_t} - |A^j|_{\phi_s} \le V_{\phi_t} - V_{\phi_s} = t - s,$$

it follows that this random time change satisfies the required condition.   □

7.3 Stochastic Differential Equations

Let us consider the stochastic differential equation (3.5.1) where instead of a Brownian motion as in Chap. 3, here $W = (W^1, W^2, \ldots, W^d)$ is a amenable semi-martingale. The growth estimate (7.2.5) enables one to conclude that in this case too, Theorem 3.30 is true and the same proof works essentially—using (7.2.5) instead of (3.4.4). Moreover, using random time change, one can conclude that the same is true even when W is any continuous semimartingale. We will prove this along with some results on approximations to the solution of an SDE.

   We are going to consider the following general framework for the SDE driven by continuous semimartingales, where the evolution from a time t_0 onwards could depend upon the entire past history of the solution rather than only on its current value as was the case in Eq. (3.5.1) driven by a Brownian motion.

   Let $Y^1, Y^2, \ldots Y^m$ be continuous semimartingales w.r.t. the filtration $(\mathcal{F}_\cdot)$. Let $Y = (Y^1, Y^2, \ldots Y^m)$. Here we will consider an SDE

$$dU_t = b(t, \cdot, U)dY_t, \quad t \ge 0, \quad U_0 = \xi_0 \tag{7.3.1}$$

where the functional b is given as follows. Recall that $\mathbb{C}_d = \mathbb{C}([0, \infty), \mathbb{R}^d)$. Let

$$a : [0, \infty) \times \Omega \times \mathbb{C}_d \to L(d, m) \tag{7.3.2}$$

be such that for all $\zeta \in \mathbb{C}_d$,

$$(t, \omega) \mapsto a(t, \omega, \zeta) \text{ is an r.c.l.l. } (\mathcal{F}_\cdot) \text{ adapted process} \tag{7.3.3}$$

and there is an increasing r.c.l.l. adapted process K such that for all $\zeta_1, \zeta_2 \in \mathbb{C}_d$,

$$\sup_{0 \le s \le t} \|a(s, \omega, \zeta_2) - a(s, \omega, \zeta_1)\| \le K_t(\omega) \sup_{0 \le s \le t} |\zeta_2(s) - \zeta_1(s)|. \tag{7.3.4}$$

Finally, $b : [0, \infty) \times \Omega \times \mathbb{C}_d \to L(d, m)$ be given by

$$b(s, \omega, \zeta) = a(s-, \omega, \zeta). \tag{7.3.5}$$

Lemma 7.17 *Suppose the functional a satisfies* (7.3.2)–(7.3.4).

(i) *For all* $t \geq 0$,

$$(\omega, \zeta) \mapsto a(t, \omega, \zeta) \text{ is } \mathcal{F}_t \otimes \mathcal{B}(\mathbb{C}_d) \text{ measurable.} \qquad (7.3.6)$$

(ii) *For any continuous* $(\mathcal{F}_.)$ *adapted process* V, Z *defined by* $Z_t = a(t, \cdot, V)$
    (i.e. $Z_t(\omega) = a(t, \omega, V(\omega))$) *is an r.c.l.l.* $(\mathcal{F}_.)$ *adapted process.*
(iii) *For any stopping time* τ,

$$(\omega, \zeta) \mapsto a(\tau(\omega), \omega, \zeta) \text{ is } \mathcal{F}_\tau \otimes \mathcal{B}(\mathbb{C}_d) \text{ measurable.} \qquad (7.3.7)$$

Proof Since for fixed t, ζ, the mapping $\omega \mapsto a(t, \omega, \zeta)$ is $\mathcal{F}_t$ measurable and in view of (7.3.4), $\zeta \mapsto a(t, \omega, \zeta)$ is continuous for fixed t, ω, it follows that (7.3.6) is true since $\mathbb{C}_d$ is separable.

For part (ii), let us define a process V^t by $V_s^t = V_{s \wedge t}$. In view of assumption (7.3.3), Z is an r.c.l.l. process. The fact that $\omega \mapsto V^t(\omega)$ is $\mathcal{F}_t$ measurable along with (7.3.6) implies that $Z_t = a(t, \cdot, V^t)$ is $\mathcal{F}_t$ measurable.

For part (iii), when τ is a simple stopping time, (7.3.7) follows from (7.3.6). For a general bounded stopping time τ, the conclusion (7.3.7) follows by approximating τ from above by simple stopping times and using right continuity of $a(t, \omega, \zeta)$. For a general stopping time τ, (7.3.7) follows by approximating τ by $\tau \wedge n$. $\qquad \square$

Let $\mathbf{0}$ denote the process that is identically equal to zero. Since $(t, \omega) \mapsto a(t, \omega, \mathbf{0})$ is an r.c.l.l. adapted process, using hypothesis (7.3.4), it follows that for $\zeta \in \mathbb{C}_d$

$$\sup_{0 \leq s \leq t} \|a(s, \omega, \zeta)\| \leq K_t'(\omega)(1 + \sup_{0 \leq s \leq t} |\zeta(s)|) \qquad (7.3.8)$$

where

$$K_t'(\omega) = K_t(\omega) + \sup_{0 \leq s \leq t} \|a(s, \omega, \mathbf{0})\|. \qquad (7.3.9)$$

K_t' is clearly an r.c.l.l. adapted process.

Here too, as in the Brownian motion case, a continuous ($\mathbb{R}^d$-valued) adapted process U is said to be a solution to the Eq. (7.3.1) if

$$U_t = \xi_0 + \int_{0+}^t b(s, \cdot, U) dY_s \qquad (7.3.10)$$

i.e. for $1 \leq j \leq d$,

$$U_t^j = \xi_0^j + \sum_{k=1}^m \int_{0+}^t b_{jk}(s, \cdot, U) dY_s^k$$

where $U = (U^1, \ldots, U^d)$ and $b = (b_{jk})$.

It is convenient to introduce matrix- and vector-valued processes and stochastic integral $\int f dX$ where f is matrix-valued and X is vector-valued. All our vectors are column vectors, though we will write as $c = (c_1, c_2, \ldots, c_m)$.

Let $X^1, X^2, \ldots X^m$ be continuous semimartingales w.r.t. the filtration $(\mathcal{F}_\cdot)$. We will say that $X = (X^1, X^2, \ldots X^m)$ is an $\mathbb{R}^m$-valued semimartingale. Similarly, for $1 \leq j \leq d$, $1 \leq k \leq m$ let f_{jk} be locally bounded predictable process. $f = (f_{jk})$ will be called an $L(d, m)$-valued locally bounded predictable process. The stochastic integral $Y = \int f dX$ is defined as follows: $Y = (Y^1, Y^2, \ldots, Y^d)$ where

$$Y^j = \sum_{k=1}^{m} \int f_{jk} dX^k.$$

Let us recast the growth estimate in matrix–vector form:

Theorem 7.18 *Let $X = (X^1, X^2, \ldots X^m)$, where X^j is a amenable semimartingale for each j, $1 \leq j \leq m$. Then for any locally bounded $L(d, m)$-valued predictable f, and a stopping time σ, one has*

$$\mathsf{E}[\sup_{0 \leq s \leq \sigma \wedge T} |\int_{0+}^{s} f dX|^2] \leq 2m(4 + T)\mathsf{E}[\int_{0+}^{\sigma \wedge T} \|f_s\|^2 ds]. \tag{7.3.11}$$

Proof

$$\mathsf{E}[\sup_{0 \leq s \leq \sigma \wedge T} |\int_{0+}^{s} f dX|^2]$$

$$\leq \sum_{j=1}^{d} \mathsf{E}[\sup_{0 \leq s \leq \sigma \wedge T} |\sum_{k=1}^{m} \int_{0+}^{s} f_{jk} dX^k|^2]$$

$$\leq m \sum_{j=1}^{d} \sum_{k=1}^{m} \mathsf{E}[\sup_{0 \leq s \leq \sigma \wedge T} |\int_{0+}^{s} f_{jk} dX^k|^2]$$

$$\leq m(8 + 2T) \sum_{j=1}^{d} \sum_{k=1}^{m} \mathsf{E}[\int_{0+}^{\sigma \wedge T} |f_{jk}(s)|^2 ds]$$

$$= 2m(4 + T)\mathsf{E}[\int_{0+}^{\sigma \wedge T} \|f_s\|^2 ds]$$

where, for the last inequality above we have used the estimate (7.2.5). $\qquad \square$

We will prove existence and uniqueness of solution of (7.3.1). When the driving semimartingale satisfies (7.2.2), the proof is almost the same as the proof when the driving semimartingale is a Brownian motion. We will prove this result without making any integrability assumptions on the initial condition ξ_0 and the uniqueness assertion is without any moment condition. For this the following simple observation is important.

Remark **7.19** Suppose M is a square integrable martingale w.r.t. a filtration $(\mathcal{F}_\cdot)$ and let $\Omega_0 \in \mathcal{F}_0$. Then $N_t = 1_{\Omega_0} M_t$ is also a square integrable martingale and further $[N, N]_t = 1_{\Omega_0}[M, M]_t$. Thus, the estimate (5.3.22) can be recast as

$$E[1_{\Omega_0} \sup_{0 \le s \le T} |M_s|^2] \le 4E[1_{\Omega_0}[M, M]_T].$$

Using this we can refine the estimate (7.2.5): for a amenable semimartingale X, any locally bounded predictable f, a stopping time σ and $\Omega_0 \in \mathcal{F}_0$, one has

$$E[1_{\Omega_0} \sup_{0 \le s \le \sigma \wedge T} |\int_{0+}^{s} f dX|^2] \le 2(4 + T)E[1_{\Omega_0} \int_{0+}^{\sigma \wedge T} |f_s|^2 ds]. \qquad (7.3.12)$$

Here is the modified estimate for vector-valued case: if $X = (X^1, X^2, \ldots X^m)$ where each X^j is a amenable semimartingale and $f = (f_{jk})$ is an $L(d, m)$-valued locally bounded predictable process, then

$$E[1_{\Omega_0} \sup_{0 \le s \le \sigma \wedge T} |\int_{0+}^{s} f dX|^2] \le 2m(4 + T)E[1_{\Omega_0} \int_{0+}^{\sigma \wedge T} \|f_s\|^2 ds]. \qquad (7.3.13)$$

We will first prove uniqueness of solution in the special case when the driving semimartingale is a amenable semimartingale.

Theorem 7.20 *Let $Y = (Y^1, Y^2, \ldots Y^m)$ where Y^j is a amenable continuous semimartingale for each j. Let the functional a satisfy conditions (7.3.2)–(7.3.4) and b be defined by (7.3.5). Let ξ_0 be any $\mathcal{F}_0$ measurable random variable. Then if $U, \tilde{U}$ are $(\mathcal{F}.)$ adapted continuous process satisfying*

$$U_t = \xi_0 + \int_{0+}^{t} b(s, \cdot, U) dY_s, \qquad (7.3.14)$$

$$\tilde{U}_t = \xi_0 + \int_{0+}^{t} b(s, \cdot, \tilde{U}) dY_s. \qquad (7.3.15)$$

then

$$P(U_t = \tilde{U}_t \ \forall t \ge 0) = 1. \qquad (7.3.16)$$

Proof For $i \ge 1$, let $\tau_i = \inf\{t \ge 0 : K'_t(\omega) \ge i \text{ or } K'_{t-}(\omega) \ge i\} \wedge i$ where $K'_t(\omega)$ is the r.c.l.l. adapted process given by (7.3.9). Thus each τ_i is a stopping time, $\tau_i \uparrow \infty$ and for $0 \le t < \tau_i(\omega)$, we have

$$0 \le K_t(\omega) \le K'_t(\omega) \le i.$$

Recalling that $b(t, \cdot, \zeta) = a(t-, \cdot, \zeta)$, we conclude that for $\zeta, \zeta_1, \zeta_2 \in \mathbb{C}_d$

$$\sup_{0 \le s \le (t \wedge \tau_i(\omega))} \|b(s, \omega, \zeta_2) - b(s, \omega, \zeta_1)\| \le i \sup_{0 \le s \le (t \wedge \tau_i(\omega))} |\zeta_2(s) - \zeta_1(s)|. \qquad (7.3.17)$$

and

$$\sup_{0 \le s \le (t \wedge \tau_i(\omega))} \|b(s, \omega, \zeta)\| \le i(1 + \sup_{0 \le s \le (t \wedge \tau_i(\omega))} |\zeta(s)|). \qquad (7.3.18)$$

We first show that if V is any solution to (7.3.14), i.e. V satisfies

$$V_t = \xi_0 + \int_{0+}^{t} b(s, \cdot, V) dY_s \qquad (7.3.19)$$

then for $k \ge 1$, $i \ge 1$,

$$E[1_{\{|\xi_0| \le k\}} \sup_{0 \le t \le \tau_i} |V_t|^2] < \infty. \qquad (7.3.20)$$

Let us fix k, i for now. For $j \ge 1$, let $\sigma_j = \inf\{t \ge 0 : |V_t| \ge j\}$. Since V is a continuous adapted process with $V_0 = \xi_0$, it follows that σ_j is a stopping time, $\lim_{j \to \infty} \sigma_j = \infty$ and

$$\sup_{0 \le t \le (\tau_i \wedge \sigma_j)} |V_t|^2 \le \max(|\xi_0|^2, j^2). \qquad (7.3.21)$$

Thus using the estimate (7.3.13) along with (7.3.18), we get for, $i, j, k \ge 1$ and $u \ge 0$

$$E[1_{\{|\xi_0| \le k\}} \sup_{0 \le t \le (u \wedge \tau_i \wedge \sigma_j)} |V_t|^2]$$

$$\le 2[E[1_{\{|\xi_0| \le k\}}(|\xi_0|^2 + 2m(4 + i) \int_{0+}^{(u \wedge \tau_i \wedge \sigma_j)} \|b(s, \cdot, V_s)\|^2 ds)]]$$

$$\le E[1_{\{|\xi_0| \le k\}}(2k^2 + 8m(4 + i)i^2 \int_0^{(u \wedge \tau_i \wedge \sigma_j)} (1 + \sup_{0 \le t \le (s \wedge \tau_i \wedge \sigma_j)} |V_s|^2)ds)].$$

Writing

$$\beta_j(u) = E[1_{\{|\xi_0| \le k\}} \sup_{0 \le t \le (u \wedge \tau_i \wedge \sigma_j)} |V_t|^2],$$

it follows that for $0 \le u \le i$,

$$\beta_j(u) \le 2k^2 + 8m(4 + i)i^3 + 8m(4 + i)i^2 \int_0^u \beta_j(s)ds$$

and further, (7.3.21) yields that β_j is a bounded function. Thus, using (Gronwall's) Lemma 3.27, we conclude

$$\beta_j(u) \le [8m(4 + i)i^3 + 2k^2] \exp\{8m(4 + i)i^2 u\}, \quad 0 \le u \le i.$$

Now letting j increase to ∞ we conclude that (7.3.20) is true.

Returning to the proof of (7.3.16), since $U, \tilde{U}$ both satisfy (7.3.19), both also satisfy (7.3.20) and hence we conclude that for each i

$$E[1_{\{|\xi_0|\leq k\}} \sup_{0\leq t\leq \tau_i} |U_t - \tilde{U}_t|^2] < \infty. \qquad (7.3.22)$$

Now

$$U_t - \tilde{U}_t = \int_{0+}^t (b(s, \cdot, U) - b(s, \cdot, \tilde{U}))dY_s$$

and hence using the Lipschitz condition (7.3.17) and the growth estimate (7.3.13), we conclude that for $0 \leq t \leq T$,

$$E[1_{\{|\xi_0|\leq k\}} \sup_{0\leq s\leq (t\wedge\tau_i)} |U_s - \tilde{U}_s|^2]$$

$$\leq 2m(4+i)i^2 E[\int_0^{(t\wedge\tau_i)} 1_{\{|\xi_0|\leq k\}} \sup_{0\leq s\leq u} |U_s - \tilde{U}_s|^2]du.$$

Fixing i, we note that the function β defined by

$$\beta(t) = E[1_{\{|\xi_0|\leq k\}} \sup_{0\leq s\leq (t\wedge\tau_i)} |U_s - \tilde{U}_s|^2]$$

satisfies, for a suitable constant C_i

$$\beta(t) \leq C_i \int_0^t \beta(u)du, \quad t \geq 0.$$

As noted above (see (7.3.22)) $\beta(t)$ is bounded. Now (Gronwall's) Lemma 3.27 implies that $\beta(t) = 0$ for all t. Thus we conclude

$$P(\{|\xi_0| \leq k\} \cap \{\sup_{0\leq s\leq (t\wedge\tau_i)} |U_s - \tilde{U}_s| > 0\}) = 0$$

for all $i \geq 1$ and $k \geq 1$. Since $\tau_i \uparrow \infty$, this proves (7.3.16) completing the proof. $\square$

We have thus seen that if Y is a amenable semimartingale then uniqueness of solution holds for the SDE (7.3.14) and the proof is on the lines of Brownian motion case. One big difference is that uniqueness is proven without a priori requiring the solution to satisfy a moment condition. This is important as while the stochastic integral is invariant under time change, moment conditions are not. And this is what enables us to prove uniqueness when the driving semimartingale may not be a amenable semimartingale.

Using random time change we extend the result on uniqueness of solutions to the SDE (7.3.14) to the case when the driving semimartingale may not be a amenable semimartingale.

Theorem 7.21 *Let $X = (X^1, X^2, \ldots X^m)$ where each X^j is a continuous semimartingale. Let the functional a satisfy conditions (7.3.2)–(7.3.4) and b be defined* by (7.3.5). *Let ξ_0 be any $\mathcal{F}_0$ measurable random variable. If $V, \tilde{V}$ are $(\mathcal{F}_\cdot)$ adapted continuous processes satisfying*

$$V_t = \xi_0 + \int_{0+}^{t} b(s, \cdot, V) dX_s, \tag{7.3.23}$$

$$\tilde{V}_t = \xi_0 + \int_{0+}^{t} b(s, \cdot, \tilde{V}) dX_s. \tag{7.3.24}$$

Then

$$P(V_t = \tilde{V}_t \ \forall t \geq 0) = 1. \tag{7.3.25}$$

Proof Let ϕ be a $(\mathcal{F}_\cdot)$ random time change such that $Y^j = \phi[X^j]$, $1 \leq j \leq m$ are amenable semimartingales (such a random time change exists as seen in Theorem 7.16). Let $(\mathcal{G}_\cdot) = (\phi\mathcal{F}_\cdot)$, $\psi = [\phi]^{-1}$ be defined via (7.1.5).

We define $c(t, \omega, \zeta), d(t, \omega, \zeta)$ as follows: fix ω and let $\theta_\omega(\zeta) \in \mathbb{C}_d$ be defined by $\theta_\omega(\zeta)(s) = \zeta(\psi_s(\omega))$ and let

$$\begin{aligned} c(t, \omega, \zeta) &= a(\phi_t(\omega), \omega, \theta_\omega(\zeta)), \\ d(t, \omega, \zeta) &= b(\phi_t(\omega), \omega, \theta_\omega(\zeta)). \end{aligned} \tag{7.3.26}$$

Since ϕ is continuous, it follows that $d(t, \omega, \zeta) = c(t-, \omega, \zeta)$.

We will first observe that for all $\zeta_1, \zeta_2 \in \mathbb{C}_d$,

$$\begin{aligned} \sup_{0 \leq u \leq s} &\|c(u, \omega, \zeta_2) - c(u, \omega, \zeta_1)\| \\ &= \sup_{0 \leq u \leq s} \|a(\phi_u(\omega), \omega, \theta_\omega(\zeta_2)) - a(\phi_u(\omega), \omega, \theta_\omega(\zeta_1))\| \\ &\leq K_{\phi_s}(\omega) \sup_{0 \leq v \leq \phi_s(\omega)} |\theta_\omega(\zeta_2)(v) - \theta_\omega(\zeta_1)(v)| \\ &\leq K_{\phi_s}(\omega) \sup_{0 \leq v \leq \phi_s(\omega)} |\zeta_2(\psi_v(\omega)) - \zeta_1(\psi_v(\omega))| \\ &\leq K_{\phi_s}(\omega) \sup_{0 \leq u \leq s} |\zeta_2(u) - \zeta_1(u)|. \end{aligned} \tag{7.3.27}$$

We now prove that for each ζ,

$$(t, \omega) \mapsto c(t, \omega, \zeta) \text{ is an r.c.l.l. } (\mathcal{G}_\cdot) \text{ adapted process.} \tag{7.3.28}$$

That the mapping is r.c.l.l. follows since a is r.c.l.l. and ϕ_t is continuous strictly increasing function. To see that it is adapted, fix t and let ζ^t be defined by $\zeta^t(s) = \zeta(s \wedge t)$. In view of (7.3.27), it follows that

$$c(t, \omega, \zeta) = c(t, \omega, \zeta^t) = a(\phi_t(\omega), \omega, \theta_\omega(\zeta^t)). \tag{7.3.29}$$

Note that

$$\theta_\omega(\zeta^t) = \zeta^t(\psi_s(\omega)) = \zeta(\psi_s(\omega) \wedge t).$$

Since ψ_s is a $(\mathcal{G}.)$ stopping time, it follows that $\psi_s \wedge t$ is $\mathcal{G}_t$ measurable and hence so is $\zeta(\psi_s(\omega) \wedge t)$. It follows that $\omega \mapsto \theta_\omega(\zeta^t)$ is $\mathcal{G}_t = \mathcal{F}_{\phi_t}$ measurable. Further, since ϕ_t is a $(\mathcal{F}.)$ stopping time, part (iii) in Lemma 7.17 gives that $(\omega, \zeta) \mapsto a(\phi_t(\omega), \omega, \zeta)$ is $\mathcal{F}_{\phi_t} \otimes \mathcal{B}(\mathbb{C}_d)$ measurable. From these we get

$$\omega \mapsto a(\phi_t(\omega), \omega, \theta_\omega(\zeta^t)) \text{ is } \mathcal{G}_t \text{ measurable.} \tag{7.3.30}$$

The conclusion (7.3.28) follows from (7.3.29) and (7.3.30). Let $H = \phi[K]$. Then H is an $(\mathcal{G}.)$ adapted increasing process and (7.3.27) can be rewritten as

$$\sup_{0 \le u \le s} \|c(u, \omega, \zeta_2) - c(u, \omega, \zeta_1)\| \le H_s(\omega) \sup_{0 \le u \le s} |\zeta_2(u) - \zeta_1(u)|. \tag{7.3.31}$$

Since $d(s, \cdot, \zeta) = c(s-, \cdot, \zeta)$, (7.3.31) implies

$$\sup_{0 \le u \le s} \|d(u, \omega, \zeta_2) - d(u, \omega, \zeta_1)\| \le H_{s-}(\omega) \sup_{0 \le u < s} |\zeta_2(u) - \zeta_1(u)|. \tag{7.3.32}$$

Let $U = \phi[V]$, $\tilde{U} = \phi[\tilde{V}]$. Then recall $V = \psi[U]$, $\tilde{V} = \psi[\tilde{U}]$. Let A, $\tilde{A}$, B, $\tilde{B}$ be defined by $A_s = b(s, \cdot, V)$, $\tilde{A}_s = b(s, \cdot, \tilde{V})$, $B_s = A_{\phi_s}$ and $\tilde{B}_s = \tilde{A}_{\phi_s}$. Then

$$\begin{aligned} B_s &= A_{\phi_s} \\ &= b(\phi_s, \cdot, \psi[U]) \\ &= d(s, \cdot, U) \end{aligned}$$

and likewise $\tilde{B}_t = d(t, \cdot, \tilde{U})$.

Thus the processes U, $\tilde{U}$ satisfy

$$U_t = \xi + \int_{0+}^{t} d(s, \cdot, U) d\tilde{Y}_s,$$

$$\tilde{U}_t = \xi + \int_{0+}^{t} d(s, \cdot, \tilde{U}) d\tilde{Y}_s.$$

Since c, d satisfy (7.3.2)–(7.3.5), Theorem 7.20 implies

$$\mathsf{P}(U_t = \tilde{U}_t \ \forall t \ge 0) = 1$$

which in turn also proves (7.3.25) since $V = \psi[U]$ and $\tilde{V} = \psi[\tilde{U}]$. $\qquad\square$

We are now ready to prove the main result on existence of solution to an SDE driven by continuous semimartingales. Our existence result is a modification of

Picard's successive approximation method. Here the approximations are explicitly constructed.

Theorem 7.22 *Let* $X^1, X^2, \ldots X^m$ *be continuous semimartingales. Let the functional* a *satisfy conditions* (7.3.2)–(7.3.4) *and* b *be defined by* (7.3.5). *Let* ξ *be any* $\mathcal{F}_0$ *measurable random variable. Then there exists a* $(\mathcal{F}_{\cdot})$ *adapted continuous process* V *that satisfies*

$$V_t = \xi_0 + \int_{0+}^{t} b(s, \cdot, V) dX_s. \tag{7.3.33}$$

Proof We will construct approximations $V^{(n)}$ that converge to a solution of (7.3.33). Let $V_t^{(0)} = \xi_0$ for all t. The processes $V^{(n)}$ are defined by induction on n. Assuming that adapted r.c.l.l. processes $V^{(0)}, \ldots, V^{(n-1)}$ have been defined, we now define $V^{(n)}$: fix n.

Let $\tau_0^{(n)} = 0$ and let $\{\tau_j^{(n)} : j \geq 1\}$ be defined inductively as follows: if $\tau_j^{(n)} = \infty$ then $\tau_{j+1}^{(n)} = \infty$ and if $\tau_j^{(n)} < \infty$ then

$$\tau_{j+1}^{(n)} = \inf\{s > \tau_j^{(n)} : \|a(s, \cdot, V^{(n-1)}) - a(\tau_j^{(n)}, \cdot, V^{(n-1)})\| \geq 2^{-n}$$
$$\text{or } \|a(s-, \cdot, V^{(n-1)}) - a(\tau_j^{(n)}, \cdot, V^{(n-1)})\| \geq 2^{-n}\}. \tag{7.3.34}$$

Since the process $s \mapsto a(s, \cdot, V^{(n-1)})$ is an adapted r.c.l.l. process, it follows that each $\tau_j^{(n)}$ is a stopping time and $\lim_{j \uparrow \infty} \tau_j^{(n)} = \infty$. Let $V_0^{(n)} = \xi_0$ and for $j \geq 0$ and $t \in (\tau_j^{(n)}, \tau_{j+1}^{(n)}]$ let

$$V_t^{(n)} = V_{\tau_j^{(n)}}^{(n)} + a(\tau_j^{(n)}, \cdot, V^{(n-1)})(X_t - X_{\tau_j^{(n)}}).$$

Equivalently,

$$V_t^{(n)} = \xi_0 + \sum_{j=0}^{\infty} a(\tau_j^{(n)}, \cdot, V^{(n-1)})(X_{t \wedge \tau_{j+1}^{(n)}} - X_{t \wedge \tau_j^{(n)}}). \tag{7.3.35}$$

Thus we have defined $V^{(n)}$ and we will show that these processes converge almost surely and the limit process V is the required solution. Let $F^{(n)}$, $Z^{(n)}$, $R^{(n)}$ be defined by

$$F_t^{(n)} = b(t, \cdot, V^{(n-1)}) \tag{7.3.36}$$

$$Z_t^{(n)} = \xi_0 + \int_{0+}^{t} F_s^{(n)} dX_s \tag{7.3.37}$$

$$R_t^{(n)} = \sum_{j=0}^{\infty} a(\tau_j^{(n)}, \cdot, V^{(n-1)}) 1_{(\tau_j^{(n)}, \tau_{j+1}^{(n)}]}(t) \tag{7.3.38}$$

$$V_t^{(n)} = \xi_0 + \int_{0+}^t R_s^{(n)} dX_s.$$

(7.3.39)

Let us note that by the definition of $\{\tau_j^{(n)} : j \geq 0\}$, we have

$$|F_t^{(n)} - R_t^{(n)}| \leq 2^{-n}.$$

(7.3.40)

We will prove convergence of $V^{(n)}$ employing the technique used in the proof of uniqueness to the SDE Theorem 7.21—namely random time change.

Let ϕ be a $(\mathcal{F}.)$ random time change such that $Y^j = \phi[X^j]$, $1 \leq j \leq m$ are amenable semimartingales (such a random time change exists as seen in Theorem 7.16). Let $(\mathcal{G}.) = (\phi \mathcal{F}.)$, $\psi = [\phi]^{-1}$ be defined via (7.1.5).

Let $c(t, \omega, \zeta), d(t, \omega, \zeta)$ be given by (7.3.26). As noted in the proof of Theorem 7.21, c, d satisfy (7.3.2)–(7.3.5). We will transform $\{\tau_i^{(n)} : n \geq 1, i \geq 1\}$, $\{V^{(n)}, F^{(n)}, Z^{(n)}, R^{(n)} : n \geq 1\}$ to the new time scale.

For $n \geq 1$, $j \geq 0$ let $\sigma_j^{(n)} = \phi[\tau_j^{(n)}]$, $U^{(n)} = \phi[V^{(n)}]$, $G^{(n)} = \phi[F^{(n)}]$, $W^{(n)} = \phi[Z^{(n)}]$, $S^{(n)} = \phi[R^{(n)}]$. Now it can be checked that

$$\sigma_{j+1}^{(n)} = \inf\{s > \sigma_j^{(n)} : \|a(s, \cdot, U^{(n-1)}) - a(\sigma_j^{(n)}, \cdot, U^{(n-1)})\| \geq 2^{-n}$$
$$\text{or } \|a(s-, \cdot, U^{(n-1)}) - a(\sigma_j^{(n)}, \cdot, U^{(n-1)})\| \geq 2^{-n}\}.$$

(7.3.41)

Each $\sigma_j^{(n)}$ is a $(\mathcal{G}.)$ stopping time and $\lim_{j \uparrow \infty} \sigma_j^{(n)} = \infty$. Further, $U_0^{(n)} = \xi_0$ and

$$U_t^{(n)} = \xi_0 + \sum_{j=0}^{\infty} c(\sigma_j^{(n)}, \cdot, U^{(n-1)})(Y_{t \wedge \sigma_{j+1}^{(n)}} - Y_{t \wedge \sigma_j^{(n)}}),$$

(7.3.42)

$$G_t^{(n)} = d(t, \cdot, U^{(n-1)}),$$

(7.3.43)

$$W_t^{(n)} = \xi_0 + \int_{0+}^t G_r^{(n)} dY_r,$$

(7.3.44)

$$S_t^{(n)} = \sum_{j=0}^{\infty} c(\sigma_j^{(n)}, \cdot, U^{(n-1)}) 1_{(\sigma_j^{(n)}, \sigma_{j+1}^{(n)}]}(t),$$

(7.3.45)

$$U_t^{(n)} = \xi_0 + \int_{0+}^t S_r^{(n)} dY_r.$$

(7.3.46)

Also, we have

$$|G_t^{(n)} - S_t^{(n)}| \leq 2^{-n}.$$

(7.3.47)

Recall that we had shown in (7.3.31) that c satisfies Lipschitz condition with coefficient $H = \phi[K]$. For $j \geq 1$, let

$$\rho_j = \inf\{t \geq 0 : |H_t| \geq j \text{ or } |H_{t-}| \geq j \text{ or } |U_t^{(1)} - \xi_0| \geq j\} \wedge j.$$

Then ρ_j are $(\mathcal{G}_.)$ stopping times and $\rho_j \uparrow \infty$. Further, for all $\zeta_1, \zeta_2 \in \mathbb{C}_d$ we have

$$\sup_{0 \leq a \leq (s \wedge \rho_j)} \| c(a, \omega, \zeta_2) - c(a, \omega, \zeta_1) \| \leq j \sup_{0 \leq a \leq (s \wedge \rho_j)} |\zeta_2(a) - \zeta_1(a)|. \qquad (7.3.48)$$

Let us fix $k \geq 1$ and $j \geq 1$. We observe that for all $n \geq 2$ (using (7.3.13) along with (7.3.43) and (7.3.44))

$$\mathsf{E}[1_{\{|\xi_0| \leq k\}} \sup_{0 \leq s \leq (t \wedge \rho_j)} |W_s^{(n)} - W_s^{(n-1)}|^2]$$

$$\leq 2m(4+j)\mathsf{E}[1_{\{|\xi_0| \leq k\}} \int_0^{(t \wedge \rho_j)} \sup_{0 \leq s \leq (r \wedge \rho_j)} |G_s^{(n)} - G_s^{(n-1)}|^2 dr]$$

$$\leq 2m(4+j)j^2 \mathsf{E}[1_{\{|\xi_0| \leq k\}} \int_0^{(t \wedge \rho_j)} \sup_{0 \leq s \leq (r \wedge \rho_j)} |U_s^{(n-1)} - U_s^{(n-2)}|^2 dr] \qquad (7.3.49)$$

Likewise, using (7.3.47) we get for $n \geq 1$,

$$\mathsf{E}[1_{\{|\xi_0| \leq k\}} \sup_{0 \leq s \leq (t \wedge \rho_j)} |W_s^{(n)} - U_s^{(n)}|^2]$$

$$\leq 2m(4+j)\mathsf{E}[1_{\{|\xi_0| \leq k\}} \int_0^{(t \wedge \rho_j)} \sup_{0 \leq s \leq (r \wedge \rho_j)} |G_s^{(n)} - S_s^{(n)}|^2 dr]$$

$$\leq 2m(4+j)4^{-n} j. \qquad (7.3.50)$$

Combining (7.3.49) and (7.3.50), we observe that for $n \geq 2$ (using for positive numbers x, y, z, $(x + y + z)^2 \leq 3(x^2 + y^2 + z^2)$)

$$\mathsf{E}[1_{\{|\xi_0| \leq k\}} \sup_{0 \leq s \leq (t \wedge \rho_j)} |U_s^{(n)} - U_s^{(n-1)}|^2]$$

$$\leq \quad 6m(4+j)(4^{-n} + 4^{-(n-1)})j$$

$$+ 6m(4+j)j^2 \mathsf{E}[1_{\{|\xi_0| \leq k\}} \int_0^{(t \wedge \rho_j)} \sup_{0 \leq s \leq (r \wedge \rho_j)} |U_s^{(n-1)} - U_s^{(n-2)}|^2 dr]$$

Let

$$f^{(n)}(t) = \mathsf{E}[1_{\{|\xi_0| \leq k\}} \sup_{0 \leq s \leq (t \wedge \rho_j)} |U_s^{(n)} - U_s^{(n-1)}|^2] \qquad (7.3.51)$$

Then, writing $C_{m,j} = 30m(4+j)j^2$, the above inequality implies for $n \geq 2$

$$f^{(n)}(t) \leq C_{m,j} + C_{m,j} \int_0^t f^{(n-1)}(r) dr.$$

Since $U_t^{(0)} = \xi_0$, by the definition of ρ_j, $f_t^{(1)} \leq j^2 \leq C_{m,j}$. Now (7.3.51) implies (via induction on n) that

$$f^{(n)}(t) \leq C_{m,j} \cdot \frac{(C_{m,j}t)^n}{n!} \qquad (7.3.52)$$

and as a consequence (writing $|\cdot|_2$ for the $\mathbb{L}^2(P)$ norm, and recalling $\rho_j \leq j$)

$$\sum_{n=1}^{\infty} \|1_{\{|\xi_0| \leq k\}} \sup_{0 \leq s \leq \rho_j} |U_s^{(n)} - U_s^{(n-1)}|\|_2 < \infty.$$

and as a consequence

$$\|1_{\{|\xi_0| \leq k\}} \sum_{n=1}^{\infty} \sup_{0 \leq s \leq \rho_j} |U_s^{(n)} - U_s^{(n-1)}|\|_2 < \infty. \qquad (7.3.53)$$

As in the proof of Theorem 3.30, it now follows that for $k \geq 1$, $j \geq 1$, $P(N_{k,j}) = 0$ where

$$N_{k,j} = \{\omega : 1_{\{|\xi_0(\omega)| \leq k\}} \sup_{0 \leq s \leq \rho_j(\omega)} (\sum_{n=1}^{\infty} |U_s^{(n)}(\omega) - U_s^{(n-1)}(\omega)|) = \infty\}.$$

Since for all $T < \infty$

$$P(\cup_{k,j=1}^{\infty} \{\omega : |\xi_0(\omega)| \leq k, \ \rho_j > T\}) = 1$$

it follows that $P(N) = 0$ where $N = \cup_{k,j=1}^{\infty} N_{k,j}$ and for $\omega \notin N$, $U_s^{(n)}(\omega)$ converges uniformly on $[0, T]$ for every $T < \infty$. So let us define U as follows:

$$U_t(\omega) = \begin{cases} \lim_{n \to \infty} U_t^{(n)}(\omega) & \text{if } \omega \in N^c \\ 0 & \text{if } \omega \in N. \end{cases}$$

By definition, U is a continuous $(\mathcal{G}_\cdot)$ adapted process (since by assumption $N \in \mathcal{F}_0 = \mathcal{G}_0$) and $U^{(n)}$ converges to U uniformly in $[0, T]$ for every T almost surely (and thus also $\mathbf{d}_{ucp}(U^{(n)}, U) \to 0$). Also, (7.3.53) yields

$$\lim_{n,r \to \infty} \|1_{\{|\xi_0| \leq k\}} \sup_{0 \leq s \leq \rho_j} |U_s^{(n)} - U_s^{(r)}|\|_2 = 0.$$

Now Fatou's Lemma implies

$$\lim_{n \to \infty} \|1_{\{|\xi_0| \leq k\}} \sup_{0 \leq s \leq \rho_j} |U_s^{(n)} - U_s|\|_2 = 0$$

which is same as

$$\lim_{n \to \infty} E[1_{\{|\xi_0| \leq k\}} \sup_{0 \leq s \leq \rho_j} |U_s^{(n)} - U_s|^2] = 0. \qquad (7.3.54)$$

Now defining $G_t = d(t, \cdot, U)$ and $W_t = \xi_0 + \int_0^t G_s dY_s$, it follows using the Lipschitz condition (7.3.48) along with (7.3.54) that

$$\lim_{n \to \infty} \mathsf{E}[1_{\{|\xi_0| \le k\}} \sup_{0 \le s \le \rho_j} |W_s^{(n)} - W_s|^2] = 0. \tag{7.3.55}$$

Now (7.3.50), (7.3.54) and (7.3.55) imply that (for all $j, k \ge 1$)

$$\mathsf{E}[1_{\{|\xi_0| \le k\}} \sup_{0 \le s \le \rho_j} |W_s - U_s|^2] = 0$$

and hence that

$$P(W_s = U_s \ \forall s \ge 0) = 1.$$

Recalling the definition of G, W, it follows that U satisfies

$$U_t = \xi_0 + \int_0^t d(s, \cdot, U) dY_s.$$

It now follows that $V = \psi[U]$ satisfies (7.3.33). $\qquad\qquad\qquad\qquad\square$

We have shown the existence and uniqueness of solution to the SDE (7.3.33). Indeed, we have explicitly constructed processes $V^{(n)}$ that converge to V. We record this in the next theorem

Theorem 7.23 *Let $X^1, X^2, \ldots X^m$ be continuous semimartingales. Let a, b satisfy conditions (7.3.2)–(7.3.5). Let ξ be any $\mathcal{F}_0$ measurable random variable. For $n \ge 1$ let $\{\tau_j^{(n)} : j \ge 1\}$ and $V^{(n)}$ be defined inductively by (7.3.34) and (7.3.35) as in the proof of Theorem 7.22. Let V be the (unique) solution to the SDE*

$$V_t = \xi_0 + \int_{0+}^t b(s, \cdot, V) dX_s.$$

Then $V^{(n)} \xrightarrow{em} V$ and $V^{(n)}$ converges to V in ucc topology almost surely.

Proof We had constructed in the proof of Theorem 7.22 a $(\mathcal{F}_\cdot)$ random time change ϕ and filtration $(\mathcal{G}_\cdot) = (\phi \mathcal{F}_\cdot)$ such that $U^{(n)} = \phi[V^{(n)}]$ converges to U in ucp metric: $\mathbf{d}_{ucp}(U^{(n)}, U) \to 0$. Now $\psi = \phi^{-1}$ is also a $(\mathcal{G}_\cdot)$ random time change, $V^{(n)} = \psi[U^{(n)}]$ and $V = \psi[U]$. It follows from Theorem 7.6 that $\mathbf{d}_{ucp}(V^{(n)}, V) \to 0$. Let $F_t = b(t, \cdot, V)$. Now the Lipschitz condition (7.3.4) on a, b implies $F^{(n)} \xrightarrow{ucp} F$, where $F^{(n)}$ is defined by (7.3.36) and then (7.3.40) implies $R^{(n)} \xrightarrow{ucp} F$. Hence Theorem 4.107 implies that $\int R^{(n)} dX$ converges in Emery topology to $\int F dX$. Now $V_t = \xi_0 + \int_0^t F_s dX_s$ and (7.3.39) together imply that $V^{(n)} \xrightarrow{em} V$. As for almost sure convergence in ucc topology, we had observed that it holds for $U^{(n)}, U$ and then we can see that same holds for $V^{(n)}, V$. $\qquad\qquad\qquad\square$

7.4 Pathwise Formula for Solution of SDE

In this section, we will consider the SDE

$$dV_t = f(t-, H, V)dX_t \tag{7.4.1}$$

for an $\mathbb{R}^d$-valued process V where $f : [0, \infty) \times \mathbb{D}_r \times \mathbb{C}_d \mapsto \mathsf{L}(d, m)$, H is an $\mathbb{R}^r$-valued r.c.l.l. adapted process, X is a $\mathbb{R}^m$-valued continuous semimartingale. Here $\mathbb{D}_r = \mathbb{D}([0, \infty), \mathbb{R}^r)$, $\mathbb{C}_d = \mathbb{C}([0, \infty), \mathbb{R}^d)$. For $t < \infty$, $\zeta \in \mathbb{C}_d$ and $\gamma \in \mathbb{D}_r$, let $\gamma^t(s) = \gamma(t \wedge s)$ and $\zeta^t(s) = \zeta(t \wedge s)$. We assume that f satisfies

$$f(t, \gamma, \zeta) = f(t, \gamma^t, \zeta^t), \quad \forall \gamma \in \mathbb{D}_r, \ \zeta \in \mathbb{C}_d, \ 0 \le t < \infty, \tag{7.4.2}$$

$$t \mapsto f(t, \gamma, \zeta) \text{ is an r.c.l.l. function } \forall \gamma \in \mathbb{D}_r, \ \zeta \in \mathbb{C}_d. \tag{7.4.3}$$

We also assume that there exists a constant $C_T < \infty$ for each $T < \infty$ such that $\forall \gamma \in \mathbb{D}_r$, $\zeta_1, \zeta_2 \in \mathbb{C}_d$, $0 \le t \le T$

$$\|f(t, \gamma, \zeta_1) - f(t, \gamma, \zeta_2)\| \le C_T(1 + \sup_{0 \le s \le t} |\gamma(s)|)(\sup_{0 \le s \le t} |\zeta_1(s) - \zeta_2(s)|). \tag{7.4.4}$$

As in Sect. 6.2, we will now obtain a mapping Ψ that yields a pathwise solution to the SDE (7.4.1).

Theorem 7.24 *Suppose f satisfies (7.4.2)–(7.4.4). Then there exists a mapping*

$$\Psi : \mathbb{R}^d \times \mathbb{D}_r \times \mathbb{C}_m \mapsto \mathbb{C}_d$$

with the following property: for an $\mathcal{F}_0$ measurable random variable ξ_0, an adapted r.c.l.l. process H and a continuous semimartingale X,

$$V = \Psi(\xi_0, H, X)$$

yields the unique solution to the SDE

$$V_t = \xi_0 + \int_0^t f(s-, H, V)dX_s. \tag{7.4.5}$$

Proof We will define mappings

$$\Psi^{(n)} : \mathbb{R}^d \times \mathbb{D}_r \times \mathbb{C}_m \mapsto \mathbb{C}([0, \infty), \mathbb{R}^d)$$

inductively for $n \ge 0$. Let $\Psi^{(0)}(u, \gamma, \zeta)(s) = u$ for all $s \ge 0$. Having defined $\Psi^{(0)}$, $\Psi^{(1)}, \ldots, \Psi^{(n-1)}$, we define $\Psi^{(n)}$ as follows. Fix n and $u \in \mathbb{R}^d$, $\gamma \in \mathbb{D}_r$ and $\zeta \in \mathbb{C}_d$.

Let $t_0^{(n)} = 0$ and let $\{t_j^{(n)} : j \geq 1\}$ be defined inductively as follows: ($\{t_j^{(n)} : j \geq 1\}$ are themselves functions of (u, γ, ζ), which are fixed for now and we will suppress writing it as a function) if $t_j^{(n)} = \infty$ then $t_{j+1}^{(n)} = \infty$ and if $t_j^{(n)} < \infty$ then writing

$$\Gamma^{(n-1)}(u, \gamma, \zeta)(s) = f(s, \gamma, \Psi^{(n-1)}(u, \gamma, \zeta)),$$

let

$$t_{j+1}^{(n)} = \inf\{s > t_j^{(n)} : \|\Gamma^{(n-1)}(u, \gamma, \zeta)(s) - \Gamma^{(n-1)}(u, \gamma, \zeta)(t_j^{(n)})\| \geq 2^{-n}$$
$$\text{or } \|\Gamma^{(n-1)}(u, \gamma, \zeta)(s-) - \Gamma^{(n-1)}(u, \gamma, \zeta)(t_j^{(n)})\| \geq 2^{-n}\}$$

(since $\Gamma^{(n-1)}(u, \gamma, \zeta)$ is an r.c.l.l. function, $t_j^{(n)} \uparrow \infty$ as $j \uparrow \infty$) and

$$\Psi^{(n)}(u, \gamma, \zeta)(s) = u + \sum_{j=0}^{\infty} \Gamma^{(n-1)}(u, \gamma, \zeta)(t_j^{(n)})(\zeta(s \wedge t_{j+1}^{(n)}) - \zeta(s \wedge t_j^{(n)})).$$

This defines $\Psi^{(n)}(u, \gamma, \zeta)$. Now we define

$$\Psi(u, \gamma, \zeta) = \begin{cases} \lim_n \Psi^{(n)}(u, \gamma, \zeta) & \text{if the limit exists in ucc topology} \\ 0 & \text{otherwise.} \end{cases} \tag{7.4.6}$$

Now it can be seen that

$$a(s, \omega, \zeta) = f(s, H(\omega), \zeta), \quad b(s, \omega, \zeta) = f(s-, H(\omega), \zeta)$$

satisfies (7.3.2)–(7.3.5) and if $V^{(n)}$ is defined inductively by (7.3.34) and (7.3.35), then

$$\Psi^{(n)}(\xi_0(\omega), H(\omega), X(\omega)) = V^{(n)}(\omega).$$

As shown in Theorem 7.23, $V^{(n)}(\omega)$ converges to $V(\omega)$ in ucc topology almost surely and hence it follows that

$$\mathsf{P}(\Psi(\xi_0(\omega), H(\omega), X(\omega))(t) = V_t(\omega) \ \forall t) = 1.$$

$\square$

This pathwise formula was obtained in [36, 40]. It was recast in [41] in the form given in this section.

7.5   Weak Solutions of SDE

Let us consider a special case of the SDE discussed in the previous section. Let
$\sigma : [0, \infty) \times \mathbb{C}_d \mapsto \mathsf{L}(d, d)$ and $h : [0, \infty) \times \mathbb{C}_d \mapsto \mathbb{R}^d$ be measurable functions.
For $\zeta \in \mathbb{C}_d$ and $t \geq 0$ let $\zeta^t(s) = \zeta(t \wedge s)$. Throughout this section, we assume that
σ, h satisfy

$$\sigma(t, \zeta) = \sigma(t, \zeta^t), \quad \forall \zeta \in \mathbb{C}_d, \ 0 \leq t < \infty, \tag{7.5.1}$$

$$h(t, \zeta) = h(t, \zeta^t), \quad \forall \zeta \in \mathbb{C}_d, \ 0 \leq t < \infty. \tag{7.5.2}$$

Let W be a d-dimensional Brownian motion adapted to $(\mathcal{F}_.)$. Consider the SDE

$$dY_t = \sigma(t, Y)dW + h(t, Y)dt. \tag{7.5.3}$$

or equivalently

$$Y_t^j = Y_0^j + \sum_{k=1}^d \int_0^t \sigma^{jk}(s, Y)dW_s^k + \int_0^t h^j(s, Y)ds. \tag{7.5.4}$$

Equation (7.5.3) is said to admit a *strong solution* if given a Brownian motion W
on some probability space $(\Omega, \mathcal{F}, \mathsf{P})$, a filtration $(\mathcal{F}_.)$ such that $(W_t, \mathcal{F}_t)_{\{t \geq 0\}}$ is a
Wiener martingale, and a $\mathcal{F}_0$ measurable random variable Y_0, there exists a process Y
adapted to $(\mathcal{F}_.)$ satisfying (7.5.4). Moreover the uniqueness of strong solution holds
if given two solution Y and Y' w.r.t. the same Brownian motion W,

$$\mathsf{P}(Y_0 = Y_0') = 1$$

implies that

$$\mathsf{P}(Y_t = Y_t' \ \forall t) = 1.$$

This is the notion of solution to an SDE that we have been considering. There is
another notion of a solution to the SDE (7.5.3), known as weak solution. It is as
follows.

We say that Eq. (7.5.3) admits a *weak solution* if for all $y_0 \in \mathbb{R}^d$ we can construct
a probability space $(\Omega, \mathcal{F}, \mathsf{P})$, a Brownian motion W adapted to a filtration $(\mathcal{F}_.)$
such that $(W_t, \mathcal{F}_t)_{\{t \geq 0\}}$ is a Wiener martingale and a $(\mathcal{F}_.)$ adapted process Y such that
$Y_0 = y_0$ satisfying (7.5.4). We say weak uniqueness of solution to Eq. (7.5.3) holds
if for all $y_0 \in \mathbb{R}^d$, given two (possibly different) probability spaces $(\Omega, \mathcal{F}, \mathsf{P})$ and
$(\widehat{\Omega}, \widehat{\mathcal{F}}, \widehat{\mathsf{P}})$, filtrations $(\mathcal{F}_.)$ on $(\Omega, \mathcal{F}, \mathsf{P})$ and $(\widehat{\mathcal{F}}_.)$ on $(\widehat{\Omega}, \widehat{\mathcal{F}}, \widehat{\mathsf{P}})$, Brownian motions
W and $\widehat{W}$ adapted to filtrations $(\mathcal{F}_.)$ and $(\widehat{\mathcal{F}}_.)$ on the two spaces, respectively, such
that $(W_t, \mathcal{F}_t)_{\{t \geq 0\}}$ and $(\widehat{W}_t, \widehat{\mathcal{F}}_t)_{\{t \geq 0\}}$ are Wiener martingales, and processes Y and $\widehat{Y}$
adapted to $(\mathcal{F}_.)$ and $(\widehat{\mathcal{F}}_.)$, respectively, such that

$$Y_t^j = y_0 + \sum_{k=1}^{d} \int_0^t \sigma^{jk}(s, Y)dW_s^k + \int_0^t h^j(s, Y)ds. \qquad (7.5.5)$$

$$\widehat{Y}_t^j = y_0 + \sum_{k=1}^{d} \int_0^t \sigma^{jk}(s, \widehat{Y})d\widehat{W}_s^k + \int_0^t h^j(s, \widehat{Y})ds, \qquad (7.5.6)$$

the distributions of Y and $\widehat{Y}$ are the same, i.e.

$$\mathsf{P} \circ Y^{-1} = \widehat{\mathsf{P}} \circ \widehat{Y}^{-1}. \qquad (7.5.7)$$

Clearly, existence of strong solution implies existence of a weak solution.

Note that the uniqueness of the weak solution requires equality in distribution of any two solutions, whereas uniqueness of the strong solution requires almost sure equality of paths. The next example illustrates the difference in these two notions.

Example **7.25** Consider the SDE

$$dX_t = \mathsf{sgn}(X_t)dW_t \qquad (7.5.8)$$

where W is a Brownian motion. Recall that for $x \in \mathbb{R}$, $\mathsf{sgn}(x) = 1$ for $x \geq 0$ and $\mathsf{sgn}(x) = -1$ for $x < 0$, so that $|x| = \mathsf{sgn}(x)x$. Let us note that if X is a solution to (7.5.8), then X is a continuous martingale (since sgn is bounded) and $[X, X]_t = t$ since $(\mathsf{sgn}(x))^2 = 1$ for all x. Thus any solution X to (7.5.8) is a Brownian motion and thus we have uniqueness of weak solution. Let us now illustrate that we can construct X, W satisfying (7.5.8) such that W is a Brownian motion. Start with a Brownian motion X and for $t \geq 0$ let

$$W_t = \int_0^t \mathsf{sgn}(X_s)dX_s.$$

Then it follows that X, W satisfy (7.5.8). Thus we have existence and uniqueness of weak solution to the SDE (7.5.8). On the other hand, easy to see that if X is a solution then so is $Y = -X$. This uses that fact that $\mathsf{P}(X_s = 0) = 0$ for all s. Thus, strong uniqueness does not hold. This example is due to Tanaka who also observed that there is no $(\mathcal{F}_\cdot^W)$ adapted process X such that (7.5.8) is true. Thus (7.5.8) does not admit a strong solution.

A general result due to Yamada–Watanabe says that strong uniqueness also implies weak uniqueness. Here we will prove that under Lipschitz conditions on the coefficients, we have strong uniqueness as well as weak uniqueness. Instead of appealing to Yamada–Watanabe result, we deduce weak uniqueness from the pathwise formula for solution to the SDE.

Suppose that there exists $C : [0, \infty) \mapsto [0, \infty)$ such that σ, h appearing in (7.5.3) satisfy

$$t \mapsto \sigma(t, \zeta) \text{ is a continuous function } \forall \; \zeta \in \mathbb{C}_d. \tag{7.5.9}$$

$$t \mapsto h(t, \zeta) \text{ is a continuous function } \forall \; \zeta \in \mathbb{C}_d. \tag{7.5.10}$$

$$\|\sigma(t, \zeta_1) - \sigma(t, \zeta_2)\| \le C_T \big(\sup_{0 \le s \le t} |\zeta_1(s) - \zeta_2(s)| \big) \tag{7.5.11}$$

$$\|h(t, \zeta_1) - h(t, \zeta_2)\| \le C_T \big(\sup_{0 \le s \le t} |\zeta_1(s) - \zeta_2(s)| \big) \tag{7.5.12}$$

Under these conditions, we can deduce the following.

Theorem 7.26 *Suppose σ, h satisfy (7.5.1), (7.5.2), (7.5.9)–(7.5.12). Let W be a d-dimensional Brownian motion adapted to $(\mathcal{F}_\cdot)$ such that $(W_t, \mathcal{F}_t)_{\{t \ge 0\}}$ is a Wiener martingale and let Y_0 be a $\mathcal{F}_0$ measurable random variable. Then*

(i) *Equation (7.5.3) admits a strong solution.*
(ii) *Strong uniqueness holds for Eq. (7.5.3).*
(iii) *Weak uniqueness also holds for Eq. (7.5.3).*

Proof For $1 \le j \le d$, defining $f^{jk}(t, \gamma, \zeta) = \sigma^{jk}(t, \zeta)$ for $1 \le k \le d$ and f^{jk} $(t, \gamma, \zeta) = h^j(t, \zeta)$ for $k = d + 1$, Eq. (7.5.3) is same as (7.4.5) with $\xi_0 = Y_0$, $X^j = W^j$ for $1 \le j \le d$ and $X_t^{d+1} = t$. Also since σ, h satisfy (7.5.9)–(7.5.12) it can be checked that f satisfies (7.4.2)–(7.4.4). Hence invoking Theorems 7.21 and 7.22, we conclude that existence of strong solution as well as uniqueness of strong solution holds for the SDE (7.5.3).

Observe that $\Psi(u, \gamma, \zeta)$ does not depend on γ, hence denoting $\Psi^*(u, \zeta) = \Psi(u, 0, \zeta)$, where 0 is the constant function, it follows that if $Y, \widehat{Y}$ satisfy (7.5.5) and (7.5.6), respectively, then

$$Y = \Psi^*(y_0, W), \quad \widehat{Y} = \Psi^*(y_0, \widehat{W}).$$

As a consequence, denoting the Wiener measure on $\mathbb{C}_d$ by μ_w and the coordinate process on $\mathbb{C}_d$ as X, $Z = \Psi^*(y_0, W)$ we have

$$\mathsf{P} \circ Y^{-1} = \mu_w \circ Z^{-1}, \quad \widehat{\mathsf{P}} \circ \widehat{Y}^{-1} = \mu_w \circ Z^{-1}.$$

Thus weak uniqueness holds. $\square$

7.6 Matrix-Valued Semimartingales

In this section, we will consider matrix-valued semimartingales. The notations introduced here will be used only in this section and in a corresponding section later. Recall that $\mathsf{L}(m, k)$ is the set of all $m \times k$ matrices. Let $L_0(d)$ denote the set of non-singular $d \times d$ matrices.

Let $X = (X^{pq})$ be an $L(m, k)$-valued process. X is said to be a semimartingale if each X^{pq} is a semimartingale. Likewise, X will be said to be a local martingale if each X^{pq} is a local martingale and we will say that $X \in \mathbb{V}$ if each $X^{pq} \in \mathbb{V}$.

If $f = (f^{ij})$ is an $L(d, m)$-valued predictable process such that $f^{ij} \in \mathbb{L}(X^{jq})$ (for all i, j, q), then $Y = \int f dX$ is defined as an $L(d, k)$-valued semimartingale as follows: $Y = (Y^{iq})$ where

$$Y^{iq} = \sum_{j=1}^{m} \int f^{ij} dX^{jq}.$$

Likewise, if $g = (g^{ij})$ is an $L(k, d)$-valued predictable process such that $g^{ij} \in \mathbb{L}(X^{pi})$ (for all i, j, p), then $Z = \int (dX)g$ is defined as follows: $Z = (Z^{pj})$ where

$$Z^{pj} = \sum_{i=1}^{k} \int g^{ij} dX^{pi}.$$

For $L(d, d)$-valued semimartingales X, Y let $[X, Y] = ([X, Y]^{ij})$ be the $L(d, d)$-valued process defined by

$$[X, Y]_t^{ij} = \sum_{k=1}^{d} [X^{ik}, Y^{kj}]_t.$$

Exercise 7.27 Let X, Y be $L(d, d)$-valued semimartingales. Show that

$$X_t Y_t = X_0 Y_0 + \int_{0+}^{t} X_{s-} dY_s + \int_{0+}^{t} (dX_s) Y_{s-} + [X, Y]_t. \qquad (7.6.1)$$

The relation (7.6.1) is the matrix analogue of the integration by parts formula (4.6.7).

Recall our terminology: we say that a $L(d, d)$-valued process h is $L_0(d)$-valued if

$$P(h_t \in \mathbb{L}_0(d) \ \forall t \geq 0) = 1.$$

Exercise 7.28 Let X, Y be $L(d, d)$-valued continuous semimartingales and let f, g, h be $L(d, d)$-valued predictable locally bounded processes. Further let h be $L_0(d)$-valued. Let $W = \int f dX$, $Z = \int (dX)g$, $U = \int (dX)h$ and $V = \int h^{-1} dY$. Show that

$$[W, Y]_t = \int_0^t f d[X, Y]. \qquad (7.6.2)$$

$$[Y, Z]_t = \int_0^t (d[Y, X])g. \qquad (7.6.3)$$

$$\int_0^t g dW = \int_0^t g f dX. \qquad (7.6.4)$$

$$\int_0^t (dZ)f = \int_0^t (dX)g f. \qquad (7.6.5)$$

$$\int_0^t f dZ \ = \int_0^t (dW)g. \tag{7.6.6}$$

$$[U, V]_t \ = [X, Y]_t. \tag{7.6.7}$$

In view of (7.6.6), with f, g, X, W, Z as in the previous exercise, we will denote $\int f dZ = \int (dW)g = \int f(dX)g$.

We can consider an analogue of the SDE (7.3.1)

$$dU_t = b(t, \cdot, U)dY_t, \quad t \geq 0, \quad U_0 = \xi_0 \tag{7.6.8}$$

where now Y is an $\mathsf{L}(m, k)$-valued continuous semimartingale, U is an $\mathsf{L}(d, k)$-valued process, ξ_0 is $\mathsf{L}(d, k)$-valued random variable and here

$$b : [0, \infty) \times \Omega \times \mathbb{C}([0, \infty), L(d, k)) \to \mathsf{L}(d, m).$$

Essentially the same arguments as given earlier in the section would give analogues of existence and uniqueness results for Eq. (7.6.8).

Exercise 7.29 Formulate and prove analogues of Theorems 7.21, 7.22 and 7.23 for Eq. (7.6.8). Make precise and do a similar analysis for $d\tilde{U}_t = (d\tilde{Y}_t)\,\tilde{b}$ $(t, \cdot, \tilde{U})$.

Exercise 7.30 Let X be an $\mathsf{L}(d, d)$-valued continuous semimartingale with $X(0) = 0$ and let I denote the $d \times d$ identity matrix. Show that the equations

$$Y_t = I + \int_0^t Y_s dX_s \tag{7.6.9}$$

and

$$Z_t = I + \int_0^t (dX_s)Z_s \tag{7.6.10}$$

admit unique solutions.

The solutions Y, Z are denoted respectively by $\mathfrak{e}(X)$ and $\mathfrak{e}'(X)$ and are the left and right exponential of X.

Exercise 7.31 Let X be an $\mathsf{L}(d, d)$-valued continuous semimartingale with $X_0 = 0$. Let $Y = -X + [X, X]$. Let $W = \mathfrak{e}(X)$ and $Z = \mathfrak{e}'(Y)$. Show that

(i) $[Y, Y] = [X, X]$.
(ii) $[X, Y] = -[X, X]$.
(iii) $[W, Z] = \int W(d[X, Y])Z$
(iv) $WZ = I$

The relation (iv) above implies that for any $\mathsf{L}(d, d)$-valued continuous semimartingale X with $X_0 = 0$, $\mathfrak{e}(X)$ is $L_0(d)$-valued and

$$[\mathfrak{e}(X)]^{-1} = \mathfrak{e}'(-X + [X, X]). \qquad (7.6.11)$$

Exercise 7.32 Let X^1, X^2, X^3, X^4 be $\mathsf{L}(d, d)$-valued continuous semimartingales with $X_0^j = 0$, $j = 1, 2, 3, 4$. Show that

(i) $\mathfrak{e}(X^1 + X^2 + [X^1, X^2]) = \mathfrak{e}(\tilde{X}^1)\mathfrak{e}(X^2)$ where $\tilde{X}^1 = \int Y^2(dX^1)(Y^2)^{-1}$
   and $Y^2 = \mathfrak{e}(X^2)$.
(ii) $\mathfrak{e}(X^1 + X^2) = \mathfrak{e}(\tilde{X}^1)\mathfrak{e}(\tilde{X}^2)$ where $\tilde{X}^1 = \int Y^2(dX^1)(Y^2)^{-1}$, $\tilde{X}^2 = X^2 - [X^1,$
   $X^2]$ and $Y^2 = \mathfrak{e}(\tilde{X}^2)$.
(iii) $\mathfrak{e}(X^3)\mathfrak{e}(X^2) = \mathfrak{e}(\tilde{X}^3 + X^2 + [\tilde{X}^3, X^2])$ where $\tilde{X}^3 = \int (Y^2)^{-1}(dX^3)Y^2$ and
   $Y^2 = \mathfrak{e}(X^2)$.
   HINT: For (i), start with right-hand side and use integration by parts
   formula (7.6.1) and simplify. For (ii), note that $X^1 + \tilde{X}^2 + [X^1, \tilde{X}^2] =$
   $X^1 + X^2$ and use (i). For (iii) note that if we let $X^1 = \tilde{X}^3$ then $\int Y^2(dX^1)$
   $(Y^2)^{-1} = X^3$.

Exercise 7.33 Let Y be an $\mathsf{L}_0(d)$-valued continuous semimartingale with $Y_0 = I$. Let $X_t = \int_{0+}^t Y^{-1}dY$. Show that

$$Y = \mathfrak{e}(X).$$

For an $\mathsf{L}_0(d)$-valued continuous semimartingale Y with $Y_0 = I$, we define $\mathfrak{log}(Y) = \int_{0+}^t Y^{-1}dY$ and $\mathfrak{log}'(Y) = \int_{0+}^t (dY)Y^{-1}$. We then have

$$\mathfrak{e}(\mathfrak{log}(Y) = Y, \quad \mathfrak{e}'(\mathfrak{log}'(Y)) = Y.$$

Likewise, for any $\mathsf{L}(d, d)$-valued continuous semimartingale X with $X_0 = 0$, we have

$$\mathfrak{log}(\mathfrak{e}(X) = X, \quad \mathfrak{log}'(\mathfrak{e}'(X)) = X.$$

Exercise 7.34 Let X be an $\mathsf{L}(d, d)$-valued continuous semimartingales with $X_0 = 0$ and Y be an $\mathsf{L}_0(d)$-valued continuous semimartingale with $Y_0 = I$. Then show that

(i) X is a local martingale if and only if $\mathfrak{e}(X)$ is a local martingale.
(ii) $X \in \mathbb{V}$ if and only if $\mathfrak{e}(X) \in \mathbb{V}$.
(iii) Y is a local martingale if and only if $\mathfrak{log}(Y)$ is a local martingale.
(iv) $Y \in \mathbb{V}$ if and only if $\mathfrak{log}(Y) \in \mathbb{V}$.

Exercise 7.35 Let Y be an $\mathsf{L}_0(d)$-valued continuous semimartingale with $Y_0 = I$. Show that Y admits a decomposition $Y = MA$ where $M_0 = I$, $A_0 = I$, M is a continuous local martingale and $A \in \mathbb{V}$. Further show that this decomposition is unique.
HINT: Let $X = \mathfrak{log}(Y)$ and use Exercise 7.32 to connect multiplicative decomposition of Y and additive decomposition of X.

The exercises given in this section are from [37].

Chapter 8
Predictable Increasing Processes

We have discussed predictable σ-field and seen the crucial role played by predictable integrands in the theory of stochastic integration. In our treatment of the integration, we have so far suppressed another role played by predictable processes. In the decomposition of semimartingales, Theorem 5.55, the process A with finite variation paths turns out to be a predictable process. Indeed, this identification played a major part in the development of the theory of stochastic integration.

In this chapter, we will make this identification and prove the Doob–Meyer decomposition theorem obtaining the predictable quadratic variation $\langle M, M \rangle$ of a square integrable martingale. We will also introduce the notion of a predictable stopping time.

An important step towards the proof of Doob–Meyer decomposition theorem is: An r.c.l.l. adapted process A with finite variation paths, $A_0 = 0$, $\mathsf{E}[\sup_{0 \le t \le T} |A_t|] < \infty$ is predictable if and only if it is natural, i.e. for all bounded r.c.l.l. martingales N, $[N, A]$ is also a martingale.

This result is usually stated assuming that the underlying filtration is right continuous. We will prove its validity without assuming this. However, some of the auxiliary results do require right continuity of σ-fields, which we state explicitly.

8.1 The σ-Field $\mathcal{F}_{\tau-}$

Recall that for a stopping time τ with respect to a filtration $(\mathcal{F}_{\cdot})$, the stopped σ-field $\mathcal{F}_{\tau}$ is defined by

$$\mathcal{F}_{\tau} = \{A \in \sigma(\cup_t \mathcal{F}_t) : \ A \cap \{\tau \le t\} \in \mathcal{F}_t \ \forall t < \infty.\}$$

We had seen that for every r.c.l.l. adapted process X and a stopping time τ, X_{τ} is $\mathcal{F}_{\tau}$ measurable. We now define the pre-stopped σ-field $\mathcal{F}_{\tau-}$ as follows.

© Springer Nature Singapore Pte Ltd. 2018
R. L. Karandikar and B. V. Rao, *Introduction to Stochastic Calculus*,
Indian Statistical Institute Series, https://doi.org/10.1007/978-981-10-8318-1_8

Definition 8.1 Let τ be a stopping time with respect to a filtration $(\mathcal{F}_\cdot)$. Then

$$\mathcal{F}_{\tau-} = \sigma(\mathcal{F}_0 \cup \{A \cap \{t < \tau\} : A \in \mathcal{F}_t, \ t < \infty\}).$$

Exercise 8.2 Let τ be the constant stopping time $\tau = t, t > 0$. Show that

$$\mathcal{F}_{t-} = \sigma(\cup_{s<t}\mathcal{F}_s).$$

Exercise 8.3 Let $0 \le s < t$. Show that $\mathcal{F}_s^+ \subseteq \mathcal{F}_{t-}$.

We note some basic properties of the pre-stopped σ-field in the next result. Recall the definition (2.34) of f_τ, (for a process f) whereby $f_\tau = f_\tau 1_{\{\tau<\infty\}}$.

Theorem 8.4 *Let τ, σ be stopping times with respect to a filtration $(\mathcal{F}_\cdot)$. Then*

  (i) *τ is $\mathcal{F}_{\tau-}$ measurable.*
 (ii) *$\mathcal{F}_{\tau-} \subseteq \mathcal{F}_\tau$.*
(iii) *Let $\sigma \le \tau$. If $\sigma < \tau$ on $\tau > 0$ (i.e. $\tau(\omega) > 0$ implies $\sigma(\omega) < \tau(\omega)$), then $\mathcal{F}_\sigma \subseteq \mathcal{F}_{\tau-}$.*
 (iv) *If $A \in \mathcal{F}_\sigma$ then $(A \cap \{\sigma < \tau\}) \in \mathcal{F}_{\tau-}$ and in particular, $\{\sigma < \tau\} \in \mathcal{F}_{\tau-}$.*
  (v) *If f is a predictable process then f_τ is $\mathcal{F}_{\tau-}$ measurable.*
 (vi) *Let W be a $\mathcal{F}_{\tau-}$ measurable random variable. Then there exists a predictable process f such that $f_\tau 1_{\{\tau<\infty\}} = W 1_{\{\tau<\infty\}}$.*
 (iv) *Let U be a $\mathcal{F}_\tau$ measurable random variable with $E[|U|] < \infty$ and $E[U \mid \mathcal{F}_{\tau-}] = 0$. Let $M_t = U 1_{[\tau,\infty)}(t)$. Then M is a martingale.*

Proof Since $\{t < \tau\} \in \mathcal{F}_{\tau-}$ by definition, (i) follows.

For (ii) note that if $A \in \mathcal{F}_t$, $t < \infty$ and $B = A \cap \{t < \tau\}$, then for any $s \in [0, \infty)$, $B \cap \{\tau \le s\}$ is empty if $s \le t$ and $B \cap \{\tau \le s\} \in \mathcal{F}_s$ if $t < s$. Thus $B \in \mathcal{F}_\tau$. This proves (ii).

For (iii), let $A \in \mathcal{F}_\sigma$. Note that writing $\mathbb{Q}^+$ to be the set of rational numbers in $[0, \infty)$,

$$A = (\cup_{r\in\mathbb{Q}^+}(A \cap \{\sigma \le r\} \cap \{r < \tau\})) \cup \{A \cap \{\sigma = \tau = 0\}\}$$

and $A \cap \{\sigma \le r\} \in \mathcal{F}_r$. Thus $A \in \mathcal{F}_{\tau-}$.

For (iv) note that for $A \in \mathcal{F}_\sigma$,

$$(A \cap \{\sigma < \tau\}) = \cup_{r\in\mathbb{Q}^+}((A \cap \{\sigma \le r\}) \cap \{r < \tau\})$$

along with $(A \cap \{\sigma \le r\}) \in \mathcal{F}_r$ implies $(A \cap \{\sigma < \tau\}) \in \mathcal{F}_{\tau-}$. Taking $A = \Omega$ we conclude $\{\sigma < \tau\} \in \mathcal{F}_{\tau-}$.

For (v), recall that $\mathcal{P}$ is the smallest σ-field generated by processes of the form (see (4.2.1))

$$f_s = a_0 1_{\{0\}}(s) + \sum_{j=0}^{m} a_{j+1} 1_{(s_j, s_{j+1}]}(s)$$

where $0 = s_0 < s_1 < s_2 < \ldots < s_{m+1} < \infty$, $m \geq 1$, a_0 is bounded $\mathcal{F}_0$ measurable and for $1 \leq j \leq (m+1)$, a_j is a bounded $\mathcal{F}_{s_{j-1}}$ measurable random variable. For such an f and $\alpha < 0$,

$$\{f_\tau \leq \alpha\} = (\{a_0 \leq \alpha\} \cap \{\tau = 0\}) \cup (\bigcup_{j=0}^m \{a_{j+1} \leq \alpha\} \cap \{s_j < \tau \leq s_{j+1}\}). \qquad (8.1.1)$$

Now a_{j+1} is $\mathcal{F}_{s_j}$ measurable implies $\{a_{j+1} \leq \alpha\} \cap \{s_j < \tau\} \in \mathcal{F}_{\tau-}$ and since τ is $\mathcal{F}_{\tau-}$ measurable, so does $\{a_{j+1} \leq \alpha\} \cap \{s_j < \tau \leq s_{j+1}\}$. This and the fact that $\mathcal{F}_0 \subseteq \mathcal{F}_{\tau-}$ together imply that $\{f_\tau \leq \alpha\} \in \mathcal{F}_{\tau-}$.

For $\alpha \geq 0$, $\{f_\tau \leq \alpha\}$ equals the expression on the right-hand side of (8.1.1) union $\{s_{j+1} < \tau\}$. Since $\{s_{j+1} < \tau\} \in \mathcal{F}_{\tau-}$, it follows that f_τ is $\mathcal{F}_{\tau-}$ measurable for simple f as given above. The result (v) follows by invoking the monotone class theorem, Theorem 2.66.

For (vi), if $W = 1_B$ where $B = A \cap \{t < \tau\}$ with $A \in \mathcal{F}_t$, then we can take $f = 1_A 1_{(t,\infty)}$ while if $B \in \mathcal{F}_0$, we can take $f = 1_B 1_{[0,\infty)}$. Thus the required result holds if $W = 1_B$ when

$$B \in \mathcal{H} = \mathcal{F}_0 \cup \{A \cap \{t < \tau\} : A \in \mathcal{F}_t, \ t \geq 0\}.$$

Thus if $\mathbb{G}$ denotes the class of simple functions over $\mathcal{H}$, if follows that the result (vi) is true if $W \in \mathbb{G}$. Note that $\mathcal{H}$ is closed under finite intersections and hence $\mathbb{G}$ is an algebra. Denoting by $\mathbb{A}$ the class of W such that (vi) is true, it follows that $\mathbb{G} \subseteq \mathbb{A}$. It is easy to check that $\mathbb{A}$ is bp-closed. Since $\mathcal{F}_{\tau-} = \sigma(\mathcal{H})$, the result (vi) follows from the monotone class theorem, Theorem 2.66.

For (vii), invoking Lemma 2.41 it follows that M is an r.c.l.l. $(\mathcal{F}_{\cdot})$ adapted stochastic process. To show that M is a martingale, suffices to show (see Theorem 2.57) that for all bounded stopping times σ,

$$\mathsf{E}[M_\sigma] = 0. \qquad (8.1.2)$$

Here $M_\sigma = U 1_{\{\tau \leq \sigma\}}$. Since $\{\tau \leq \sigma\} = \{\tau > \sigma\}^c \in \mathcal{F}_{\tau-}$ (by part (iv) above) and $\mathsf{E}[U \mid \mathcal{F}_{\tau-}] = 0$ by assumption, (8.1.2) follows. $\square$

The next result is a stopping time analogue of Exercise 8.3.

Theorem 8.5 *Let σ be an $(\mathcal{F}_{\cdot}^+)$ stopping time and τ be a $(\mathcal{F}_{\cdot})$ stopping time. Then*

$$A \in \mathcal{F}_\sigma^+ \Rightarrow A \cap \{\sigma < \tau\} \in \mathcal{F}_{\tau-}.$$

As a consequence, if $\mathcal{F}_0^+ = \mathcal{F}_0$, $\{\tau > 0\} \subseteq \{\sigma < \tau\}$ and $\{\tau = 0\} \subseteq \{\sigma = 0\}$ then

$$\mathcal{F}_\sigma^+ \subseteq \mathcal{F}_{\tau-}.$$

Proof Fix $A \in \mathcal{F}_\sigma^+$. For $t > 0$, note that

$$A \cap \{\sigma < \tau\} = \bigcup \{A \cap \{\sigma \leq r\} \cap \{s < \tau\} \ : \ r < s \text{ rationals in } [0, t]\}$$

Now for $r < s$, $A \cap \{\sigma \leq r\} \in \mathcal{F}_r^+ \subseteq \mathcal{F}_s$. Hence $A \cap \{\sigma < \tau\}$ is a countable union of sets in $\mathcal{F}_{\tau-}$ and thus belongs to $\mathcal{F}_{\tau-}$. On the other hand $A \cap \{\sigma = 0\} \in \mathcal{F}_0^+ = \mathcal{F}_0 \subseteq \mathcal{F}_{\tau-}$. Thus, $A \in \mathcal{F}_{\tau-}$.   □

Remark 8.6 Note that the conditions $\{\tau > 0\} \subseteq \{\sigma < \tau\}$ and $\{\tau = 0\} \subseteq \{\sigma = 0\}$ are the same as $\{\sigma \leq \tau\}$ and on $\{\tau > 0\}$, $\{\sigma < \tau\}$.

8.2 Predictable Stopping Times

For stopping times σ, τ we define *stochastic intervals* as follows. Recall that $\widetilde{\Omega} = [0, \infty) \times \Omega$

$$(\sigma, \tau] = \{(t, \omega) \in \widetilde{\Omega} \ : \ \sigma(\omega) < t \leq \tau(\omega)\}$$

$$[\sigma, \tau] = \{(t, \omega) \in \widetilde{\Omega} \ : \ \sigma(\omega) \leq t \leq \tau(\omega)\}$$

and likewise, $[\sigma, \tau)$, (σ, τ) are also defined. The graph $[\tau]$ of a stopping time is defined by

$$[\tau] = \{(\tau(\omega), \omega) \in \widetilde{\Omega}\}.$$

With this notation, $[\tau, \tau] = [\tau]$. Note that for any σ, τ, the processes f, g, h defined by $f_t = 1_{[\sigma, \tau)}(t)$, $g_t = 1_{(\sigma, \tau]}(t)$ and $h_t = 1_{[0, \tau]}(t)$ are adapted processes. While f is r.c.l.l., g is l.c.r.l. and thus g is predictable. h is also l.c.r.l. except at $t = 0$ and is predictable. As a consequence we get that for stopping times σ, τ with $0 \leq \sigma \leq \tau$, we have

$$[0, \tau] \in \mathcal{P}, \quad (\sigma, \tau] \in \mathcal{P}. \tag{8.2.1}$$

On the other hand if τ is a $[0, \infty)$-valued random variable such that $f_t = 1_{[0, \tau)}(t)$ is adapted, then τ is a stopping time, since in that case $\{f_t = 0\} = \{\tau \leq t\} \in \mathcal{F}_t$.

Exercise 8.7 Let X be a continuous adapted process such that $X_T = 0$. Let

$$\tau = \inf\{t \geq 0 \ : \ |X_t| = 0\} \tag{8.2.2}$$

and for $n \geq 1$, let

$$\sigma_n = \inf\{t \geq 0 \ : \ |X_t| \leq 2^{-n}\}. \tag{8.2.3}$$

Let Y be an r.c.l.l. process such that $Y_0 = 0$ and M be a martingale such that $M_0 = 0$. Show that

(i) τ and σ_n for $n \geq 1$ are bounded stopping times with $\sigma_n \leq \tau$.
(ii) For all $n \geq 1$, $\{\tau > 0\} \subseteq \{\sigma_n < \tau\}$.
(iii) $\sigma_n \uparrow \tau$ as $n \to \infty$.

(iv) $[\tau] \in \mathcal{P}$.
(v) Y_{σ_n} converges to $Y_{\tau-}$ pointwise.
(vi) M_{σ_n} converges to $M_{\tau-}$ in $\mathbb{L}^1(\mathrm{P})$.
(vii) $\mathrm{E}[(\varDelta M)_\tau \mid \mathcal{F}_{\tau-}] = 0$.

The stopping time τ in the exercise above has some special properties. Such stopping times are called predictable.

Definition 8.8 A stopping time τ is said to be predictable if

$$[\tau] \in \mathcal{P}. \tag{8.2.4}$$

We have noted that for every stopping time τ, $(\tau, \infty) \in \mathcal{P}$. Thus τ is predictable if and only if

$$[\tau, \infty) \in \mathcal{P}. \tag{8.2.5}$$

It follows that maximum as well as minimum of finitely many predictable stopping times is predictable. Indeed, supremum of countably many predictable stopping times $\{\tau_k : k \geq 1\}$ is predictable since

$$[\sup_{1 \leq k < \infty} \tau_k, \infty) = \cap_{k=1}^\infty [\tau_k, \infty).$$

Also, it follows that if τ is predictable, then so is $\tau \wedge k$ for all k.

Exercise 8.9 Let σ be any stopping time and $a \in [0, \infty)$ be a constant. Let $\tau = \sigma + a$. Show that τ is predictable.

We will be proving that predictable stopping times are characterized by properties (ii), (iii) as well as by (vii) in the Exercise 8.7 above (when the underlying filtration is right continuous).

Towards this goal, we need the following result from Metivier [50] on the predictable σ-field, interesting in its own right. This is analogous to the result that every finite measure on the Borel σ-field of a complete separable metric space is regular. Even the proof is very similar, with continuous adapted processes playing the role of bounded continuous functions and zero sets of such processes playing the role of closed sets. See [18].

Theorem 8.10 *Let μ be a finite measure on $(\widetilde{\Omega}, \mathcal{P})$ and let*

$$\mathcal{C} = \{\{(t, \omega) \in \widetilde{\Omega} : X_t(\omega) = 0\} : X \text{ is a bounded continuous adapted process.}\}$$

Then for all $\epsilon > 0$ and for all $\Gamma \in \mathcal{P}$ there exist Λ_0, Λ_1 such that $\Lambda_0 \in \mathcal{C}$, $\Lambda_1^c \in \mathcal{C}$,

$$\Lambda_0 \subseteq \Gamma \subseteq \Lambda_1$$

and

$$\mu(\Lambda_1 \cap (\Lambda_0)^c) < \epsilon.$$

Proof Easy to see that C is closed under finite unions and finite intersections: if X^1, X^2 are bounded continuous processes, so are $Y = X^1 X^2$ and $Z = |X^1| + |X^2|$. Indeed, C is closed under countable intersections as $X^j, j \geq 1$ bounded continuous (w.l.g. bounded by 1) yields that $Z = \sum_{j=1}^{\infty} 2^{-j} |X^j|$ is a bounded continuous adapted process and

$$\{Z = 0\} = \cap_j \{X^j = 0\}.$$

Let G be the class of sets Γ in $\mathcal{P}$ for which the desired conclusion holds. Clearly it is closed under complements. Now it can be checked using properties of C noted in the previous paragraph that G is a σ-field.

For any continuous adapted process X and $\alpha \in \mathbb{R}$

$$\{X \leq \alpha\} = \{Y = 0\} \quad \text{where } Y = \max(X, \alpha) - \alpha$$

and hence $\{X \leq \alpha\} \in C$. Since

$$\{X = 0\} = \cap_n \{|X| < \frac{1}{n}\} = \cap_n \{-|X| \leq -\frac{1}{n}\}^c$$

it follows that $C \subseteq G$. Invoking Proposition 4.1, part (iii) we now conclude that $G = \mathcal{P}$. $\square$

The following result gives some insight as to why stopping times satisfying (8.2.4) are called predictable.

Theorem 8.11 *Let τ be $(\mathcal{F}.)$ stopping time. Then τ is predictable if and only if there exist $(\mathcal{F}^+.)$ stopping times τ^n such that $\tau^n \leq \tau^{n+1} \leq \tau$, $\tau^n \uparrow \tau$ and $\tau^n < \tau$ on $\tau > 0$.*

Proof Let us take the easy part first. If $\{\tau^n : n \geq 1\}$ as in the statement exist then noting that

$$[\tau, \infty) \cap \{(0, \infty) \times \Omega\} = \cap_{n=1}^{\infty} (\tau^n, \infty)$$

it follows that $[\tau, \infty) \cap \{(0, \infty) \times \Omega\} \in \mathcal{P}(\mathcal{F}^+.)$. Thus using Corollary 4.6 we conclude

$$[\tau, \infty) \cap \{(0, \infty) \times \Omega\} \in \mathcal{P}(\mathcal{F}.).$$

Since τ is a $(\mathcal{F}.)$ stopping time, $\{\tau = 0\} \in \mathcal{F}_0$ and thus $[0, \infty) \times \{\tau = 0\} \in \mathcal{P}(\mathcal{F}.)$. We can thus conclude that

$$[\tau, \infty) \in \mathcal{P}(\mathcal{F}.).$$

Thus τ is predictable.

For the other part, suppose (8.2.4) holds. Consider the finite measure μ on $(\tilde{\Omega}, \mathcal{P})$ defined by

$$\mu(\Gamma) = \mathsf{P}(\{\omega : (\tau(\omega), \omega) \in \Gamma\})$$

or equivalently for a positive bounded predictable f

$$\int f d\mu = \mathsf{E}[f_\tau].$$

For $m \geq 1$, get bounded continuous adapted processes X^m such that

$$\{X^m = 0\} \subseteq [\tau] \tag{8.2.6}$$

and

$$\mu([\tau] \cap (\{X^m = 0\}^c)) \leq 2^{-m}. \tag{8.2.7}$$

Let $\alpha^m = \inf\{t \geq 0 : |X_t^m| = 0\}$. Then (8.2.6) and (8.2.7) together imply that α^m is either equal to τ or ∞ and

$$\mathsf{P}(\tau \neq \alpha^m) \leq 2^{-m}. \tag{8.2.8}$$

Let $\sigma_{m,n} = \inf\{t : |X_t^m| \leq 2^{-n}\}$. Easy to see that $\sigma_{m,n} \leq \alpha^m$ and if $0 < \alpha^m < \infty$ then $\sigma_{m,n} < \alpha^m$ and $\sigma_{m,n} \uparrow \alpha^m$ as $n \to \infty$. For $u, v \in [0, \infty]$, let $\mathbf{d}^*(u, v) = |\tan^{-1}(u) - \tan^{-1}(v)|$. Then $\mathbf{d}^*$ is a metric for the usual topology on $[0, \infty]$. Since $\sigma_{m,n} \uparrow \alpha^m$ as $n \to \infty$, we can choose $n = n_m$ large enough so that denoting $\sigma_{m,n_m} = \sigma^m$ we have

$$\mathsf{P}(\mathbf{d}^*(\sigma^m, \alpha^m) \geq 2^{-m}) \leq 2^{-m}. \tag{8.2.9}$$

Clearly, $\sigma^m \leq \alpha^m$ and further $\sigma^m < \alpha^m$ on $0 < \alpha^m < \infty$.

Let $N_m = \{\mathbf{d}^*(\sigma^m, \alpha^m) \geq 2^{-m}\} \cup \{\tau \neq \alpha^m\}$ and $N = \limsup_{m \to \infty} N_m = \bigcap_{m=1}^{\infty} \bigcup_{n=m}^{\infty} N_n$. By Borel–Cantelli Lemma and the probability estimates (8.2.8) and (8.2.9), we conclude $\mathsf{P}(N) = 0$. For $\omega \notin N$, $\exists m_0 = m_0(\omega)$ such that for $m \geq m_0$, $\alpha^m(\omega) = \tau(\omega)$, and $\mathbf{d}^*(\sigma^m(\omega), \alpha^m(\omega)) \leq 2^{-m}$. Recall that for all $m \geq 1$, $\sigma^m(\omega) \leq \alpha^m(\omega)$ for all ω. Further, for any ω such that $0 < \alpha^m(\omega) < \infty$, by construction $\sigma^m(\omega) < \alpha^m(\omega)$.

Let $\tau^m = \inf\{\sigma^k : k \geq m\}$. Then τ^m are $(\mathcal{F}_\cdot^+)$ stopping times such that for $\omega \notin N$,

$$\tau^m(\omega) \leq \tau^{m+1}(\omega) \leq \tau(\omega), \quad \forall m \geq 1,$$

$$\tau^m(\omega) < \tau(\omega) \text{ if } \tau(\omega) < \infty$$

and

$$\mathbf{d}^*(\tau^m(\omega), \tau(\omega)) \leq 2^{-m}.$$

Thus $\{\tau^n\}$ so constructed satisfy required properties. $\qquad\square$

Remark **8.12** The stopping times $\{\tau^n : n \geq 1\}$ in the theorem above are said to *announce the predictable stopping time* τ. Indeed, this characterization was the definition of predictability of stopping times in most treatments.

Here are two observations linking the filtrations $(\mathcal{F}_\cdot)$, $(\mathcal{F}_\cdot^+)$ with stopping times and martingales.

Lemma 8.13 *Let τ be a $(\mathcal{F}_.)$ stopping time. Then*

$$[\tau] \in \mathcal{P}(\mathcal{F}_.^+) \implies [\tau] \in \mathcal{P}(\mathcal{F}_.).$$

Proof Let $f = 1_{[\tau]}$. Then $f_0 = 1_{\{\tau=0\}}$ is $\mathcal{F}_0$ measurable as τ is $(\mathcal{F}_.)$ stopping time. Now the conclusion, namely f being $(\mathcal{F}_.)$ predictable, follows from Corollary 4.5. □

Lemma 8.14 *Let M be an r.c.l.l. $(\mathcal{F}_.)$-martingale. Then M is also a $(\mathcal{F}_.^+)$-martingale.*

Proof Fix $s < t$. Let $u_n = s + 2^{-n}$. Note that

$$\mathbb{E}[M_t \mid \mathcal{F}_{u_n}] = M_{u_n \wedge t}.$$

Since $\mathcal{F}_s^+ = \cap_n \mathcal{F}_{u_n}$ using Theorem 1.38 we conclude

$$\mathbb{E}[M_t \mid \mathcal{F}_s^+] = M_s.$$

□

We reiterate here that when the underlying filtration is required to be right continuous, we will state it explicitly. Otherwise martingales, stopping times, predictable, etc., refer to the filtration $(\mathcal{F}_.)$.

Theorem 8.15 *Let τ be a $(\mathcal{F}_.^+)$ predictable stopping time and let τ^n be as in Theorem 8.11 announcing τ. Then*

$$\sigma(\bigcup_{n=1}^{\infty} \mathcal{F}_{\tau^n}^+) = \mathcal{F}_{\tau-}^+. \tag{8.2.10}$$

Proof As seen in Theorem 8.4 (applied to the filtration $(\mathcal{F}_.^+)$) $\mathcal{F}_{\tau^n}^+ \subseteq \mathcal{F}_{\tau-}^+$. To see the other inclusion, let $B \in \mathcal{F}_{\tau-}^+$. If $B \in \mathcal{F}_0^+$, then $B \in \mathcal{F}_{\sigma_n}^+$ for each $n \geq 1$. If $B = A \cap \{t < \tau\}, A \in \mathcal{F}_t^+$. Then $B_n = A \cap \{t < \tau^n\} \in \mathcal{F}_{\tau^n-}^+ \subseteq \mathcal{F}_{\tau^n}^+$. Thus

$$B_n \in \sigma(\bigcup_{m=1}^{\infty} \mathcal{F}_{\tau^m}^+)$$

and of course easy to see that $B = \cup_{n=1}^{\infty} B_n$ completing the proof. □

Exercise 8.16 Let τ be a $(\mathcal{F}_.)$ predictable stopping time and let τ^n be as in Theorem 8.11 announcing τ. Suppose $\mathcal{F}_0 = \mathcal{F}_0^+$. Show that

$$\sigma(\bigcup_{n=1}^{\infty} \mathcal{F}_{\tau^n}^+) = \mathcal{F}_{\tau-}. \tag{8.2.11}$$

Thus conclude $\mathcal{F}_{\tau-} = \mathcal{F}_{\tau-}^+$.

Here is a consequence of predictability of stopping times.

Theorem 8.17 *Let τ be a bounded predictable stopping time. Then for all r.c.l.l. martingales M with $M_0 = 0$ we have*

$$\mathsf{E}[(\Delta M)_\tau \mid \mathcal{F}_{\tau-}] = 0 \tag{8.2.12}$$

Proof Let T be a bound for τ and let τ^n be a sequence of $(\mathcal{F}^+)$ stopping times announcing τ as in Theorem 8.11 above. If $\tau > 0$, then M_{τ^n} converges to $M_{\tau-}$ almost surely whereas if $\tau = 0$ then $\tau^n = 0$ for all n and $M_{\tau^n} = 0$ and $M_{\tau-} = 0$ by definition of $M_{0-} = 0$. Thus we conclude that M_{τ^n} converges to $M_{\tau-}$ almost surely. On the other hand $\tau^n \leq \tau$ and the martingale property of M gives $\mathsf{E}[M_\tau \mid \mathcal{F}_{\tau^n}^+] = M_{\tau^n}$—see Lemma 8.14 and Corollary 2.56. Now Theorem 1.37 along with (8.2.10) yields

$$\mathsf{E}[M_\tau \mid \mathcal{F}_{\tau-}^+] = M_{\tau-}. \tag{8.2.13}$$

Now (8.2.12) follows from this as $M_{\tau-}$ is $\mathcal{F}_{\tau-}$ measurable and $\mathcal{F}_{\tau-} \subseteq \mathcal{F}_{\tau-}^+$. □

We now observe that (8.2.12) characterizes predictable stopping times when the filtration is right continuous.

Theorem 8.18 *Let τ be a bounded stopping time for the filtration $(\mathcal{F}_\cdot)$. Then the following are equivalent.*

(i) τ *is predictable, i.e.* $[\tau] \in \mathcal{P}(\mathcal{F}_\cdot)$.
(ii) *For all bounded $(\mathcal{F}_\cdot^+)$ martingales M with $M_0 = 0$, one has*

$$\mathsf{E}[(\Delta M)_\tau \mid \mathcal{F}_{\tau-}^+] = 0. \tag{8.2.14}$$

(iii) *For all bounded $(\mathcal{F}_\cdot^+)$ martingales M with $M_0 = 0$, one has*

$$\mathsf{E}[(\Delta M)_\tau] = 0. \tag{8.2.15}$$

Proof If $[\tau] \in \mathcal{P}(\mathcal{F}_\cdot)$ then of course $[\tau] \in \mathcal{P}(\mathcal{F}_\cdot^+)$ and hence (*ii*) holds as seen in Theorem 8.17 (invoked here for the filtration $(\mathcal{F}_\cdot^+)$). Thus (*i*) implies (*ii*).

That (*ii*) implies (*iii*) is obvious.

Let us now assume that (8.2.15) holds for all bounded $(\mathcal{F}_\cdot^+)$-martingales M. We will show that there exists a sequence of stopping times announcing τ. In view of Theorem 8.11, this will prove (*i*) completing the proof. Since τ is bounded, let T be such that $\tau \leq T$.

Let N be the r.c.l.l. version of the martingale $\mathsf{E}[\tau \mid \mathcal{F}_t^+]$. Let $M_t = N_t - N_0$ and $Z_t = N_t - t$. Noting that $N_\tau = \mathsf{E}[\tau \mid \mathcal{F}_\tau^+] = \tau$ by Theorem 2.55, we have $Z_\tau = 0$. We will first prove that

$$\mathsf{P}(Z_t 1_{\{t \leq \tau\}} \geq 0 \quad \forall t \leq T) = 1. \tag{8.2.16}$$

To see this note that for $A \in \mathcal{F}_t^+$, $A \cap \{t \leq \tau\} \in \mathcal{F}_t^+$ and using that N is a martingale with $N_T = \tau$, we have

$$\mathsf{E}[1_A 1_{\{t \leq \tau\}} N_t] = \mathsf{E}[1_A 1_{\{t \leq \tau\}} \tau].$$

Hence

$$\mathsf{E}[1_A Z_t 1_{\{t \leq \tau\}}] = \mathsf{E}[1_A (\tau - t) 1_{\{t \leq \tau\}}] \geq 0.$$

Since this holds for all $A \in \mathcal{F}_t^+$ and $Z_t 1_{\{t \leq \tau\}}$ is $\mathcal{F}_t^+$ measurable, it follows that for each t

$$Z_t 1_{\{t \leq \tau\}} \geq 0 \quad a.s.$$

Now right continuity of $t \mapsto Z_t$ shows the validity of (8.2.16). It now follows that $Z_{\tau-} \geq 0$ $a.s.$ and hence

$$N_{\tau-} \geq \tau \quad a.s. \tag{8.2.17}$$

On the other hand, M is a bounded martingale with $M_0 = 0$ and hence in view of the assumption (8.2.15) on τ, it follows that $\mathsf{E}[(\Delta M)_\tau] = 0$. Noting that

$$(\Delta M)_\tau = (\Delta N)_\tau 1_{\{\tau > 0\}}$$

we have $\mathsf{E}[(\Delta N)_\tau 1_{\{\tau > 0\}}] = 0$. Now using $N_\tau = \tau$ we conclude

$$\mathsf{E}[N_{\tau-} 1_{\{\tau > 0\}}] = \mathsf{E}[N_\tau 1_{\{\tau > 0\}}] = \mathsf{E}[\tau 1_{\{\tau > 0\}}]. \tag{8.2.18}$$

In view of (8.2.17) and (8.2.18) we conclude

$$N_{\tau-} 1_{\{\tau > 0\}} = \tau 1_{\{\tau > 0\}} \quad a.s. \tag{8.2.19}$$

Let

$$\sigma_n = \inf\{t \geq 0 : Z_t < 2^{-n}\}.$$

We will show that σ_n are $(\mathcal{F}_\cdot^+)$ stopping times and announce τ. Clearly, $\sigma_n \leq \sigma_{n+1} \leq \tau$ for all n. As $(\mathcal{F}_\cdot^+)$ is right continuous, σ_n is a $(\mathcal{F}_\cdot^+)$ stopping time by Lemma 2.48.

By definition of σ_n we have $Z_{\sigma_n} \leq 2^{-n}$, i.e. $N_{\sigma_n} - \sigma_n \leq 2^{-n}$. Further, if $\sigma_n > 0$, then by left continuity of paths of Z^-, we also have

$$Z_{\sigma_n-} \geq 2^{-n}. \tag{8.2.20}$$

Since $\tau \leq T$ and hence $\mathcal{F}_T^+$ measurable , we conclude that $N_T = \tau$. Now N being a martingale, we have $\mathsf{E}[N_{\sigma_n}] = \mathsf{E}[N_0] = \mathsf{E}[N_T]$ and hence $\mathsf{E}[N_{\sigma_n}] = \mathsf{E}[\tau]$. Since $N_{\sigma_n} - \sigma_n \leq 2^{-n}$, we conclude

$$\mathsf{E}[\tau - \sigma_n] \leq 2^{-n}. \tag{8.2.21}$$

Since $\sigma_n \leq \sigma_{n+1} \leq \tau$ for all n we conclude that

$$\lim_{n\to\infty} \sigma_n = \tau. \tag{8.2.22}$$

Remains to prove that on $\{\tau > 0\}$, $\sigma_n < \tau$.

If $\tau > 0$ and $\sigma_n > 0$, then as seen in (8.2.20), $Z_{\sigma_n-} \geq 2^{-n}$ and so $N_{\sigma_n-} > \sigma_n$ a.s. In view of (8.2.19), this implies $\sigma_n < \tau$ a.s. on $\tau > 0$. Thus $\{\sigma_n\}$ announces τ and as a consequence τ is predictable. $\qquad\square$

Let τ be a stopping time and ξ be a $\mathcal{F}_\tau$ measurable random variable. Then $h = \xi 1_{(\tau,\infty)}$ is an l.c.r.l. adapted process and hence predictable.

The next result refines this observation when τ is predictable.

Theorem 8.19 *Let τ be a predictable stopping time and let ξ be a $\mathcal{F}_{\tau-}$ measurable random variable. Then $f = \xi 1_{[\tau,\infty)}$ and $g = \xi 1_{[\tau]}$ are predictable processes.*

Proof Note that $f = g + h$ where $h = \xi 1_{(\tau,\infty]}$ and that h is predictable being l.c.r.l. adapted. Thus suffices to show that g is predictable.

For $A \in \mathcal{F}_s$, $B = A \cap \{s < \tau\}$, observe that

$$1_B(\omega)1_{[\tau]}(t,\omega) = 1_B(\omega)1_{(s,\infty)}(t)1_{[\tau]}(t,\omega).$$

The process $1_B(\omega)1_{(s,\infty)}(t)$ is l.c.r.l. adapted and hence predictable while $1_{[\tau]}(t,\omega)$ is predictable because τ is predictable. Thus, the desired conclusion namely g is predictable is true when $\xi = 1_B$, $B = A \cap \{s < \tau\}$ and $A \in \mathcal{F}_s$. It is easily seen to be true when $\xi = 1_B$, $B \in \mathcal{F}_0$. Since the class of bounded $\mathcal{F}_{\tau-}$ measurable random variables ξ for which the desired conclusion is true is a linear space and is bp-closed, the conclusion follows by invoking the monotone class theorem—Theorem 2.66. $\qquad\square$

We will next show that the jump times of an r.c.l.l. adapted process X are stopping times and if the process is predictable, then the stopping times can also be chosen to be predictable.

Lemma 8.20 *Let X be an r.c.l.l. $(\mathcal{F}_.)$ adapted process with $X_0 = 0$. For $\alpha > 0$ let*

$$\tau = \inf\{t > 0 : |\Delta X|_t \geq \alpha\}.$$

Then τ is a stopping time with $\tau(\omega) > 0$ for all ω. Further, $\tau < \infty$ implies $|\Delta X|_\tau \geq \alpha$.

Proof Note that for any $\omega \in \Omega$ and $T < \infty$

$$\{t \in [0,T] : |\Delta X(\omega)|_t \geq \alpha\}$$

is a finite set since X has r.c.l.l. paths. Thus $\tau(\omega) < \infty$ implies $|\Delta X(\omega)|_{\tau(\omega)} \geq \alpha$. Moreover,

$$\{\omega : \tau(\omega) \leq t\} = \{\omega : \exists s \in [0, t] : |\Delta X|_s(\omega) \geq \alpha\}. \qquad (8.2.23)$$

It now follows that (writing $Q^t = \{r \in [0, t] : r \text{ is rational}\} \cup \{t\}$) for $\omega \in \Omega, \tau(\omega) \leq t$ if and only if $\forall n \geq 1, \exists s_n, r_n \in Q^t, 0 < s_n < r_n < s_n + \frac{1}{n}$,

$$|X_{r_n}(\omega) - X_{s_n}(\omega)| \geq \alpha - \frac{1}{n}. \qquad (8.2.24)$$

To see this, if such s_n, r_n exist, then choose a subsequence n_k such that

$$s_{n_k}, r_{n_k}, X_{s_{n_k}}(\omega), X_{r_{n_k}}(\omega)$$

converge. Using $s_n < r_n < s_n + \frac{1}{n}$ and (8.2.24), it follows that

$$s_{n_k} \to u, \ r_{n_k} \to u, \ \text{for some } u, 0 < u \leq t.$$

Since $|X_{s_{n_k}} - X_{r_{n_k}}| \geq \alpha - \frac{1}{n_k}$ and $X_{s_{n_k}}, X_{r_{n_k}}$ are converging, the only possibility is that $X_{s_{n_k}}(\omega) \to X_{u-}, X_{r_{n_k}}(\omega) \to X_u$ and $|X_u(\omega) - X_{u-}(\omega)| \geq \alpha$. Hence $\tau(\omega) \leq t$. For the other part, if $\tau(\omega) = s \leq t$, using Q^t is dense in $[0, t]$ and $t \in Q^t$, we can get $s_n, r_n \in Q^t, 0 < s_n < s \leq r_n < s_n + \frac{1}{n}$,

$$|X_{s-}(\omega) - X_{s_n}(\omega)| \leq \frac{1}{2n}, \ |X_{r_n}(\omega) - X_s(\omega)| \leq \frac{1}{2n}$$

and hence using $|\Delta X(\omega)|_s \geq \alpha$, we get $|X_{r_n}(\omega) - X_{s_n}(\omega)| \geq \alpha - \frac{1}{n}$. Thus

$$\{\tau \leq t\} = \cap_{n=1}^{\infty} (\cup_{\{s, r \in Q^t, 0 < s < r \leq s + \frac{1}{n}\}} \{|X_s - X_r| \geq \alpha - \frac{1}{n}\}).$$

Thus τ is a stopping time. Since $X_0 = 0$ implies $(\Delta X)_0 = 0$, (8.2.23) implies $\tau > 0$. $\square$

The next result shows that the jumps of an r.c.l.l. process can be covered by a countable sequence of stopping times.

Lemma 8.21 *Let A be an r.c.l.l. $(\mathcal{F}_\cdot)$ adapted process with $A_0 = 0$. For $n \geq 1$ and $\omega \in \Omega$, let $\sigma_0^n = 0$ and for $i \geq 1$ let $\sigma_{i+1}^n(\omega) = \infty$ if $\sigma_i^n(\omega) = \infty$ and*

$$\sigma_{i+1}^n(\omega) = \inf\{t > \sigma_i^n(\omega) : |(\Delta A)|_t(\omega) \geq \frac{1}{n}\}. \qquad (8.2.25)$$

Then

(i) *For all $n \geq 1, i \geq 1$, σ_i^n is a stopping time and $\sigma_i^n > 0$.*

(ii) *$\forall \omega$, $\sigma_i^n(\omega) < \infty$ implies $|(\Delta A)|_{\sigma_i^n(\omega)}(\omega) \geq \frac{1}{n}$ and $\sigma_i^n(\omega) < \sigma_{i+1}^n(\omega)$.*

(iii) *$\forall \omega$, $\lim_{i \to \infty} \sigma_i^n(\omega) = \infty$.*

(iv) *(Recall $\tilde{\Omega} = [0, \infty) \times \Omega$)*

$$\{(t, \omega) \in \tilde{\Omega} : |(\Delta A)_t(\omega)| \geq \frac{1}{n}\} = \{(\sigma_i^n(\omega), \omega) : i \geq 1\} \cap \tilde{\Omega}. \qquad (8.2.26)$$

(v) If A is also predictable, then $\{\sigma_i^n : i \geq 1, n \geq 1\}$ are predictable stopping times.

Proof Fix $n \geq 1$. Lemma 8.20 implies that σ_1^n is a stopping time. We will prove that σ_i^n are stopping times by induction. Assuming this to be the case for $i = j$, consider the process $Y_t = A_t - A_{t \wedge \sigma_j^n}$. Applying Lemma 8.20 to the r.c.l.l. adapted process Y, we conclude that σ_{j+1}^n is a stopping time. Moreover, if $\sigma_j^n < \infty$ then $\sigma_{j+1}^n > \sigma_j^n$. This completes the induction step proving (i), (ii). Since an r.c.l.l. function can have only finitely many jumps larger than $\frac{1}{n}$ in any finite interval, it follows that $\lim_{i \to \infty} \sigma_i^n(\omega) = \infty$ proving (iii). Hence for a fixed ω, the set $\{t : |(\Delta A)_t(\omega)| \geq \frac{1}{n}\} \cap [0, T]$ is finite for every T and is thus contained in $\{\sigma_i^n(\omega) : 1 \leq i \leq m\}$ for a sufficiently large m and thus

$$\{(t, \omega) \in \widetilde{\Omega} : |(\Delta A)_t(\omega)| \geq \tfrac{1}{n}\} \subseteq \{(\sigma_i^n(\omega), \omega) : i \geq 1\} \cap \widetilde{\Omega}.$$

Part (ii) proven above implies that the equality holds proving (8.2.26). For (v), if A is predictable, then $\Delta A = A - A^-$ is also predictable (since A^- is l.c.r.l. adapted). Also $(\sigma_{i-1}^n, \sigma_i^n] \in \mathcal{P}$ as its indicator is a l.c.r.l. adapted process. In view of (8.2.26) for $i \geq 1$

$$[\sigma_i^n] = \{(t, \omega) \in \widetilde{\Omega} : |(\Delta A)_t(\omega)| \geq \tfrac{1}{n}\} \cap (\sigma_{i-1}^n, \sigma_i^n]$$

and thus if A is predictable, $[\sigma_i^n] \in \mathcal{P}$; i.e. σ_i^n is predictable. $\qquad\square$

The previous result shows that the jumps of an r.c.l.l. adapted process can be covered by countably many stopping times. We now show that one can choose finite or countably many stopping times that cover jumps and have mutually disjoint graphs. Note that stopping times are allowed to take ∞ as its value and the graph of a stopping time is a subset of $[0, \infty) \times \Omega$ and thus several (or all!) may take value ∞ for an ω without violating the requirement that the graphs are mutually disjoint.

Theorem 8.22 *Let X be an r.c.l.l. $(\mathcal{F}.)$ adapted process with $X_0 = 0$. Then there exists a sequence of stopping times $\{\tau_m : m \geq 1\}$, such that*

$$\{(\Delta X) \neq 0\} = \bigcup_{m=1}^{\infty} [\tau_m] \tag{8.2.27}$$

and further that for $m \neq n$, $[\tau_m] \cap [\tau_n] = \emptyset$. As a consequence

$$(\Delta X) = \sum_{m=1}^{\infty} (\Delta X)_{\tau_m} 1_{[\tau_m]}. \tag{8.2.28}$$

Further, if the process X is also predictable, then the stopping times $\{\tau_m : m \geq 1\}$ can be chosen to be predictable and then $(\Delta X)_{\tau_m}$ are $\mathcal{F}_{\tau_m -}$ measurable.

Proof Let $\{\sigma_i^n : i \geq 0, n \geq 1\}$ be defined by (8.2.25). For $k = 2^{(n-1)}(2i-1)$, $i \geq 1, n \geq 1$ let

$$\xi_k = \sigma_i^n.$$

Then the sequence of stopping times $\{\xi_k : k \geq 1\}$ is just an enumeration of $\{\sigma_i^n : i \geq 1, n \geq 1\}$ and thus in view of (8.2.26) we have

$$\{(t, \omega) \in \widetilde{\Omega} : |(\Delta X)_t(\omega)| > 0\} = \{(\xi_k(\omega), \omega) : k \geq 1\} \cap \widetilde{\Omega}. \qquad (8.2.29)$$

However, the graphs of $\{\xi_k : k \geq 1\}$ may not be disjoint. Define $\tau_1 = \xi_1$ and define stopping times $\tau_k : k \geq 2$ inductively as follows. Having defined stopping times τ_k, $1 \leq k \leq m$ let

$$\tau_{m+1}(\omega) = \begin{cases} \xi_{m+1}(\omega) & \text{if } \omega \in \cap_{j=1}^m \{\xi_{m+1}(\omega) \neq \tau_j(\omega), \ \tau_j(\omega) < \infty\} \\ \infty & \text{otherwise .} \end{cases} \qquad (8.2.30)$$

Fix t. Note that

$$\{\tau_{m+1} \leq t\} = \{\xi_{m+1} \leq t\} \cap (\cap_{j=1}^m \{\xi_{m+1} \neq \tau_j\})$$

and $A = \cap_{j=1}^m \{\xi_{m+1} \neq \tau_j\} \in \mathcal{F}_{\xi_{m+1}}$ by Theorem 2.54. Thus,

$$\{\tau_{m+1} \leq t\} = \{\xi_{m+1} \leq t\} \cap A \in \mathcal{F}_t.$$

and hence each τ_{m+1} is also a stopping time. Thus $\{\tau_k : k \geq 1\}$ is a sequence of stopping times.

In view of (8.2.29) and the definition (8.2.30) of τ_{m+1}, we can check that the sequence $\{\tau_m : m \geq 1\}$ satisfies the required conditions, (8.2.27) and (8.2.28).

When the process X is predictable, we have seen that the stopping times $\{\sigma_i^n : n \geq 1, i \geq 1\}$ are predictable and thus here $\{\xi_k : k \geq 1\}$ are predictable. Since

$$[\tau_{m+1}] = [\xi_{m+1}] \cap (\cup_{j=1}^m [\tau_j])^c$$

it follows inductively that $\{\tau_m\}$ are also predictable. Predictability of X implies that ΔX is also predictable and then part (v) Theorem 8.4 now shows that $(\Delta X)_{\tau_m}$ are $\mathcal{F}_{\tau_m-}$ measurable   □

The sequence of stopping times $\{\tau_k : k \geq 1\}$ satisfying (8.2.27) and (8.2.28) is said to be *enumerating the jumps* of X.

Remark 8.23 It is possible that in the construction given above $P(\tau_m = \infty) = 1$ for some m. Of course, such a τ_m can be removed without altering the conclusion.

Here is an observation.

Corollary 8.24 *Let A be a $(\mathcal{F}_\cdot)$ predictable r.c.l.l. process with finite variation paths. Let $|A| = \text{VAR}(A)$ denote the total variation of A. Then $|A|$ is $(\mathcal{F}_\cdot)$ predictable.*

Proof As seen earlier, predictability of A implies that of ΔA. Since $(\Delta|A|) = |(\Delta A)|$ it follows that $(\Delta|A|)$ is predictable and hence it follows that $|A|$ is predictable. Remember, $|A|_t$ is the total variation of $s \mapsto A_s$ on $[0, t]$. $\qquad \square$

The following result is essentially proven above.

Lemma 8.25 *Let H be an r.c.l.l. adapted process. Then H is predictable if and only if (ΔH) admits a representation*

$$(\Delta H) = \sum_{m=1}^{\infty} (\Delta H)_{\tau_m} 1_{[\tau_m]} \tag{8.2.31}$$

where $\{\tau_m : m \geq 1\}$ is a sequence of predictable stopping times and $(\Delta H)_{\tau_m}$ is $\mathcal{F}_{\tau_m-}$ measurable.

Proof One part is proven in Theorem 8.22. For the other part, if H admits such a representation then by Theorem 8.19, (ΔH) is predictable. Since $H = H^- + (\Delta H)$ and H^- is predictable being left continuous, it follows that H is predictable. $\qquad \square$

The following result would show that an r.c.l.l. predictable process is locally bounded—in a sense proving that *since we can predict the jumps, we can stop just before a big jump.*

Lemma 8.26 *Let A be an r.c.l.l. predictable process with $A_0 = 0$. Then for every n, the stopping time*

$$\tau_n = \inf\{t > 0 : |A_t| \geq n \text{ or } |A_{t-}| \geq n\}$$

is predictable. As a consequence, A is locally bounded as a $(\mathcal{F}_\cdot^+)$ adapted process.

Proof We have shown in Lemma 2.42 that τ_n is a stopping time, $\tau_n > 0$ and if $\tau_n < \infty$ then $|A_{\tau_n}| \geq n$ or $|A_{\tau_n-}| \geq n$. Further, it follows that $\lim_n \tau_n = \infty$. Let

$$\Gamma_n = \{(t, \omega) \in \widetilde{\Omega} : |A_t| \geq n \text{ or } |A_{t-}| \geq n\}.$$

Since A is predictable and so is A^- (being l.c.r.l.), it follows that $\Gamma_n \in \mathcal{P}$ and hence

$$[\tau_n] = [0, \tau_n] \cap \Gamma_n \in \mathcal{P}.$$

This shows τ_n is predictable.

For the second part, fix $n \geq 1$. By Theorem 8.11 there exist $(\mathcal{F}_\cdot^+)$ stopping times $\{\tau_{n,m} : m \geq 1\}$ that increase to τ_n strictly from below. Thus we can get m_n such that $\tau_n^* = \tau_{n,m_n}$ satisfies

$$P(\tau_n^* < \tau_n) = 1, \quad P(\tau_n^* \leq \tau_n - 2^{-n}) \leq 2^{-n}. \tag{8.2.32}$$

Let $\sigma_n = \max\{\tau_1^*, \tau_2^*, \ldots, \tau_n^*\}$. Then $\{\sigma_n : n \geq 1\}$ are $(\mathcal{F}_\cdot^+)$ stopping times and

$$P(\sigma_n < \tau_n) = 1, \quad P(\sigma_n \leq \sigma_{n+1}) = 1, \quad P(\sigma_n \leq \tau_n - 2^{-n}) \leq 2^{-n}. \qquad (8.2.33)$$

Since $\tau_n \uparrow \infty$, it follows that $\sigma_n \uparrow \infty$. And $\sigma_n < \tau_n$ implies that $|A_t| \leq n$ for $t \leq \sigma_n$. Thus $A^{[\sigma_n]}$ is bounded (by n) and so A is locally bounded as a $(\mathcal{F}_\cdot^+)$ adapted process. $\qquad \square$

Here is another important consequence of the preceding discussion on predictable stopping times.

Theorem 8.27 *Let M be an r.c.l.l. martingale. If M is also predictable, then M has continuous paths almost surely.*

Proof Let τ be a bounded predictable stopping time. By Lemma 8.14, M is also a $(\mathcal{F}_\cdot^+)$-martingale. Now using Theorem 8.18, we have

$$E[(\Delta M)_\tau \mid \mathcal{F}_{\tau-}^+] = 0.$$

On the other hand, as seen in Theorem 8.4, part (v), M_τ is $\mathcal{F}_{\tau-}^+$ measurable and thus so is $(\Delta M)_\tau$ and so $(\Delta M)_\tau = 0$ a.s. and hence we get $M_\tau = M_{\tau-}$ (a.s.). Now if σ is any predictable stopping time, $\sigma \wedge k$ is also predictable and hence we conclude

$$(\Delta M)_{\sigma \wedge k} = 0 \quad a.s. \ \forall k \geq 1$$

and hence passing to the limit

$$(\Delta M)_\sigma = 0 \quad a.s.$$

By Theorem 8.22, the jumps of M can be covered by countably many predictable stopping times. This shows

$$P((\Delta M)_t = 0 \ \forall t) = 1$$

and hence paths of M are continuous almost surely. $\qquad \square$

By localizing, we can deduce the following:

Corollary 8.28 *Let M be an r.c.l.l. $(\mathcal{F}_\cdot)$ local martingale. If M is also $(\mathcal{F}_\cdot)$ predictable, then M has continuous paths almost surely.*

Here is an interesting result.

Theorem 8.29 *Let A be an r.c.l.l. $(\mathcal{F}_\cdot)$ predictable process with finite variation paths with $A_0 = 0$. If A is also a $(\mathcal{F}_\cdot)$-martingale, then*

$$P(A_t = 0 \ \forall t) = 1. \qquad (8.2.34)$$

Proof Theorem 8.27 implies that A is continuous. Now Theorem 4.74 implies

$$P([A, A]_t = 0 \ \forall t) = 1.$$

Now part (v) of Theorem 5.19 and its Corollary 5.20 together imply that (8.2.34) is true. $\qquad\square$

Once again, we can localize and obtain the following.

Corollary 8.30 *Let A be an r.c.l.l. $(\mathcal{F}_.)$ predictable process with finite variation paths with $A_0 = 0$. If A is also a $(\mathcal{F}_.)$ local martingale, then*

$$P(A_t = 0 \ \forall t) = 1.$$

8.3 Natural FV Processes Are Predictable

The main result of this section is to show that a process $A \in \mathbb{V}$ is natural if and only if it is predictable. To achieve this, we need to consider the right continuous filtration $(\mathcal{F}_.^+)$ along with the given filtration.

Recall that we had observed in Corollary 4.5 that a $(\mathcal{F}_.^+)$ predictable process f such that f_0 is $\mathcal{F}_0$ measurable is $(\mathcal{F}_.)$ predictable.

In his work on decomposition of submartingales, P. A. Meyer had introduced a notion of *natural* increasing process. It was an ad hoc definition, given with the aim of showing uniqueness in the Doob–Meyer decomposition.

Definition 8.31 Let $A \in \mathbb{V}_0$; i.e. A is an adapted process with finite variation paths and $A_0 = 0$. Suppose $|A|$ is locally integrable where $|A|_t = \text{VAR}_{[0,t]}(A)$. A is said to be natural if for all bounded r.c.l.l. martingales M

$$[M, A] \text{ is a local martingale.} \tag{8.3.1}$$

Let $\mathbb{W} = \{V \in \mathbb{V}_0 : |V| \text{ is locally integrable where } |V|_t = \text{VAR}_{[0,t]}(V)\}$.

Remark 8.32 Let $A \in \mathbb{V}_0$ be such that $[A, A]$ is locally integrable. Since $\Delta[A, A] = (\Delta A)^2$, it follows that (ΔA) is locally integrable and as a consequence A is locally integrable and thus $A \in \mathbb{W}$.

Theorem 8.33 *Let $A \in \mathbb{W}$ be natural. Let $\sigma_n \uparrow \infty$ be such that $|A|^{[\sigma_n]}$ is integrable. Then $[M, A^{[\sigma_n]}]$ is a martingale for all bounded martingales M and $n \geq 1$.*

Proof Let $A^n = A^{[\sigma_n]}$. Let M be a bounded martingale. As seen in (4.6.9), $[M, A]_t^{[\sigma_n]} = [M, A^n]_t$ and since A is natural, $[M, A^n]_t$ is also a local martingale. Invoking Theorem 4.74 we have

$$[M, A^n]_t = \sum_{0 < s \leq t} (\Delta M)_s (\Delta A^n)_s. \tag{8.3.2}$$

Thus if the martingale M is bounded by a constant C, (ΔM) is bounded by $2C$ and then we have

$$|[M,A^n]_t| \le 2C \sum_{0<s\le t} |(\Delta A^n)_s| \le 2C|A|_{t\wedge\sigma_n}. \qquad (8.3.3)$$

By choice of σ_n, $\mathsf{E}[|A|_{t\wedge\sigma_n}] < \infty$. Thus

$$\mathsf{E}[\sup_{s\le t} |[M,A^n]_s|] \le 2C\mathsf{E}[|A|_{t\wedge\sigma_n}] < \infty.$$

Now Lemma 5.5 implies that the local martingale $[M,A^n]$ is indeed a martingale. □

We will first show that if $A \in \mathbb{W}$ is predictable then it is natural. The converse is also true and would be taken up subsequently.

Theorem 8.34 *Let $A \in \mathbb{W}$ be predictable. Then A is natural.*

Proof Let M be a r.c.l.l. martingale bounded by C and let $\sigma_n \uparrow \infty$ be such that $|A|^{[\sigma_n]}$ is integrable. Let us write $A^n = A^{[\sigma_n]}$. Note that A predictable implies A^n is predictable for all n. Fix n.

Predictability of A^n implies that (ΔA^n) is predictable since $(A^n)^-$ is predictable being l.c.r.l. adapted. Let $\{\tau_m\}$ be predictable stopping times covering jumps of A^n as constructed in Lemma 8.21. Now predictability of (ΔA^n) and part (v) of Theorem 8.4 implies that $(\Delta A^n)_{\tau_m}$ is $\mathcal{F}_{\tau_m-}$ measurable for all $m \ge 1$. Since $\{\tau_m\}$ cover the jumps of A^n, it follows from (8.3.2) that

$$[M,A^n]_t = \sum_{m=0}^{\infty} (\Delta A^n)_{\tau_m}(\Delta M)_{\tau_m} 1_{[\tau_m,\infty)}(t). \qquad (8.3.4)$$

For each m, τ_m as well as $(\Delta A^n)_{\tau_m}$ are $\mathcal{F}_{\tau_m-}$ measurable by Theorem 8.4. Hence

$$\mathsf{E}[(\Delta A^n)_{\tau_m}(\Delta M)_{\tau_m} 1_{[\tau_m,\infty)}(t)] = \mathsf{E}[\mathsf{E}[(\Delta A^n)_{\tau_m}(\Delta M)_{\tau_m} 1_{[\tau_m,\infty)}(t) \mid \mathcal{F}_{\tau_m-}]]$$
$$= \mathsf{E}[(\Delta A^n)_{\tau_m} 1_{[\tau_m,\infty)}(t)\mathsf{E}[(\Delta M)_{\tau_m} \mid \mathcal{F}_{\tau_m-}]]$$
$$= 0$$

as M is a martingale and τ_m is predictable. Since M is bounded by C, ΔM is bounded by $2C$ and then we have

$$\mathsf{E}[\sum_{m=1}^{\infty} |(\Delta M)_{\tau_m}(\Delta A^n)_{\tau_m}| 1_{[\tau_m,\infty)}(t)] \le 2C\mathsf{E}[\sum_{m=1}^{\infty} |(\Delta A^n_{\tau_m})| 1_{[\tau_m,\infty)}(t)$$
$$\le 2C|A^n|_t$$
$$< \infty.$$

The dominated convergence theorem implies that the series in (8.3.4) converges in $\mathbb{L}^1(\mathsf{P})$ and as a consequence, $\mathsf{E}[[M,A^n]_t] = 0$ for all $t < \infty$.

Now given a stopping time σ bounded by T, apply the above to $N = M^{[\sigma]}$ to get

$$E[[M, A^n]_\sigma] = E[[M, A^n]_{\sigma \wedge T}] = E[[N, A^n]_T] = 0$$

where we have used (4.6.9) for the last equality. Invoking Theorem 2.55 we conclude that $[M, A^n]$ is a martingale. Thus $[M, A]$ is a local martingale for every bounded martingale M and so A is natural.   □

Our next observation will play an important role in the converse that we will take up later.

Lemma 8.35 *Let $A \in \mathbb{W}$ be natural. Then for all stopping times τ, $(\Delta A)_\tau$ is $\mathcal{F}_{\tau-}$ measurable.*

Proof To see this, first let τ be bounded, say by k and let U be any bounded $\mathcal{F}_\tau$ measurable random variable and let

$$W_t = (U - E[U \mid \mathcal{F}_{\tau-}]) 1_{[\tau,\infty)}(t).$$

As seen in Theorem 8.4, part (vii), W is a martingale with r.c.l.l. paths. Since W is bounded and A is natural

$$[W, A]_t = (\Delta A)_\tau (U - E[U \mid \mathcal{F}_{\tau-}]) 1_{[\tau,\infty)}(t)$$

is a local martingale.

Let $\sigma_n \uparrow \infty$ be stopping times such that $|A|_{\sigma_n}$ is integrable for all n. As seen in Theorem 8.33, $[W, A^{[\sigma_n]}]$ is a martingale and since $\tau \leq k$ we have $E[[W, A^{[\sigma_n]}]_\tau] = 0$. Thus

$$E[(\Delta A^{[\sigma_n]})_\tau U] = E[(\Delta A^{[\sigma_n]})_\tau E[U \mid \mathcal{F}_{\tau-}]]$$

for all bounded $\mathcal{F}_\tau$ measurable random variables U. Thus by Lemma 5.49, it follows that $(\Delta A^{[\sigma_n]})_\tau$ is $\mathcal{F}_{\tau-}$ measurable. Since $\sigma_n \uparrow \infty$, we conclude that ΔA_τ is $\mathcal{F}_{\tau-}$ measurable.

For a general stopping time τ, noting that $(\Delta A)_{\tau \wedge n}$ is $\mathcal{F}_{\tau \wedge n-}$ measurable and hence $\mathcal{F}_{\tau-}$ measurable. As seen in Theorem 8.4 part (i), $\{\tau < \infty\} \in \mathcal{F}_{\tau-}$. Since

$$(\Delta A)_\tau = \lim_{n \to \infty} (\Delta A)_{\tau \wedge n} 1_{\{\tau < \infty\}}$$

it follows that

$$(\Delta A)_\tau \text{ is } \mathcal{F}_{\tau-} \text{ measurable.} \tag{8.3.5}$$

□

Next we will show that if A is natural for $(\mathcal{F}_\cdot)$ then it is so for $(\mathcal{F}_\cdot^+)$ filtration as well.

Theorem 8.36 *Let A be an* $(\mathcal{F}_\cdot)$ *adapted r.c.l.l. process with finite variation paths such that* $A_0 = 0$ *and* $|A|$ *is locally integrable where* $|A|_t = \text{VAR}_{[0,t]}(A)$. *Suppose that A is natural for the filtration* $(\mathcal{F}_\cdot)$. *Then it is also natural for the filtration* $(\mathcal{F}_\cdot^+)$.

Proof Let $\sigma_n \uparrow \infty$ be such that $|A|^{[\sigma_n]}$ is integrable and let $A^n = A^{[\sigma_n]}$. Let us fix a $(\mathcal{F}_\cdot^+)$-martingale N bounded by C with r.c.l.l. paths. To show $[N, A]$ is a local martingale, we will prove that for all n

$$[N, A^n] \text{ is a}(\mathcal{F}_\cdot^+) - \text{martingale.} \qquad (8.3.6)$$

We carry out the proof in 3 steps.
Step 1:. We will first prove that for $r \geq 0$ fixed,

$$U = (\Delta N)_r 1_{[r,\infty)}(t) \qquad (8.3.7)$$

is a $(\mathcal{F}_\cdot^+)$-martingale and $[U, A^n]$ is given by

$$[U, A^n]_t = (\Delta N)_r (\Delta A^n)_r 1_{[r,\infty)}(t), \qquad (8.3.8)$$

and is a $(\mathcal{F}_\cdot^+)$-martingale.

To see this, note that $\mathsf{E}[(\Delta N)_r \mid \mathcal{F}_{r-}^+] = 0$ as N is a $(\mathcal{F}_\cdot^+)$-martingale. Thus U is a $(\mathcal{F}_\cdot^+)$-martingale. The process U has a single jump at $t = r$ and thus (8.3.8) holds. Invoking Lemma 8.35 we observe that $(\Delta A^n)_r$ is $\mathcal{F}_{r-}$ measurable. Since $\mathcal{F}_{r-} \subseteq \mathcal{F}_{r-}^+$, we have

$$(\Delta A^n)_r \text{ is } \mathcal{F}_{r-}^+ \text{ measurable.} \qquad (8.3.9)$$

As a consequence

$$
\begin{aligned}
\mathsf{E}[(\Delta[U, A^n])_r \mid \mathcal{F}_{r-}^+] &= \mathsf{E}[(\Delta N)_r (\Delta A^n)_r \mid \mathcal{F}_{r-}^+] \\
&= (\Delta A^n)_r \mathsf{E}[(\Delta N)_r \mid \mathcal{F}_{r-}^+] \qquad (8.3.10) \\
&= 0
\end{aligned}
$$

and thus $[U, A^n]$ is a $(\mathcal{F}_\cdot^+)$-martingale. This completes step 1.
Step 2: Let $D = \{t \in [0, \infty) : \mathsf{P}((\Delta N)_t \neq 0) > 0\}$. Then D is countable.

To see this, for $t \geq 0$ let $h(t) = \mathsf{E}[[N, N]_t]$. Then h is an $[0, \infty)$-valued function since N is a bounded martingale. Clearly, h is increasing. Since

$$h(t) - h(t-) = \mathsf{E}[[N, N]_t - [N, N]_{t-}] = \mathsf{E}[((\Delta N)_t)^2],$$

it follows that t is a continuity point of h if and only if $\mathsf{P}((\Delta N)_t \neq 0) = 0$. Thus D is precisely the set of discontinuity points of h which is countable. This completes step 2.

Let $t_1, t_2, \ldots t_m \ldots$ be an enumeration of D. Let us write

$$Z_t^m = \sum_{k=1}^{m} (\Delta N)_{t_k} 1_{[t_k,\infty)}(t)$$

and $M^m = N - Z^m$. Then Z^m and $[Z^m, A^n]$ are $(\mathcal{F}_\cdot^+)$-martingales as seen in step 1. Note that

$$[Z^m, A^n]_t = \sum_{k=1}^{m} (\Delta N)_{t_k} (\Delta A^n)_{t_k} 1_{[t_k,\infty)}(t). \tag{8.3.11}$$

From the definition of Z^m, M^m it follows that

$$\{t : \ \mathsf{P}((\Delta M^m)_t \neq 0) > 0\} = \{t_k : k \geq (m+1)\}. \tag{8.3.12}$$

Next we will prove
Step 3: Z^m converges to a $(\mathcal{F}_\cdot^+)$-martingale Z in the Emery topology and $[Z, A^n]$ is a $(\mathcal{F}_\cdot^+)$-martingale for each $n \geq 1$.

To see this, for $j < m$ we have

$$[Z^m - Z^j, Z^m - Z^j]_t = \sum_{k=j+1}^{m} ((\Delta N)_{t_k})^2 1_{[t_k,\infty)}(t).$$

Since N is a bounded martingale, it is square integrable and thus using Corollary 5.20, $\mathsf{E}[[N, N]_t] < \infty$. Hence

$$\mathsf{E}[\sum_{k=1}^{\infty} ((\Delta N)_{t_k})^2 1_{[t_k,\infty)}(t)] \leq \mathsf{E}[[N, N]_t] < \infty.$$

It follows that $\mathsf{E}[[Z^m - Z^j, Z^m - Z^j]_t] \to 0$ for all $t < \infty$ as $j, m \to \infty$. Hence using Lemma 5.41, we conclude that Z^m converges to a $(\mathcal{F}_\cdot^+)$-martingale Z such that

$$\mathsf{E}[[Z^m - Z, Z^m - Z]_t] \to 0 \text{ for all } t < \infty \tag{8.3.13}$$

and

$$Z^m \xrightarrow{em} Z. \tag{8.3.14}$$

Let us note that since N is bounded by C we have

$$\sum_{k=1}^{\infty} |(\Delta N)_t (\Delta A^n)_t| 1_{[t_k,\infty)}(t) \leq 2C \sum_{k=1}^{\infty} |(\Delta A^n)_t| 1_{[t_k,\infty)}(t) \tag{8.3.15}$$

$$\leq 2C |A^n|_t$$

Let V^n be defined by

$$V_t^n = \sum_{k=1}^{\infty} (\Delta N)_{t_k} (\Delta A^n)_{t_k} 1_{[t_k, \infty)}(t). \tag{8.3.16}$$

The series defining V^n converges almost surely and in $\mathbb{L}^1(\mathsf{P})$ in view of (8.3.15). Then

$$\mathsf{E}[\,|V_t^n - [Z^m, A^n]_t|\,] \le \sum_{k=m+1}^{\infty} 1_{[t_k, \infty)}(t) \mathsf{E}[\,|(\Delta N)_{t_k}(\Delta A^n)_{t_k}|\,] \tag{8.3.17}$$
$$\to 0 \text{ as } m \to \infty$$

in view of (8.3.15). Since $[Z^m, A^n]$ is a $(\mathcal{F}_{\cdot}^{+})$-martingale, (8.3.17) implies that V^n is a $(\mathcal{F}_{\cdot}^{+})$-martingale. On the other hand (8.3.14) implies that

$$[Z^m, A^n] \to [Z, A^n] \text{ as } m \to \infty$$

in the Emery topology (see Theorem 4.1.1.1). Thus $V^n = [Z, A^n]$ is a $(\mathcal{F}_{\cdot}^{+})$-martingale. This completes step 3.

Since $M^m = N - Z^m$, we get $M^m \xrightarrow{em} M$ where $M = N - Z$ is also a $(\mathcal{F}_{\cdot}^{+})$-martingale. Also, $M^m \xrightarrow{ucp} M$. It then follows from (8.3.12) that

$$\mathsf{P}((\Delta M)_t \ne 0) = 0 \quad \forall t \ge 0. \tag{8.3.18}$$

This observation (8.3.18) and the assumption that $\mathcal{F}_0$ contains all P null sets together imply that M_t is $\mathcal{F}_{t-}^{+}$ measurable. Since $\mathcal{F}_{t-}^{+} \subseteq \mathcal{F}_t \subseteq \mathcal{F}_t^{+}$, we conclude that M is a $(\mathcal{F}_{\cdot})$-martingale. In view of the assumption that A is natural for the filtration $(\mathcal{F}_{\cdot})$, we conclude that $[M, A]$ is a $(\mathcal{F}_{\cdot})$-local martingale. The process $[M, A]$ is r.c.l.l. and thus is also a $(\mathcal{F}_{\cdot}^{+})$-local martingale.

Since $N = M + Z$, $[N, A] = [M, A] + [Z, A]$. We have already shown in step 3 that $[Z, A]$ is a $(\mathcal{F}_{\cdot}^{+})$-local martingale and thus it follows that $[N, A]$ is a $(\mathcal{F}_{\cdot}^{+})$-local martingale. $\qquad\square$

We now come to one of the main results of this chapter.

Theorem 8.37 *Let $B \in \mathbb{W}$ be natural for the filtration $(\mathcal{F}_{\cdot})$, i.e. for all bounded $(\mathcal{F}_{\cdot})$-martingales M*

$$[M, B] \text{ is a local martingale.} \tag{8.3.19}$$

Then B is $(\mathcal{F}_{\cdot})$ predictable.

Proof Let $\tau_m \uparrow \infty$ be stopping times such that $B^m = B^{[\tau_m]}$ satisfies $|B^m|_t$ is integrable for all m, t. Suffices to prove that for all m, B^m is predictable. So fix $m \ge 1$ and let us write $A = B^m$. As seen in Theorem 8.33, for all bounded martingales M,

$$[M, A] \text{ is a martingale.} \tag{8.3.20}$$

By Theorem 8.36, (8.3.20) is true for all bounded $(\mathcal{F}_\cdot^+)$-martingales M.

Let $\{\sigma_i^n : n \geq 1, i \geq 0\}$ be the stopping times defined by (8.2.25). We are going to prove that each of these is predictable. Note that (8.2.26) can be rewritten as

$$(\Delta A)1_{\{|\Delta A| \geq \frac{1}{n}\}} = \sum_{i=0}^{\infty} (\Delta A)_{\sigma_i^n} 1_{[\sigma_i^n]}. \tag{8.3.21}$$

Here, $|\Delta A|$ denotes the process $s \mapsto |(\Delta A)_s|$. It follows from Lemma 8.35 that

$$W^{n,i} = (\Delta A)_{\sigma_i^n} \text{ is } \mathcal{F}_{\sigma_i^n-} \text{ measurable.} \tag{8.3.22}$$

Invoking part (vi) of Theorem 8.4, we obtain predictable processes $f^{n,i}$ such that

$$f_{\sigma_i^n}^{n,i} 1_{\{\sigma_i^n < \infty\}} = W^{n,i} 1_{\{\sigma_i^n < \infty\}}.$$

Let

$$g^n = \sum_{i=1}^{\infty} f^{n,i} 1_{(\sigma_{i-1}^n, \sigma_i^n]}.$$

Then g^n is predictable since $f^{n,i}$ is predictable and $1_{(\sigma_{i-1}^n, \sigma_i^n]}$ is l.c.r.l. adapted process and hence predictable. Further, note that by definition

$$g_{\sigma_i^n}^n = (\Delta A)_{\sigma_i^n}.$$

In view of the definition of $\{\sigma_i^n\}$ as seen in (8.2.26), it follows that

$$g^n 1_{\{|\Delta A| \geq \frac{1}{n}\}} = (\Delta A)1_{\{|\Delta A| \geq \frac{1}{n}\}}. \tag{8.3.23}$$

In particular, for all $m \geq n$

$$g^n 1_{\{|\Delta A| \geq \frac{1}{n}\}} = g^m 1_{\{|\Delta A| \geq \frac{1}{n}\}}$$

and hence defining $g = \limsup_{n \to \infty} g^n$, it follows that g is predictable and

$$g 1_{\{|\Delta A| > 0\}} = (\Delta A)1_{\{|\Delta A| > 0\}}. \tag{8.3.24}$$

Now fix $m \geq 1$ and let $h^m = 1_{\{|g| > 0\}} g^{-1} 1_{\{|\Delta A| \geq \frac{1}{m}\}}$. Then h is bounded predictable and

$$(\Delta A)h^m = 1_{\{|\Delta A| \geq \frac{1}{m}\}}. \tag{8.3.25}$$

Now given a bounded $(\mathcal{F}_\cdot^+)$-martingale M with $M_0 = 0$, $N^m = \int h^m dM$ is also a $(\mathcal{F}_\cdot^+)$-martingale since h^m is a bounded predictable process. Thus $[N^m, A]$ is a $(\mathcal{F}_\cdot^+)$-martingale—see Theorem 8.36. On the other hand

$$[N^m, A]_t = \int_0^t h^m d[M, A] = \sum_{0 \le s \le t} h_s^m (\Delta M)_s (\Delta A)_s$$

$$= \sum_{0 \le s \le t} (\Delta M)_s 1_{\{|(\Delta A)_s| \ge \frac{1}{m}\}}$$

$$= \sum_{j=0}^{\infty} (\Delta M)_{\sigma_j^m \wedge t}.$$

Since $[N^m, A]$ is a martingale, $\mathsf{E}[[N^m, A]_{\sigma_k^m \wedge t}] = 0$ for all k. Thus

$$\mathsf{E}[\sum_{j=0}^{k} (\Delta M)_{\sigma_j^m \wedge t}] = 0$$

for all k. Hence for all $m \ge 1$, $i \ge 1$ and $t < \infty$

$$\mathsf{E}[(\Delta M)_{\sigma_i^m \wedge t}] = 0. \tag{8.3.26}$$

Since (8.3.26) holds for all bounded $(\mathcal{F}_\cdot^+)$-martingales M with $M_0 = 0$, invoking Theorem 8.18 we conclude that for all $m \ge 1, i \ge 1$ and $t < \infty$, the bounded stopping time $\sigma_i^m \wedge t$ is $(\mathcal{F}_\cdot^+)$ predictable.

Since $[\sigma_i^m] = \cup_j \cap_{k \ge j} [\sigma_i^m \wedge k]$, it follows that σ_i^m is $(\mathcal{F}_\cdot^+)$ predictable for every m, i. This along with the observation (8.3.22) gives us

$$(\Delta A)_{\sigma_i^n} 1_{[\sigma_i^n]}$$

is $(\mathcal{F}_\cdot^+)$ predictable for each n, i. As a consequence

$$\xi^n = \sum_{i=0}^{\infty} (\Delta A)_{\sigma_i^n} 1_{[\sigma_i^n]}$$

is $(\mathcal{F}_\cdot^+)$ predictable. It can be seen using (8.3.21) that

$$\lim_{n \to \infty} \xi^n = \Delta A$$

and thus ΔA is $(\mathcal{F}_\cdot^+)$ predictable. Since

$$A_t = A_{t-} + (\Delta A)_t$$

and since A_{t-} is $(\mathcal{F}_\cdot^+)$ predictable, being an l.c.r.l. adapted process, we conclude that A is $(\mathcal{F}_\cdot^+)$ predictable. Since $B \in \mathbb{W}$, it follows that $B_0 = A_0 = 0$ and then Corollary 4.5 implies that A is $(\mathcal{F}_\cdot)$ predictable. $\qquad\square$

8.4 Decomposition of Semimartingales Revisited

In view of this identification of natural FV processes as predictable, we can recast Theorem 5.50 as follows.

Theorem 8.38 *Let X be a stochastic integrator such that*

(i) $X_t = X_{t \wedge T}$ *for all t.*
(ii) $E[\sup_{s \leq T} |X_s|] < \infty$.
(iii) $E[[X, X]_T] < \infty$.

Then X admits a decomposition

$$X = M + A, \ M \in \mathbb{M}^2, \ A \in \mathbb{V}, \ A_0 = 0 \text{ and } A \text{ is predictable}. \tag{8.4.1}$$

Further, the decomposition (8.4.1) *is unique.*

Proof Let $X = M + A$ be the decomposition in Theorem 5.50. As seen in Corollary 5.50, the process A satisfies $A_t = A_{t \wedge T}$. Since $E[[M, A]_T] = 0$, we have

$$E[[X, X]_T] = E[[M, M]_T] + E[[A, A]_T]$$

and hence $E[[A, A]_T] < \infty$ and so $A \in \mathbb{W}$. Thus A satisfies conditions of Theorem 8.37 and hence A is predictable. For uniqueness, if $X = N + B$ is another decomposition with $N \in \mathbb{M}^2$ and $B \in \mathbb{V}$ and B being predictable, then $M - N = B - A$ is a predictable process with finite variation paths which is also a martingale and hence by Theorem 8.29, $M = N$ and $B = A$. $\qquad\square$

Remark **8.39** We note that in Theorem 8.38 we have not assumed that the underlying filtration is right continuous.

We need two more results before we can deduce the Doob–Meyer decomposition result. First, we need to extend Theorem 8.38 to all locally integrable semimartingales. Then we need to show that when X is a submartingale, then the FV process appearing in the decomposition is an increasing process. We begin with the second result.

Theorem 8.40 *Let U be an r.c.l.l. ($\mathcal{F}_{\cdot}$) predictable process with finite variation paths with $U_0 = 0$ and*

$$E[\text{VAR}_{[0,T]}(U)] < \infty \text{ for all } T < \infty. \tag{8.4.2}$$

If U is also a ($\mathcal{F}_{\cdot}$) submartingale, then U is an increasing process, i.e.

$$P(U_t \geq U_s \ \forall \ 0 \leq s \leq t < \infty) = 1. \tag{8.4.3}$$

Proof Let $V = |U|$ denote the total variation process of U ($V_t = \text{VAR}_{[0,t]}(U)$). As seen in Corollary 8.24, V is also predictable and of course, V is an increasing process.

Fix $T < \infty$ and define measures μ and λ on $(\widetilde{\Omega}, \mathcal{P})$ as follows. For a bounded predictable process f

$$\int f d\mu = \mathsf{E}[\int_0^T f_s dU_s] \tag{8.4.4}$$

$$\int f d\lambda = \mathsf{E}[\int_0^T f_s dV_s]. \tag{8.4.5}$$

Since V is an increasing process, λ is a positive finite measure. We shall show that μ is also a positive measure. Note that if $f = a1_{(s_1, s_2]}$ with a being $\mathcal{F}_{s_1}$ measurable, $a \geq 0$, $s_1 < s_2 \leq T$ then

$$\begin{aligned}
\int f d\mu &= \mathsf{E}[\int_0^T f_s dU_s] \\
&= \mathsf{E}[a(U_{s_2} - U_{s_1})] \\
&= \mathsf{E}[a\mathsf{E}[(U_{s_2} - U_{s_1}) \mid \mathcal{F}_{s_1}]] \\
&\geq 0
\end{aligned}$$

as $\mathsf{E}[(U_{s_2} - U_{s_1}) \mid \mathcal{F}_{s_1}] \geq 0$ $a.s.$ since U is a submartingale and $a \geq 0$. Since a simple predictable process is a linear combination of such functions, it follows that for a simple predictable f given by (4.2.1) such that $f \geq 0$, $\int f d\mu \geq 0$. Since § generates the predictable σ-field, it follows that μ is a positive measure (see Exercise 4.2). If f is a non-negative bounded predictable process, then

$$|\int_0^T f_s dU_s| \leq \int_0^T f_s dV_s$$

and thus for such an f,

$$\int f d\mu \leq \int f d\lambda.$$

Thus μ is absolutely continuous w.r.t. λ and thus denoting the Radon–Nikodym derivative by ξ, it follows that ξ is a $[0, \infty)$-valued predictable process such that

$$\int f d\mu = \int f\xi d\lambda. \tag{8.4.6}$$

Let us define a process B by

$$B_t(\omega) = \int_0^t 1_{[0,T]}(s)\xi_s(\omega)dV_s(\omega). \tag{8.4.7}$$

Since ξ is a $[0, \infty)$-valued and V is an r.c.l.l. increasing process, it follows that B is increasing as well. Since ξ and V are $(\mathcal{F}.)$ adapted, it follows that B is also $(\mathcal{F}.)$

adapted. Further, let us note that

$$(\Delta B) = \xi(\Delta V) \tag{8.4.8}$$

and hence predictability of V implies that (ΔV) and as a consequence (ΔB) is predictable. Since $B = B^- + (\Delta B)$, and B^- is l.c.r.l. and hence predictable, it follows that B is predictable. Let $C = B - U$. Then C is a predictable process with finite variation paths and $C_0 = 0$.

We will show that C is a martingale. For this, using (8.4.6) and (8.4.7), we have for any bounded predictable f

$$\begin{aligned} \mathsf{E}[\int_0^T f_s dB_s] &= \mathsf{E}[\int_0^T f_s \xi_s dV_s] \\ &= \int f \xi d\lambda \tag{8.4.9} \\ &= \int f d\mu. \end{aligned}$$

Using (8.4.4) and (8.4.9) and recalling that $C = B - U$ we conclude that for bounded predictable processes f we have

$$\mathsf{E}[\int_0^T f_s dC_s] = 0.$$

Taking $f = a1_{(s_1,s_2]}$ with a being $\mathcal{F}_{s_1}$ measurable, $s_1 < s_2 \leq T$, we conclude

$$\mathsf{E}[a(C_{s_2} - C_{s_1})] = 0.$$

This being true for all $T < \infty$, we conclude that C is a martingale.

By Theorem 8.29, it follows that $C = 0$ and as a consequence, $U = B$. Since by construction, B is an increasing process, this completes the proof. $\qquad\square$

The preceding result also contains a proof of the following:

Corollary 8.41 *Let U be an r.c.l.l. $(\mathcal{F}_{\cdot})$ predictable process with finite variation paths with $U_0 = 0$ such that (8.4.2) is true. Let $V = |U|$ denote the total variation process of U ($V_t = \mathrm{VAR}_{[0,t]}(U)$). Then there exists a predictable process ξ such that $|\xi| = 1$ and*

$$U_t = \int_0^t \xi_s dV_s. \tag{8.4.10}$$

Here are some auxiliary results.

Lemma 8.42 *Let M be a local martingale. Then M is locally integrable.*

Proof Let σ^n be stopping times increasing to ∞ such that $M^{[\sigma^n]}$ is a martingale. Let α_n be the stopping times defined by

$$\alpha_n = \inf\{t \ge 0 : |M_t| \ge n \text{ or } |M_{t-}| \ge n\}$$

and let $\tau_n = \alpha_n \wedge \sigma_n \wedge n$. Note that τ^n also increases to ∞. Let $M^n = M^{[\tau_n]}$. Then M^n is a martingale and for any $T < \infty$

$$\sup_{0 \le t \le \tau_n} |M_t| \le n + |M_{\tau_n}| = n + |M_n^n|.$$

Since $\mathsf{E}[M_n^n] < \infty$, it follows that M is locally integrable. $\square$

Lemma 8.43 *Let τ be a stopping time and let ξ be a $\mathcal{F}_\tau$ measurable $[0, \infty)$-valued integrable random variable. Let*

$$X_t = \xi I_{[\tau, \infty)}(t).$$

Then X admits a decomposition

$$X = M + A, \quad M \in \mathbb{M}, \ A \text{ predictable}, \ A \in \mathbb{V}^+. \tag{8.4.11}$$

The decomposition (8.4.11) is unique. Further, for all $T < \infty$,

$$\mathsf{E}[A_T] \le \mathsf{E}[\xi] \tag{8.4.12}$$

$$\mathsf{E}[|M_T|] \le 2\mathsf{E}[\xi]. \tag{8.4.13}$$

Proof Let $\{c_n : n \ge 1\}$ be such that $\mathsf{E}[\xi 1_{\{\xi \ge c_n\}}] \le 2^{-n}$. Such a sequence exists as $\xi \ge 0$ and $\mathsf{E}[\xi] < \infty$. Let $\xi^m = \xi 1_{\{\xi \le c_m\}}$ and $X_t^m = \xi^m 1_{[\tau, \infty)}(t)$. Then, X^m is bounded FV processes and hence a stochastic integrator. Thus by Theorem 8.38, it admits a decomposition

$$X^m = M^m + A^m, \ M^m \in \mathbb{M}^2, \ A_0^m = 0 \text{ and } A^m \text{ is predictable}.$$

Also, $\xi^m \ge 0$ implies that X^m is a submartingale. Since M^m is a martingale, it follows that A^m is a submartingale and hence by Theorem 8.40, A^m is an increasing process.

Let us note that $X_t^{n+1} - X_t^n = \xi 1_{\{c_n < \xi \le c_{n+1}\}} 1_{[\tau, \infty)}$ and clearly it is an increasing process. Thus, we have

$$\mathsf{E}[\sup_{0 \le t \le T} |X_t^{n+1} - X_t^n|] \le \mathsf{E}[X_T^{n+1} - X_T^n]$$

$$\le \mathsf{E}[\xi 1_{\{c_n < \xi \le c_{n+1}\}}] \tag{8.4.14}$$

$$\le 2^{-n}.$$

Moreover $X^{n+1} - X^n$ is a submartingale. Noting that

$$X_t^{n+1} - X_t^n = M_t^{n+1} - M_t^n + A_t^{n+1} - A_t^n, \quad \forall t \ge 0 \tag{8.4.15}$$

and the fact that $M^{n+1} - M^n$ is a martingale, it follows that $A^{n+1} - A^n$ is a submartingale. As a consequence, $A^{n+1} - A^n$ is an increasing process in view of Theorem 8.40. Thus

$$\mathsf{E}[\sup_{0 \leq t \leq T} |A_t^{n+1} - A_t^n|] = \mathsf{E}[A_T^{n+1} - A_T^n] \qquad (8.4.16)$$

Using $\mathsf{E}[M_0^m] = \mathsf{E}[X_0^m] = \mathsf{E}[\xi 1_{\{\xi \leq c_m\}} 1_{\{\tau=0\}}]$, we conclude that

$$\begin{aligned} \mathsf{E}[M_T^{n+1} - M_T^n] &= \mathsf{E}[M_0^{n+1} - M_0^n] \\ &= \mathsf{E}[\xi 1_{\{c_n < \xi \leq c_n+1\}} 1_{\{\tau=0\}}] \qquad (8.4.17) \\ &\geq 0. \end{aligned}$$

As a result we have

$$\begin{aligned} \mathsf{E}[A_T^{n+1} - A_T^n] &= \mathsf{E}[X_T^{n+1} - X_T^n] - \mathsf{E}[M_T^{n+1} - M_T^n] \\ &\leq \mathsf{E}[X_T^{n+1} - X_T^n] \qquad (8.4.18) \\ &\leq 2^{-n}. \end{aligned}$$

Using (8.4.16) and (8.4.18) we conclude

$$\begin{aligned} \mathsf{E}[\sum_{n=1}^{\infty} \sup_{0 \leq t \leq T} |A_t^{n+1} - A_t^n|] &\leq \sum_{n=1}^{\infty} \mathsf{E}[\sup_{0 \leq t \leq T} |A_t^{n+1} - A_t^n|] \\ &\leq \sum_{n=1}^{\infty} \mathsf{E}[A_T^{n+1} - A_T^n] \qquad (8.4.19) \\ &\leq \sum_{n=1}^{\infty} 2^{-n} \\ &< \infty. \end{aligned}$$

As seen in the proof of Theorem 2.71, we conclude that outside a null set, A_t^n converges to A_t uniformly in $t \in [0, T]$ for every $T < \infty$. On the exceptional set, A_t is defined to be zero. Since each A^n is predictable and is an increasing process, it follows that so is A, thus $A \in \mathbb{V}^+$. The estimate (8.4.19) also implies

$$\mathsf{E}[|A_t^n - A_t|] \to 0 \quad \forall t. \qquad (8.4.20)$$

Defining $M = X - A$, using (8.4.20) and the easily checked fact that $\mathsf{E}[|X_t^n - X_t|] \to 0$, it follows that

$$\mathsf{E}[|M_t^n - M_t|] \to 0 \quad \forall t. \qquad (8.4.21)$$

Since each M^n is a martingale, using (8.4.21) and Theorem 2.23 we conclude that M is a martingale.

The uniqueness of the decomposition follows from Theorem 8.29—if $X = N + B$ is another decomposition with a martingale N and $B \in \mathbb{V}$ with $B_0 = 0$, then $U = M - N = B - A$ is a martingale as well as a predictable process with $U \in \mathbb{V}$ with $U_0 = 0$ and hence $U_t = 0$ for all t by Theorem 8.29.

Note that $A_0 = 0$ implies $M_0 = \xi 1_{\{\tau=0\}}$ and hence $\mathsf{E}[M_T] = \mathsf{E}[M_0] \geq 0$. Thus,

$$\mathsf{E}[A_T] = \mathsf{E}[X_T] - \mathsf{E}[M_T] \leq \mathsf{E}[X_T] \leq \mathsf{E}[\xi].$$

This proves (8.4.12). For (8.4.13), note that $|M_T| \leq X_T + A_T$ and hence

$$\mathsf{E}[\,|M_T|\,] \leq \mathsf{E}[A_T] + \mathsf{E}[X_T] \leq 2\mathsf{E}[\xi].$$

$\square$

Corollary 8.44 *The processes A, M constructed in the previous result also satisfy*

$$\mathsf{E}[\sup_{0 \leq t \leq T} |A_t|] \leq \mathsf{E}[\xi] \tag{8.4.22}$$

$$\mathsf{E}[\sup_{0 \leq t \leq T} |M_t|] \leq 2\mathsf{E}[\xi]. \tag{8.4.23}$$

Proof Since A and X are increasing processes, $\sup_{0 \leq t \leq T} |A_t| = A_T$ and $\sup_{0 \leq t \leq T} |X_t| = X_T$. Thus

$$\sup_{0 \leq t \leq T} |M_t| \leq A_T + X_T.$$

The inequalities (8.4.22) and (8.4.23) follow from these observations. $\square$

Corollary 8.45 *Let τ_n be a sequence of stopping times and let ξ_n be a sequence of $[0, \infty)$-valued random variables, with ξ_n being $\mathcal{F}_{\tau_n}$ measurable and*

$$\sum_{n=1}^{\infty} \mathsf{E}[\xi_n] < \infty. \tag{8.4.24}$$

Let

$$Z_t = \sum_{n=1}^{\infty} \xi_n I_{[\tau_n, \infty)}. \tag{8.4.25}$$

Then there exists a unique predictable increasing process B with $B_0 = 0$ and a martingale N such that

$$Z = N + B.$$

Proof In view of (8.4.24), it follows that

$$\sum_{n=1}^{\infty} \xi_n < \infty \ \ a.s$$

and hence Z is an r.c.l.l. process. Let $X_t^n = \xi_n 1_{[\tau_n, \infty)}$ and let A^n be the predictable increasing process with $A_0^n = 0$ given by Lemma 8.43 such that $M_t^n = X_t^n - A_t^n$ is a martingale. Then as seen in Lemma 8.43 and Corollary 8.44 we have

$$\mathsf{E}[\sup_{0 \le t \le T} |A_t^n|] \le \mathsf{E}[\xi_n] \tag{8.4.26}$$

and hence the assumption (8.4.24) implies that $B_t = \sum_{n=1}^{\infty} A_t^n$ defines a integrable predictable process. Indeed, (8.4.26) implies that the series defining B converges uniformly in $t \in [0, T]$ for every $T < \infty$ almost surely. Thus, B is an r.c.l.l. increasing predictable process. Further $\sum_{n=1}^{k} M_t^n$ converges in $L^1(\mathsf{P})$ to $N_t = Z_t - B_t$. By Theorem 2.23 N is a martingale. This proves existence part. The uniqueness again follows from Theorem 8.29. □

We can now extend the decomposition of semimartingales where the FV process is predictable to a wider class of semimartingales.

Theorem 8.46 *Let X be a semimartingale that is locally integrable. Then X admits a decomposition*

$$X = M + A, \ \ M \in \mathbb{M}_{loc}, \ \ A \in \mathbb{V}, A_0 = 0 \ and \ A \ is \ predictable. \tag{8.4.27}$$

The decomposition as in (8.4.27) is unique. Conversely, if a semimartingale X admits a decomposition (8.4.27), then X is locally integrable.

The process A appearing in (8.4.27) is called the compensator of the semimartingale X.

Proof Let σ_n be stopping times increasing to ∞ such that

$$\sup_{0 \le t \le \sigma_n} |X_t| \ \text{is integrable.} \tag{8.4.28}$$

Let α_n be the stopping times defined by

$$\alpha_n = \inf\{t \ge 0 : |X_t| \ge n \text{ or } |X_{t-}| \ge n \text{ or } [X, X]_{t-} \ge n\} \tag{8.4.29}$$

and let $\tau_n = \alpha_n \wedge \sigma_n \wedge n$. Let $X^n = X^{[\tau_n]}$, $\xi^n = (\Delta X)_{\tau_n}$, $U^n = (\xi^n)^+ 1_{[\tau_n, \infty)}$, $V^n = (\xi^n)^- 1_{[\tau_n, \infty)}$ and $Z^n = X^n - U^n + V^n$. Then

$$Z_t^n = \begin{cases} X_t & \text{if } t < \tau_n \\ X_{\tau_n -} & \text{if } t \ge \tau_n. \end{cases}$$

Hence Z^n is a bounded semimartingale. Note that $[Z^n, Z^n]$ is also bounded. Thus by Theorem 8.38, Z^n admits a decomposition as in (8.4.1). In view of (8.4.28) $(\xi^n)^+$ and $(\xi^n)^-$ are integrable and hence by Lemma 8.43, U^n, V^n also admit decomposition as in (8.4.11). Thus X^n also admits a decomposition

$$X^n = M^n + A^n \qquad (8.4.30)$$

with $A_0^n = 0$, A^n being a predictable FV process. This decomposition is unique in view of the Corollary 8.30. For $n < m$,

$$X_t^n = X_t^m \text{ for } t \le \tau_n$$

The uniqueness of the decompositions (8.4.30) then shows that

$$M_t^n = M_t^m, \quad A_t^n = A_t^m \text{ for } t \le \tau_n.$$

Thus defining $M_t = \lim_n M_t^n$ and $A_t = \lim_n A_t^n$, it follows that

$$M_t = M_t^n, \quad A_t = A_t^n \text{ for } t \le \tau_n.$$

Thus M is a local martingale and A is a predictable process with finite variation paths with $A_0 = 0$ and by construction, $X = M + A$. Thus a decomposition as in (8.4.27) exists. On the other hand if a semimartingale X admits a decomposition (8.4.27), then A, being a predictable FV process, is locally bounded (see Lemma 8.26) and hence locally integrable and M being a local martingale is locally integrable (see Lemma 8.42).

The uniqueness part once again follows from Corollary 8.30. □

Remark **8.47** A locally integrable semimartingale is called a *Special Semimartingale*. The previous result says that a semimartingale admits a decomposition as in (8.4.27) if and only if it is special.

Exercise 8.48 Let X be a semimartingale such that $X_0 = 0$ and

$$|(\Delta X)| \le a$$

for a constant a. Show that X is locally integrable. Further, if $X = M + A$ is the canonical decomposition with $M \in \mathbb{M}_{\text{loc}}, A \in \mathbb{V}, A_0 = 0$ and A predictable. Show that

$$|(\Delta A)| \le a.$$

HINT: If α_n are defined by (8.4.29), then $X^{[\alpha_n]}$ are bounded. Observe that for any predictable stopping time τ,

$$(\Delta A^{[\alpha_n]})_\tau = \mathsf{E}[(\Delta A^{[\alpha_n]})_\tau \mid \mathcal{F}_{\tau-}].$$

When M, N are local martingales such that the quadratic variation process $[M, N]$ is locally integrable, then the unique predictable process $A \in \mathbb{V}$ such that $A_0 = 0$ and $[M, N]_t - A_t$ is a local martingale (in other words the process A appearing in the decomposition (8.4.27) is denoted by $\langle M, N \rangle$).

Definition 8.49 For local martingales M, N such that $[M, N]$ is locally integrable, the predictable cross-quadratic variation $\langle M, N \rangle$ is the unique predictable process with FV paths which is zero at $t = 0$ and such that

$$[M, N]_t - \langle M, N \rangle_t$$

is a local martingale. When $M = N$ and M is locally square integrable then $\langle M, M \rangle$ is called the predictable quadratic variation of M.

If M, N are locally square integrable, $[M, N]$ is locally integrable as seen in (4.6.17) and hence the predictable quadratic variation $\langle M, N \rangle$ is defined. It is easy to see that $(M, N) \mapsto \langle M, N \rangle$ is a bilinear mapping from $\mathbb{M}^2_{\text{loc}} \times \mathbb{M}^2_{\text{loc}}$ into $\mathbb{V}$.

8.5 Doob–Meyer Decomposition

As was mentioned earlier, the Doob–Meyer decomposition was the starting point of the theory of stochastic integration. For a square integrable martingale M, the Doob–Meyer decomposition of the submartingale M^2 gives an increasing process $\langle M, M \rangle$ (also called the predictable quadratic variation of M) which gave an estimate on the growth of stochastic integral w.r.t. M. In this book, we have developed the theory of stochastic integration via the quadratic variation $[M, M]$. Nonetheless, the predictable quadratic variation of a (locally) square integrable martingale M plays an important role in the theory and now we will show that if $M \in \mathbb{M}^2_{\text{loc}}$ then $\langle M, M \rangle$ is an increasing process. We start with an auxiliary result.

Lemma 8.50 *Let A be an adapted increasing integrable process with $A_0 = 0$ and let U be a predictable process, $U \in \mathbb{V}$ such that $M = A - U$ is a martingale. Then $U \in \mathbb{V}^+$; i.e. U is an increasing process.*

Proof Let $\{\tau_m : m \in F\}$ be the sequence of stopping times given by Theorem 8.22 (F is a subset of natural numbers) so that (8.2.27) and (8.2.28) are true. Let

$$C_t = \sum_{m \in F} (\Delta A)_{\tau_m} 1_{[\tau_m, \infty)}$$

and

$$D_t = A_t - C_t.$$

It follows that C and D are adapted increasing processes, $C_0 = 0$, $D_0 = 0$ and D is continuous. Since

$$0 \le C_t \le A_t \ \forall t$$

it follows that C is also integrable and thus by Corollary 8.45 we can get a predictable increasing process B such that $N_t = C_t - B_t$ is a martingale. Thus $N_t = A_t - D_t - B_t$. Thus $N - M = U - D - B$. Now $N - M$ is a martingale and at the same time $U - D - B$ is an FV process that is predictable. Thus by Theorem 8.29, we have

$$U = D + B.$$

By construction, D and B are increasing processes and this shows U is increasing. $\square$

Corollary 8.51 *Let M be a locally square integrable martingale. Then the predictable quadratic variation $\langle M , M \rangle$ (see Definition 8.49) is an increasing process.*

Proof The result follows from Lemma 8.50 since $[M, M]$ is an increasing process. $\square$

We are now ready to prove the classical result.

Theorem 8.52 *(The Doob–Meyer Decomposition Theorem) Let N be a locally square integrable martingale with $N_0 = 0$ w.r.t. a filtration $(\mathcal{F}_.)$. Then $\langle N, N \rangle$ is the unique r.c.l.l. $(\mathcal{F}_.)$ predictable increasing process A such that $A_0 = 0$ and M defined by*

$$M_t = N_t^2 - A_t \tag{8.5.1}$$

is a local martingale. Further, for any stopping time σ

$$\mathsf{E}[\,\langle N, N \rangle_\sigma] \le \mathsf{E}[\sup_{0 \le s \le \sigma} |N_s|^2] \le 4\mathsf{E}[\,\langle N, N \rangle_\sigma]. \tag{8.5.2}$$

Proof Since $N_t^2 - [N, N]_t$ is a local martingale (see Theorem 5.19), for a predictable process A, $N^2 - A$ is a local martingale if and only if $[N, N] - A$ is a local martingale. Now the first part follows from Corollary 8.51.

For the remaining part, let τ_n be a localizing sequence for the local martingale $M = N^2 - \langle N, N \rangle$. Then for $n \ge 1$

$$Y_t^n = N_{\tau_n \wedge t}^2 - \langle N, N \rangle_{\tau_n \wedge t}$$

is a martingale. Hence for any bounded stopping time σ we have,

$$\mathsf{E}[N_{\tau_n \wedge \sigma}^2] = \mathsf{E}[\langle N, N \rangle_{\tau_n \wedge \sigma}] \tag{8.5.3}$$

Now by Doob's maximal inequality, Theorem 2.26, we get

$$\mathsf{E}[\langle N, N \rangle_{\tau_n \wedge \sigma}] \le \mathsf{E}[\sup_{0 \le s \le \tau_n \wedge \sigma} |N_s|^2] \le 4\mathsf{E}[\langle N, N \rangle_{\tau_n \wedge \sigma}].$$

Now taking limit as $n \to \infty$ and using monotone convergence theorem, we conclude

$$E[\langle N, N \rangle_\sigma] \leq E[\sup_{0 \leq s \leq \sigma} |N_s|^2] \leq 4E[\langle N, N \rangle_\sigma]. \qquad (8.5.4)$$

$\square$

Remark **8.53** Using (8.5.3), it follows that if N is a locally square integrable martingale then for all stopping times σ,

$$E[[N, N]_\sigma] = E[N_\sigma^2] = E[\langle N, N \rangle_\sigma]. \qquad (8.5.5)$$

All the quantities can be ∞, but if one is finite, so are the others and they are equal.

Lemma 8.54 *Let M be a locally square integrable martingale and let f be a predictable process such that for all t*

$$B_t = \int_0^t |f_s|^2 d\langle M, M \rangle_s < \infty \quad a.s. \qquad (8.5.6)$$

Then B is a predictable increasing process.

Proof Let us note that B is an r.c.l.l. increasing adapted process and for any stopping time τ,

$$(\Delta B)_\tau = |f_\tau|^2 (\Delta \langle M, M \rangle)_\tau.$$

Now the result follows from Lemma 8.25 and part (v) of Theorem 8.4. $\square$

Exercise 8.55 Let $V \in \mathbb{V}^+$ be predictable and f be a predictable process such that $\int_0^t |f| dV < \infty$ for all $t \geq 0$. Show that the process U defined by $U_t = \int_0^t f dV$ is predictable.

Lemma 8.56 *Let M be a locally square integrable martingale and let f be a locally bounded predictable process and let $N = \int f dM$. Then*

$$\langle N, N \rangle_t = \int_0^t |f_s|^2 d\langle M, M \rangle_s < \infty. \qquad (8.5.7)$$

Proof That B_t defined by (8.5.6) is predictable has been noted above. One can show that $N_t^2 - B_t$ is a local martingale starting with f simple and then by approximation. Thus $\langle N, N \rangle_t = B_t$. $\square$

Remark **8.57** For f, M, N as above, we have seen that

$$[N, N]_t = \int_0^t |f_s|^2 d[M, M]_s < \infty \qquad (8.5.8)$$

and hence it follows that for a stopping time σ,

$$\mathsf{E}[\int_0^\sigma |f_s|^2 d\langle M, M\rangle_s] = \mathsf{E}[\int_0^\sigma |f_s|^2 d[M, M]_s]. \tag{8.5.9}$$

The estimate (5.4.10) on the growth of the stochastic integral can be recast as:

Theorem 8.58 *Let M be a locally square integrable martingale with $M_0 = 0$. For a locally bounded predictable process f, the processes $Y_t = \int_0^t f dM$ and $Z_t = Y_t^2 - \int_0^t f_s^2 d\langle M, M\rangle_s$ are local martingales and for any stopping time σ such that $\mathsf{E}[\int_0^\sigma f_s^2 d\langle M, M\rangle_s] < \infty$,*

$$\mathsf{E}[\sup_{0 \le t \le \sigma} |\int_0^t f dM|^2] \le 4\mathsf{E}[\int_0^\sigma f_s^2 d\langle M, M\rangle_s]. \tag{8.5.10}$$

For locally square integrable martingales M, N, the *predictable cross-quadratic variation* $\langle M, N\rangle$ is the unique predictable process in $\mathbb{V}_0$ such that $[M, N]_t - \langle M, N\rangle_t$ is a local martingale. Since $M_t N_t - [M, N]_t$ is a local martingale, it follows that $\langle M, N\rangle$ is the unique predictable process in $\mathbb{V}_0$ such that

$$Z_t = M_t N_t - \langle M, N\rangle_t \tag{8.5.11}$$

is a local martingale. We can see that the predictable cross-quadratic variation also satisfies the polarization identity (for locally square integrable martingales M, N)

$$\langle M, N\rangle_t = \frac{1}{4}(\langle M + N, M + N\rangle_t - \langle M - N, M - N\rangle_t) \tag{8.5.12}$$

We had seen that $(M, N) \mapsto \langle M, N\rangle$ is bilinear in M and N. This yields an analogue of Theorem 4.78 which we note here.

Theorem 8.59 *Let M, N be locally square integrable martingales. Then for any $s \le t$*

$$\mathrm{VAR}_{(s,t]}(\langle M, N\rangle) \le \sqrt{(\langle M, M\rangle_t - \langle M, M\rangle_s).(\langle N, N\rangle_t - \langle N, N\rangle_s)} \tag{8.5.13}$$

and

$$\mathrm{VAR}_{[0,t]}(\langle M, N\rangle) \le \sqrt{\langle M, M\rangle_t \langle N, N\rangle_t}, \tag{8.5.14}$$

Proof Let

$$\Omega_{a,b,s,r} = \{\omega \in \Omega : \langle aM + bN, aM + bN\rangle_r(\omega) \ge \langle aM + bN, aM + bN\rangle_s(\omega)\}$$

and

$$\Omega_0 = \cup\{\Omega_{a,b,s,r} : s, r, a, b \in \mathbb{Q}, r \ge s\}.$$

Then it follows that $\mathsf{P}(\Omega_0) = 1$ (since for any locally square integrable martingale Z, $\langle Z, Z\rangle$ is an increasing process) and that for $\omega \in \Omega_0$, for $0 \le s \le r, s, r, a, b \in \mathbb{Q}$

$$(a^2(\langle M,M\rangle_r - \langle M,M\rangle_s) + b^2(\langle N,N\rangle_r - \langle N,N\rangle_s)$$
$$+ 2ab(\langle M,N\rangle_r - \langle M,N\rangle_s))(\omega) \geq 0,$$

and since the quadratic form above remains positive, we conclude

$$|(\langle M,N\rangle_r(\omega) - \langle M,N\rangle_s(\omega))|$$
$$\leq \sqrt{(\langle M,M\rangle_r(\omega) - \langle M,M\rangle_s(\omega))(\langle N,N\rangle_r(\omega) - \langle N,N\rangle_s(\omega))}.$$
$$(8.5.15)$$

Since all the processes occurring in (8.5.15) are r.c.l.l., it follows that (8.5.15) is true for all $s \leq r$, $s, r \in [0, \infty)$. Now given $s < t$ and $s = t_0 < t_1 < \ldots < t_m = t$, we have

$$\sum_{j=0}^{m-1} |\langle M,N\rangle_{t_{j+1}} - \langle M,N\rangle_{t_j}|$$

$$\leq \sum_{j=0}^{m-1} \sqrt{(\langle M,M\rangle_{t_{j+1}} - \langle M,M\rangle_{t_j})(\langle N,N\rangle_{t_{j+1}} - \langle N,N\rangle_{t_j})} \qquad (8.5.16)$$

$$\leq \sqrt{(\langle M,M\rangle_t - \langle M,M\rangle_s)(\langle N,N\rangle_t - \langle N,N\rangle_s)}$$

where the last step follows from Cauchy–Schwarz inequality and the fact that $\langle M,M\rangle$, $\langle N,N\rangle$ are increasing processes. Now taking supremum over partitions of $[0, t]$ in (8.5.16) we get (8.5.13). Now (8.5.14) follows from (8.5.13) taking $s = 0$ since $\langle M,M\rangle_0 = 0$, $\langle N,N\rangle_0 = 0$ and $\langle M,N\rangle_0 = 0$. □

And here is an analogue of (4.77). However, the proof is different from that given earlier.

Lemma 8.60 *Let U, V be locally square integrable martingales. Then for any t we have*

$$|\langle U,U\rangle_t - \langle V,V\rangle_t| \leq \sqrt{2\langle U - V, U - V\rangle_t(\langle U,U\rangle_t + \langle V,V\rangle_t)}. \qquad (8.5.17)$$

Proof Using bilinearity of $(M,N) \mapsto \langle M,N\rangle$, note that

$$\langle U,U\rangle_t - \langle V,V\rangle_t = \langle U + V, U - V\rangle_t. \qquad (8.5.18)$$

Invoking (8.5.13) with $s = 0$, we get

$$\langle U,U\rangle_t - \langle V,V\rangle_t \leq \sqrt{\langle U - V, U - V\rangle_t(\langle U + V, U + V\rangle_t)}$$

$$= \sqrt{\langle U - V, U - V\rangle_t(\langle U,U\rangle_t + \langle V,V\rangle_t + 2\langle U,V\rangle_t)}.$$
$$(8.5.19)$$

Since

$$\langle U - V, U - V\rangle_t = \langle U,U\rangle_t + \langle V,V\rangle_t - 2\langle U,V\rangle_t \geq 0$$

we have

$$2\langle U, V\rangle_t \le \langle U, U\rangle_t + \langle V, V\rangle_t. \tag{8.5.20}$$

Now (8.5.17) follows from (8.5.19) and (8.5.20). $\square$

We will now prove that $[M, M]$ and $\langle M, M\rangle$ depend continuously on $M \in \mathbb{M}_{\text{loc}}^2$.

Theorem 8.61 *Let $M^n, M \in \mathbb{M}_{\text{loc}}^2$ be such that for a sequence $\{\sigma_j\}$ of stopping times increasing to ∞, one has for each $j \ge 1$,*

$$\mathsf{E}[\sup_{t \le \sigma_j} |M_t^n - M_t|^2] \to 0. \tag{8.5.21}$$

Then we have

$$[M^n - M, M^n - M] \xrightarrow{ucp} 0 \tag{8.5.22}$$

$$\langle M^n - M, M^n - M\rangle \xrightarrow{ucp} 0, \tag{8.5.23}$$

$$[M^n, M^n] \xrightarrow{ucp} [M, M], \tag{8.5.24}$$

$$\langle M^n, M^n\rangle \xrightarrow{ucp} \langle M, M\rangle. \tag{8.5.25}$$

Proof Using (5.3.24) and (8.5.2) it follows from (8.5.21) that

$$\mathsf{E}[\,[M^n - M, M^n - M]_{\sigma_j}\,] \to 0$$

and

$$\mathsf{E}[\langle M^n - M, M^n - M\rangle_{\sigma_j}] \to 0.$$

Since $[M^n - M, M^n - M]$ and $\langle M^n - M, M^n - M\rangle$ are increasing processes, (8.5.22) and (8.5.23) follow from these observations and Lemma 2.75. Using (4.6.17), (8.5.14) along with (8.5.22) and (8.5.23), we conclude that

$$[M^n - M, M] \xrightarrow{ucp} 0 \tag{8.5.26}$$

and

$$\langle M^n - M, M\rangle \xrightarrow{ucp} 0. \tag{8.5.27}$$

Using

$$[M^n, M^n] - [M, M] = [M^n - M, M^n - M] + 2[M^n - M, M]$$

and

$$\langle M^n, M^n \rangle - \langle M, M \rangle = \langle M^n - M, M^n - M \rangle + 2\langle M^n - M, M \rangle,$$

the remaining conclusions (8.5.24) and (8.5.25) follow from (8.5.22), (8.5.26) and (8.5.23), (8.5.27), respectively. □

Using Lemma 8.25, one can show that if A is a predictable r.c.l.l. FV process and f is a $[0, \infty)$-valued predictable process, then B defined by

$$B_t = \int_0^t f_s dA_s$$

is predictable. Further, it can be checked (first for simple integrands and then by limiting arguments) that for f, g predictable and locally bounded, M, N locally square integrable martingales, $X = \int f dM$ and $Y = \int g dN$,

$$U_t = X_t Y_t - \int_0^t f g d\langle M, N \rangle$$

is a local martingale. Thus

$$\langle \int f dM, \int g dN \rangle = \int f g d\langle M, N \rangle. \tag{8.5.28}$$

These observations lead us to the following analogue of Theorem 5.31 which we record here for use later.

Theorem 8.62 *Let M be a locally square integrable martingale with $M_0 = 0$ and σ be a stopping time. For a locally bounded predictable process f such that $\mathrm{E}[\int_0^\sigma f_s^2 d\langle M, M \rangle_s] < \infty$, we have*

$$\mathrm{E}[\sup_{0 \le t \le \sigma} |\int_0^t f dM|^2] \le 4\mathrm{E}[\int_0^\sigma f_s^2 d\langle M, M \rangle_s]. \tag{8.5.29}$$

Proof We had observed that $Y = \int f dM$ is a local martingale above and that

$$\langle Y, Y \rangle_t = \int_0^t f_s^2 d\langle M, M \rangle_s.$$

Now the estimate (8.5.29) follows from (8.5.2). □

Definition 8.63 Two locally square integrable martingales M, N are said to be strongly orthogonal if $\langle M, N \rangle_t = 0$ for all $t < \infty$.

Equivalently, M, N are strongly orthogonal if $Z_t = M_t N_t$ is a local martingale.

Exercise 8.64 Construct an example of martingales M, N such that $\langle M, N \rangle_t = 0$ for all $t < \infty$ but for some $T < \infty$, $[M, N]_T \ne 0$.

The following interesting observation is due to Kunita–Watanabe who studied structure of square integrable martingales.

Theorem 8.65 *Let M, N be locally square integrable martingales. Then N admits
a decomposition*

$$N_t = \int_0^t f dM + U_t \tag{8.5.30}$$

*where $f \in \mathbb{L}_m^2(M)$ and U is a locally square integrable martingale strongly orthog-
onal to M.*

Proof Let τ_n be a sequence of stopping times, $\tau_n \le n$, such that $N_{t \wedge \tau_n}$ is a square
integrable martingale. Let $a_n = \mathsf{E}[N_{\tau_n}^2]$. Let

$$\alpha = \inf\{\textstyle\sum_{n=1}^\infty \frac{1}{2^n(1+a_n)} \mathsf{E}[(N_{\tau_n} - \int_0^{\tau_n} g dM)^2] \; : \; g \in \mathbb{L}_m^2(M)\}.$$

It can be seen that $\alpha < \infty$, indeed, $\alpha \le 1$ since $g = 0 \in \mathbb{L}_m^2(M)$. Now it can be
shown (proceeding as in the proof of existence of orthogonal projections onto a
closed subspace in a Hilbert space as used proof of Theorem 5.44) that the infimum
is attained, say for $f \in \mathbb{L}_m^2(M)$ and then for every n

$$\mathsf{E}[(N_{\tau_n} - \int_0^{\tau_n} f dM)(\int_0^{\tau_n} g dM)] = 0 \; \forall g \in \mathbb{L}_m^2(M).$$

Thus defining U by (8.5.30) with this f we have

$$\mathsf{E}[U_{\tau_n}(\int_0^{\tau_n} g dM)] = 0 \; \forall g \in \mathbb{L}_m^2(M).$$

Given a stopping time σ, taking $g = 1_{[0,\sigma]}$, this yields

$$\mathsf{E}[U_{\tau_n} N_{\sigma \wedge \tau_n}] = 0$$

which in turn yields

$$\mathsf{E}[U_{\sigma \wedge \tau_n} N_{\sigma \wedge \tau_n}] = 0.$$

Writing $Z_t = U_t N_t$, we conclude that $Z_t^n = Z_{t \wedge \tau_n}$ is a martingale and thus Z is a local
martingale completing the proof. $\qquad\square$

For $N, M \in \mathbb{M}_{\mathrm{loc}}$, if $[M, N] = 0$ then it follows that MN is a local martingale and
hence $\langle M, N \rangle = 0$. This observation has an important consequence.

Lemma 8.66 *Let $N \in \mathbb{V} \cap \mathbb{M}_{loc}^2$ and $M \in \mathbb{M}_{loc}^2$ be such that for all stopping
times τ,*

$$(\Delta N)_\tau (\Delta M)_\tau = 0 \quad a.s.$$

Then $\langle M, N \rangle = 0$. In particular, $\mathbb{V} \cap \mathbb{M}_{loc}^2 \subseteq \mathbb{M}_{d,loc}^2$.

Proof As seen in Theorem 4.74, $N \in \mathbb{V}$ implies

$$[M, N]_t = \sum_{0 < s \le t} (\Delta M)_s (\Delta N)_s.$$

Using Theorem 8.22, we can get stopping times $\{\tau_n : n \in F\}$ (F is a finite or countable set) that cover the jumps of N and then it follows that

$$[M, N]_t = \sum_{n \in F} (\Delta M)_{\tau_m} (\Delta N)_{\tau_m} 1_{\{\tau_m \le t\}}.$$

Now in view of the assumption, $(\Delta M)_{\tau_m} (\Delta N)_{\tau_m} = 0$ (*a.s.*) for each m and as a consequence, $[M, N]_t = 0$ (*a.s.*) for all t. Thus $Z_t = M_t N_t$ is a local martingale and as a consequence $\langle M, N \rangle = 0$. The last statement follows easily. □

8.6 Square Integrable Martingales

In this section we will obtain a decomposition of square integrable martingales into a martingale with continuous paths and a martingale with jumps.

Theorem 8.67 *Let τ be a predictable stopping time and let ξ be a $\mathcal{F}_\tau$ measurable square integrable random variable with $E[\xi \mid \mathcal{F}_{\tau-}] = 0$. Then*

$$M_t = \xi 1_{[\tau, \infty)}(t) \tag{8.6.1}$$

is a martingale and $\langle M, M \rangle = A$ where

$$A_t = E[\xi^2 \mid \mathcal{F}_{\tau-}] 1_{[\tau, \infty)}(t). \tag{8.6.2}$$

Proof From part (*vii*) in Theorem 8.4 it follows that M is a martingale. Since τ is predictable, by Theorem 8.19 it follows that A is predictable and clearly A is an increasing process. Noting that

$$M_t^2 - A_t = (\xi^2 - E[\xi^2 \mid \mathcal{F}_{\tau-}]) 1_{[\tau, \infty)}(t)$$

it follows, again invoking part (*vii*) in Theorem 8.4, that $N_t = M_t^2 - A_t$ is a martingale.

It is clear from the definition of A that $A_0 = 0$ on the set $\tau > 0$. On the other hand, ξ is $\mathcal{F}_\tau$ measurable and hence $\xi 1_{\{\tau = 0\}}$ is $\mathcal{F}_0$ measurable. Thus $E[\xi \mid \mathcal{F}_{\tau-}] = 0$ implies $\xi 1_{\{\tau = 0\}} = 0$. Since $\mathcal{F}_0 \subseteq \mathcal{F}_{\tau-}$, we conclude that $A_0 = 0$ on $\{\tau = 0\}$ as well. Thus $A_0 = 0$ and hence

$$\langle M, M \rangle_t = E[\xi^2 \mid \mathcal{F}_{\tau-}] 1_{[\tau, \infty)}(t) \tag{8.6.3}$$

follows from the uniqueness part of the Doob–Meyer decomposition (Theorem 8.52). $\qquad\square$

We are going to prove a structural decomposition result for square integrable martingales. This would play an important role in the proof of Metivier–Pellaumail inequality in the next chapter. Here are some preparatory results. Recall the Definition 5.68—a locally square integrable martingale M is purely discontinuous if

$$[M, N] = 0 \quad \forall N \in \mathbb{M}_{c, \text{loc}}.$$

The class of purely discontinuous square integrable martingales is denoted by $\mathbb{M}_d^2$ and purely discontinuous locally square integrable martingales is denoted by $\mathbb{M}_{d, \text{loc}}^2$.

Exercise 8.68 Show that a locally square integrable martingale is purely discontinuous if and only if

$$\langle M, N \rangle_t = 0 \quad \forall t \geq 0, \ \forall N \in \mathbb{M}_{c, \text{loc}}^2.$$

Exercise 8.69 Let $M^n \in \mathbb{M}_d^2$ and $M \in \mathbb{M}^2$ be such that

$$\mathsf{E}[[M^n - M, M^n - M]_T] \to 0 \quad \forall T < \infty$$

then $M \in \mathbb{M}_d^2$.

Lemma 8.70 *Let X be a square integrable martingale with $X_t = X_{t \wedge T}$ for all t, for some $T < \infty$ and let τ be a predictable stopping time such that*

$$\mathsf{P}((\Delta X)_\tau \neq 0) > 0.$$

Let N be defined by
$$N_t = (\Delta X)_\tau I_{[\tau, \infty)}(t).$$

Then $N \in \mathbb{M}_d^2$. Let $Y = X - N$. Then $\mathsf{P}((\Delta Y)_\tau \neq 0) = 0$, Y is a square integrable martingale,

$$\langle Y, N \rangle = 0$$

and
$$\langle N, N \rangle_t = \mathsf{E}[(\Delta X)_\tau^2 \mid \mathcal{F}_{\tau-}] I_{[\tau, \infty)}(t). \tag{8.6.4}$$

Proof Using $X_t = X_{t \wedge T}$, we have

$$|(\Delta X)_\tau| \leq 2 \sup_{t \leq T} |X_t|$$

and hence in view of Doob's maximal inequality, Theorem 2.26, we have that $\xi = (\Delta X)_\tau$ is square integrable. Since τ is predictable, $\mathsf{E}[(\Delta X)_\tau \mid \mathcal{F}_{\tau-}] = 0$. So Theorem

8.67 implies that N is a square integrable martingale and that (8.6.4) is true. Since $N \in \mathbb{V}$, using Lemma 8.66 it follows that $N \in \mathbb{M}_d^2$.

It follows that Y is a square integrable martingale and by construction, $P((\Delta Y)_\tau \neq 0) = 0$. Hence

$$(\Delta Y)_t(\omega)(\Delta N)_t(\omega) = 0, \quad \forall t, \; \forall \omega$$

and hence by Lemma 8.66, $\langle Y, N \rangle = 0$. $\qquad\qquad\qquad\qquad\qquad\qquad\qquad\qquad \Box$

For $X \in \mathbb{M}_{loc}^2$, applying the previous result to $X^{[\sigma_m]} \in \mathbb{M}^2$ where $\{\sigma_m : m \geq 1\}$ is a localizing sequence we get:

Corollary 8.71 *Let X be a locally square integrable martingale and let τ be a predictable stopping time such that*

$$P((\Delta X)_\tau \neq 0) > 0.$$

Let N, Y be defined by

$$N_t = (\Delta X)_\tau 1_{[\tau, \infty)}(t)$$

and $Y = X - N$. Then $P((\Delta Y)_\tau \neq 0) = 0, N, Y$ are locally square integrable martingales with

$$\langle Y, N \rangle = 0.$$

Further, $\langle N, N \rangle$ has a single jump at τ and if $\sigma_m \uparrow \infty$ are bounded stopping times such that $X^{[\sigma_m]} \in \mathbb{M}^2$ then

$$\langle N, N \rangle_\tau 1_{\{\tau \leq \sigma_m\}} = E[(\Delta X^{[\sigma_m]})_\tau^2 \mid \mathcal{F}_{\tau-}]. \tag{8.6.5}$$

The following technical result will be used in the decomposition of $M \in \mathbb{M}_{loc}^2$.

Theorem 8.72 *Let $M \in \mathbb{M}_{loc}^2$ be a locally square integrable martingale with $M_0 = 0$. Let $\{\tau_k : k \geq 1\}$ be a sequence of stopping times with disjoint graphs. Let*

$$V_t^j = (\Delta M)_{\tau_j} 1_{[\tau_j, \infty)} \tag{8.6.6}$$

and let A^j be the compensator of V^j. Let $W^j = V^j - A^j$. Then

(i) *For all $j \geq 1$, $W^j \in \mathbb{M}_{d,loc}^2$.*
(ii) *$S^m = \sum_{j=1}^m W^j$ converges in Emery topology to W.*
(iii) *$W \in \mathbb{M}_{d,loc}^2$ and if $\{\sigma_i : i \geq 1\}$ is a sequence of bounded stopping times increasing to ∞ with $W^{[\sigma_i]} \in \mathbb{M}_d^2$, then for each i*

$$E[\sup_{t \leq \sigma_i} |S_t^m - W_t|^2] \to 0 \text{ as } m \to \infty. \tag{8.6.7}$$

(vi) *$[S^n, S^n] \xrightarrow{ucp} [W, W]$ and $\langle S^n, S^n \rangle \xrightarrow{ucp} \langle W, W \rangle$.*

Proof Since graphs of $\{\tau_k : k \geq 1\}$ are disjoint,

$$\sum_{j=1}^{\infty} [V^j, V^j]_t = \sum_{j=1}^{\infty} (\Delta M)^2_{\tau_j} 1_{[\tau_j, \infty)}(t) \leq \sum_{0 < s \leq t} (\Delta M)^2_s \leq [M, M]_t. \qquad (8.6.8)$$

Also, easy to see that for $j \neq k$ $\mathsf{P}(\tau_j = \tau_k) = 0$ implies

$$[V^j, V^k]_t = 0. \qquad (8.6.9)$$

Let $\sigma_i \uparrow \infty$ be bounded stopping times such that $M^{[\sigma_i]} \in \mathbb{M}^2$ so that

$$\mathsf{E}[[M, M]_{\sigma_i}] < \infty \quad \forall i \geq 1. \qquad (8.6.10)$$

It follows that for every i,

$$\lim_{m \to \infty} \mathsf{E}[\sum_{j=m}^{\infty} [V^j, V^j]_{\sigma_i}] = 0. \qquad (8.6.11)$$

Since A^j is compensator of V^j and $W^j = V^j - A^j$, it follows that $\sum_{j=1}^{n} A^j$ is the compensator of $\sum_{j=1}^{n} V^j$ and $S^n = \sum_{j=1}^{n} V^j - \sum_{j=1}^{n} A^j$. Using (5.6.37) along with (8.6.8) and (8.6.9) we have

$$\mathsf{E}[[S^n, S^n]_{\sigma_i}] \leq \mathsf{E}[[\sum_{j=1}^{n} V^j, \sum_{j=1}^{n} V^j]_{\sigma_i}]$$

$$= \sum_{j=1}^{n} \mathsf{E}[[V^j, V^j]_{\sigma_i}]$$

$$\leq \mathsf{E}[[M, M]_{\sigma_i}]$$

$$< \infty.$$

Thus $S^n \in \mathbb{M}^2_{\text{loc}}$. Since $S^n \in \mathbb{V}$ by construction, using Lemma 8.66 it follows that $S^n \in \mathbb{M}^2_{d,\text{loc}}$. Similarly, for $m \leq n$ we have

$$\mathsf{E}[[S^n - S^m, S^n - S^m]_{\sigma_i}] \leq \mathsf{E}[\sum_{j=m+1}^{n} [V^j, V_j]_{\sigma_i}]. \qquad (8.6.12)$$

Now (8.6.11) and (8.6.12) imply that S^n is Cauchy in $\mathbf{d}_{em}$ and thus converges to say W such that $W \in \mathbb{M}^2_{\text{loc}}$ (see Corollary 5.42) and further (8.6.7) holds. Since $(S^n)^{[\sigma_i]} \in \mathbb{M}^2_d$, the relation (8.6.7) implies that $W^{[\sigma_i]} \in \mathbb{M}^2_d$ and hence $W \in \mathbb{M}^2_{d,\text{loc}}$. As seen in Theorem 8.61, (8.6.7) implies (*iv*). $\square$

Lemma 8.73 *Let $M \in \mathbb{M}^2$ be such that $M_0 = 0$, $M_t = M_{t \wedge T}$ for $t \geq 0$ for some T and let τ be a predictable stopping time. Then*

$$\Delta \langle M, M \rangle_\tau = \mathsf{E}[(\Delta M)^2_\tau \mid \mathcal{F}_{\tau-}] \tag{8.6.13}$$

and

$$\mathsf{E}[(\Delta M)^2_\tau] = \mathsf{E}[(\Delta M^2)_\tau]. \tag{8.6.14}$$

Proof Let $\tau_n \uparrow \tau$ be a sequence of $(\mathcal{F}^{+} \cdot)$ stopping time announcing τ so that $\tau_n < \tau$ on $\tau > 0$. Note that since $M^2_t - \langle M, M \rangle_t$ is a martingale, we have for $m \leq n$

$$\mathsf{E}[M^2_\tau - M^2_{\tau_n} \mid \mathcal{F}^{+}_{\tau_m}] = \mathsf{E}[\langle M, M \rangle_\tau - \langle M, M \rangle_{\tau_n} \mid \mathcal{F}^{+}_{\tau_m}].$$

On the other hand, easy to see that

$$\mathsf{E}[(M_\tau - M_{\tau_n})^2 \mid \mathcal{F}^{+}_{\tau_m}] = \mathsf{E}[M^2_\tau - M^2_{\tau_n} \mid \mathcal{F}^{+}_{\tau_m}].$$

Taking limit as $n \to \infty$ and using that $M_{\tau_n} \to M_{\tau-}$ (we need to use $M_0 = 0$), it follows that

$$\mathsf{E}[(\Delta M)^2_\tau \mid \mathcal{F}^{+}_{\tau_m}] = \mathsf{E}[(\Delta M^2)_\tau \mid \mathcal{F}^{+}_{\tau_m}] = \mathsf{E}[\Delta \langle M, M \rangle_\tau \mid \mathcal{F}^{+}_{\tau_m}].$$

Now taking limit as $m \to \infty$ and using Theorem 8.15 along with Theorem 1.37, we conclude

$$\mathsf{E}[(\Delta M)^2_\tau \mid \mathcal{F}^{+}_{\tau-}] = \mathsf{E}[(\Delta M^2)_\tau \mid \mathcal{F}^{+}_{\tau-}] = \mathsf{E}[\Delta \langle M, M \rangle_\tau \mid \mathcal{F}^{+}_{\tau-}].$$

Since $\langle M, M \rangle$ is predictable, $\Delta \langle M, M \rangle_\tau$ is $\mathcal{F}_{\tau-} \subseteq \mathcal{F}^{+}_{\tau-}$ measurable, we conclude

$$\mathsf{E}[(\Delta M)^2_\tau \mid \mathcal{F}_{\tau-}] = \mathsf{E}[(\Delta M^2)_\tau \mid \mathcal{F}_{\tau-}] = \Delta \langle M, M \rangle_\tau.$$

Both the required relations (8.6.13) and (8.6.14) follow from this. $\square$

Corollary 8.74 *Let $M \in \mathbb{M}^2_{loc}$ and let τ be a predictable stopping time. Then for all bounded stopping times σ such that $M^{[\sigma]} \in \mathbb{M}^2$*

$$(\Delta \langle M, M \rangle_\tau) 1_{\{\tau \leq \sigma\}} = \mathsf{E}[(\Delta M^{[\sigma]})^2_\tau \mid \mathcal{F}_{\tau-}]. \tag{8.6.15}$$

Proof Follows from (8.6.13) by observing that $\langle M^{[\sigma]}, M^{[\sigma]} \rangle_t = \langle M, M \rangle_{t \wedge \sigma}$ and as a consequence,

$$(\Delta \langle M^{[\sigma]}, M^{[\sigma]} \rangle)_\tau = (\Delta \langle M, M \rangle)_{\tau \wedge \sigma}).$$

$\square$

Theorem 8.75 *Let $Y \in \mathbb{M}^2_{loc}$ be a locally square integrable martingale with $Y_0 = 0$. Then there exists a sequence $\{\tau_k : k \geq 1\}$ of predictable stopping times such that the* *local martingale Y admits a decomposition*

$$Y = Z + U \tag{8.6.16}$$

satisfying the following.

(i) $U_t^k = (\Delta Y)_{\tau_k} I_{[\tau_k, \infty)}(t)$ *satisfies* $U^k \in \mathbb{M}_{d, loc}^2$.
(ii) $U = \sum_{k=1}^{\infty} U^k \in \mathbb{M}_{d, loc}^2$ *(here the sum converges in the Emery topology).*
(iii) $[U^j, U^k] = 0$ *and* $\langle U^j, U^k \rangle = 0$ *for all* $j, k \geq 1$.
(iv) $[U, U] = \sum_{k=1}^{\infty} [U^k, U^k]$.
(v) $\langle U, U \rangle = \sum_{k=1}^{\infty} \langle U^k, U^k \rangle$.
(vi) $[Z, U] = 0$ *and* $\langle Z, U \rangle = 0$.
(vii) Z *is a locally square integrable martingale,*

$$[Y, Y] = [Z, Z] + \sum_{k=1}^{\infty} [U^k, U^k] \tag{8.6.17}$$

and

$$\langle Y, Y \rangle = \langle Z, Z \rangle + \sum_{k=1}^{\infty} \langle U^k, U^k \rangle. \tag{8.6.18}$$

(viii) $\langle Z, Z \rangle$ *is a continuous process.*

Further, $\langle U^k, U^k \rangle$ *is a process with a single jump at* τ_k *and if* $\sigma_m \uparrow \infty$ *are bounded stopping times such that* $Y^{\lfloor \sigma_m \rfloor} \in \mathbb{M}^2$

$$\langle U^k, U^k \rangle_{\tau_k} I_{\{\tau_k \leq \sigma_m\}} = \mathsf{E}[(\Delta Y^{\lfloor \sigma_m \rfloor})_{\tau_k}^2 \mid \mathcal{F}_{\tau_k -}]. \tag{8.6.19}$$

Proof Let $A = \langle Y, Y \rangle$. By definition A is an increasing predictable process with $A_0 = 0$. Using Theorem 8.22, we can get predictable stopping times $\{\tau_m : m \geq 1\}$ with disjoint graphs that enumerate the jumps of A, i.e.

$$(\Delta A) = \sum_{m=1}^{\infty} (\Delta A)_{\tau_m} 1_{[\tau_m]}. \tag{8.6.20}$$

For $k \geq 1$, let

$$U_t^k = (\Delta Y)_{\tau_k} 1_{[\tau_k, \infty)}(t). \tag{8.6.21}$$

As observed in Lemma 8.70 and Corollary 8.71, U^k is a locally square integrable martingale and (8.6.19) holds. Since graphs of $\{\tau_m : m \geq 1\}$ are disjoint, it follows that

$$[U^k, U^j] = 0, \quad \text{for } j, k \geq 1 \tag{8.6.22}$$

and as a consequence

$$\langle U^j, U^k \rangle = 0 \quad \text{for } j, k \geq 1. \tag{8.6.23}$$

We invoke Theorem 8.72. In the notation of that theorem, $V^j = U^j$ and $A^j = 0$ (since U^j is local martingale) and hence $W^j = U^j$. Thus it follows from Theorem 8.72 that

$$S^k = \sum_{k=1}^{m} U^k \xrightarrow{em} U$$

where $U \in \mathbb{M}^2_{d, \text{loc}}$. In view of (8.6.22) and (8.6.23) we have $[S^n, S^n] = \sum_{k=1}^{n} [U^k, U^k]$ and $\langle S^n, S^n \rangle = \sum_{k=1}^{n} \langle U^k, U^k \rangle$. Part (iv) in Theorem 8.72 now implies

$$[U, U] = \sum_{k=1}^{\infty} [U^k, U^k] \tag{8.6.24}$$

$$\langle U, U \rangle = \sum_{k=1}^{\infty} \langle U^k, U^k \rangle. \tag{8.6.25}$$

Since $Y - S^k$ and S^k do not have any common jump and $S^k \in \mathbb{V}$,

$$[Y - S^k, S^k] = 0.$$

Now using that $S^m \xrightarrow{em} U$ we get

$$[Y - U, U] = 0. \tag{8.6.26}$$

We define $Z = Y - U$ so that (8.6.16) holds and conclude using (8.6.26) that

$$[Y, Y] = [Z, Z] + [U, U].$$

This along with (8.6.24) implies (8.6.17). Finally $[Z, U] = 0$ also gives $\langle Z, U \rangle = 0$ and thus

$$\langle Y, Y \rangle = \langle Z, Z \rangle + \langle U, U \rangle$$

and in turn (8.6.18) follows.

Remains to prove that $\langle Z, Z \rangle$ is continuous. By the choice of the stopping times $\{\tau_k : k \geq 1\}$, it follows that the $\langle Y, Y \rangle$ does not have jumps other than at $\{\tau_k : k \geq 1\}$. From Corollary 8.74 we get

$$(\Delta \langle Y, Y \rangle_{\tau_k}) 1_{\{\tau_k \leq \sigma_m\}} = \mathsf{E}[(\Delta Y^{[\sigma_m]})^2_{\tau_k} \mid \mathcal{F}_{\tau_k-}]$$

while from (8.6.19) it follows that

$$(\Delta \langle U^k, U^k \rangle_{\tau_k}) 1_{\{\tau_k \leq \sigma_m\}} = \mathsf{E}[(\Delta Y^{[\sigma_m]})^2_{\tau_k} \mid \mathcal{F}_{\tau_k-}].$$

Since $\langle U^j, U^j \rangle$ has a single jump at τ_j and since the graphs of $\{\tau_k : k \geq 1\}$ are disjoint, it follows that

$$\Delta \langle U, U \rangle_{\tau_k} = \Delta \langle U^k, U^k \rangle_{\tau_k}.$$

Thus we conclude that for all $k \geq 1$

$$\Delta \langle Y, Y \rangle_{\tau_k} = \Delta \langle U, U \rangle_{\tau_k}.$$

This implies $\langle Z, Z \rangle$ is continuous.   □

Remark **8.76** Note that in the decomposition (8.6.16), Z and U are such that $\langle U, U \rangle$ is purely discontinuous and $\langle Z, Z \rangle$ is continuous.

We will explore further structure of elements in $\mathbb{M}^2_{d, \mathrm{loc}}$ and conclude with identifying the continuous part $[X, X]^{(c)}$ of the quadratic variation of a semimartingale X as the quadratic variation of its continuous local martingale part, to be defined below.

Lemma 8.77 *Let $Z \in \mathbb{M}^2_{loc}$ be such that $\langle Z, Z \rangle$ is continuous. Then for any predictable stopping time σ,*

$$(\Delta Z)_\sigma = 0. \qquad (8.6.27)$$

Proof Let $\tau_k \uparrow \infty$ be stopping times such that $Z^k = Z^{[\tau_k]} \in \mathbb{M}^2$ and let $Y^k_t = (Z^k_t)^2 - \langle Z, Z \rangle_{t \wedge \tau_k}$. Then Y^k is a martingale and hence using Theorem 8.17 it follows that

$$\mathsf{E}[(\Delta Y^k)_{\sigma \wedge n}] = 0.$$

Since $\langle Z, Z \rangle$ is continuous, we conclude

$$\mathsf{E}[(Z^k_{\sigma \wedge n})^2 - (Z^k_{(\sigma \wedge n)-})^2] = 0.$$

Since $Z^k \in \mathbb{M}^2$, Lemma 8.73 now yields

$$\mathsf{E}[(Z^k_{\sigma \wedge n} - Z^k_{(\sigma \wedge n)-})^2] = 0.$$

Thus

$$\mathsf{E}[(\Delta Z)^2_{\sigma \wedge n \wedge \tau_k}] = 0$$

so that

$$\mathsf{P}((\Delta Z)^2_{\sigma \wedge n \wedge \tau_k} = 0) = 1.$$

Since this holds for all n and for all k and $\tau_k \uparrow \infty$, this completes the proof.   □

Lemma 8.78 *Let $Z \in \mathbb{M}^2_{loc}$ be such that $\langle Z, Z \rangle$ is continuous. Let σ be a stopping time and let $D = (\Delta Z)_\sigma I_{[\sigma, \infty)}$. Let A be the compensator of D, namely A is the unique predictable process in $\mathbb{V}$ such that $R = D - A$ is a local martingale. Then A is a continuous process, $R \in \mathbb{M}^2_{d, \mathrm{loc}}$ and*

$$[R, R]_t = (\Delta Z)_\sigma^2 I_{[\sigma, \infty)}(t) \tag{8.6.28}$$

Proof Since A is predictable and the discontinuities of A can be covered by countably many predictable stopping times (as seen in Theorem 8.22), it suffices to show that for any bounded predictable stopping time τ,

$$P((\Delta A)_\tau = 0) = 1.$$

So, fix a predictable stopping time τ. By Lemma 8.77, we have $(\Delta Z)_\tau = 0$. Since $D = R + A$ this gives

$$(\Delta A)_\tau = -(\Delta R)_\tau.$$

Thus $(\Delta R)_\tau$ is $\mathcal{F}_{\tau-}$ measurable (as A is predictable). On the other hand

$$E[(\Delta R)_\tau \mid \mathcal{F}_{\tau-}] = 0$$

by Theorem 8.17. Thus we get

$$(\Delta A)_\tau = 0$$

for all bounded predictable stopping times. Since A is predictable, this shows that A is continuous. Continuity of A gives $[R, R] = [D, D]$. $\square$

Theorem 8.79 *Let $Z \in \mathrm{M}^2_{loc}$ be such that $\langle Z, Z \rangle$ is continuous. Then Z admits a decomposition*

$$Z = M + R \tag{8.6.29}$$

where $M \in \mathrm{M}_{c, loc}$ and $R \in \mathrm{M}^2_{d, loc}$. Further, there exist stopping times $\{\sigma_j : j \geq 1\}$ such that

$$[Z, Z]_t = [M, M]_t + \sum_{j=1}^\infty (\Delta Z)_{\sigma_j}^2 I_{[\sigma_j, \infty)}(t). \tag{8.6.30}$$

Proof By Theorem 8.22, we can get stopping times $\{\sigma_j : j \geq 1\}$ that enumerates the jumps of Z. Let

$$V_t^j = (\Delta Z)_{\sigma_j} 1_{[\sigma_j, \infty)}(t), \tag{8.6.31}$$

and let A^j be the compensator of V^j and $R^j = Z^j - A^j$. By Lemma 8.78, A^j is a continuous process, $R^j \in \mathrm{M}^2_{d, loc}$ and

$$[R^j, R^j]_t = (\Delta Z)_{\sigma_j}^2 1_{[\sigma_j, \infty)}(t).$$

Since A^j, A^k are continuous, $[R^j, R^k] = [Z^j, Z^k] = 0$ for $j \neq k$. Let $W^n = \sum_{j=1}^n R^j$. It follows that

$$[W^n, W^n] = \sum_{j=1}^{n} [R^j, R^j]_t = \sum_{j=1}^{n} (\Delta Z)^2_{\sigma_j} 1_{[\sigma_j, \infty)}(t) \tag{8.6.32}$$

and

$$[Z - W^n, W^n] = 0. \tag{8.6.33}$$

Further, as seen in Theorem 8.72, W^n converges in Emery topology to $R \in \mathbb{M}^2_{d,\,loc}$ and

$$[W^n, W^n] \to [R, R]. \tag{8.6.34}$$

Theorem 4.1.1.1 along with $W^n \xrightarrow{em} R$ and (8.6.33) implies

$$[Z - R, R] = 0.$$

Defining $M = Z - R$, we conclude

$$[Z, Z] = [M, M] + [R, R]. \tag{8.6.35}$$

The relations (8.6.32) and (8.6.34) yield

$$[R, R] = \sum_{j=1}^{\infty} (\Delta Z)^2_{\sigma_j} 1_{[\sigma_j, \infty)}(t). \tag{8.6.36}$$

In turn, (8.6.35) and (8.6.36) together yield the validity of (8.6.30). Since $(\Delta[Z, Z])_s = (\Delta Z)^2_s$ and as the stopping times $\{\sigma_j : j \geq 1\}$ cover jumps of Z, (8.6.30) implies that $[M, M]$ is continuous and hence M is a continuous local martingale. This completes the proof. $\square$

Based on general considerations such as projections in a Hilbert space, we had shown that every square integrable local martingale X can be written as a sum of $M \in \mathbb{M}_{c,loc}$ and $N \in \mathbb{M}^2_{d,loc}$. Now we have a more concrete description of this decomposition.

Theorem 8.80 *Let $Y \in \mathbb{M}^2_{loc}$ be a locally square integrable martingale with $Y_0 = 0$. There exist predictable stopping times $\{\tau_k : k \geq 1\}$ and a sequence of stopping times $\{\sigma_j : j \geq 1\}$ such that U^k and V^j defined by*

$$U^k_t = (\Delta Y)_{\tau_k} I_{[\tau_k, \infty)}(t)$$

$$V^j_t = (\Delta Y)_{\sigma_j} I_{[\sigma_j, \infty)}(t)$$

satisfy the following.

(i) $U^k \in \mathbb{M}^2_{d,\,loc}$ for all $k \geq 1$.
(ii) $\sum_{k=1}^{m} U^k \xrightarrow{em} U \in \mathbb{M}^2_{d,\,loc}$.

(iii) For all $j \geq 1$, $\exists$ continuous process $A^j \in \mathbb{V}$ such that $R^j = V^j - A^j \in \mathbb{M}^2_{d, loc}$.

(iv) $\sum_{j=1}^{m} R^j \xrightarrow{em} R \in \mathbb{M}^2_{d, loc}$.

(v) $[U^j, U^k] = 0$ and $\langle U^j, U^k \rangle = 0$ for all $j, k \geq 1, j \neq k$.

(vi) $[R^j, R^k] = 0$ and $\langle R^j, R^k \rangle = 0$ for all $j, k \geq 1, j \neq k$.

(vii) $[R^j, U^k] = 0$ and $\langle R^j, U^k \rangle = 0$ for all $j, k \geq 1$.

(viii) $[U, U] = \sum_{k=1}^{\infty} [U^k, U^k]$.

(ix) $\langle U, U \rangle = \sum_{k=1}^{\infty} \langle U^k, U^k \rangle$.

(x) $[R, R] = \sum_{j=1}^{\infty} [R^j, R^j]$.

(xi) $\langle R, R \rangle = \sum_{j=1}^{\infty} \langle R^j, R^j \rangle$.

(xii) $N = R + U \in \mathbb{M}^2_{d, loc}$ and $M = Y - N \in \mathbb{M}^2_{c, loc}$.

(xii) $[N, N]_t = \sum_{0 < s \leq t} (\Delta Y)^2_s$.

(xiii) $[M, M]$ is continuous.

Thus $Y = M + N$ is a decomposition with $M \in \mathbb{M}^2_{c, loc}$ and $N \in \mathbb{M}^2_{d, loc}$.

Proof The proof is just putting together various parts proven in Theorem 8.75 and Theorem 8.79. First get $\{\tau_k : k \geq 1\}$ as in Theorem 8.75 that cover jumps of $\langle Y, Y \rangle$ and define U^k, U as above. Writing $Z = Y - U$, we conclude $\langle Z, Z \rangle$ is continuous. Now we get σ_j to cover jumps of Z. Since $\{\tau_k : k \geq 1\}$ are predictable, for all k, j,

$$\mathsf{P}(\tau_k = \sigma_j, \ \sigma_j < \infty) = 0$$

and hence $(\Delta U)_{\sigma_j} = 0$ and as a consequence, $(\Delta Z)_{\sigma_j} = (\Delta Y)_{\sigma_j}$. The rest now follows from Theorem 8.79. $\qquad\square$

Corollary 8.81 *Let $N \in \mathbb{M}^2_{d, loc}$. Then*

$$[N, N] = \sum_{0 < s \leq t} (\Delta N)^2_s.$$

Remark 8.82 In the decompositions $Y = M + N$ and $N = R + U$, with M, N, R, $U \in \mathbb{M}^2_{loc}$, various parts are characterized by the following :

$[M, M]$ and $\langle R, R \rangle$ are continuous, $[N, N]$ and $\langle U, U \rangle$ are purely discontinuous.

We now come to identify the continuous part $[X, X]^{(c)}$ of quadratic variation $[X, X]$ of a semimartingale X as the quadratic variation $[X^{(c)}, X^{(c)}]$ of the continuous local martingale part $X^{(c)}$ of X.

Theorem 8.83 *Let X be a semimartingale. Then X admits a decomposition*

$$X = X^{(c)} + S$$

such that $X^{(c)}$ is a continuous local martingale with $X_0^{(c)} = 0$ and

$$[U, S] = 0 \quad \text{for all continuous local martingales } U.$$

Further, such a decomposition is unique and

$$[X, X] = [X^{(c)}, X^{(c)}] + \sum_{0 < s \leq t} (\Delta X)_s^2.$$

As a consequence, $[X, X]^{(c)} = [X^{(c)}, X^{(c)}]$.

Proof First, we invoke Corollary 5.60 and decompose $X = Y + A$ where $Y \in \mathbb{M}^2_{\text{loc}}$ and $A \in \mathbb{V}$. Then we use Theorem 8.80, we get a decomposition $Y = M + N$ with $M \in \mathbb{M}^2_{c, \text{loc}}$ and $N \in \mathbb{M}^2_{d, \text{loc}}$. Let $X^{(c)} = M$ and $S = A + N$. Since $N \in \mathbb{M}^2_{d, \text{loc}}$,

$$[U, N] = 0 \quad \forall U \in \mathbb{M}^2_{c, \text{loc}}.$$

Of course $[U, A] = 0$ by Theorem 4.74.

Uniqueness follows easily : if $X = Z + R$ is any decomposition with $Z \in \mathbb{M}_{c, \text{loc}}$ such that $[U, R] = 0$ for all $U \in \mathbb{M}_{c, \text{loc}}$, then $W = X^{(c)} - Z = R - S$ is a continuous local martingale. Since $[W, S] = 0$ and $[W, R] = 0$ it follows that $[W, W] = 0$ and hence $W = 0$. $\qquad\square$

Chapter 9
The Davis Inequality

In this chapter, we would give the continuous-time version of the Burkholder–Davis–Gundy inequality $-p = 1$ case. This is due to Davis. This plays an important role in answering various questions on the stochastic integral w.r.t. a martingale M—including condition on $f \in \mathbb{L}(M)$ under which $\int f \, dM$ is a local martingale. This naturally leads us to the notion of a sigma-martingale which we discuss.

We will begin with a result on martingales obtained from process with a single jump.

9.1 Preliminaries

Lemma 9.1 *Let τ be a stopping time and let ξ be a $\mathcal{F}_\tau$ measurable $[0, \infty)$-valued integrable random variable. Let*

$$X_t = \xi \, 1_{[\tau, \infty)}(t)$$

Let A be the compensator of X and $M = X - A$.
Then for all $T < \infty$ we have

$$\mathsf{E}[\sqrt{[M, M]_T}] \leq 3\mathsf{E}[\xi]. \tag{9.1.1}$$

Proof Note that if σ is a bounded predictable stopping time, then

$$(\Delta A)_\sigma = \mathsf{E}[(\Delta X)_\sigma \mid \mathcal{F}_{\sigma-}]$$

© Springer Nature Singapore Pte Ltd. 2018
R. L. Karandikar and B. V. Rao, *Introduction to Stochastic Calculus*,
Indian Statistical Institute Series, https://doi.org/10.1007/978-981-10-8318-1_9

since $\mathsf{E}[(\Delta M)_\sigma \mid \mathcal{F}_{\sigma-}] = 0$ by Theorem 8.17 and $(\Delta A)_\sigma$ is $\mathcal{F}_{\sigma-}$ measurable by Theorem 8.4. Thus, for a bounded predictable stopping time σ we have (recall that by definition, $(\Delta X)_\tau = \xi \, 1_{\{\tau < \infty\}}$)

$$
\begin{aligned}
\mathsf{E}[\,|(\Delta A)_\sigma|\,] &= \mathsf{E}[\,|\mathsf{E}[(\Delta X)_\sigma \mid \mathcal{F}_{\sigma-}]|\,] \\
&\leq \mathsf{E}[\mathsf{E}[\,|(\Delta X)_\sigma|\mid \mathcal{F}_{\sigma-}]] \\
&= \mathsf{E}[\,|(\Delta X)_\sigma|\,] \\
&= \mathsf{E}[\,|\xi|\,1_{\{\tau<\infty\}}\,1_{\{\sigma=\tau\}}].
\end{aligned}
\tag{9.1.2}
$$

Given a $[0, \infty]$-valued stopping time σ, note that the set

$$
A = \{a \in \mathbb{R} : \mathsf{P}(\tau = a) > 0 \text{ or } \mathsf{P}(\sigma = a) > 0\}
$$

is countable. Thus, for $k \geq 1$, we choose $a_k \in (k, k+1) \cap A^c$. Let $\tau_k = \sigma \wedge a_k$. Then (9.1.2) gives us, for each $k \geq 1$

$$
\mathsf{E}[\,|(\Delta A)_{\tau_k}|\,] \leq \mathsf{E}[\,|\xi|\,1_{\{\tau<\infty\}}\,1_{\{\tau_k=\tau\}}].
\tag{9.1.3}
$$

Since

$$
|(\Delta A)_\sigma| \leq \liminf_{k\to\infty} (\Delta A)_{\tau_k}
$$

and

$$
|\xi|\,1_{\{\tau<\infty\}}\,1_{\{\tau_k=\tau\}} \uparrow |\xi|\,1_{\{\tau<\infty\}}\,1_{\{\sigma=\tau\}} \quad a.s.
$$

we can take limit as $k \to \infty$ in (9.1.3) and use Fatou's lemma on left-hand side and monotone convergence theorem on right-hand side of (9.1.3) to conclude

$$
\mathsf{E}[\,|(\Delta A)_\sigma|\,] \leq \mathsf{E}[\,|\xi|\,1_{\{\tau<\infty\}}\,1_{\{\sigma=\tau\}}].
\tag{9.1.4}
$$

Let σ_n be predictable stopping times with disjoint graphs such that

$$
\{(\Delta A) \neq 0\} = \cup_{m\geq 1}[\sigma_m]
$$

(existence of such stopping times was proven in Theorem 8.22). Recall that the graphs being disjoint means

$$
\mathsf{P}(\sigma_n = \sigma_m, \ \sigma_n < \infty) = 0 \quad \forall n, m, \ n \neq m.
\tag{9.1.5}
$$

Thus

$$
\begin{aligned}
\sqrt{[A, A]_T} &= \sqrt{\textstyle\sum_{m=1}^{\infty} (\Delta A)_{\sigma_m}^2} \\
&= \textstyle\sum_{m=1}^{\infty} |(\Delta A)_{\sigma_m}|.
\end{aligned}
$$

Noting that by Lemma 4.75, $[M, M]_T \leq 2([X, X]_T + [A, A]_T)$ and by definition of X, $[X, X]_T = \xi^2 \, 1_{\{\tau \leq T\}}$ we conclude that

$$
\begin{aligned}
\mathsf{E}[\sqrt{[M, M]_T}] &\leq \sqrt{2}\mathsf{E}[\sqrt{[X, X]_T} + \sqrt{[A, A]_T}] \\
&\leq \sqrt{2}\mathsf{E}[\xi] + \sqrt{2}\mathsf{E}[\sum_{m=1}^{\infty} |(\Delta A)_{\sigma_m}|] \\
&\leq \sqrt{2}\mathsf{E}[\xi] + \sqrt{2}\mathsf{E}[\sum_{m=1}^{\infty} 1_{\{\tau < \infty\}}\xi \, 1_{\{\sigma_m = \tau\}}] \\
&\leq \sqrt{2}\mathsf{E}[\xi] + \sqrt{2}\mathsf{E}[\xi]
\end{aligned}
\tag{9.1.6}
$$

where we have used (9.1.4) and (9.1.5). This completes the proof of (9.1.1). □

9.2 Burkholder–Davis–Gundy Inequality—Continuous Time

We will prove the $p = 1$ case of the Burkholder–Davis–Gundy inequality: for $1 \leq p < \infty$, there exist universal constants c_p^1, c_p^2 such that for all martingales M and $T < \infty$,

$$
c_p^1 \mathsf{E}[([M, M]_T)^{\frac{p}{2}}] \leq \mathsf{E}[\sup_{0 \leq t \leq T} |M_t|^p]
$$
$$
\leq c_p^2 \mathsf{E}[([M, M]_T)^{\frac{p}{2}}].
$$

We have given a proof for $p = 1$ in the discrete case, and here we will approximate the continuous-time martingale by its restriction to a discrete skeleton and then pass to the limit.

One inequality follows easily from the discrete case. For the other we first note it for the case of square integrable martingale and then later we will prove the same without this restriction.

Theorem 9.2 *Let c^1, c^2 be the universal constants appearing in Theorem 1.45. Let M be a martingale with $M_0 = 0$. Then*

$$
c^1 \mathsf{E}[([M, M]_T)^{\frac{1}{2}}] \leq \mathsf{E}[\sup_{0 \leq t \leq T} |M_t|].
\tag{9.2.1}
$$

Further, if $\mathsf{E}[M_T^2] < \infty$, then

$$
\mathsf{E}[\sup_{0 \leq t \leq T} |M_t|] \leq c^2 \mathsf{E}[([M, M]_T)^{\frac{1}{2}}].
\tag{9.2.2}
$$

Proof For $0 \leq k \leq 2^n$ and $n \geq 1$, let $t_{k,n} = \frac{Tk}{2^n}$ and let

$$Q^n = [\sum_{k=0}^{2^n-1} (M_{t_{k+1,n}} - M_{t_{k,n}})^2]^{\frac{1}{2}}$$

and

$$Z^n = \max_{1 \leq k \leq 2^n} |M_{t_{k,n}}|.$$

Also let

$$Q = \sqrt{[M, M]_T}.$$

Applying the discrete version of the inequality proven in Theorem 1.45, we have

$$c^1 \mathsf{E}[Q^n] \leq \mathsf{E}[Z^n] \leq c^2 \mathsf{E}[Q^n]. \tag{9.2.3}$$

We have seen in Theorem 4.64 that Q^n converges to Q in probability, and hence a subsequence Q^{n_k} converges almost surely. Then applying Fatou's lemma we conclude from (9.2.3) that

$$c^1 \mathsf{E}[Q] \leq c^1 \liminf_{k \to \infty} \mathsf{E}[Q^{n_k}] \leq \liminf_{k \to \infty} \mathsf{E}[Z^{n_k}].$$

Since Z^n increases to Z, $\mathsf{E}[Z^{n_k}]$ converges to $\mathsf{E}[Z]$ and thus (9.2.1) follows. Also we get from (9.2.3)

$$\mathsf{E}[Z] \leq c^2 \liminf_{k \to \infty} \mathsf{E}[Q^{n_k}].$$

If $\mathsf{E}[M_T^2] < \infty$, it follows that

$$\mathsf{E}[(Q^n)^2] = \mathsf{E}[M_T^2] < \infty.$$

Hence $\{Q^n : n \geq 1\}$ is uniformly integrable and thus Q^n converges to Q in $\mathbb{L}^1(\mathsf{P})$ and thus

$$\liminf_{k \to \infty} \mathsf{E}[Q^{n_k}] = \mathsf{E}[Q].$$

The inequality (9.2.2) follows.   □

Remark 9.3 If M is a locally square integrable martingale, then it satisfies (9.2.2). To see this let τ_n be stopping times increasing to ∞ such that $M_t^n = M_{t \wedge \tau_n}$ is a square integrable martingale. Then we have from the previous theorem that (9.2.2) holds for M^n. The desired conclusion follows by passing to the limit and invoking monotone convergence theorem.

Our aim is to show that (9.2.2) holds for all martingales. We first consider a special case and show that (9.2.2) holds in this case.

Lemma 9.4 *Let M be a martingale with $M_0 = 0$ and let σ be a stopping time bounded by T such that*

$$|M_t| \leq K \quad \forall t < \sigma. \tag{9.2.4}$$

and

$$M_t = M_{t \wedge \sigma} \quad \forall t. \tag{9.2.5}$$

Then $\mathsf{E}[\sqrt{[M, M]_T}\,] < \infty$ and (9.2.2) holds for M.

Proof Let $\xi = (\Delta M)_\sigma$. Since M is a martingale and σ is bounded, it follows that ξ is integrable. Further,

$$\sup_{0 \leq t \leq T} |M_t| \leq K + |\xi|$$

and hence $\sup_{0 \leq t \leq T} |M_t|$ is integrable. Hence $\mathsf{E}[\sqrt{[M, M]_T}\,] < \infty$ by Theorem 9.2. Let

$$A_t^j = \xi^+ \, 1_{\{|\xi| \geq j\}} \, 1_{[\sigma, \infty)}(t)$$

$$B_t^j = \xi^- \, 1_{\{|\xi| \geq j\}} \, 1_{[\sigma, \infty)}(t)$$

and let C^j, D^j be the compensators of A^j, B^j, respectively, i.e. predictable increasing processes (see Lemmas 8.43 and 9.1) such that

$$U_t^j = A_t^j - C_t^j$$

$$V_t^j = B_t^j - D_t^j$$

are martingales. Since $A_t^j = A_{t \wedge \sigma}^j$, it follows that $C_t^j = C_{t \wedge \sigma}^j$ and hence $U_t^j = U_{t \wedge \sigma}^j$. Likewise, $V_t^j = V_{t \wedge \sigma}^j$. As seen in Lemma 9.1, we have

$$\mathsf{E}[\sqrt{[U^j, U^j]_T}\,] \leq 3\mathsf{E}[\xi^+ \, 1_{\{|\xi| \geq j\}}] \tag{9.2.6}$$

$$\mathsf{E}[\sqrt{[V^j, V^j]_T}\,] \leq 3\mathsf{E}[\xi^- \, 1_{\{|\xi| \geq j\}}]. \tag{9.2.7}$$

Also, using Lemma 8.43, it follows that for all t

$$\mathsf{E}[\,|U_t^j|\,] \leq 2\mathsf{E}[\xi^+ \, 1_{\{|\xi| \geq j\}}] \tag{9.2.8}$$

$$\mathsf{E}[\,|V_t^j|\,] \leq 2\mathsf{E}[\xi^- \, 1_{\{|\xi| \geq j\}}] \tag{9.2.9}$$

Let

$$M_t^j = M_t - U_t^j + V_t^j.$$

Then $M_\sigma^j = M_\sigma - U_\sigma^j + V_\sigma^j$. Now

$$M_t^j = M_t - (A_t^j - B_t^j) + (C_t^j - D_t^j).$$

Now $|M_t - (A_t^j - B_t^j)| \leq K$ for $t < \sigma$ (as in that case $A_t^j = B_t^j = 0$). Also

$$M_\sigma - (A_\sigma^j - B_\sigma^j) = M_{\sigma-} + \xi - \xi^+ \, 1_{\{|\xi| \geq j\}} + \xi^- \, 1_{\{|\xi| \geq j\}}$$
$$= M_{\sigma-} + \xi \, 1_{\{|\xi| < j\}}.$$

Further,

$$M_t - (A_t^j - B_t^j) = M_{t \wedge \sigma} - (A_{t \wedge \sigma}^j - B_{t \wedge \sigma}^j).$$

Hence

$$|M_t - (A_t^j - B_t^j)| \leq K + j \quad \forall t.$$

For each j, C^j and D^j, being predictable r.c.l.l. processes, are locally bounded and hence it follows that M^j is locally bounded. Thus M^j is locally square integrable for each j. Thus by Remark 9.3, we have

$$\mathsf{E}[\sup_{0 \leq t \leq T} |M_t^j|] \leq c^2 \mathsf{E}[([M^j, M^j]_T)^{\frac{1}{2}}]. \tag{9.2.10}$$

In view of (9.2.8) and (9.2.9), $U_T^j \to 0$ and $V_T^j \to 0$ in $\mathbb{L}^1(\mathsf{P})$ and hence $M_T^j \to M_T$ in $\mathbb{L}^1(\mathsf{P})$. Thus, by Doob's maximal inequality, (2.3.6)

$$\mathsf{P}(\sup_{0 \leq t \leq T} |M_t^j - M_t| > \epsilon) \to 0.$$

By going through a subsequence and using Fatou's lemma, we conclude from (9.2.10) that

$$\mathsf{E}[\sup_{0 \leq t \leq T} |M_t|] \leq c^2 \liminf_{j \to \infty} \mathsf{E}[([M^j, M^j]_T)^{\frac{1}{2}}]. \tag{9.2.11}$$

Now $M - M^j = U^j - V^j$ and hence (using (4.6.13))

$$[M - M^j, M - M^j]_t \leq 2([U^j, U^j]_t + [V^j, V^j]_t)$$

Thus

$$\mathsf{E}[([M - M^j, M - M^j]_T)^{\frac{1}{2}}] \leq \sqrt{2}\mathsf{E}[([U^j, U^j]_T)^{\frac{1}{2}}] + \sqrt{2}\mathsf{E}[([V^j, V^j]_T)^{\frac{1}{2}}]$$
$$\leq 3\sqrt{2}c^2 \mathsf{E}[|\xi| \, 1_{\{|\xi| \geq j\}}] \tag{9.2.12}$$

where we have used (9.2.6), (9.2.7) and hence

$$\lim_{j \to \infty} \mathsf{E}[([M - M^j, M - M^j]_T)^{\frac{1}{2}}] = 0. \tag{9.2.13}$$

Using (4.6.21), we see that

$$[M - M^j, M - M^j]_T = [M, M]_T + [M^j, M^j]_T - 2[M, M^j]_T$$
$$\geq [M, M]_T + [M^j, M^j]_T - 2\sqrt{[M, M]_T[M^j, M^j]_T}$$
$$\geq (\sqrt{[M, M]_T} - \sqrt{[M^j, M^j]_T})^2.$$

(9.2.14)

Hence in view of (9.2.13), we conclude

$$\lim_{j \to \infty} \mathsf{E}[\sqrt{[M^j, M^j]_T}] = \mathsf{E}[\sqrt{[M, M]_T}].$$

This and (9.2.11) together imply that (9.2.2) holds for M. $\qquad\square$

We are now in a position to prove $p = 1$ case of Burkholder–Davis–Gundy inequality.

Theorem 9.5 *There exist universal constants c^1, c^2 such that for all local martingales M with $M_0 = 0$ and for all $T > 0$ one has*

$$c^1 \mathsf{E}[([M, M]_T)^{\frac{1}{2}}] \leq \mathsf{E}[\sup_{0 \leq t \leq T} |M_t|] \leq c^2 \mathsf{E}[([M, M]_T)^{\frac{1}{2}}].$$

(9.2.15)

Proof Let $\{\tau_n : n \geq 1\}$ be stopping times increasing to ∞ such that $M_{t \wedge \tau_n}$ is a martingale. For $n \geq 1$ let

$$\theta_n = \inf\{t \geq 0 : |M_t| \geq n \text{ or } |M_{t-}| \geq n\}$$

and let $\sigma_n = \tau_n \wedge \theta_n \wedge n$. Let

$$N_t^n = M_{t \wedge \sigma_n}.$$

Then N^n is a martingale and satisfies the conditions of Lemma 9.4 with $\sigma = \sigma_n$, $K = n$, $T = n$ and hence N^n satisfies (9.2.2). We have already noted that (9.2.1) holds for N^n in Theorem 9.2. Thus we have

$$c^1 \mathsf{E}[([N^n, N^n]_T)^{\frac{1}{2}}] \leq \mathsf{E}[\sup_{0 \leq t \leq T} |N_t^n|] \leq c^2 \mathsf{E}[([N^n, N^n]_T)^{\frac{1}{2}}] < \infty.$$

(9.2.16)

As $n \to \infty$, $([N^n, N^n]_T)^{\frac{1}{2}}$ increases to $([M, M]_T)^{\frac{1}{2}}$ and $\sup_{0 \leq t \leq T} |N_t^n|$ increases to $\sup_{0 \leq t \leq T} |M_t|$ and thus (9.2.15) follows from (9.2.16) using monotone convergence theorem. The constants c^1, c^2 are the universal constants as in Theorem 9.2 and do not depend upon n or M. $\qquad\square$

Definition 9.6 A martingale M is said to be a $\mathcal{H}^1$-martingale if

$$\mathsf{E}[\sup_{0 \leq t < \infty} |M_t|] < \infty$$

(9.2.17)

Remark **9.7** In view of the Burkholder–Davis–Gundy inequality, it follows that M is a $\mathcal{H}^1$-martingale if and only if

$$\mathsf{E}[\sup_{0 \le t < \infty} \sqrt{[M, M]_t}] < \infty \qquad (9.2.18)$$

It also follows that if M is a $\mathcal{H}^1$-martingale and if f is a bounded predictable process, then $N = \int f \, dM$ is also a $\mathcal{H}^1$-martingale.

During the proof of Theorem 9.5 above we have shown the following:

Corollary 9.8 *Let M be a local martingale. Then there exist stopping times σ_n increasing to ∞ such that for all $n \ge 1$*

$$M^{[\sigma_n]} \in \mathcal{H}^1, \qquad (9.2.19)$$

$$\mathsf{E}[\sqrt{[M, M]_{\sigma_n}}] < \infty \quad \forall n \ge 1 \qquad (9.2.20)$$

and

$$\mathsf{E}[\sup_{0 \le t \le \sigma_n} |M_t|] < \infty \quad \forall n \ge 1. \qquad (9.2.21)$$

Corollary 9.9 *Let M be a local martingale. For any stopping time σ, one has*

$$c^1 \mathsf{E}[([M, M]_\sigma)^{\frac{1}{2}}] \le \mathsf{E}[\sup_{0 \le t \le \sigma} |M_t|] \le c^2 \mathsf{E}[([M, M]_\sigma)^{\frac{1}{2}}]. \qquad (9.2.22)$$

Corollary 9.10 *If M is a local martingale and σ is a stopping time such that*

$$\mathsf{E}[([M, M]_\sigma)^{\frac{1}{2}}] < \infty] \qquad (9.2.23)$$

then it follows that $\mathsf{E}[\sup_{0 \le t \le \sigma} |M_t|] < \infty$ and hence $N_t = M_{t \wedge \sigma}$ is a martingale.

Here is a consequence of Theorem 9.5 that will be needed later.

Theorem 9.11 *Let X be a martingale such that*

$$\mathsf{E}[\sup_{0 \le t \le T} |X_t|] < \infty \quad \forall T < \infty \qquad (9.2.24)$$

or equivalently, such that

$$\mathsf{E}[([X, X]_T)^{\frac{1}{2}}] < \infty \quad \forall T < \infty.$$

Then there exists a sequence of bounded martingales Z^k such that

$$\lim_{k \to \infty} \mathsf{E}[\sup_{0 \le t \le T} |Z_t^k - X_t|] = 0 \quad \forall T < \infty. \qquad (9.2.25)$$

Proof We will show that for all $k \geq 1$, there exists a bounded martingale Z^k such that

$$E[\sup_{0 \leq t \leq k} |Z_t^k - X_t|] \leq \frac{1}{k}. \tag{9.2.26}$$

The required result follows from this.

Thus we fix an integer $k < \infty$. Let σ_n be the stopping times constructed in the proof of Theorem 9.5 with $M = X$ and N^n denote the martingale X stopped at σ_n. Since $\sup_{0 \leq t \leq T} |N_t^n|$ increases to $\sup_{0 \leq t \leq T} |X_t|$, we can get integer n such that $Y = N^n$ satisfies

$$E[\sup_{0 \leq t \leq k} |Y_t - X_t|] \leq \frac{1}{3k}. \tag{9.2.27}$$

As noted in the proof of Theorem 9.5, N^n and hence Y satisfy the conditions of Lemma 9.4. Thus for $M = Y$, we can get locally bounded martingales M^j such that (9.2.13) holds, i.e.

$$\lim_{j \to \infty} E[([Y - M^j, Y - M^j]_k)^{\frac{1}{2}}] = 0. \tag{9.2.28}$$

Now using Burkholder–Davis–Gundy inequality Theorem 9.5, we can get j such that $W = M^j$ satisfies

$$E[\sup_{0 \leq t \leq k} |W_t - Y_t|] \leq \frac{1}{3k}. \tag{9.2.29}$$

Finally, W being locally bounded, we can get stopping times τ_n increasing to ∞ such that U^n given by $U_t^n = W_{t \wedge \tau_n}$ is a bounded martingale and

$$E[\sup_{0 \leq t \leq k} |U_t^n - W_t|] \to 0 \text{ as } n \to \infty$$

and hence can get n such that $Z = U^n$ satisfies

$$E[\sup_{0 \leq t \leq k} |Z_t - W_t|] \leq \frac{1}{3k}. \tag{9.2.30}$$

Now (9.2.28)–(9.2.30) together imply (9.2.26) with $Z^k = Z$. $\square$

Remark **9.12** The martingales Z^k obtained in the theorem also satisfy

$$\lim_{k \to \infty} E[([Z^k - X, Z^k - X]_T)^{\frac{1}{2}}] \to 0 \quad \forall T < \infty. \tag{9.2.31}$$

This follows from the Burkholder–Davis–Gundy inequality (Theorem 9.5).

9.3   On Stochastic Integral w.r.t. a Martingale

For a local martingale M, the stochastic integral $Y = \int f \, dM$ for $f \in \mathbb{L}(M)$ is defined (since M is also a stochastic integrator), and we have only observed that when M is locally square integrable martingale and $f \in \mathbb{L}_m^2(M)$, Y is also a locally square integrable martingale. We now explore as to when is Y a martingale or a local martingale. We begin with an observation.

Theorem 9.13 *Let M be a martingale such that $\mathsf{E}[([M, M]_T)^{\frac{1}{2}}] < \infty \, \forall T < \infty$ and f be a bounded predictable process. Then $N = \int f \, dM$ is also a martingale and $\mathsf{E}[([N, N]_T)^{\frac{1}{2}}] < \infty \, \forall T < \infty$.*

Proof If f is bounded by c, and $N = \int f \, dM$ (interpreted as a stochastic integral w.r.t. stochastic integrator M), then N satisfies

$$[N, N]_T = \int_0^T |f|^2 d[M, M]_s \le c^2 [M, M]_T$$

and hence $\mathsf{E}[([N, N]_T)^{\frac{1}{2}}] < \infty \, \forall T < \infty$. Let $\mathbb{A}$ be the class of bounded predictable process f such that $N = \int f \, dM$ is a martingale. It is easy to see that simple predictable processes belong to $\mathbb{A}$. If $g^n \in \mathbb{A}$ and $g^n \xrightarrow{bp} g$, then writing $N^n = \int g^n \, dM$ and $N = \int g \, dM$, we see that

$$\mathsf{E}[\,(\int_0^T |g_s^n - g_s|^2 d[M, M]_s)^{\frac{1}{2}}\,] \to 0 \text{ as } n \to \infty.$$

Hence

$$\mathsf{E}[\sqrt{[N^n - N, N^n - N]_T}\,] \to 0 \text{ as } n, \to \infty$$

and as a consequence, (using (9.2.15))

$$\mathsf{E}[\sup_{0 \le t \le T} |N_s^n - N_s|\,] \to 0 \text{ as } n \to \infty.$$

Thus N is a martingale. Thus $\mathbb{A}$ is closed under bp-convergence, and hence by Theorem 2.66, it follows that $\mathbb{A}$ is the class of all bounded predictable processes completing the proof.   □

By localizing, we immediately conclude that

Corollary 9.14 *Let M be a local martingale and f be a locally bounded predictable process. Then $Y = \int f \, dM$ is also a local martingale.*

As noted earlier, for an r.c.l.l. adapted process X, X^- defined by $X_s^- = X_{s-}$ is a locally bounded predictable process. Hence we conclude from the corollary above

Corollary 9.15 *Let M be a local martingale and X be an r.c.l.l. adapted process. Then $Y = \int X^- dM$ is also a local martingale.*

Earlier we have defined $\mathbb{L}_m^2(M)$ for a locally square integrable martingale M. We now define $\mathbb{L}_m^1(M)$ for a local martingale M.

Definition 9.16 For a local martingale M, $\mathbb{L}_m^1(M)$ is the class of predictable processes f such that there exist stopping times σ_n increasing to ∞ with

$$\mathsf{E}[(\int_0^{\sigma_n} f_s^2 d[M, M]_s)^{\frac{1}{2}}] < \infty. \tag{9.3.1}$$

Theorem 9.17 *Let M be a local martingale. Then $\mathbb{L}_m^1(M) \subseteq \mathbb{L}(M)$ and for $f \in \mathbb{L}_m^1(M)$, $N = \int f dM$ is a local martingale.*

Proof Let $f \in \mathbb{L}_m^1(M)$ be such that (9.3.1) holds. We first show that $f \in \mathbb{L}(M)$. Let g^k be bounded predictable processes converging pointwise to g such that $|g^k| \le |f|$.

For $k \ge 1$, let $Y^k = \int g^k dM$. Then we have seen that Y^k is a local martingale. From properties of stochastic integrators, we have

$$[Y^k, Y^k]_t = \int_0^t (g^k)^2 d[M, M]$$

and hence

$$\mathsf{E}[\sup_{0 \le t \le \sigma_n} |Y_s^k|] \le c^2 \mathsf{E}[(\int_0^{\sigma_n} (f_s)^2 d[M, M]_s)^{\frac{1}{2}}] < \infty \tag{9.3.2}$$

where σ_n is as in (9.3.1). It thus follows that $U_t^{k,n} = Y_{t \wedge \sigma_n}^k$ is a martingale for each k, n. Moreover, it follows that for $k \ge j \ge 1$

$$[Y^k - Y^j, Y^k - Y^j]_t = \int_0^t (g^k - g^j)^2 d[M, M].$$

Hence, using (9.2.22), we get

$$\mathsf{E}[\sup_{0 \le t \le \sigma_n} |Y^k - Y^j|] \le c^2 \mathsf{E}[(\int_0^{\sigma_n} (g_s^k - g_s^j)^2 d[M, M]_s)^{\frac{1}{2}}].$$

The right-hand side above goes to 0 as k, j tend to ∞ in view of the assumption (9.3.1) and choice of g^k (using Lebesgue's dominated convergence theorem). Thus we have for each $n \ge 1$

$$\lim_{m \to \infty} (\sup_{j,k \ge m} \mathsf{E}[\sup_{0 \le t \le \sigma_n} |Y_t^k - Y_t^j|]) = 0. \tag{9.3.3}$$

Thus $\{Y^k\}$ are Cauchy in $\mathbf{d}_{ucp}$ and hence $f \in \mathbb{L}(M)$. Let Y be the limit of Y^k. By dominated convergence theorem, we get $Y = \int g dM$. We also get from (9.3.3) that

$$\lim_{k\to\infty} \mathsf{E}[\sup_{0\le t\le\sigma_n} |Y_t^k - Y_t|] = 0. \tag{9.3.4}$$

Noting that $U_t^{k,n} = Y_{t\wedge\sigma_n}^k$ is a martingale for each k, n, (9.3.4) implies that $U_t^n = Y_{t\wedge\sigma_n}$ is a martingale. Thus Y is a local martingale.

To show that N is a local martingale, let us take $g^k = f \, 1_{\{|f|\le k\}}$. The process Y for this choice of $\{g^k\}$ equals N which has been shown to be a local martingale completing the proof. $\qquad\square$

Corollary 9.18 *Let M be a local martingale and $f \in \mathbb{L}(M)$ be such that*

$$\mathsf{E}[\,|f_\sigma(\Delta M)_\sigma|\,] < \infty \ \forall \ bounded\ stopping\ times\ \sigma. \tag{9.3.5}$$

Then $f \in \mathbb{L}_m^1(M)$ and $Z = \int f\,dM$ is a local martingale.

Proof For $n \ge 1$, let

$$\sigma_n = \inf\{s : (s + \int_{[0,s)} f_u^2 d[M, M]_u) \ge n\}.$$

Then $f \in \mathbb{L}(M)$ implies that $\sigma_n \uparrow \infty$. Of course $\sigma_n \le n$ and for $t < \sigma_n$, $\int_0^t f_s^2 d[M, M]_s \le n$. Thus,

$$\sqrt{\int_0^{\sigma_n} f_s^2 d[M, M]_s} \le \sqrt{n + f_{\sigma_n}^2 (\Delta M)_{\sigma_n}^2} \le \sqrt{n} + |f_{\sigma_n}(\Delta M)_{\sigma_n}|$$

and thus in view of the assumption (9.3.5) on M,

$$\mathsf{E}[\sqrt{\int_0^{\sigma_n} f_s^2 d[M, M]_s}] < \infty$$

and thus $f \in \mathbb{L}_m^1(M)$. The second part follows from Theorem 9.17. $\qquad\square$

Corollary 9.19 *Let M be a local martingale and $f \in \mathbb{L}(M)$ be such that*

$$Z_t = \int_0^t f_s\,dM_s$$

is bounded. Then Z is a martingale.

Proof Since Z is bounded, f satisfies (9.3.5). Thus by Corollary 9.18, Z is a local martingale. Since it is bounded, it follows that Z is a martingale. $\qquad\square$

Corollary 9.20 *Let M be a continuous local martingale. Then*

$$\mathbb{L}(M) = \mathbb{L}_m^1(M) \tag{9.3.6}$$

Proof Since M is continuous, (9.3.5) is trivially satisfied and hence the result follows from Theorem 9.17 and Corollary 9.18. □

The Burkholder–Davis–Gundy inequality helps us to conclude the converse to Theorem 9.17:

Theorem 9.21 *Let M be a local martingale and $f \in \mathbb{L}(M)$ and let $N = \int f dM$ be a local martingale. Then $f \in \mathbb{L}_m^1(M)$.*

Proof Since N is a local martingale and $[N, N]_t = \int_0^t f_s^2 d[M, M]_s$, Corollary 9.8 implies $f \in \mathbb{L}_m^1(M)$. □

9.4 Sigma-Martingales

We have seen that if M is a local martingale and $f \in \mathbb{L}_m^1(M)$, then $X = \int f dM$ is a local martingale. On the other hand if $f \in \mathbb{L}(M)$ but does not belong to $\mathbb{L}_m^1(M)$ then X is defined and is a semimartingale but it is not a local martingale. Nonetheless, it shares some properties of a local martingale and is called a sigma-martingale.

Definition 9.22 A semimartingale X is said to be a sigma-martingale if there exists a local martingale N and $f \in \mathbb{L}(N)$ such that $X = \int f dN$.

If X is a sigma-martingale with f, N as in the definition above and $g \in \mathbb{L}(X)$, then $Y = \int g dX = \int g f dN$ and hence Y is also a sigma-martingale. Here is an elementary observation.

Lemma 9.23 *Let X be a semimartingale. Then X is a sigma-martingale if and only if there exists a $(0, \infty)$-valued predictable process ϕ such that $\phi \in \mathbb{L}(X)$ and $M = \int \phi dX$ is a $\mathcal{H}^1$-martingale.*

Proof If such a M, ϕ exist, then $\psi = \frac{1}{\phi} \in \mathbb{L}(M)$ and $X = \int \psi dM$.

For the converse part, suppose N is a local martingale, $X = \int f dN$ with $f \in \mathbb{L}(N)$. Then taking $g = (1 + |f|)^{-1}$, we observe that $\int g dX = \int f g dN$. Since N is a local martingale and fg is bounded by 1, invoking Corollary 9.14 we conclude that $Y = \int f g dN$ is itself a local martingale. As seen in Corollary 9.8, there exist stopping times σ_n increasing to ∞ such that

$$a_n = E[\sqrt{[Y, Y]_{\sigma_n}}] < \infty.$$

Let h be the predictable process defined by

$$h_s = \frac{1}{1 + |Y_0|} 1_{\{0\}}(s) + \sum_{n=1}^{\infty} 2^{-n} \frac{1}{1 + a_n} 1_{(\sigma_{n-1}, \sigma_n]}(s).$$

Then h is $(0, 1)$ valued and thus $M = \int h\,dY$ is a local martingale. Since

$$
\begin{aligned}
[M, M]_{\sigma_n} &= \int_0^{\sigma_n} h_s^2 d[Y, Y]_s \\
&= \sum_{j=1}^n \int_{\sigma_{j-1}}^{\sigma_j} h_s^2 d[Y, Y]_s \\
&\leq \sum_{j=1}^n 4^{-j} \frac{1}{(1 + a_j)^2} [Y, Y]_{\sigma_j}
\end{aligned}
$$

Thus

$$
\sqrt{[M, M]_{\sigma_n}} \leq \sum_{j=1}^n 2^{-j} \frac{1}{(1 + a_j)} \sqrt{[Y, Y]_{\sigma_j}}.
$$

From the choice of a_j, it now follows that

$$
E[\sqrt{[M, M]_{\sigma_n}}] \leq 1 \ \forall n \geq 1
$$

and as a consequence

$$
E[\sup_{t < \infty} \sqrt{[M, M]_t}] \leq 1.
$$

Hence, M is a martingale and $M \in \mathcal{H}^1$. Let $\phi = hg$. Then ϕ is $(0, \infty)$ valued and $\int \phi\,dX = M$.   □

From the definition, it is not obvious that sum of sigma-martingales is also a sigma-martingale, but this is so as the next result shows.

Theorem 9.24 Let X^1, X^2 be sigma-martingales and a_1, a_2 be real numbers. Then $Y = a_1 X^1 + a_2 X^2$ is also a sigma-martingale.

Proof Let ϕ^1, ϕ^2 be $(0, \infty)$-valued predictable processes such that

$$
M_t^i = \int_0^t \phi_s^i dX_s^i, \quad i = 1, 2
$$

are martingales. Then, writing $\xi = \min(\phi^1, \phi^2)$ and $\eta_s^i = \frac{\xi_s}{\phi_s^i}$, it follows that

$$
N_t^i = \int_0^t \eta_s^i dM_s^i = \int_0^t \xi_s dX_s^i
$$

are martingales since η^i is bounded by one. Clearly, $Y = a_1 X^1 + a_2 X^2$ is a semi-martingale and $\xi \in \mathbb{L}(X^i)$ for $i = 1, 2$ implies $\xi \in \mathbb{L}(Y)$ and

$$\int_0^t \xi_s \, dY_s = a_1 N_t^1 + a_2 N_t^2$$

is a martingale. Since ξ is $(0, \infty)$-valued predictable process, it follows that Y is a sigma-martingale. □

The following result gives conditions under which a sigma-martingale is a local martingale.

Lemma 9.25 *Let X be a sigma-martingale with $X_0 = 0$. Suppose there exists a sequence of stopping times $\tau_n \uparrow \infty$ such that*

$$\mathsf{E}[\sqrt{[X, X]_{\tau_n}}] < \infty \ \forall n. \tag{9.4.1}$$

Then X is a local martingale.

Proof Let N be a local martingale and $f \in \mathbb{L}(N)$ be such that $X = \int f \, dN$. Note that

$$[X, X]_t = \int_0^t (f_s)^2 d[N, N]_s. \tag{9.4.2}$$

Let

$$X_t^k = \int_0^t f_s \, \mathbf{1}_{\{|f_s| \le k\}} dN_s. \tag{9.4.3}$$

Noting that $f_s \, \mathbf{1}_{\{|f_s| \le k\}}$ is bounded, it follows that X^k is a local martingale. Since

$$\mathsf{E}[\sqrt{[X^k, X^k]_{t \wedge \tau_n}}] \le \mathsf{E}[\sqrt{\int_0^{\tau_n} f_s^2 \, \mathbf{1}_{\{|f_s| \le k\}} d[N, N]_s}]$$
$$\le \mathsf{E}[\sqrt{[X, X]_{\tau_n}}]$$
$$< \infty$$

we conclude that for k, n fixed, $Z_t^{k,n} = X_{t \wedge \tau_n}^k$ is a martingale. Let $Z_t^n = X_{t \wedge \tau_n}$. Clearly, for $t > 0$

$$\mathsf{E}[\sqrt{[X - X^k, X - X^k]_{t \wedge \tau_n}}] \le \mathsf{E}[\sqrt{\int_0^{\tau_n} f_s^2 \, \mathbf{1}_{\{|f_s| > k\}} d[N, N]_s}]. \tag{9.4.4}$$

The assumption (9.4.1) and the estimate (9.4.4) imply that for n fixed,

$$\lim_{k \to \infty} \mathsf{E}[\sqrt{[X - X^k, X - X^k]_{t \wedge \tau_n}}] = 0. \tag{9.4.5}$$

The Burkholder–Davis–Gundy inequality Corollary 9.9 now gives

$$\lim_{k \to \infty} \mathsf{E}[\sup_{s \le t} |Z_t^{k,n} - Z_t^n|] = 0. \tag{9.4.6}$$

Since $Z^{k,n}$ is a martingale for each k, n, (9.4.6) implies Z^n is a martingale and thus X is a local martingale.   □

Corollary 9.26 *Let X be a sigma-martingale. Suppose that X_0 is integrable and that there exists a sequence of stopping times $\tau_n \uparrow \infty$ such that (9.4.1) holds then X is a local martingale.*

Proof Let $Y_t = X_t - X_0$. Observe that $[Y, Y]_t = [X, X]_t$ for all t. Now the previous result gives that Y is a local martingale and hence X is a local martingale.   □

Corollary 9.27 *A bounded sigma-martingale X is a martingale.*

Proof Since X is bounded, say by K, it follows that jumps of X are bounded by $2K$. Thus jumps of the increasing process $[X, X]$ are bounded by $4K^2$ and thus X satisfies (9.4.1) for

$$\tau_n = \inf\{t \geq 0 : [X, X]_{t-} \geq n\}.$$

Hence X is a local martingale and being bounded, it is a martingale.   □

Exercise 9.28 Let X be a sigma-martingale. Suppose $|X_t| \leq \xi$ where ξ is an integrable random variable. Show that X is a martingale.

Here is a variant of the example given by Emery [17] of a sigma-martingale that is not a local martingale.

Example **9.29** Let τ, ξ be independent random variables with τ having exponential distribution and $P(\xi = 1) = P(\xi = -1) = 0.5$. Without loss of generality, we assume that $0 < \tau(\omega) < \infty$ for all ω. Let

$$M_t = \xi \, \mathbf{1}_{[\tau, \infty)}(t)$$

and $\mathcal{F}_t = \sigma(M_s : s \leq t)$. Easy to see that M is a martingale. Let $f_t = \frac{1}{t} \mathbf{1}_{(0,\infty)}(t)$ and $X_t = \int_0^t f \, dM$. Then X is a sigma-martingale and

$$[X, X]_t = \frac{1}{\tau^2} \, \mathbf{1}_{[\tau, \infty)}(t).$$

For any stopping time σ, it can be checked that σ is a constant on $\sigma < \tau$ and thus if σ is not identically equal to 0, $\sigma \geq (\tau \wedge a)$ for some $a > 0$. Thus, $\sqrt{[X, X]_\sigma} \geq \frac{1}{\tau} \mathbf{1}_{\{\tau < a\}}$. It follows that for any stopping time σ, not identically zero, $E[\sqrt{[X, X]_\sigma}] = \infty$ and so X is not a local martingale.

9.5   Auxiliary Results

We have seen that if M, N are locally square integrable martingales, then $MN - [M, N]$ is a local martingale. We now show that the same is true for all local martingales M, N.

Theorem 9.30 *Let M, N be local martingales with $M_0 = 0$. Let $X_t = M_t N_t - [M, N]_t$. Then X is a local martingale.*

Proof The integration by parts formula (4.6.7) gives

$$X_t = \int_0^t M_{s-} dN_s + \int_0^t N_{s-} dM_s$$

and by Corollary 9.15 it follows that X is a local martingale. $\qquad\square$

Lemma 9.31 *Let N be martingale such that $E[\sup_{0 \le t \le T} |N_t|] < \infty$ for all $T < \infty$. Let $A \in \mathbb{V}$ be a predictable process with $A_0 = 0$. Suppose for each $T < \infty$, there is $K_T < \infty$ such that $|A|_T \le K_T$. Here $|A|_T$ is the variation of A on $[0, T]$. Then $Y_t = A_t N_t - \int_0^t N_{s-} dA_s$ is a martingale.*

Proof Invoking Theorem 9.11, obtain bounded martingales N^k such that

$$\lim_{k \to \infty} E[\sup_{0 \le t \le T} |N_t^k - N_t|] = 0 \quad \forall T < \infty.$$

Let $Y_t^k = A_t N_t^k - \int_0^t N_{s-}^k dA_s$. Note that

$$E[\sup_{0 \le t \le T} |Y_t^k|] \le 2 K_T C_k \qquad (9.5.1)$$

where C_k is a bound for N^k. Also,

$$E[\sup_{0 \le t \le T} |Y_t^k - Y_t|] \le 2 K_T E[\sup_{0 \le t \le T} |N_t^k - N_t|] \qquad (9.5.2)$$

By integration by parts formula (4.6.7),

$$Y_t^k = \int_0^t A_{s-} dN_s^k + [A, N^k].$$

The integral appearing above is a local martingale by Corollary 9.15, and $[A, N^k]$ is a martingale by Theorem 8.34. Thus Y^k is a local martingale. Lemma 5.5 along with the observation (9.5.1) implies that Y^k is a martingale for each k and then (9.5.2) along with Theorem 2.23 shows that Y is a martingale. $\qquad\square$

Theorem 9.32 *Let M be local martingale such that $M_0 = 0$ and let $A \in \mathbb{V}$ be a predictable process with $A_0 = 0$. Then $Y_t = A_t M_t - \int_0^t M_{s-} dA_s$ is a local martingale.*

Proof Since $A \in \mathbb{V}$ is predictable, so is its total variation process $B = |A|$ (see Corollary 8.24). Thus B is locally bounded, and we can get stopping times $\tau^n \uparrow \infty$ such that B^{τ^n} is bounded. Invoking Corollary 9.8, get stopping times $\sigma^n \uparrow \infty$ such that $M^n = M^{[\sigma^n]}$ are martingales satisfying, for each $n \ge 1$

$$E[\sup_{0 \le t \le T} |M_t^n|] < \infty \ \forall T < \infty.$$

Without loss of generality (by replacing σ^n by $\min(\sigma^n, \tau^n)$ if necessary), we can assume that $\sigma^n \le \tau^n$ and so $B^n = B^{\sigma^n}$ is bounded for each n. Let $A^n = A^{\sigma^n}$. It follows that $Y^n = Y^{[\sigma^n]}$ satisfies

$$Y_t^n = A_t^n M_t^n - \int_0^t M_{s-}^n dA_s^n.$$

By Lemma 9.31, Y^n is a martingale for each n and thus Y is a local martingale. $\square$

As a consequence of Theorem 9.17, we have the following observation.

Lemma 9.33 *Let M, N be local martingales such that $M_0 N_0$ is integrable. Then the process Z defined by $Z_t = M_t N_t$ is locally integrable if and only if $[M, N]$ is locally integrable.*

Proof The integration by parts formula gives

$$M_t N_t = M_0 N_0 + \int_0^t M_{s-} dN_s + \int_0^t N_{s-} dM_s + [M, N]_s.$$

By corollary 9.15, the stochastic integral terms in the right-hand side above are local martingales and thus the result follows. $\square$

Remark **9.34** If M, N are local martingales such that MN is locally integrable, then so is $[M, N]$ and thus $\langle M, N \rangle$ exists and is the unique predictable process in $\mathbb{V}_0$ such that

$$M_t N_t - \langle M, N \rangle_t$$

is a local martingale.

Chapter 10
Integral Representation of Martingales

In this chapter we will consider the question as to when do all martingales adapted to a filtration $(\mathcal{F}_.)$ admit a representation as a stochastic integral with respect to a given local martingale M. This result was proved by Ito's when the underlying filtration is the filtration generated by a multidimensional Wiener process. Ito's had proven the integral representation property for square integrable martingales and this was extended to all martingales by Clark.

Jacod and Yor investigated this aspect and proved that the integral representation property holds if and only if there does not exist any other probability measure Q equivalent to the underlying probability measure with the property that M is a Q—local martingale. Such a measure Q is called an *Equivalent Martingale Measure* (EMM). In other words, martingale representation property holds if and only if EMM is unique. Jacod–Yor proved this result in one dimension and in a special case for multidimensional local martingale, which was subsequently extended.

This result is important from the point of view of mathematical finance. We will give brief introduction to the same and prove the second fundamental theorem of asset pricing.

10.1 Preliminaries

Throughout this chapter, we will be working with one fixed filtration $(\mathcal{F}_.)$ such that $\mathcal{F}_0$ contains all null sets. We do not assume that the filtration is right continuous. All notions—martingale, local martingale, stopping time, adapted process, predictable process—are with reference to this fixed filtration. Since in this chapter, we need to deal with martingales which may not have r.c.l.l. paths a priori, we will explicitly assume r.c.l.l. paths when it is needed.

© Springer Nature Singapore Pte Ltd. 2018   321
R. L. Karandikar and B. V. Rao, *Introduction to Stochastic Calculus*,
Indian Statistical Institute Series, https://doi.org/10.1007/978-981-10-8318-1_10

For r.c.l.l.semimartingales $X^1, X^2, \ldots, X^d$, we introduce the class of semimartingales that admit integral representation w.r.t. $X^1, X^2, \ldots, X^d$:

$$\mathbb{I}(X^1, X^2, \ldots, X^d)$$

$$= \{Y : \exists g^j \in \mathbb{L}(X^j), \ 1 \leq j \leq d \ \text{with} \ Y_t = Y_0 + \sum_{j=1}^{d} \int_0^t g^j \, dX^j \ \forall t\}$$

Let us note that if $Y \in \mathbb{I}(X^1, X^2, \ldots, X^d)$ then for any stopping time τ, $\tilde{Y}$ defined by $\tilde{Y}_t = Y_{t \wedge \tau}$ also belongs to $\mathbb{I}(X^1, X^2, \ldots, X^d)$. Also, if $Y \in \mathbb{I}(X^1, X^2, \ldots, X^d)$ then we can always choose $g^j \in \mathbb{L}(X^j)$ with $g_0^j = 0$ for $1 \leq j \leq d$ such that

$$Y_t = Y_0 + \sum_{j=1}^{d} \int_0^t g^j \, dX^j \ \forall t < \infty.$$

If $Y \in \mathbb{I}(X^1, X^2, \ldots, X^d)$, the semimartingale Y is said to have an integral representation w.r.t. semimartingales $X^1, X^2, \ldots, X^d$. Here is an elementary observation on the class $\mathbb{I}(X^1, X^2, \ldots, X^d)$.

Lemma 10.1 *Let Y be a semimartingale such that for a sequence of stopping times $\tau_n \uparrow \infty$, Y^n defined by $Y_t^n = Y_{t \wedge \tau_n}$ admits an integral representation w.r.t. r.c.l.l. semimartingales $X^1, X^2, \ldots, X^d$ for each $n \geq 1$. Then Y also admits an integral representation w.r.t. $X^1, X^2, \ldots, X^d$.*

Proof Let $f^{n,j} \in \mathbb{L}(X^j), 1 \leq j \leq d, n \geq 1$ be such that for all n,

$$Y_t^n = Y_0^n + \sum_{j=1}^{d} \int_0^t f^{n,j} \, dX^j.$$

Define f^j by

$$f^j = \sum_{n=1}^{\infty} 1_{(\tau_{n-1}, \tau_n]} f^{n,j}.$$

Then it is easy to check (using Theorem 4.43) that $f^j \in \mathbb{L}(X^j)$ and

$$Y_t = Y_0 + \sum_{j=1}^{d} \int_0^t f^j \, dX^j.$$

This completes the proof.   □

Exercise 10.2 Let Y be an r.c.l.l.process such that it is a local martingale under probability measure Q_1 as well as Q_2. Show that Y is a local martingale under $Q = \frac{1}{2}(Q_1 + Q_2)$.

Given r.c.l.l. adapted processes $X^1, X^2, \ldots, X^d$, let $\mathbb{E}(X^1, X^2, \ldots, X^d)$ denote the class of probability measures Q on $(\Omega, \mathcal{F})$ such that $X^1, X^2, \ldots, X^d$ are Q-local martingales.

For a probability measure P on $(\Omega, \mathcal{F})$, let $\mathbb{E}_P(X^1, X^2, \ldots, X^d)$ denote the class of measures $Q \in \mathbb{E}(X^1, X^2, \ldots, X^d)$ such that Q is equivalent to P and let $\widetilde{\mathbb{E}}_P(X^1, X^2, \ldots, X^d)$ denote the class of measure $Q \in \mathbb{E}(X^1, X^2, \ldots, X^d)$ such that Q is absolutely continuous w.r.t. P. It is easy to see that for any $X^1, X^2, \ldots, X^d$, the sets $\mathbb{E}(X^1, X^2, \ldots, X^d)$, $\mathbb{E}_P(X^1, X^2, \ldots, X^d)$ and $\widetilde{\mathbb{E}}_P(X^1, X^2, \ldots, X^d)$ are convex.

Elements of $\mathbb{E}_P(X^1, X^2, \ldots, X^d)$ are referred to as EMM- equivalent martingale measures, though they should be called equivalent local martingale measures.

Likewise let $\mathbb{E}^\sigma(X^1, X^2, \ldots, X^d)$ be the class of probability measures Q on $(\Omega, \mathcal{F})$ such that $X^1, X^2, \ldots, X^d$ are sigma-martingales on $(\Omega, \mathcal{F}, Q)$ and $\mathbb{E}_P^\sigma(X^1, X^2, \ldots, X^d)$ denote the class of measure $Q \in \mathbb{E}^\sigma(X^1, X^2, \ldots, X^d)$ such that Q is equivalent to P. $\mathbb{E}_P^\sigma(X^1, X^2, \ldots, X^d)$ is the class of equivalent σ-martingale measures.

Jacod and Yor discovered a connection between extreme points P of $\mathbb{E}(X)$ and the martingale representation property w.r.t. X. This was later generalized to multidimensions under suitable conditions. We will first deal with the one-dimensional case and then take up multidimensional case and prove integral representation theorem for multidimensional σ-martingales. This necessitates definition of *vector stochastic integral*. We will also discuss relevance of integral representation theorem to mathematical finance.

10.2 One-Dimensional Case

In this section, we will fix a local martingale M and explore as to when $\mathbb{I}(M)$ contains all martingales. The next lemma gives an important property of $\mathbb{I}(M)$.

Lemma 10.3 *Let M be an r.c.l.l. local martingale and let $N^n \in \mathbb{I}(M)$ be martingales such that $\mathsf{E}[\,|N^n_t - N_t|\,] \to 0$ for all t. Then $N \in \mathbb{I}(M)$.*

Proof The assumptions imply that N is a martingale (see Theorem 2.23). In view of Theorem 5.39 and the assumptions on N^n, N, it follows that

$$N^n \text{ converges to } N \text{ in Emery topology.}$$

Thus, invoking Theorem 4.111, we conclude that

$$[N^n - N, N^n - N]_T \to 0 \text{ in probability as } n \to \infty. \tag{10.2.1}$$

Hence, using $[N^n - N^m, N^n - N^m]_T \le 2([N^n - N, N^n - N]_T + [N^m - N, N^m - N]_T)$ (see (4.6.13)), we have

$$[N^n - N^m, N^n - N^m]_T \to 0 \text{ in probability as } n, m \to \infty. \tag{10.2.2}$$

Since $N^n \in \mathbb{I}(M)$, there exists predictable process $g^n \in \mathbb{L}(M)$ such that

$$N_t^n = N_0^n + \int_0^t g^n \, dM. \tag{10.2.3}$$

As a consequence, for all $T < \infty$,

$$[N^n - N^m, N^n - N^m]_T = \int_0^T (g_s^n - g_s^m)^2 d[M, M]_s.$$
$$\to 0 \quad \text{in probability as } n, m \to \infty.$$

By taking a subsequence, if necessary and relabelling, we assume that for $1 \leq k \leq n$,

$$P((\int_0^k (g_s^n - g_s^k)^2 d[M, M]_s)^{\frac{1}{2}} \geq \frac{1}{2^k}) \leq \frac{1}{2^k}. \tag{10.2.4}$$

Then by Borel–Cantelli Lemma, we conclude

$$\sum_{k=1}^{\infty} (\int_0^T (g_s^{k+1} - g_s^k)^2 d[M, M]_s)^{\frac{1}{2}} < \infty \quad a.s. \tag{10.2.5}$$

for all $T < \infty$. Since

$$[\int_0^T (\sum_{k=1}^{\infty} |g_s^{k+1} - g_s^k|)^2 d[M, M]_s]^{\frac{1}{2}} \leq \sum_{k=1}^{\infty} (\int_0^T (g_s^{k+1} - g_s^k)^2 d[M, M]_s)^{\frac{1}{2}}$$

we conclude that for all $T < \infty$

$$[\int_0^T (\sum_{k=1}^{\infty} |g_s^{k+1} - g_s^k|)^2 d[M, M]_s] < \infty \quad a.s. \tag{10.2.6}$$

and also

$$\lim_{k \to \infty} \int_0^T \sup_{m, n \geq k} |g_s^m - g_s^n|^2 d[M, M]_s = 0 \quad a.s. \tag{10.2.7}$$

Let $\widetilde{\Omega} = [0, \infty) \times \Omega$ and $\widetilde{\mathcal{F}}$ be the product of $\mathcal{F}$ and the Borel σ-field on $[0, \infty)$. Let Γ be the (σ-finite) measure on $(\widetilde{\Omega}, \widetilde{\mathcal{F}})$ defined by, for $E \in \widetilde{\mathcal{F}}$

$$\Gamma(E) = \int [\int_0^{\infty} 1_E(s, \omega) d[M, M]_s(\omega)] dP(\omega). \tag{10.2.8}$$

Now (10.2.6) implies that

$$\sum_{k=1}^{\infty} |g_s^{k+1}(\omega) - g_s^k(\omega)| < \infty \quad a.e. \ \Gamma.$$

Thus g^k converges *a.e.* Γ. Let

$$g_s(\omega) = \limsup_{k \to \infty} g_s^k(\omega).$$

Using (10.2.7), one can conclude that for all $t < \infty$, for all $\epsilon > 0$

$$\lim_{k \to \infty} P\left(\left(\int_0^t (g_s^k - g_s)^2 d[M, M]_s\right)^{\frac{1}{2}} \geq \epsilon\right) = 0. \qquad (10.2.9)$$

and

$$\int_0^t (g_s^k)^2 d[M, M]_s \to \int_0^t (g_s)^2 d[M, M]_s \text{ in probability.}$$

On the other hand convergence of N^n to N in Emery topology and Theorem 4.111 imply for all $t < \infty$

$$\int_0^t (g_s^k)^2 d[M, M]_s = [N^k, N^k]_t \to [N, N]_t \text{ in probability.}$$

Thus

$$[N, N]_t = \int_0^t (g_s)^2 d[M, M]_s. \qquad (10.2.10)$$

Now N being a martingale, as seen in Corollary 9.8, there exist stopping times σ_n increasing to ∞ such that

$$E[\sqrt{[N, N]_{\sigma_n}}] < \infty \ \forall n \geq 1 \qquad (10.2.11)$$

and hence (10.2.10) implies that $g \in \mathbb{L}_m^1(M)$. Let $Y_t = N_0 + \int_0^t g \, dM$. By definition, Y is a local martingale with $Y_0 = N_0$. Now

$$[N^n - Y, N^n - Y]_T = \int_0^T (g_s^n - g_s)^2 d[M, M]_s$$

and thus as seen in (10.2.9),

$$[N^n - Y, N^n - Y]_T \to 0 \text{ in probability} \qquad (10.2.12)$$

On the other hand N^n converges to N in Emery topology and thus invoking Theorem 4.111 we conclude that

$$[N^n - Y, N^n - Y]_T \to [N - Y, N - Y]_T \text{ in probability.} \qquad (10.2.13)$$

Thus $[N - Y, N - Y]_T = 0$ and recalling that N, Y are local martingales such that $N_0 = Y_0$, it follows (once again invoking Burkholder-Davis-Gundy inequality for

$p = 1$) that $N = Y$. Hence $N_t = N_0 + \int_0^t f \, dM$ with $f \in \mathbb{L}(M)$. This proves $N \in \mathbb{I}(M)$.   □

Let us recall that $\mathbb{M}$ denotes the class of r.c.l.l. martingales.

Theorem 10.4 *For an r.c.l.l. local martingale M and $T < \infty$ let*

$$\mathbb{K}_T(M) = \{N_T : \ N \in \mathbb{I}(M) \cap \mathbb{M}\}. \tag{10.2.14}$$

Then $\mathbb{K}_T(M)$ is a closed linear subspace of $\mathbb{L}^1(\Omega, \mathcal{F}_T, \mathsf{P})$.

Proof Let $N^n \in \mathbb{I}(M) \cap \mathbb{M}$ be such that N_T^n converges in $\mathbb{L}^1(\Omega, \mathcal{F}_T, \mathsf{P})$ to ξ. Without loss of generality, we assume that $N_t^n = N_{t \wedge T}^n$ for all $n \geq 1$ and for all $t < \infty$. Now for each n, N^n is a uniformly integrable martingale. It follows that for all t

$$\mathsf{E}[|N_t^n - N_t^m|] \to 0 \text{ as } n, m \to \infty. \tag{10.2.15}$$

Thus, by Theorem 5.39, it follows that $\{N^n\}$ is Cauchy in the $\mathbf{d}_{em}$ metric for the Emery topology. Since this metric is complete, the sequence N^n converges in the Emery topology to say N. Then N^n also converges in $\mathbf{d}_{ucp}$ to N and in view of (10.2.15), we conclude $N_T = \xi$ and for each t

$$\mathsf{E}[|N_t^n - N_t|] \to 0 \text{ as } n \to \infty. \tag{10.2.16}$$

Thus Lemma 10.3 implies $N \in \mathbb{I}(M)$ and thus $N_T = \xi \in \mathbb{K}_T(M)$.   □

Exercise 10.5 Show that $\xi \in \mathbb{K}_T(M)$ if and only if $\xi \in \mathbb{L}^1(\Omega, \mathcal{F}_T, \mathsf{P})$ and there exist $\eta \in \mathbb{L}^1(\Omega, \mathcal{F}_0, \mathsf{P})$ and $g \in \mathbb{L}_m^1(M)$ such that $\xi = \eta + \int_0^T g \, dM$.

We now come to the main result of this section, due to Jacod–Yor [29]. This characterizes martingales M with property that all martingales N admit an integral representation w.r.t. M.

Definition 10.6 A process Y is said to admit an integral representation w.r.t. an r.c.l.l. semimartingale X if

$$\exists f \in \mathbb{L}(X) \text{ such that } Y_t = Y_0 + \int_0^t f_s \, dX_s \ \text{ a.s. } \forall t. \tag{10.2.17}$$

Note that if Y is a process that admits a representation w.r.t. an r.c.l.l. semimartingale X, then Y has r.c.l.l. modification since the stochastic integral is by definition an r.c.l.l. process.

Here is an important observation on integral representation.

Lemma 10.7 *Let M be an r.c.l.l. local martingale. Then all martingales N admit a representation w.r.t. M if and only if*

$$\mathbb{K}_T(M) = \mathbb{L}^1(\Omega, \mathcal{F}_T, \mathsf{P}) \ \ \forall T < \infty. \tag{10.2.18}$$

Proof Suppose all martingales N admit a representation w.r.t. M. Given $\xi \in \mathbb{L}^1(\Omega, \mathcal{F}_T, \mathsf{P})$, consider the martingale $N_t = \mathsf{E}[\xi \mid \mathcal{F}_t]$. Note that N may not be r.c.l.l. to begin with. In view of our assumption, get $f \in \mathbb{L}(M)$ such that

$$N_t = N_0 + \int_0^t f \, dM \quad a.s. \ \forall t.$$

This implies that $V_t = N_0 + \int_0^t f \, dM$ is an r.c.l.l. martingale and is a version of N. Thus by definition of $K_T(M)$, it follows that $N_T = \xi \in \mathbb{K}_T(M)$.

Conversely, if (10.2.18) holds, then given a martingale N, fix n and let $\xi = N_n$. Then $\xi \in \mathbb{K}_n(M)$ and so we get $f^n \in \mathbb{L}(X)$ such that $\xi = N_n = N_0 + \int_0^n f^n \, dM$ and further that $Z^n = \int f^n \, dM$ is a martingale. It follows that $Z_t^n = N_t - N_0$ $a.s.$ for $t \le n$. Let

$$f_s = \sum_n f_s^n \, 1_{(n-1, n]}.$$

Then one can check that $f \in \mathbb{L}(M)$, $Z = \int f \, dM$ is a martingale and $Z_t = N_t - N_0$ $a.s.$ for all t. Hence N admits a representation w.r.t. M. $\qquad\square$

Essentially the same proof also gives us the following.

Corollary 10.8 *Let M be an r.c.l.l. local martingale. Then all bounded martingales N admit a representation w.r.t. M if and only if*

$$\mathbb{L}^\infty(\Omega, \mathcal{F}_T, \mathsf{P}) \subseteq \mathbb{K}_T(M) \quad \forall \, T < \infty. \tag{10.2.19}$$

Theorem 10.9 *Let M be an r.c.l.l. local martingale on $(\Omega, \mathcal{F}, \mathsf{P})$ with a filtration $(\mathcal{F}_.)$. Suppose that $\mathcal{F}_0$ is trivial and $\mathcal{F} = \sigma(\cup_t \mathcal{F}_t)$. Then the following are equivalent.*

(i) Every bounded martingale N admits an integral representation w.r.t. M.
(ii) Every martingale N admits an integral representation w.r.t. M.
(iii) P is an extreme point of the convex set $\mathbb{E}(M)$.
(iv) $\widetilde{\mathbb{E}}_\mathsf{P}(M) = \{\mathsf{P}\}$.
(v) $\mathbb{E}_\mathsf{P}(M) = \{\mathsf{P}\}$.

Proof We have seen that (i) is same as $\mathbb{L}^\infty(\Omega, \mathcal{F}_T, \mathsf{P}) \subseteq \mathbb{K}_T(M) \, \forall T \in (0, \infty)$ and (ii) is same as $\mathbb{L}^1(\Omega, \mathcal{F}_T, \mathsf{P}) = \mathbb{K}_T(M) \, \forall T \in (0, \infty)$. As seen in Theorem 10.4, $\mathbb{K}_T(M)$ is a closed subspace of $\mathbb{L}^1(\Omega, \mathcal{F}_T, \mathsf{P})$. Since $\mathbb{L}^\infty(\Omega, \mathcal{F}_T, \mathsf{P})$ is dense in $\mathbb{L}^1(\Omega, \mathcal{F}_T, \mathsf{P})$, it follows that (i) and (ii) are equivalent.

On the other hand, suppose (iv) holds and suppose $\mathsf{Q}_1, \mathsf{Q}_2 \in \mathbb{E}(M)$ and $\mathsf{P} = \alpha \mathsf{Q}_1 + (1 - \alpha)\mathsf{Q}_2$. It follows that $\mathsf{Q}_1, \mathsf{Q}_2$ are absolutely continuous w.r.t. P and hence $\mathsf{Q}_1, \mathsf{Q}_2 \in \widetilde{\mathbb{E}}_\mathsf{P}(M)$. In view of (iv), $\mathsf{Q}_1 = \mathsf{Q}_2 = \mathsf{P}$ and thus P is an extreme point of $\mathbb{E}(M)$ and so (iii) is true. Thus $(iv) \Rightarrow (iii)$.

Since $\{\mathsf{P}\} \subseteq \mathbb{E}_\mathsf{P}(M) \subseteq \widetilde{\mathbb{E}}_\mathsf{P}(M)$, it follows that (iv) implies (v).

On the other hand, suppose (v) is true and $\mathsf{Q} \in \widetilde{\mathbb{E}}_\mathsf{P}(M)$. Then $\mathsf{Q}_1 = \frac{1}{2}(\mathsf{Q} + \mathsf{P}) \in \mathbb{E}_\mathsf{P}(M)$. Then (v) implies $\mathsf{Q}_1 = \mathsf{P}$ and hence $\mathsf{Q} = \mathsf{P}$. Thus $(v) \Rightarrow (iv)$ holds.

Till now we have proved $(i) \Longleftrightarrow (ii)$ and $(iii) \Longleftarrow (iv) \Longleftrightarrow (v)$. To complete the proof, we will show $(i) \Rightarrow (v)$ and $(iii) \Rightarrow (ii)$.

First we come to the proof of $(iii) \Rightarrow (ii)$. Suppose P is an extreme point of $\mathbb{E}(M)$ but (ii) is not true. We will show that this leads to a contradiction. Since (ii) is not true, it follows that $\mathbb{K}_T(M)$ is a closed proper subspace of $\mathbb{L}^1(\Omega, \mathcal{F}, P)$. Since $\mathbb{K}_T(M)$ is not equal to $\mathbb{L}^1(\Omega, \mathcal{F}_T, P)$, by the Hahn–Banach Theorem (see [55]), there exists $\xi \in \mathbb{L}^\infty(\Omega, \mathcal{F}_T, P)$, $P(\xi \neq 0) > 0$ such that

$$\int \theta \xi dP = 0 \quad \forall \theta \in \mathbb{K}_T(M).$$

Then for $c \in \mathbb{R}$, we have

$$\int \theta(1 + c\xi)dP = \int \theta dP \quad \forall \theta \in \mathbb{K}_T(M). \tag{10.2.20}$$

Since ξ is bounded, we can choose a $c > 0$ such that

$$P(c|\xi| < 0.5) = 1.$$

Now, let Q be the measure with density $\eta = (1 + c\xi)$. Then Q is a probability measure. Thus (10.2.20) yields

$$\int \theta dQ = \int \theta dP \quad \forall \theta \in \mathbb{K}_T(M). \tag{10.2.21}$$

Let $\sigma_n \uparrow \infty$ be bounded stopping times such that $M_t^n = M_{t \wedge \sigma_n}$ is a P-martingale. For any bounded stopping time τ, $M_{\tau \wedge T}^n = M_{\tau \wedge \sigma_n \wedge T} \in \mathbb{K}_T$ and hence (remembering that $\mathcal{F}_0$ is trivial)

$$E_Q[M_{\tau \wedge T}^n] = E_P[M_{\tau \wedge T}^n] = M_0 \tag{10.2.22}$$

On the other hand,

$$\begin{aligned}
E_Q[M_{\tau \vee T}^n] &= E_P[\eta M_{\tau \vee T}^n] \\
&= E_P[E_P[\eta M_{\tau \vee T}^n \mid \mathcal{F}_T]] \\
&= E_P[\eta E_P[M_{\tau \vee T}^n \mid \mathcal{F}_T]] \\
&= E_P[\eta M_T^n] \\
&= E_Q[M_T^n] \\
&= E_P[M_T^n] \\
&= M_0.
\end{aligned} \tag{10.2.23}$$

where we have used the facts that η is $\mathcal{F}_T$ measurable, M^n is a P-martingale and that (10.2.22) holds for $\tau = T$. Now noting that $M_\tau^n = M_{\tau \wedge T}^n + M_{\tau \vee T}^n - M_T^n$, we conclude

$$E_Q[M_\tau^n] = E_Q[M_{\tau \wedge T}^n] + E_Q[M_{\tau \vee T}^n] - E_Q[M_T^n] = M_0.$$

Thus $M_t^n = M_{t \wedge \sigma_n}$ is a Q-martingale for every n so that M is a Q-local martingale and thus $Q \in \mathbb{E}(M)$. Similarly, if $\tilde{Q}$ is the measure with density $\eta = (1 - c\xi)$, we can prove that $\tilde{Q} \in \mathbb{E}(M)$. Here $P = \frac{1}{2}(Q + \tilde{Q})$ and $P \neq Q$ (since $P(\xi \neq 0) > 0$). This contradicts the assumption that P is an extreme point of $\mathbb{E}(M)$. Thus $(iii) \Rightarrow (ii)$.

To complete the proof we will show that (i) implies (v). Suppose (i) is true and let $Q \in \mathbb{E}_P(M)$. Fix $T < \infty$ and let η be any $\mathcal{F}_T$ measurable bounded random variable. Since $\mathbb{L}^\infty(\Omega, \mathcal{F}_T, P) \subseteq \mathbb{K}_T(M)$ and $\mathcal{F}_0$ is trivial, we can get $g \in \mathbb{L}(M)$ with

$$\eta = c + \int_0^T g \, dM$$

such that $\int_0^t g \, 1_{[0,T]} dM$ is a P-martingale and $c = \mathbb{E}_P[\eta]$.

Let $Z_t = \int_0^t g_s \, 1_{(0,T]}(s) dM_s$. Then $Z_t = \mathbb{E}_P[(\eta - c) \mid \mathcal{F}_t]$ and since η is bounded, it follows that Z is bounded. As noted earlier, since P and Q are equivalent, the stochastic integrals under P and Q are identical. Under Q, M being a local martingale and $Z = \int f \, dM$ being bounded, we conclude invoking Corollary 9.19 that Z is also a martingale under Q. Thus, $\mathbb{E}_Q[Z_T] = 0 = \mathbb{E}_P[Z_T]$ and thus using $\eta = c + Z_T$ we get $\mathbb{E}_Q[\eta] = c = \mathbb{E}_P[\eta]$. Since this holds for all $\mathcal{F}_T$ measurable bounded random variables η, we conclude Q and P agree on $\mathcal{F}_T$. In view of the assumption $\mathcal{F} = \sigma(\cup_t \mathcal{F}_t)$, we get $Q = P$ proving (v). This completes the proof. $\qquad\square$

We can now deduce the integral representation property for Brownian motion, due to Ito's [25] and Clark [10].

Theorem 10.10 *Let W be one-dimensional Brownian motion and let $\mathcal{F}_t = \mathcal{F}_t^W$ and $\mathcal{F} = \sigma(\cup_t \mathcal{F}_t^W)$. Then every martingale M w.r.t. the filtration $(\mathcal{F}_.)$ admits an integral representation*

$$M_t = M_0 + \int_0^t f \, dW, \quad \forall t \geq 0 \qquad (10.2.24)$$

for some $f \in \mathbb{L}(W)$.

Proof We will prove that $\mathbb{E}_P(W) = \{P\}$. The conclusion then would follow from Theorem 10.9. If $Q \in \mathbb{E}_P(W)$, then by definition, W is a Q-local martingale and $[W, W]_t^Q = t$ since Q is equivalent to P and $[W, W]_t^P = t$, see Remark 4.81. Now part (v) in Theorem 5.19 implies that W_t and $W_t^2 - t$ are Q-local martingales and then Levy's characterization of Brownian motion, Theorem 3.7 implies that W is a Brownian motion under Q. Thus for $t_1, t_2, \ldots, t_m \in [0, \infty)$ and $B \in \mathcal{B}(\mathbb{R}^m)$,

$$Q((W_{t_1}, W_{t_2}, \ldots, W_{t_m}) \in B) = P((W_{t_1}, W_{t_2}, \ldots, W_{t_m}) \in B). \qquad (10.2.25)$$

Hence $P = Q$ since $\mathcal{F} = \sigma(W_s : s \in [0, \infty))$. Thus $\mathbb{E}_P(W) = \{P\}$. $\qquad\square$

10.3   Quasi-elliptical Multidimensional Semimartingales

The d-dimensional version of Theorem 10.9 is not true in general—the implication (iii) implies (ii) may not be true. The difficulty is this: given a sequence of martingales $\{N^n : n \geq 1\}$ such that N_T^n is converging in $\mathbb{L}^1$ (say to N_T) for every $T < \infty$, and $\{g^{n,j} : n \geq 1\}$ such that

$$N_t^n = N_0^n + \sum_{j=1}^{d} \int_0^t g^{n,j} dM^j$$

we cannot conclude that the sequence of integrands $g^{n,j}$ is converging as was the case in one-dimensional. See counter example in [29]. This prompts us to introduce a condition under which the class of martingales that admit integral representation is closed under $\mathbb{L}^1$ convergence.

For r.c.l.l. semimartingales $X^1, X^2, \ldots, X^d$ and $\lambda^1, \ldots, \lambda^d \in \mathbb{R}$, defining $Y = \sum_{j=1}^{d} \lambda^j X^j$, note that

$$[Y, Y]_t = \sum_{i,j=1}^{d} \lambda^i \lambda^j [X^i, X^j]_t$$

and hence

$$\sum_{i,j=1}^{d} \lambda^i \lambda^j ([X^i, X^j]_t - [X^i, X^j]_s) \geq 0 \quad a.s. \tag{10.3.1}$$

In other words, for $s < t$ fixed, the matrix $(([X^i, X^j]_t - [X^i, X^j]_s))$ is non-negative definite.

Definition 10.11 A d-dimensional r.c.l.l. semimartingale $X = (X^1, \ldots, X^d)$ is said to be *quasi-elliptic* if there exists a sequence of stopping times $\tau_n \uparrow \infty$ and constants $\alpha_n > 0$ such that $\forall \lambda^1, \ldots, \lambda^d \in \mathbb{R}$, $\forall s < t \leq \tau_n$, one has

$$\sum_{i,j=1}^{d} \lambda^i \lambda^j ([X^i, X^j]_t - [X^i, X^j]_s) \geq \alpha_n^2 \sum_{i=1}^{d} (\lambda^i)^2 ([X^i, X^i]_t - [X^i, X^i]_s) \quad a.s.$$

$$\tag{10.3.2}$$

Remark **10.12** Note that if $X = (X^1, X^2, \ldots, X^d)$ is a quasi-elliptic semimartingale on $(\Omega, \mathcal{F}, \mathsf{P})$ and Q is a probability measure absolutely continuous w.r.t. P, then X continues to be a quasi-elliptic semimartingale on $(\Omega, \mathcal{F}, \mathsf{Q})$.

Example **10.13** Let $X = (X^1, X^2, \ldots, X^d)$ be a semimartingale such that $[X^i, X^j] = 0$ for $i \neq j$. Then trivially, X is a quasi-elliptic semimartingale. This is the case when X is d-dimensional standard Brownian motion.

Example **10.14** Let $X = (X^1, X^2, \ldots, X^d)$ be the solution to the SDE (3.51) with $m = d$ where σ, b satisfy (3.5.3) and (3.5.4). Further, suppose that $\exists \alpha > 0$ such that for all $t \geq 0$, $x \in \mathbb{R}^d$, $\lambda_1, \lambda_2, \ldots, \lambda_d$

$$\sum_{i,j=1}^{d} \lambda_i \lambda_j \sigma_{ij}(t, x) \geq \alpha \sum_{i=1}^{d} \lambda_i^2. \qquad (10.3.3)$$

Then it is easy to verify that $X = (X^1, X^2, \ldots, X^d)$ is a quasi-elliptic semimartingale.

Lemma 10.15 *Let $X = (X^1, X^2, \ldots, X^d)$ be a quasi-elliptic semimartingale. Then for all $h^j \in \mathbb{L}(X^j)$, $1 \leq j \leq d$, one has*

$$\sum_{j=1}^{d} \int_0^t (h_s^j)^2 d[X^j, X^j]_s \leq \frac{1}{\alpha_n^2} \sum_{i,j=1}^{d} \int_0^t h_s^i h_s^j d[X^i, X^j]_s \ \forall t \leq \tau_n \ a.s. \qquad (10.3.4)$$

where τ_n and α_n are as in (10.3.2).

Proof Clearly, the assumption (10.3.2) implies that (10.3.4) is true for simple predictable processes $h^1, h^2, \ldots, h^d \in \mathbb{S}$. Now fixing $h^2, \ldots, h^d \in \mathbb{S}$, the class of h^1 for which (10.3.4) is true is seen to be bp-closed and hence by monotone class theorem, (Theorem 2.66) contains all bounded predictable processes. Similarly, assuming that (10.3.4) is true for $h^1, \ldots, h^j$ bounded predictable and $h^{j+1}, \ldots, h^d \in \mathbb{S}$, we can show that the same is true for $h^1, \ldots, h^{j+1}$ bounded predictable and $h^{j+2}, \ldots, h^d \in \mathbb{S}$. Thus by induction we conclude that (10.3.4) holds when $h^1, h^2, \ldots, h^d$ are bounded.

Now note that by the Kunita–Watanabe inequality (Theorem 4.80) and Remark 4.87, the right-hand side in (10.3.4) is finite *a.s.* for $h^j \in \mathbb{L}(X^j)$, $1 \leq j \leq d$. Let $\eta = \sum_{j=1}^{d} |h^j|^2$ and for $1 \leq j \leq d$ and $n \geq 1$, let

$$h^{n,j} = h^j \, 1_{\{\eta \leq n\}}.$$

Using (10.3.1), it follows that

$$\sum_{i,j=1}^{d} \int_0^t h_s^{n,i} h_s^{n,j} d[X^i, X^j]_s \text{ increases to } \sum_{i,j=1}^{d} \int_0^t h_s^i h_s^j d[X^i, X^j]_s$$

and also easy to see that

$$\sum_{i=1}^{d} \int_0^t (h_s^{n,j})^2 d[X^j, X^j]_s \text{ increases to } \sum_{i=1}^{d} \int_0^t (h_s^j)^2 d[X^j, X^j]_s.$$

Thus validity of (10.3.4) for $\{h^{n,j} : 1 \leq j \leq d\}$ for all $n \geq 1$ implies validity of (10.3.4) for $\{h^j \in \mathbb{L}(X^j) : 1 \leq j \leq d\}$.   □

Corollary 10.16 *Let $X = (X^1, X^2, \ldots, X^d)$ be a quasi-elliptic semimartingale. Suppose $g^{n,j} \in \mathbb{L}(X^j)$, $1 \leq j \leq d$, $n \geq 1$ are such that $\forall t < \infty$*

$$\sum_{i,j=1}^{d} \int_0^t (g_s^{n,i} - g_s^{m,i})(g_s^{n,j} - g_s^{m,j}) d[X^i, X^j]_s \to 0 \quad \text{in probability as } n, m \to \infty.$$

Then $\forall t < \infty$

$$\sum_{i,j=1}^{d} \int_0^t (g_s^{n,i} - g_s^{m,i})^2 d[X^i, X^i]_s \to 0 \quad \text{in probability as } n, m \to \infty.$$

Proof This follows from Lemmas 2.75 and 10.15.   □

We have noted that if $X^1, X^2, \ldots, X^d$ are r.c.l.l. semimartingales such that $[X^i, X^j] = 0$ for $i \neq j$ then $X = (X^1, X^2, \ldots, X^d)$ is quasi-elliptic semimartingale. In particular, if $X^1, X^2, \ldots, X^d$ are continuous local martingales such that $X^i X^j$ is also a local martingale for $i \neq j$, then $X = (X^1, X^2, \ldots, X^d)$ is quasi-elliptic local martingale.

Here is the analogue of Lemma 10.3 in multidimensional case for a quasi-elliptic semimartingale.

Lemma 10.17 *Let $(M^1, M^2, \ldots, M^d)$ be a quasi-elliptic semimartingale such that each component is a local martingale. Let $N^n \in \mathbb{I}(M^1, M^2, \ldots, M^d)$ be martingales such that $\mathsf{E}[\, |N_t^n - N_t|\,] \to 0$ for all t.*
    Then $N \in \mathbb{I}(M^1, M^2, \ldots, M^d)$.

Proof The proof follows that of Lemma 10.3. First we get $g^{n,j} \in \mathbb{L}(M^j)$ for $1 \leq j \leq d$, $n \geq 1$ such that

$$N_t^n = N_0^n + \sum_{j=1}^{d} \int_0^t g^{n,j} dM^j.$$

We choose $g^{n,j}$ such that $g_0^{n,j} = 0$. We then conclude that N^n is converging in Emery topology and as a consequence, for all $T < \infty$,

$$[N^n - N^m, N^n - N^m]_T \to 0 \quad \text{in probability as } n, m \to \infty. \tag{10.3.5}$$

Here note that

$$[N^n - N^m, N^n - N^m]_T = \sum_{j,k=1}^{d} \int_0^T (g^{n,j} - g^{m,j})(g^{n,k} - g^{m,k}) d[M^j, M^k]. \tag{10.3.6}$$

Now (10.3.5), (10.3.6), the assumption that $(M^1, M^2, \ldots, M^d)$ is a quasi-elliptic local martingale and Corollary 10.16 implies that for each j, $1 \leq j \leq d$

$$\int_0^T (g_s^{n,j} - g_s^{m,j})^2 d[M, M]_s \to 0 \quad \text{in probability as } n, m \to \infty. \tag{10.3.7}$$

Now, by taking a subsequence, if necessary and relabelling, we assume that for $1 \leq k \leq n, 1 \leq j \leq d$

$$\mathsf{P}((\int_0^k (g_s^{n,j} - g_s^{k,j})^2 d[M, M]_s)^{\frac{1}{2}} \geq \frac{1}{2^k}) \leq \frac{1}{2^k}. \tag{10.3.8}$$

Proceeding as in the proof of Lemma 10.3, defining

$$g_s^j(\omega) = \limsup_{k \to \infty} g_s^{k,j}(\omega)$$

we can conclude that

$$[N, N]_t = \sum_{j,k=1}^d \int_0^t g_s^j g_s^k d[M^j, M^k]_s.$$

Getting $\sigma_n \uparrow \infty$ such that (10.2.11) holds and using that $(M^1, \ldots, M^d)$ is quasi-elliptic, we conclude, for a suitable sequence of stopping times $\tau_n \uparrow \infty$ (as in definition of quasi-elliptic semimartingales),

$$\mathsf{E}[(\int_0^{\sigma_n \wedge \tau_n} (g_s^j)^2 d[M^j, M^j]_s)^{\frac{1}{2}}] < \infty \quad 1 \leq j \leq d, \ n \geq 1.$$

Thus $g^j \in \mathbb{L}(M^j)$. Now defining $Y_t = N_0 + \sum_{j=1}^d \int_0^t g^j dM^j$, we can show that (10.2.12) and (10.2.13) hold and thus $N = Y$ completing the proof that $N \in \mathbb{I}(M^1, M^2, \ldots, M^d)$. $\square$

Now the same proof as that of Theorem 10.4 gives us the following.

Theorem 10.18 *For r.c.l.l. local martingales $M^1, M^2, \ldots, M^d$ and $T < \infty$ let*

$$\mathbb{K}_T(M^1, M^2, \ldots, M^d) = \{N_T : \ N \in \mathbb{I}(M^1, M^2, \ldots, M^d) \cap \mathbb{M}\}. \tag{10.3.9}$$

Suppose $(M^1, M^2, \ldots, M^d)$ is quasi-elliptic semimartingale such that each M^j is a local martingale. Then $\mathbb{K}_T(M^1, M^2, \ldots, M^d)$ is a closed linear subspace of $\mathbb{L}^1(\Omega, \mathcal{F}_T, \mathsf{P})$.

We are now ready to prove the multidimensional version of Theorem 10.9.

Theorem 10.19 *Let $M = (M^1, M^2, \ldots, M^d)$ be a quasi-elliptic semimartingale such that each component is a local martingale on a probability space $(\Omega, \mathcal{F}, \mathsf{P})$*

with a filtration $(\mathcal{F}_.)$. Suppose that $\mathcal{F}_0$ is trivial and $\mathcal{F} = \sigma(\cup_t \mathcal{F}_t)$. Then the following are equivalent.

(i) For every bounded martingale N, $\exists f^j \in \mathbb{L}(M^j)$, $1 \leq j \leq d$ such that

$$N_t = N_0 + \sum_{j=1}^{d} \int_0^t f_s^j \, dM_s^j \quad a.s. \ \forall t. \tag{10.3.10}$$

(ii) For every martingale N, $\exists f^j \in \mathbb{L}(M^j)$, $1 \leq j \leq d$ such that (10.3.10) is true.
(iii) P is an extreme point of the convex set $\mathbb{E}(M^1, M^2, \ldots, M^d)$.
(iv) $\widetilde{\mathbb{E}}_{\mathsf{P}}(M^1, M^2, \ldots, M^d) = \{\mathsf{P}\}$.
(v) $\mathbb{E}_{\mathsf{P}}(M^1, M^2, \ldots, M^d) = \{\mathsf{P}\}$.

Proof The proof closely follows that of Theorem 10.9. Once again we can observe that (i) is same as $\mathbb{L}^\infty(\Omega, \mathcal{F}_T, \mathsf{P}) \subseteq \mathbb{K}_T(M^1, M^2, \ldots, M^d)$, $\forall T \in (0, \infty)$ and (ii) is same as $\mathbb{L}^1(\Omega, \mathcal{F}_T, \mathsf{P}) = \mathbb{K}_T(M^1, M^2, \ldots, M^d)$, $\forall T \in (0, \infty)$.

Proofs of $(i) \Longleftrightarrow (ii)$ and $(iii) \Longleftarrow (iv) \Longleftrightarrow (v)$ are exactly the same.

The proof of $(i) \Rightarrow (v)$ is also on similar lines, invoking Theorem 10.18 in place of Theorem 10.4 to conclude that the class of $\mathcal{F}_T$ measurable random variables that admit representation is a closed subspace of $\mathbb{L}^1(\Omega, \mathcal{F}, \mathsf{P})$.

For the proof of the last part, namely (i) implies (v), assume (i) is true and let $\mathsf{Q} \in \mathbb{E}_{\mathsf{P}}(M^1, M^2, \ldots, M^d)$.

Fix $T < \infty$ and let η be a $\mathcal{F}_T$ measurable bounded random variable. Since $\mathbb{L}^\infty(\Omega, \mathcal{F}_T, \mathsf{P})$ is a subset of $\mathbb{K}_T(M^1, M^2, \ldots, M^d)$ and $\mathcal{F}_0$ is trivial, we can get $g^j \in \mathbb{L}(M^j)$ for $1 \leq j \leq d$ with

$$\eta = c + \sum_{j=1}^{d} \int_0^T g^j \, dM^j$$

such that $V_t = c + \sum_{j=1}^{d} \int_0^t g^j \, dM^j$ is a P-martingale.

Let $Z_t = \sum_{j=1}^{d} \int_0^t g_s^j \, 1_{[0,T]}(s) dM_s^j$. Then $Z_t = \mathsf{E}[(\eta - c) \mid \mathcal{F}_t]$ and thus Z is a bounded P-martingale.

Since $M^1, M^2, \ldots, M^d$ are Q-local martingales and $g^j \in \mathbb{L}(M^j)$, it follows that Z is a Q-sigma-martingale. But Z is a bounded process and now invoking Corollary 9.27 we conclude that Z is a Q-martingale. The rest of the proof that $\mathsf{Q} = \mathsf{P}$ is exactly as in Theorem 10.9. $\square$

We can now deduce the integral representation property for d-dimensional Brownian motion, due to Ito's [25] and Clark [10].

Theorem 10.20 *Let $W = (W^1, W^2, \ldots, W^d)$ be d-dimensional Brownian motion. Thus each W^j is a one-dimensional Brownian motion and moreover $W^1, W^2, \ldots, W^d$ are independent. Let $\mathcal{F}_t = \mathcal{F}_t^W$ and $\mathcal{F} = \sigma(\cup_t \mathcal{F}_t^W)$. Then every martingale M w.r.t. the filtration $(\mathcal{F}_.)$ admits a representation*

$$M_t = M_0 + \sum_{j=1}^{d} \int_0^t f^j dW^j \quad \forall t \tag{10.3.11}$$

where $f^j \in \mathbb{L}(W^j)$ for $1 \leq j \leq d$.

Proof The proof is on the same lines as in the case of one-dimensional version, Theorem 10.10. First we note that $[W^j, W^k] = 0$ for $j \neq k$ implies that W is a quasi-elliptic semimartingale so that we can use Theorem 10.19. We will show that if $Q \in \mathbb{E}_P(W^1, W^2, \ldots, W^d)$ then $Q = P$. Once again as in Theorem 10.10, we deduce that for each j, W^j is a square integrable martingale with $[W^j, W^j]_t^Q = t$ and for $j \neq k$, $[W^j, W^k]_t^Q = 0$. Thus Levy's characterization Theorem 3.8 implies that W is a d-dimensional Brownian motion on $(\Omega, \mathcal{F}, Q)$. The assumption that $\mathcal{F}$ is generated by $\{W_t : t \geq 0\}$ yields $P = Q$ completing the proof. □

Example **10.21** Let $W = (W^1, W^2, \ldots, W^d)$ be d-dimensional Brownian motion. Thus each W^j is a real-valued Brownian motion and $W^1, W^2, \ldots, W^d$ are independent. Let $\mathcal{F}_t = \mathcal{F}_t^W$ and $\mathcal{F} = \sigma(\cup_t \mathcal{F}_t)$. Let $X = (X^1, X^2, \ldots, X^d)$ be the solution to the SDE (3.5.1) with $m = d$ and $b = 0$ where σ satisfies (3.5.3), (3.5.4). Further, suppose that $\exists \alpha > 0$ such that for all $t \geq 0$, $x \in \mathbb{R}^d$, $\lambda_1, \lambda_2, \ldots, \lambda_d$

$$\sum_{i,j=1}^{d} \lambda_i \lambda_j \sigma_{ij}(t, x) \geq \alpha \sum_{i=1}^{d} \lambda_i^2. \tag{10.3.12}$$

Then as noted earlier, $X = (X^1, X^2, \ldots, X^d)$ is a quasi-elliptic semimartingale. Moreover, the condition (10.3.12) implies that $\sigma(t, x)$ is invertible and then

$$W_t = \int_0^t \sigma^{-1}(s, X_s) dX_s. \tag{10.3.13}$$

Thus, W_t is $\mathcal{F}_t^X$ measurable and as a consequence, $\mathcal{F}_t^X = \mathcal{F}_t$. Hence every martingale M admits a representation

$$M_t = M_0 + \sum_{j=1}^{d} \int_0^t g^j dX^j \quad \forall t \tag{10.3.14}$$

where $g^j \in \mathbb{L}(X^j)$ for $1 \leq j \leq d$- just define $g_s = f_s \sigma^{-1}(s, X_s)$ where f is as in (10.3.11). Since $(\mathcal{F}_.^X) = (\mathcal{F}_.)$, g above is also $(\mathcal{F}_.^X)$ predictable. As a consequence, we also get that

$$\mathbb{E}_P(X^1, X^2, \ldots, X^d) = \{P\}.$$

10.4　Continuous Multidimensional Semimartingales

We will show that for continuous semimartingales $X^1, X^2, \ldots, X^d$, we can get bounded predictable processes f^{ij} such that $Y = (Y^1, \ldots, Y^d)$ defined by

$$Y^i_t = \sum_{j=1}^d \int_0^t f^{ij} \, dX^j$$

satisfies for $j \neq k$

$$[Y^j, Y^k]_t = 0 \; \forall t$$

and thus is quasi-elliptic. Thus, Theorem 10.19 would hold for $(Y^1, \ldots, Y^d)$ if $X^1, \ldots, X^d$ were local martingales.

We will first show that such a transformation is always possible. In order to achieve this, we need some auxiliary results.

Lemma 10.22 *Let $\mathbb{N}^{[d]}, \mathbb{O}^{[d]}, \mathbb{D}^{[d]}$ be the class of $d \times d$ symmetric non-negative definite matrices, $d \times d$ Orthogonal matrices and $d \times d$ diagonal matrices, respectively. Then there exists a Borel measurable mapping $\theta : \mathbb{N}^{[d]} \mapsto \mathbb{O}^{[d]} \times \mathbb{D}^{[d]}$ such that*

$$\theta(C) = (B, D) \text{ satisfies } C = B^T D B.$$

Proof Given a non-negative definite C, the eigenvalue-eigenvector decomposition gives existence of orthogonal B and diagonal D such that $C = B^T D B$. Since for all $C \in \mathbb{N}^{[d]}$, the set

$$\{(B, D) \in \mathbb{O}^{[d]} \times \mathbb{D}^{[d]} \; : \; C = B^T D B\}$$

is compact, measurable selection theorem (See [20] or Corollary 5.2.6 of [57]) yields the existence of Borel measurable θ.　　□

Lemma 10.23 *Let $\mathcal{D}$ be a σ-field on a non-empty set Γ and for $1 \leq i, j \leq d$, λ_{ij} be σ-finite signed measures on $(\Gamma, \mathcal{D})$ such that for all $E \in \mathcal{D}$, the matrix$((\lambda_{ij}(E)))$ is a symmetric non-negative definite matrix. Let $\Lambda(E) = \sum_{i=1}^d \lambda_{ii}(E)$. Then for $1 \leq i, j \leq d$ there exists a version c^{ij} of the Radon-Nikodym derivative $\frac{d\lambda_{ij}}{d\Lambda}$ such that for all $\alpha \in \Gamma$, the matrix $((c^{ij}(\alpha)))$ is non-negative definite.*

Proof For $1 \leq i \leq j \leq d$ let f^{ij} be a version of the Radon-Nikodym derivative $\frac{d\lambda_{ij}}{d\Lambda}$ and let $f^{ji} = f^{ij}$. For rational numbers $r_1, r_2, \ldots, r_d$, let

$$A_{r_1, r_2, \ldots, r_d} = \{\alpha : \sum_{ij} r_i r_j f^{ij}(\alpha) < 0\}.$$

Then $\Lambda(A_{r_1, r_2, \ldots, r_d}) = 0$ and hence $\Lambda(A) = 0$ where

$$A = \cup\{A_{r_1, r_2, \ldots, r_d} : r_1, r_2, \ldots, r_d \text{ rationals}\}.$$

The required version is now given by

$$c^{ij}(\alpha) = f^{ij}(\alpha)\, 1_{A^c}(\alpha).$$

$\square$

We are now ready to prove

Theorem 10.24 *Let $M^1, M^2, \ldots, M^d$ be continuous semimartingales. We can get predictable processes b^{ij} bounded by 1 such that $N = (N^1, \ldots, N^d)$ defined by*

$$N_t^j = \sum_{i=1}^d \int_0^t b^{ji}\, dM^i \tag{10.4.1}$$

satisfies

$$[N^j, N^k]_t = 0 \quad \forall t. \tag{10.4.2}$$

Further,

$$M_t^k = \sum_{j=1}^d \int_0^t b^{jk}\, dN^j. \tag{10.4.3}$$

Proof First let us assume that $[M^k, M^k]_t \le C$ for all t and $1 \le k \le d$. Recall that the predictable σ-field $\mathcal{P}$ is the smallest σ-field on $\widetilde{\Omega} = [0, \infty) \times \Omega$ with respect to which all continuous adapted processes are measurable.

Let signed measures Γ_{ij} on $\mathcal{P}$ be defined as follows: for $E \in \mathcal{P}, 1 \le i, j \le d$,

$$\Gamma_{ij}(E) = \int_\Omega \int_0^\infty 1_E(s, \omega)\, d[M^i, M^j]_s(\omega)\, d\mathrm{P}(\omega).$$

Let $\Lambda = \sum_{j=1}^d \Gamma_{jj}$. From the properties of quadratic variation $[M^i, M^j]$, it follows that for all $E \in \mathcal{P}$, the matrix $((\Gamma_{ij}(E)))$ is non-negative definite. In particular, for all i, j

$$|\Gamma_{ij}(E)| \le \Gamma_{ii}(E) + \Gamma_{jj}(E)$$

Hence, Γ_{ij} is absolutely continuous w.r.t. Λ, $\forall i, j$. It follows that we can get predictable processes c^{ij} such that

$$\frac{d\Gamma_{ij}}{d\Lambda} = c^{ij} \tag{10.4.4}$$

and that $C = ((c^{ij}))$ is a non-negative definite matrix (see Lemma 10.23). By construction $|c^{ij}| \le 1$. Using Lemma 10.22, we can obtain predictable processes b^{ij}, d^j such that for all i, k, (writing $\delta_{ik} = 1$ if $i = k$ and $\delta_{ik} = 0$ if $i \ne k$),)

$$\sum_{j=1}^{d} b_s^{ij} b_s^{kj} = \delta_{ik} \tag{10.4.5}$$

$$\sum_{j=1}^{d} b_s^{ji} b_s^{jk} = \delta_{ik} \tag{10.4.6}$$

$$\sum_{j,l=1}^{d} b_s^{ij} c_s^{jl} b_s^{kl} = \delta_{ik} d_s^i \tag{10.4.7}$$

Since $C = ((c_s^{ij}))$ is non-negative definite, it follows that $d_s^i \geq 0$. Further, $B = ((b^{ij}))$ being orthogonal matrix the process b^{ij} is bounded by one and hence is in $\mathbb{L}(M)$ for all $1 \leq i, j \leq d$.

For $1 \leq j \leq d$, let N^j be defined by (10.4.1). Using (10.4.6), it follows that

$$\sum_{j=1}^{d} \int_0^t b^{jk} dN^j = \int_0^t \sum_{j=1}^{d} \sum_{i=1}^{d} b^{jk} b^{ji} dM^i \tag{10.4.8}$$
$$= M_t^k$$

Note that

$$[N^i, N^k]_t = \sum_{j,l=1}^{d} \int_0^t b_s^{ij} b_s^{kl} d[M^j, M^l]_s$$

and hence for any bounded predictable process h such that $|h_s| \leq C \, 1_{[0,T]}(s)$ for some $T < \infty$ and $C < \infty$ and $i \neq k$

$$\mathsf{E}_\mathsf{P}[\int_0^\infty h_s d[N^i, N^k]_s] = \int_\Omega \int_0^\infty h_s \sum_{j,l=1}^{d} b_s^{ij} b_s^{kl} d[M^j, M^l]_s d\mathsf{P}(\omega)$$
$$= \int_{\tilde{\Omega}} h \sum_{j,l=1}^{d} b^{ij} b^{kl} d\Gamma_{jl} \tag{10.4.9}$$
$$= \int_{\tilde{\Omega}} h \sum_{j,l=1}^{d} b^{ij} b^{kl} c^{jl} d\Lambda$$
$$= 0$$

where the last step follows from (10.4.7). Given a bounded stopping time σ, taking $h = 1_{[0,\sigma]}$, it follows that h is predictable and thus using (10.4.9) we conclude from (10.4.9) that for $i \neq k$

$$\mathsf{E}[[N^i, N^k]_\sigma] = 0.$$

Thus $[N^i, N^k]$ is a martingale for $i \neq k$ (see Theorem 2.57). Also, it is a continuous process (as each N^j is continuous) and $[N^i, N^k] \in \mathbb{V}$ with $[N^i, N^k]_0 = 0$ by definition. Hence it follows (see Corollary 5.24) that $[N^i, N^k]_t = 0$ for $i \neq k$. Thus we have proved the result for the case when $[M^k, M^k]_t$ are bounded. For the general case, let

$$\sigma_n = \inf\{t \geq 0 : \sum_{j=1}^{d} [M^j, M^j]_t \geq n\}.$$

Then $\sigma_n \uparrow \infty$ and for each n,

$$[M^{k,[\sigma_n]}, M^{k,[\sigma_n]}]_t \text{ is bounded for } 1 \leq k \leq d$$

where $M^{k,[\sigma_n]}$ is defined by $M_t^{k,[\sigma_n]} = M_{t \wedge \sigma_n}^k$.

Let $((b^{[n],ij}))$ be the predictable processes obtained in the preceding paragraphs for $M^{1,[\sigma_n]}, M^{2,[\sigma_n]}, \ldots, M^{d,[\sigma_n]}$. Then defining

$$b_s^{ij} = \sum_{n=1}^{\infty} b^{[n],ij} 1_{(\sigma_{n-1}, \sigma_n]}(s)$$

we can verify that N defined by (10.4.1) satisfies (10.4.2) and (10.4.3). $\qquad\square$

Remark **10.25** Let M, N be as in Theorem 10.24.
Then it follows that $M^1, \ldots, M^d$ are local martingales if and only if $N^1, \ldots, N^d$ are local martingales since b^{ik} are bounded predictable processes. Further, (10.4.1), (10.4.3) imply that

$$\mathbb{E}(M^1, M^2, \ldots, M^d) = \mathbb{E}(N^1, N^2, \ldots, N^d),$$
$$\mathbb{E}_\mathsf{P}(M^1, M^2, \ldots, M^d) = \mathbb{E}_\mathsf{P}(N^1, N^2, \ldots, N^d),$$
$$\widetilde{\mathbb{E}}_\mathsf{P}(M^1, M^2, \ldots, M^d) = \widetilde{\mathbb{E}}_\mathsf{P}(N^1, N^2, \ldots, N^d).$$

In view of Remark 10.25, we have the following result as a direct consequence of Theorem 10.19.

Theorem 10.26 *Let $M^1, M^2, \ldots, M^d$ be continuous local martingales on $(\Omega, \mathcal{F}, \mathsf{P})$. Suppose that $\mathcal{F}_0$ is trivial and $\mathcal{F} = \sigma(\cup_t \mathcal{F}_t)$. Let $N^1, N^2, \ldots, N^d$ be as in Theorem 10.24 so that (10.4.1), (10.4.2) and (10.4.3) hold. Then the following are equivalent.*

(i) For every bounded martingale U, $\exists f^j \in \mathbb{L}(N^j)$, $1 \leq j \leq d$ such that

$$U_t = U_0 + \sum_{j=1}^{d} \int_0^t f_s^j \, dN_s^j \quad \forall t. \qquad (10.4.10)$$

(ii) For every martingale N, $\exists f^j \in \mathbb{L}(N^j)$, $1 \leq j \leq d$ such that (10.4.10) is true.

(iii) P *is an extreme point of the convex set* $\mathbb{E}(M^1, M^2, \ldots, M^d)$.
(iv) $\widetilde{\mathbb{E}}_P(M^1, M^2, \ldots, M^d) = \{P\}$.
 (v) $\mathbb{E}_P(M^1, M^2, \ldots, M^d) = \{P\}$.

10.5   General Multidimensional Case

We have commented that in general, the multidimensional version of Theorem 10.9
is not true. In view of this, we had given a version in case of quasi-ellipticity, Theorem
10.19 and a version in the case of continuous local martingales, Theorem 10.26. We
now come to the general multidimensional case.

For r.c.l.l. semimartingales $X^1, \ldots, X^d$ and $h^j \in \mathbb{L}(X^j)$, we can define the *vector
stochastic integral* as

$$\int_0^t \langle h, dX \rangle = \sum_{j=1}^d \int h^j \, dX^j.$$

In order to discuss the general case of integral representation theorem, we need
to extend the notion of vector stochastic integral. See Jacod [27], Cherny and
Shiryaev [7].

Definition 10.27 For r.c.l.l. semimartingales $X^1, \ldots, X^d$, let $\mathbb{L}_v(X^1, \ldots, X^d)$
denote the class of $\mathbb{R}^d$-valued predictable processes $h = (h^1, \ldots h^d)$ such
that for any sequence of predictable processes ϕ^n satisfying

  (i) $|\phi^n| \leq 1$,
 (ii) $h^j \phi^n$ is bounded for all $j, n, 1 \leq j \leq d, n \geq 1$,
(iii) $\phi^n \to 0$ pointwise

the processes $Z^n = \sum_{j=1}^d \int h^j \phi^n \, dX^j$ converge to 0 in $\mathbf{d}_{ucp}$ metric.

Here is an observation.

Lemma 10.28 *Let* $X = (X^1, \ldots, X^d)$ *be r.c.l.l. semimartingales and let* $h = (h^1, \ldots, h^d)$ *be an* $\mathbb{R}^d$-*valued predictable process. Let*

$$X^{<h>} = \sum_{j=1}^d \int h^j \, 1_{\{|h|>0\}} \frac{1}{|h|} dX^j \tag{10.5.1}$$

where $|h| = \sqrt{\sum_{j=1}^d (h^j)^2}$. *Then* $h^j \frac{1}{|h|} \in \mathbb{L}(X^j)$ *for* $1 \leq j \leq d$ *and*

$$h \in \mathbb{L}_v(X^1, \ldots, X^d) \text{ if and only if } |h| \in \mathbb{L}(X^{<h>}) \tag{10.5.2}$$

Proof Since $h^j \, 1_{\{|h|>0\}} \frac{1}{|h|}$ is bounded, clearly, $h^j \, 1_{|h|>0} \frac{1}{|h|} \in \mathbb{L}(X^j)$ for $1 \leq j \leq d$.
Let $h \in \mathbb{L}_v(X^1, \ldots, X^d)$. Let f^n be bounded predictable processes with $|f^n| \leq |h|$

and $f^n \to 0$ pointwise. Let $\phi^n = f^n \, 1_{\{|h|>0\}} \frac{1}{|h|}$. Then ϕ^n are predictable processes with $|\phi^n| \leq 1$ and $\phi^n \to 0$ pointwise. Also, $f^n = \phi^n |h|$ and so $\int f^n dX^{<h>} = \sum_{j=1}^d \int h^j \phi^n dX^j$. Thus, $\int f^n dX^{<h>} \to 0$ in $\mathbf{d}_{ucp}$ metric and thus $|h| \in \mathbb{L}(X^{<h>})$ (see Theorem 4.18).

Conversely, suppose $|h| \in \mathbb{L}(X^{<h>})$. Given predictable ϕ^n as in Definition 10.27, let $f^n = |h| \phi^n$. Then $|f^n| \leq |h|$ and $f^n \to 0$ pointwise and hence

$$W^n = \int f^n dX^{<h>} \to 0 \ \mathbf{d}_{ucp} \text{ metric.}$$

Noting that

$$\int f^n dX^{<h>} = \sum_{j=1}^d \int h^j \phi^n dX^j$$

we conclude that $h \in \mathbb{L}_v(X^1, \ldots, X^d)$.   □

Definition 10.29 For $h \in \mathbb{L}_v(X^1, \ldots, X^d)$, where $X^1, \ldots, X^d$ are r.c.l.l. semimartingales, the vector stochastic integral $\int_0^t \langle h, dX \rangle$ is defined by

$$\int_0^t \langle h, dX \rangle = \int_0^t |h_s| dX_s^{<h>}$$

where $X^{<h>}$ is defined in (10.5.1).

Note that if $Z_t = \int_0^t \langle h, dX \rangle$, then

$$[Z, Z]_t = \int_0^t |h_s|^2 d[X^{<h>}, X^{<h>}]_s$$

$$= \sum_{j,k=1}^d \int_0^t |h_s|^2 \frac{1}{|h_s|^2} (h_s^j h_s^k) d[X^j, X^k]_s \qquad (10.5.3)$$

$$= \sum_{j,k=1}^d \int_0^t (h_s^j h_s^k) d[X^j, X^k]_s.$$

Likewise, for $h \in \mathbb{L}_v(X^1, \ldots, X^d)$ and $g \in \mathbb{L}_v(U^1, \ldots, U^d)$, $Z_t = \int_0^t \langle h, dX \rangle$ and $W_t = \int_0^t \langle g, dU \rangle$, we have

$$[Z, W]_t = \sum_{j,k=1}^d \int_0^t (h_s^j g_s^k) d[X^j, U^k]_s. \qquad (10.5.4)$$

Remark **10.30** Let $X^1, \ldots, X^d$ be r.c.l.l. semimartingales and let $h^j \in \mathbb{L}(X^j)$ for $1 \leq j \leq d$. Then it is easy to see that $h = (h^1, \ldots, h^d) \in \mathbb{L}_v(X^1, \ldots, X^d)$

and

$$\int \langle h, dX \rangle = \sum_{j=1}^{d} \int h^j \, dX^j.$$

However, $h = (h^1, \dots, h^d) \in \mathbb{L}_v(X^1, \dots, X^d)$ does not imply that $h^j \in \mathbb{L}(X^j)$ for $1 \le j \le d$.

Remark **10.31** Let $h \in \mathbb{L}_v(X^1, \dots, X^d)$, where $X^1, \dots, X^d$ are r.c.l.l. semi-martingales. For any predictable process ψ such that $\psi > 0$ and $|h| \le \psi$, we have

$$\int_0^t \langle h, dX \rangle = \int_0^t \psi_s \, dZ \qquad (10.5.5)$$

where

$$Z_t = \sum_{j=1}^{d} \int_0^t \frac{h^j}{\psi} \, dX^j. \qquad (10.5.6)$$

We introduce the class of semimartingales that admit a representation as vector integral w.r.t. an $\mathbb{R}^d$-valued r.c.l.l. semimartingale $X = (X^1, \dots, X^d)$. Let

$$\mathbb{I}^v(X^1, \dots, X^d)$$
$$= \{Z : \exists g \in \mathbb{L}_v(X^1, \dots, X^d), \text{ with } Z_t = Z_0 + \int_0^t \langle g, dX \rangle \; \forall t < \infty\}$$

When $X = (X^1, \dots, X^d)$, we will also write $\mathbb{L}_v(X) = \mathbb{L}_v(X^1, \dots, X^d)$ and $\mathbb{I}^v(X) = \mathbb{I}^v(X^1, \dots, X^d)$.

Here is an observation about vector integral, an analogue of Theorem 4.33.

Theorem 10.32 *Let $X^1, \dots, X^d$ be r.c.l.l. semimartingales and let ϕ be a $(0, \infty)$-valued predictable process such that $\phi \in \mathbb{L}(X^j)$ for $1 \le j \le d$. Let $Y^j = \int \phi \, dX^j$ and $h = (h^1, h^2, \dots, h^d)$ be an $\mathbb{R}^d$-valued predictable process. Let $f^j = \phi h^j$ and $f = (f^1, f^2, \dots, f^d)$. Writing $X = (X^1, \dots, X^d)$ and $Y = (Y^1, \dots, Y^d)$. Then*

$$f \in \mathbb{L}_v(X) \text{ if and only if } h \in \mathbb{L}_v(Y) \qquad (10.5.7)$$

and then

$$\int \langle f, dX \rangle = \int \langle h, dY \rangle. \qquad (10.5.8)$$

As a consequence

$$\mathbb{I}^v(X) = \mathbb{I}^v(Y). \qquad (10.5.9)$$

Proof Recall that $f \in \mathbb{L}_v(X)$ if and only if $|f| \in \mathbb{L}(X^{<f>})$ where $X^{<f>} = \sum_{j=1}^{d} \int f_j \, 1_{\{|f|>0\}} \frac{1}{|f|} dX^j$ and $h \in \mathbb{L}_v(Y)$ if and only if $|h| \in \mathbb{L}(Y^{<h>})$ where $Y^{<h>} = \sum_{j=1}^{d} \int h_j \, 1_{\{|h|>0\}} \frac{1}{|h|} dY^j$. Since ϕ is $(0, \infty)$-valued, it follows that for all j

$$h^j \, 1_{\{|h|>0\}} \frac{1}{|h|} = f^j \, 1_{\{|f|>0\}} \frac{1}{|f|}$$

and as a consequence

$$\int \phi \, dX^{<f>} = \sum_{j=1}^{d} \int f^j \, 1_{\{|f|>0\}} \frac{1}{|f|} \phi \, dX^j$$

$$= \sum_{j=1}^{d} \int h^j \, 1_{\{|h|>0\}} \frac{1}{|h|} \phi \, dX^j$$

$$= \sum_{j=1}^{d} \int h^j \, 1_{\{|h|>0\}} \frac{1}{|h|} \, dY^j$$

$$= Y^{<h>}.$$

Thus, using Theorem 4.33, we have

$$|h| \in \mathbb{L}(Y^{<h>}) \text{ if and only if } |f| = |h|\phi \in \mathbb{L}(X^{<f>})$$

and

$$\int |h| \, dY^{<h>} = \int |f| \, dX^{<f>}.$$

Since $\int \langle f, dX \rangle = \int |f| \, dX^{<f>}$ and $\int \langle h, dY \rangle = \int |h| \, dY^{<h>}$, this completes the proof. □

We observe that an analogue of Theorem 4.43 holds for vector integral as well.

Theorem 10.33 *Let $X^1, X^2, \ldots, X^d$ be r.c.l.l. stochastic integrators and let $h^1, h^2, \ldots, h^d$ be predictable processes such that there exist stopping times τ_m increasing to ∞ with*

$$(h^1 \, 1_{[0,\tau_m]}, \ldots, h^d \, 1_{[0,\tau_m]}) \in \mathbb{L}_v(X^1, \ldots, X^d) \quad \forall n \geq 1. \tag{10.5.10}$$

Then $(h^1, h^2, \ldots, h^d) \in \mathbb{L}_v(X^1, \ldots, X^d)$.

Proof Let ϕ^n be predictable processes, $|\phi^n| \leq 1$, $h^j \phi^n$ is bounded for all $j, n, 1 \leq j \leq d, n \geq 1$, $\phi^n \to 0$ pointwise. Then in view of (10.5.10), it follows that for each m,

$$Z^{m,n} = \sum_{j=1}^{d} \int h^j \, 1_{[0,\tau_m]} \phi^n \, dX^j \to 0 \text{ in } \mathbf{d}_{ucp} \text{ metric.} \tag{10.5.11}$$

We need to show that

$$Z^n = \sum_{j=1}^{d} \int h^j \phi^n dX^j \to 0 \text{ in } \mathbf{d}_{ucp} \text{ metric.} \qquad (10.5.12)$$

By (4.4.2) in Lemma 4.36, it follows that

$$Z^{m,n}_t = Z^n_{t \wedge \tau_m}.$$

Now the required conclusion, namely (10.5.12) follows from (10.5.11) and Lemma 2.75.   □

Exercise 10.34 Show that the mapping $(h, X) \mapsto \int \langle h, dX \rangle$ is linear in h and X.

In analogy with Lemma 10.1, here we have the following result, with very similar proof.

Lemma 10.35 *Let $X^1, X^2, \ldots, X^d$ be semimartingales and Y be a semimartingale such that for a sequence of stopping times $\tau_n \uparrow \infty$, Y^n defined by $Y^n_t = Y_{t \wedge \tau_n}$ satisfies*

$$Y^n \in \mathbb{I}^v(X^1, X^2, \ldots, X^d).$$

Then

$$Y \in \mathbb{I}^v(X^1, X^2, \ldots, X^d).$$

Proof Let $f^{n,j} \in \mathbb{L}(X^j), 1 \le j \le d, n \ge 1$ be such that for all n,

$$Y^n_t = Y^n_0 + \int_0^t \langle f^n, dX \rangle.$$

Define f^j by

$$f^j = \sum_{n=1}^{\infty} 1_{(\tau_{n-1}, \tau_n]} f^{n,j}.$$

Then it is easy to check (using Theorem 10.33) that $f^j \in \mathbb{L}(X^j)$ and

$$Y_t = Y_0 + \int_0^t \langle f, dX \rangle.$$

This completes the proof.   □

With the introduction of vector integral, we can now prove the multidimensional analogue of Lemma 10.3.

Lemma 10.36 *Let $M^1, M^2, \ldots, M^d$ be r.c.l.l. local martingales. For $n \geq 1$, let X^n be martingales such that $X^n \in \mathbb{I}^v(M^1, M^2, \ldots, M^d)$. Suppose that $\mathbb{E}[\,|X_t^n - X_t|\,] \to 0$ as $n \to \infty$ $\forall t$. Then $X \in \mathbb{I}^v(M^1, M^2, \ldots, M^d)$.*

Proof The proof follows that of Lemma 10.3. In view of Lemma 10.35, suffices to consider the case when for some $T < \infty$,

$$X_t = X_{t \wedge T}, \quad X_t^n = X_{t \wedge T}^n, \quad \forall t \geq 0, \ n \geq 1. \tag{10.5.13}$$

So let us fix a $T < \infty$ such that (10.5.13) holds. Without loss of generality, we can assume that $M_t^j = M_{t \wedge T}^j$ for $1 \leq j \leq d$. We note that Theorem 5.39 implies that X^n converges to X in Emery topology and as a consequence,

$$[X^n - X, X^n - X]_T \to 0 \quad \text{in probability as } n \to \infty, \tag{10.5.14}$$

$$[X^n, X^n]_T \to [X, X]_T \quad \text{in probability as } n \to \infty \tag{10.5.15}$$

and

$$[X^n - X^m, X^n - X^m]_T \to 0 \quad \text{in probability as } n, m \to \infty. \tag{10.5.16}$$

By taking a subsequence and relabelling if necessary, we assume that

$$P([X^n - X, X^n - X]_T \geq 2^{-k}) \leq 2^{-k}, \quad \forall n \geq k. \tag{10.5.17}$$

Using this estimate and invoking Borel–Cantelli Lemma it follows that

$$\sum_{n=1}^{\infty} \sqrt{[X^n - X, X^n - X]_T} < \infty \quad a.s.$$

Let

$$B = \sum_{j=1}^{d} \sqrt{[M^j, M^j]_T} + \sum_{n=1}^{\infty} \sqrt{[X^n - X, X^n - X]_T} + \sqrt{[X, X]_T}$$

Using (4.6.22), it follows that $\sqrt{[X^n, X^n]_T} \leq 2B$ and also

$$\sqrt{[X^n - X^m, X^n - X^m]_T} \leq 2B \tag{10.5.18}$$

We are going to carry out a orthogonalization as in Theorem 10.24. However, this time $[M^j, M^k]$ are not continuous and thus we cannot assume them to be locally integrable. Thus we introduce an equivalent measure Q as follows: let

$$\alpha = \mathbb{E}_P[\exp\{-B\}],$$

$$\xi = \frac{1}{\alpha} \exp\{-B\}$$

and Q be the probability measure on $(\Omega, \mathcal{F})$ defined by

$$\frac{dQ}{dP} = \xi.$$

Since $E_Q[B^k] < \infty$ for all t and for all k, and $[X^n - X^m, X^n - X^m]_T$ converges to 0 in P and hence in Q probability as $n, m \to \infty$, it follows using (10.5.18) that

$$E_Q[\,[X^n - X^m, X^n - X^m]_T] \to 0 \quad \text{as } n, m \to \infty. \tag{10.5.19}$$

Likewise

$$E_Q[\,[X^n - X, X^n - X]_T] \to 0 \quad \text{as } n \to \infty \tag{10.5.20}$$

and

$$E_Q[\,[X^n, X^n]_T] \to E_Q[\,[X, X]_T] \quad \text{as } n \to \infty. \tag{10.5.21}$$

Since $X^n \in \mathbb{I}^v(M^1 \ldots, M^d)$, we can get $f^n = (f^{n,1}, \ldots, f^{n,d})$ such that $f^n \in \mathbb{L}_v(M^1 \ldots, M^d)$ and

$$X^n_t = X^n_0 + \int_0^t \langle f^n, dM \rangle. \tag{10.5.22}$$

We repeat the construction that we carried out in proof of Theorem 10.24, with a subtle difference. Here we do not have continuity of $[M^i, M^j]$ but do have integrability of $[M^j, M^j]_T$ under probability measure Q for each j. Let signed measures Γ_{ij} on $\mathcal{P}$ be defined as follows: for $E \in \mathcal{P}$, $1 \leq i, j \leq d$,

$$\Gamma_{ij}(E) = \int_\Omega \int_0^T 1_E(s, \omega) d[M^i, M^j]_s(\omega) dQ(\omega).$$

Let $\Lambda = \sum_{j=1}^d \Gamma_{jj}$. From the properties of quadratic variation $[M^i, M^j]$, it follows that for all $E \in \mathcal{P}$, the matrix $((\Gamma_{ij}(E)))$ is non-negative definite. Further, Γ_{ij} is absolutely continuous w.r.t. Λ $\forall i, j$. It follows that we can get predictable processes c^{ij} such that

$$\frac{d\Gamma_{ij}}{d\Lambda} = c^{ij} \tag{10.5.23}$$

and that $C = ((c^{ij}))$ is a non-negative definite matrix (see Lemma 10.23). By construction $|c^{ij}| \leq 1$. Using Lemma 10.22, we can obtain predictable processes b^{ij}, d^j such that for all i, k, (writing $\delta_{ik} = 1$ if $i = k$ and $\delta_{ik} = 0$ if $i \neq k$),

$$\sum_{j=1}^d b_s^{ij} b_s^{kj} = \delta_{ik} \tag{10.5.24}$$

$$\sum_{j=1}^{d} b_s^{ji} b_s^{jk} = \delta_{ik} \tag{10.5.25}$$

$$\sum_{j,l=1}^{d} b_s^{ij} c_s^{jl} b_s^{kl} = \delta_{ik} d_s^{i} \tag{10.5.26}$$

Since $((c_s^{ij}))$ is non-negative definite, it follows that $d_s^i \geq 0$. For $1 \leq j \leq d$, let N^j be defined by

$$N_t^j = \sum_{i=1}^{d} \int_0^t b^{ji} dM^i \tag{10.5.27}$$

Using (10.5.25), it follows that for $t \leq T$

$$\sum_{j=1}^{d} \int_0^t b^{jk} dN^j = \int_0^t \sum_{j=1}^{d} \sum_{i=1}^{d} b^{jk} b^{ji} dM^i \tag{10.5.28}$$
$$= M_t^k$$

Note that

$$[N^i, N^k]_t = \sum_{j,l=1}^{d} \int_0^t b_s^{ij} b_s^{kl} d[M^j, M^l]_s$$

and hence for any bounded predictable process h and for $i \neq k$

$$\mathsf{E}_\mathsf{Q}[\int_0^T h_s d[N^i, N^k]_s] = \int_\Omega \int_0^T h_s \sum_{j,l=1}^{d} b_s^{ij} b_s^{kl} d[M^j, M^l]_s dQ(\omega)$$
$$= \int_{\bar\Omega} h \sum_{j,l=1}^{d} b^{ij} b^{kl} d\Gamma_{jl} \tag{10.5.29}$$
$$= \int_{\bar\Omega} h \sum_{j,l=1}^{d} b^{ij} b^{kl} c^{jl} d\Lambda$$
$$= 0$$

where the last step follows from (10.5.26). As a consequence, for bounded predictable h^i

$$\mathsf{E}_\mathsf{Q}[\sum_{i,k=1}^{d} \int_0^T h_s^i h_s^k d[N^i, N^k]_s] = \mathsf{E}_\mathsf{Q}[\sum_{k=1}^{d} \int_0^T (h_s^k)^2 d[N^k, N^k]_s] \tag{10.5.30}$$

Let us observe that (10.5.30) holds for any predictable processes $\{h^i : 1 \leq i \leq d\}$ provided the right-hand side is finite: we can first note that it holds for $\tilde{h}^i = h^i 1_{\{|h| \leq k\}}$ where $|h| = \sum_{i=1}^d |h^i|$ and then let $k \uparrow \infty$. Let us define

$$g^{n,k} = \sum_{j=1}^d f^{n,j} b^{kj} \qquad (10.5.31)$$

and let $\psi^n = 1 + \sum_{k=1}^d (|g^{n,k}| + |f^{n,k}|)$. Then note that

$$\int \langle g^n, dN \rangle = \int \psi^n dW^n$$

where

$$W^n = \sum_{k=1}^d \int \frac{g^{n,k}}{\psi^n} dN^k.$$

Note that

$$W^n = \sum_{k=1}^d \sum_{j=1}^d \int \frac{1}{\psi^n} f^{n,j} b^{kj} dN^k$$

$$= \sum_{k=1}^d \sum_{j=1}^d \sum_{l=1}^d \int \frac{1}{\psi^n} f^{n,j} b^{kj} b^{kl} dM^l \qquad (10.5.32)$$

$$= \sum_{j=1}^d \int \frac{1}{\psi^n} f^{n,j} dM^j$$

and hence

$$\int \langle f^n, dM \rangle = \int \psi^n dW^n.$$

Thus we have

$$\int \langle g^n, dN \rangle = \int \langle f^n, dM \rangle \qquad (10.5.33)$$

and as a consequence, recalling (10.5.3), we have

$$[X^n - X^m, X^n - X^m]_T = \sum_{j,k=1}^d \int_0^T (g^{n,j} - g^{m,j})(g^{n,k} - g^{m,k}) d[N^j, N^k].$$

$$(10.5.34)$$

Now invoking (10.5.29) we get

$$\mathsf{E}_Q[\,[X^n - X^m, X^n - X^m]_T] = \mathsf{E}_Q[\sum_{k=1}^{d} \int_0^T (g_s^{n,k} - g_s^{m,k})^2 d[N^k, N^k]_s]$$

$$(10.5.35)$$

$$= \sum_{k=1}^{d} \int (g^{n,k} - g^{m,k})^2 d\Gamma_{kk}$$

Since left-hand side in (10.5.35) converges to 0 (see (10.5.19)), using completeness of $\mathbb{L}^2(\bar{\Omega}, \mathcal{P}, \Gamma_{kk})$, we can get predictable processes g^k such that

$$\int (g^{n,k} - g^k)^2 d\Gamma_{kk} \to 0.$$

As a consequence,

$$\sum_{j,k=1}^{d} \int_0^T (g^{n,j} - g^j)(g^{n,k} - g^k) d[N^j, N^k] \to 0 \quad \text{in } Q\text{-probability as } n \to \infty,$$

$$(10.5.36)$$

and thus for any bounded stopping time τ

$$\sum_{j,k=1}^{d} \int_0^\tau (g^{n,j})(g^{n,k}) d[N^j, N^k] \to \sum_{j,k=1}^{d} \int_0^\tau (g^j)(g^k) d[N^j, N^k] \qquad (10.5.37)$$

in Q probability as $n \to \infty$. Noting that Q and P are equivalent and

$$[X^n, X^n]_\tau = \sum_{j,k=1}^{d} \int_0^\tau (g^{n,j})(g^{n,k}) d[N^j, N^k],$$

we conclude that

$$[X^n, X^n]_t \to \sum_{j,k=1}^{d} \int_0^t (g^j)(g^k) d[N^j, N^k] \quad \text{in } P\text{ - probability as } n \to \infty$$

and as a consequence

$$[X, X]_t = \sum_{j,k=1}^{d} \int_0^t (g^j)(g^k) d[N^j, N^k] \qquad (10.5.38)$$

Let us define bounded predictable processes ϕ^j and predictable process h and a P-martingale Z as follows:

$$h_s = 1 + \sum_{i=1}^{d} |g_s^i| \qquad (10.5.39)$$

$$\phi_s^j = \frac{g_s^j}{h_s} \tag{10.5.40}$$

$$Z_t = \sum_{j=1}^{d} \int_0^t \phi_s^j \, dN_s^j \tag{10.5.41}$$

Then

$$Z_t = \sum_{i=1}^{d} \sum_{j=1}^{d} \int_0^t \phi_s^j b_s^{ji} \, dM^i \tag{10.5.42}$$

Since ϕ^j, b^{ji} are predictable and bounded by 1, it follows that Z is a P-local martingale. Let us note that

$$\int_0^t h_s^2 d[Z, Z]_s = \sum_{j=1}^{d} \sum_{k=1}^{d} \int_0^t h_s^2 \phi_s^j \phi_s^k \, d[N^j, N^k]_s$$

$$= \sum_{j=1}^{d} \sum_{k=1}^{d} \int_0^t g_s^k g_s^j \, d[N^j, N^k]_s. \tag{10.5.43}$$

Putting together (10.5.38) and (10.5.43), we conclude

$$[X, X]_t = \int_0^t h_s^2 d[Z, Z]_s \tag{10.5.44}$$

We now forget Q and focus only on P. Since X is a martingale, we can get stopping times $\sigma_n \uparrow \infty$ such that $\mathsf{E}_\mathsf{P}[\,[X, X]_{\sigma_n}] < \infty$ and thus using (10.5.44), we conclude that $h \in \mathbb{L}_m^1(Z)$. Defining $Y_t = X_0 + \int_0^t h \, dZ$, we note that $Y_t = X_0 + \int_0^t < g, dN >$. Thus Y is a local martingale and further

$$[X^n - Y, X^n - Y]_t = \int_0^t \sum_{j,k=1}^{d} (g^{n,j} - g^j)(g^{n,k} - g^k) d[N^j, N^k]. \tag{10.5.45}$$

Using (10.5.36) and the observation that Q and P are equivalent, we conclude

$$[X^n - Y, X^n - Y]_t \to 0 \text{ in Pprobability as } n \to \infty. \tag{10.5.46}$$

Since for all $n \geq 1$

$$\sqrt{[X - Y, X - Y]_t} \leq \sqrt{[X^n - X, X^n - X]_t} + \sqrt{[X^n - Y, X^n - Y]_t} \tag{10.5.47}$$

using (10.5.46) and (10.5.14), we conclude

$$[X - Y, X - Y]_t = 0 \quad \forall t. \tag{10.5.48}$$

Since X, Y are local martingales and $X_0 = Y_0$, (10.5.48) implies $X_t = Y_t$ for all t. To complete the proof, we will show that

$$X_t = X_0 + \int_0^t \langle g, dN \rangle = X_0 + \int_0^t \langle f, dM \rangle$$

for a suitably defined f. So let

$$f^i = \sum_{k=1}^d g^k b^{ki}$$

and let $\psi = \sum_{j=1}^d (1 + |f^j| + |g^j|)$. Then

$$
\begin{aligned}
W &= \sum_{i=1}^d \int_0^t \frac{1}{\psi} f^i dM^i \\
&= \sum_{i=1}^d \sum_{j=1}^d \int_0^t \frac{1}{\psi} f^i b^{ji} dN^j \\
&= \sum_{i=1}^d \sum_{k=1}^d \sum_{j=1}^d \int_0^t \frac{1}{\psi} g^k b^{ki} b^{ji} dN^j \\
&= \sum_{k=1}^d \sum_{j=1}^d \int_0^t \frac{1}{\psi} g^k \delta_{jk} dN^j \\
&= \sum_{j=1}^d \int_0^t \frac{1}{\psi} g^j dN^j
\end{aligned}
$$

where we have used (10.5.28), definition of f^i and (10.5.24). Thus

$$\int \langle f, dM \rangle = \int \psi dW = \int \langle g, dN \rangle.$$

Hence

$$X_t = X_0 + \int_0^t \langle f, dM \rangle.$$

$\square$

For semimartingales $X^1, X^2, \ldots, X^d$ let $\mathbb{K}_T^v(X^1, \ldots, X^d)$ be defined by

$$\mathbb{K}_T^v(X^1, \ldots, X^d) = \{N_T : N \in \mathbb{I}^v(X^1, \ldots, X^d) \cap \mathbb{M}\}. \tag{10.5.49}$$

Note that if Q is equivalent to P the class $\mathbb{K}_T^v$ under Q may not be the same as the one under P defined above as the class of martingales $\mathbb{M}$ need not be the same under the two measures.

As an immediate consequence of Lemma 10.36 we have

Theorem 10.37 *For local martingales* $M^1, \ldots, M^d$, $\mathbb{K}_T^v(M^1, \ldots, M^d)$ *is a closed linear subspace of* $\mathbb{L}^1(\Omega, \mathcal{F}_T, P)$ *for every* $T < \infty$.

Now using Theorem 10.37 instead of Theorem 10.18, we can obtain the integral representation property for general multidimensional local martingales—rest of the argument is essentially same as in the proof of Theorem 10.19, but we will give it here for the reader's convenience.

Theorem 10.38 *Let* $M^1, M^2, \ldots, M^d$ *be local martingales on* $(\Omega, \mathcal{F}, P)$. *Suppose that* $\mathcal{F}_0$ *is trivial and* $\mathcal{F} = \sigma(\cup_t \mathcal{F}_t)$. *Then the following are equivalent.*

(i) *For every bounded martingale* S, $\exists f \in \mathbb{L}_v(M^1, M^2, \ldots, M^d)$ *such that*

$$S_t = S_0 + \int_0^t \langle f, dM \rangle \ \ a.s. \ \ \forall t. \tag{10.5.50}$$

(ii) *For every martingale* S, $\exists f \in \mathbb{L}_v(M^1, M^2, \ldots, M^d)$ *such that* (10.5.50) *is true.*
(iii) P *is an extreme point of the convex set* $\mathbb{E}(M^1, M^2, \ldots, M^d)$.
(iv) $\widetilde{\mathbb{E}}_P(M^1, M^2, \ldots, M^d) = \{P\}$.
(v) $\mathbb{E}_P(M^1, M^2, \ldots, M^d) = \{P\}$.

Proof It can be seen that (i) is same as $\mathbb{L}^\infty(\Omega, \mathcal{F}_T, P) \subseteq \mathbb{K}_T^v(M^1, \ldots, M^d) \ \forall T \in (0, \infty)$ and (ii) is same as $\mathbb{L}^1(\Omega, \mathcal{F}_T, P) = \mathbb{K}_T^v(M^1, \ldots, M^d) \ \forall T \in (0, \infty)$. As seen in Theorem 10.37, $\mathbb{K}_T^v(M^1, \ldots, M^d)$ is a closed subspace of $\mathbb{L}^1(\Omega, \mathcal{F}_T, P)$. Since $\mathbb{L}^\infty(\Omega, \mathcal{F}_T, P)$ is dense in $\mathbb{L}^1(\Omega, \mathcal{F}_T, P)$, it follows that (i) and (ii) are equivalent.

On the other hand, suppose (iv) holds and let $P = \alpha Q_1 + (1 - \alpha)Q_2$ where $Q_1, Q_2 \in \mathbb{E}(M^1, M^2, \ldots, M^d), 0 \le \alpha \le 1$. Then Q_1, Q_2 are absolutely continuous w.r.t. P and hence $Q_1, Q_2 \in \widetilde{\mathbb{E}}_P(M^1, M^2, \ldots, M^d)$. In view of (iv), $Q_1 = Q_2 = P$ and thus P is an extreme point of $\mathbb{E}(M^1, M^2, \ldots, M^d)$ and so (iii) is true. Thus (iv) $\Rightarrow (iii)$.

Since $\{P\} \subseteq \mathbb{E}_P(M^1, M^2, \ldots, M^d) \subseteq \widetilde{\mathbb{E}}_P(M^1, M^2, \ldots, M^d)$, it follows that (iv) implies (v).

If (v) is true and $Q \in \widetilde{\mathbb{E}}_P(M^1, M^2, \ldots, M^d)$, then $Q_1 = \frac{1}{2}(Q + P) \in \mathbb{E}_P(M^1, M^2, \ldots, M^d)$. Then (v) implies $Q_1 = P$ and hence $Q = P$. Thus $(v) \Rightarrow (iv)$ holds.

To see that $(iii) \Rightarrow (ii)$ let P be an extreme point of $\mathbb{E}(M^1, M^2, \ldots, M^d)$ but (ii) is not true. Then $\mathbb{K}_T^v(M^1, \ldots, M^d)$ is a closed proper subspace of $\mathbb{L}^1(\Omega, \mathcal{F}, P)$ and by the Hahn–Banach Theorem (see [55]), there exists $\xi \in \mathbb{L}^\infty(\Omega, \mathcal{F}_T, P), P(\xi \ne 0) > 0$ such that

$$\int \theta \xi dP = 0 \ \ \forall \theta \in \mathbb{K}_T^v(M^1, \ldots, M^d).$$

Then for $c \in \mathbb{R}$, we have

$$\int \theta(1 + c\xi)d\mathsf{P} = \int \theta d\mathsf{P} \quad \forall \theta \in \mathbb{K}_T^v(M^1, \ldots, M^d). \tag{10.5.51}$$

Since ξ is bounded, we can choose a $c > 0$ such that

$$\mathsf{P}(c|\xi| < 0.5) = 1.$$

Now, let Q be the measure with density $\eta = (1 + c\xi)$. Then Q is a probability measure. Thus (10.5.51) yields

$$\int \theta d\mathsf{Q} = \int \theta d\mathsf{P} \quad \forall \theta \in \mathbb{K}_T^v(M^1, \ldots, M^d). \tag{10.5.52}$$

Let $\sigma_n \uparrow \infty$ be bounded stopping times such that $M_t^{j,n} = M_{t \wedge \sigma_n}^j$ is a P-martingale. For any bounded stopping time τ, $M_{\tau \wedge T}^{j,n} = M_{\tau \wedge \sigma_n \wedge T}^j \in \mathbb{K}_T$ and hence

$$\mathsf{E}_\mathsf{Q}[M_{\tau \wedge T}^{j,n}] = \mathsf{E}_\mathsf{P}[M_{\tau \wedge T}^{j,n}] = M_0^j \tag{10.5.53}$$

On the other hand,

$$\begin{aligned}
\mathsf{E}_\mathsf{Q}[M_{\tau \vee T}^{j,n}] &= \mathsf{E}_\mathsf{P}[\eta M_{\tau \vee T}^{j,n}] \\
&= \mathsf{E}_\mathsf{P}[\mathsf{E}_\mathsf{P}[\eta M_{\tau \vee T}^{j,n} \mid \mathcal{F}_T]] \\
&= \mathsf{E}_\mathsf{P}[\eta \mathsf{E}_\mathsf{P}[M_{\tau \vee T}^{j,n} \mid \mathcal{F}_T]] \\
&= \mathsf{E}_\mathsf{P}[\eta M_T^{j,n}] \\
&= \mathsf{E}_\mathsf{Q}[M_T^{j,n}] \\
&= M_0^j.
\end{aligned} \tag{10.5.54}$$

where we have used the facts that η is $\mathcal{F}_T$ measurable, $M^{j,n}$ is a P-martingale and (10.5.53). Now noting that $M_\tau^{j,n} = M_{\tau \wedge T}^{j,n} + M_{\tau \vee T}^{j,n} - M_T^{j,n}$, we conclude

$$\mathsf{E}_\mathsf{Q}[M_\tau^{j,n}] = \mathsf{E}_\mathsf{Q}[M_{\tau \wedge T}^{j,n}] + \mathsf{E}_\mathsf{Q}[M_{\tau \vee T}^{j,n}] - \mathsf{E}_\mathsf{Q}[M_T^{j,n}] = M_0^j.$$

Thus $M_t^{j,n} = M_{t \wedge \sigma_n}^j$ is a Q-martingale for every n so that M^j is a Q-local martingale and thus $\mathsf{Q} \in \mathbb{E}(M^1, M^2, \ldots, M^d)$. Similarly, if $\tilde{\mathsf{Q}}$ is the measure with density $\eta = (1 - c\xi)$, we can prove that $\tilde{\mathsf{Q}} \in \mathbb{E}(M^1, M^2, \ldots, M^d)$. Here $\mathsf{P} = \frac{1}{2}(\mathsf{Q} + \tilde{\mathsf{Q}})$ and $\mathsf{P} \neq \mathsf{Q}$ (since $\mathsf{P}(\xi \neq 0) > 0$). This contradicts the assumption that P is an extreme point of $\mathbb{E}(M^1, M^2, \ldots, M^d)$. Thus $(iii) \Rightarrow (ii)$.

To complete the proof, we need to show that (i) implies (v). Suppose (i) is true and let $\mathsf{Q} \in \mathbb{E}_\mathsf{P}(M^1, M^2, \ldots, M^d)$. Fix $T < \infty$ and let η be any $\mathcal{F}_T$ measurable bounded random variable. Since $\mathbb{L}^\infty(\Omega, \mathcal{F}_T, \mathsf{P}) \subseteq \mathbb{K}_T^v(M^1, M^2, \ldots, M^d)$ and $\mathcal{F}_0$ is trivial, we can get $g = (g^1, \ldots, g^d) \in \mathbb{L}_v(M^1, M^2, \ldots, M^d)$ with

$$\eta = c + \int_0^T \langle g, dM \rangle$$

such that $\int_0^t \langle g, dM \rangle$ is a martingale. Let $h_s = g_s \, 1_{[0,T]}(s)$ and $Z_t = \int_0^t \langle h, dM \rangle$. It follows that $Z_t = \mathbb{E}_{\mathsf{P}}[(\eta - c) \mid \mathcal{F}_t]$ and since η is bounded, it follows that Z is bounded. As noted earlier, since P and Q are equivalent, the stochastic integrals under P and Q are identical. Under Q, $M^1, M^2, \ldots, M^d$ being local martingales, $Z = \int \langle h, dM \rangle$ is a local martingale. Since it is also bounded, we conclude invoking Corollary 9.19 that Z is also a martingale under Q. Thus, $\mathbb{E}_{\mathsf{Q}}[Z_T] = 0 = \mathbb{E}_{\mathsf{P}}[Z_T]$ and thus using $\eta = c + Z_T$ we get $\mathbb{E}_{\mathsf{Q}}[\eta] = c = \mathbb{E}_{\mathsf{P}}[\eta]$. Since this holds for all $\mathcal{F}_T$ measurable bounded random variables η, we conclude Q and P agree on $\mathcal{F}_T$. In view of the assumption $\mathcal{F} = \sigma(\cup_t \mathcal{F}_t)$, we get $\mathsf{Q} = \mathsf{P}$ proving (v). This completes the proof. $\square$

10.6   Integral Representation w.r.t. Sigma-Martingales

In this section, we will prove an analogue of Theorem 10.38 for a multidimensional sigma-martingale.

Let us note that if $X^1, X^2, \ldots, X^d$ are sigma-martingales, then we can choose predictable $(0, \infty)$-valued process ϕ such that $\int \phi \, dX^j$ is a local martingale for each j. First for each j we choose ϕ^j and then take $\phi = \min(\phi^1, \ldots, \phi^d)$.

Here are two observations on sigma-martingales.

Lemma 10.39 *For sigma-martingales $X^1, X^2, \ldots, X^d$, $\mathbb{E}^\sigma(X^1, X^2, \ldots, X^d)$ and $\mathbb{E}_{\mathsf{P}}^\sigma(X^1, X^2, \ldots, X^d)$ are convex sets.*

Proof Let $\mathsf{Q}_1, \mathsf{Q}_2 \in \mathbb{E}^\sigma(X^1, X^2, \ldots, X^d)$ and let $\mathsf{Q}_0 = \alpha \mathsf{Q}_1 + (1 - \alpha) \mathsf{Q}_2$ for $0 < \alpha < 1$. For $k = 1, 2$, let ψ_k be $(0, \infty)$-valued predictable processes such that $\int \psi_k \, dX^j$ is a Q_k-martingale for each j. Then taking $\phi = \min(\psi_1, \psi_2)$ and $N^j = \int \phi \, dX^j$, it follows that for each j, N^j is a martingale under Q_1 as well as under Q_2 and thus under Q_0 as well. Hence $\mathsf{Q}_0 \in \mathbb{E}^\sigma(X^1, X^2, \ldots, X^d)$. Similarly, it can be shown that $\mathbb{E}_{\mathsf{P}}^\sigma(X^1, X^2, \ldots, X^d)$ is a convex set. $\square$

Lemma 10.40 *Let $X^1, X^2, \ldots, X^d$ be sigma-martingales. Then there exists a predictable $(0, \infty)$-valued process ϕ such that*

(i) *$N^j = \int \phi \, dX^j$ is a martingale for each j.*
(ii) *$\mathbb{K}_T^v(X^1, X^2, \ldots, X^d) = \mathbb{K}_T^v(N^1, N^2, \ldots, N^d)$.*
(iii) *$\mathbb{E}_P(N^1, N^2, \ldots, N^d) \subseteq \mathbb{E}_{\mathsf{P}}^\sigma(X^1, X^2, \ldots, X^d)$.*
(iv) *Suppose that P is an extreme point of $\mathbb{E}^\sigma(X^1, X^2, \ldots, X^d)$. Then P is also an extreme point of $\mathbb{E}(N^1, N^2, \ldots, N^d)$.*

Proof We have seen in Lemma 9.23 that we can choose $(0, \infty)$-valued predictable processes $\phi^j \in \mathbb{L}(X^j)$ such that $M^j = \int \phi^j \, dX^j$ are martingales. Let $\phi = \min(\phi^1, \ldots, \phi^d)$. Let $\psi^j = \frac{\phi}{\phi^j}$. Note that ψ^j is bounded by 1. Then $N^j = \int \phi \, dX^j =$

$\int \psi^j \, dM^j$ and then using Theorem 9.13 it follows that N^j is a martingale. This proves (i). For (ii), we have seen in Theorem 10.32 that $g = (g^1, g^2, \ldots, g^d) \in \mathbb{L}_v(N^1, N^2, \ldots, N^d)$ if and only if $g\phi = (g^1\phi, g^2\phi, \ldots, g^d\phi) \in \mathbb{L}_v(X^1, X^2, \ldots, X^d)$ and then

$$\int \langle g, dN \rangle = \int \langle \phi g, dX \rangle.$$

The assertion (ii) follows from this.

For (iii), if $Q \in \mathbb{E}_P(N^1, N^2, \ldots, N^d)$, then $X^j = \int \frac{1}{\phi} N^j$ is a Q-sigma-martingale for $1 \le j \le d$ and thus $Q \in \mathbb{E}_P^\sigma(X^1, X^2, \ldots, X^d)$.

For (iv), if $Q_1, Q_2 \in \mathbb{E}(N^1, N^2, \ldots, N^d)$ with $P = \frac{1}{2}(Q_1 + Q_2)$, then by part (iii)

$$Q_1, Q_2 \in \mathbb{E}_P(N^1, N^2, \ldots, N^d) \subseteq \mathbb{E}_P^\sigma(X^1, X^2, \ldots, X^d) \subseteq \mathbb{E}^\sigma(X^1, X^2, \ldots, X^d).$$

Since P is an extreme point of $\mathbb{E}^\sigma(X^1, X^2, \ldots, X^d)$, we conclude $Q_1 = Q_2 = P$ proving (iv) □

Part (ii) above along with Lemma 10.36 yields the following.

Corollary 10.41 *Let $X^1, X^2, \ldots, X^d$ be sigma-martingales. Then $\forall T < \infty$, $\mathbb{K}_T^v$ $(X^1, X^2, \ldots, X^d)$ is a closed subspace of $\mathbb{L}^1(\Omega, \mathcal{F}, P)$*

The proof of the next result is on the lines of corresponding results given in previous sections. The proof closely follows the proof of Theorem 10.38.

Theorem 10.42 *Let $X^1, X^2, \ldots, X^d$ be sigma-martingales on $(\Omega, \mathcal{F}, P)$. Suppose that $\mathcal{F}_0$ is trivial and $\mathcal{F} = \sigma(\cup_t \mathcal{F}_t)$. Then the following are equivalent.*

(i) *For every bounded martingale S, $\exists g \in \mathbb{L}_v(X^1, X^2, \ldots, X^d)$ such that*

$$S_t = S_0 + \int_0^t \langle g, dX \rangle \quad a.s. \ \forall t. \tag{10.6.1}$$

(ii) *For every martingale S, $\exists g \in \mathbb{L}_v(X^1, X^2, \ldots, X^d)$ such that $(10.6.1)$ is true.*
(iii) *P is an extreme point of the convex set $\mathbb{E}^\sigma(X^1, X^2, \ldots, X^d)$.*
(iv) *$\tilde{\mathbb{E}}_P^\sigma(X^1, X^2, \ldots, X^d) = \{P\}$.*
(v) *$\mathbb{E}_P^\sigma(X^1, X^2, \ldots, X^d) = \{P\}$.*

Proof Once again it can be seen that (i) is same as

$$\mathbb{L}^\infty(\Omega, \mathcal{F}_T, P) \subseteq \mathbb{K}_T^v(X^1, \ldots, X^d) \ \forall T \in (0, \infty)$$

and (ii) is same as

$$\mathbb{L}^1(\Omega, \mathcal{F}_T, P) = \mathbb{K}_T^v(X^1, \ldots, X^d) \ \forall T \in (0, \infty).$$

Also, as seen in Theorem 10.18, $\mathbb{K}_T^v(X^1, \ldots, X^d)$ is a closed subspace of $\mathbb{L}^1(\Omega,$ $\mathcal{F}_T, \mathsf{P})$. Since $\mathbb{L}^\infty(\Omega, \mathcal{F}_T, \mathsf{P})$ is dense in $\mathbb{L}^1(\Omega, \mathcal{F}_T, \mathsf{P})$, it follows that (i) and (ii) are equivalent. The proofs of $(iv) \Rightarrow (iii)$, (iv) implies (v) and $(v) \Rightarrow (iv)$ are exactly the same as that given in Theorem 10.38.

To see that $(iii) \Rightarrow (ii)$ let P be an extreme point of $\mathbb{E}(X^1, X^2, \ldots, X^d)$. Let ϕ and N^j be as in Lemma 10.40. Part (iv) in Lemma 10.40 now implies that P is an extreme point of $\mathbb{E}(N^1, N^2, \ldots, N^d)$ and then Theorem 10.38 implies that $\mathbb{L}^1(\Omega, \mathcal{F}_T, \mathsf{P}) = \mathbb{K}_T^v(N^1, \ldots, N^d)$. Thus part (ii) of Lemma 10.40 gives $\mathbb{L}^1(\Omega, \mathcal{F}_T, \mathsf{P}) = \mathbb{K}_T^v(X^1, \ldots, X^d)$ $\forall T \in (0, \infty)$ which is same as (ii).

To complete the proof, we will show that (i) implies (v). This is exactly as in Theorem 10.38. Suppose (i) is true and let $\mathsf{Q} \in \mathbb{E}_{\mathsf{P}}^\sigma(X^1, X^2, \ldots, X^d)$. Fix $T < \infty$ and let η be any $\mathcal{F}_T$ measurable bounded random variable. Since $\mathbb{L}^\infty(\Omega, \mathcal{F}_T, \mathsf{P}) \subseteq$ $\mathbb{K}_T^v(X^1, X^2, \ldots, X^d)$ and $\mathcal{F}_0$ is trivial, we can get $g \in \mathbb{L}_v(X^j)$ with

$$\eta = c + \int_0^T \langle g, dX \rangle$$

such that $\int_0^t \langle g, dX \rangle$ is a martingale. Let $h_s = g_s \, 1_{[0, T]}(s)$ and $Z_t = \int_0^t \langle h, dX \rangle$. It follows that $Z_t = \mathbb{E}_{\mathsf{P}}[(\eta - c) \mid \mathcal{F}_t]$ and since η is bounded, it follows that Z is bounded. As noted earlier, since P and Q are equivalent, the stochastic integrals under P and Q are identical. Under Q, $X^1, X^2, \ldots, X^d$ are sigma-martingales and thus $Z = \int \langle h, dX \rangle$ is also a sigma-martingale. Since it is also bounded, we conclude invoking Corollary 9.27 that Z is also a martingale under Q. Thus, $\mathbb{E}_{\mathsf{Q}}[Z_T] = 0 = \mathbb{E}_{\mathsf{P}}[Z_T]$ and thus using $\eta = c + Z_T$ we get $\mathbb{E}_{\mathsf{Q}}[\eta] = c = \mathbb{E}_{\mathsf{P}}[\eta]$. Since this holds for all $\mathcal{F}_T$ measurable bounded random variables η, we conclude Q and P agree on $\mathcal{F}_T$. In view of the assumption $\mathcal{F} = \sigma(\cup_t \mathcal{F}_t)$, we get $\mathsf{Q} = \mathsf{P}$ proving (v). This completes the proof. $\square$

This result has strong connections to mathematical finance and in particular to the theory of asset pricing. We will give a brief background in the next section.

10.7 Connections to Mathematical Finance

Connections of stochastic processes and mathematical finance go back to 1900 when Bachelier [1] studied the question of option pricing in his Doctoral Thesis. Here he had modelled the stock price movement as a Brownian motion. This was before Einstein used Brownian motion in the context of physics and movement of particles. Samuelson, Merton worked extensively on this question [49, 56]. The paper by Black–Scholes brought the connection to the forefront. The papers by Harrison and Pliska around 1980 built the formal connection between mathematical finance and stochastic calculus [22, 23]. The fundamental papers by Kreps [45], Yan [61], Stricker [58] laid the foundation for the so-called First Fundamental Theorem of Asset Pricing.

The final version of this result is due to Delbaen and Schachermayer [11, 12]. Also see [2, 30, 32, 43, 54].

We will give a brief account of the framework. We consider a market with d stocks, whose prices are modelled as stochastic processes $X^1, X^2, \ldots, X^d$, assumed to be processes with r.c.l.l. paths. The market is assumed to be ideal where there are no transaction costs and rate of interest r on deposits is same as rate of interest on loans, with instantaneous compounding, so that deposit of 1\$ is worth e^{rt} at time t. Let $S_t^j = X_t^j e^{-rt}$, $1 \le j \le d$ denote the discounted stock prices. Let $\mathcal{F}_t = \sigma(S_u^j : 0 \le u \le t, 1 \le j \le d)$.

A simple trading strategy is where an investor trades stock at finitely many time points and at a time s she/he can use information available up to time s. Then it can be seen that the strategy can be represented as follows: the times where the stock holdings change should be a stopping time and thus the strategy $f = (f^1, f^2, \ldots, f^d)$ can be seen to be representable as

$$f^j = \sum_{k=0}^{m-1} a_k^j \, 1_{(\sigma_k, \sigma_{k+1}]} \tag{10.7.1}$$

where σ_k are stopping times and a_k^j are $\mathcal{F}_{\sigma_k}$ measurable bounded random variables. For such a trading strategy, the value function (representing gain or loss from the strategy) is given by

$$V_f(f) = \sum_{j=1}^{d} \sum_{k=0}^{m-1} a_k^j (S_{\sigma_{k+1} \wedge t}^j - S_{\sigma_k \wedge t}^j). \tag{10.7.2}$$

When $S^1, \ldots S^j$ are semimartingales, then we see that $V_t(f) = \int_0^t \langle f, dS \rangle$.

A simple trading strategy f is said to be an arbitrage opportunity if for some T the following two conditions hold : (i) $\mathsf{P}(V_T(f) \ge 0) = 1$ and (ii) $\mathsf{P}(V_T(f) > 0) > 0$. One of the economic principles is that such a strategy cannot exist in a market in equilibrium, i.e. there are many buyers and sellers at that price for if it existed, all investors will follow the strategy as it gives an investor a shot at making money without taking any risk, thus disturbing the equilibrium. This is referred to as the *no arbitrage* principle or simply NA.

If each S^j is a martingale, then $V_t(f)$ is a martingale for every simple strategy f and thus $\mathsf{E}[V_T(f)] = 0$ and thus NA holds. It can be seen that NA is true even when each S^j is martingale under a probability measure Q that is equivalent to P. The converse to this statement is not true. However, it was recognized that if one rules out approximate arbitrage (in an appropriate sense) then indeed the converse is true. We will not trace the history of this line of thought (see references given above for the same) but give three results on this theme. The following result is Theorem 7.2 in [11].

Theorem 10.43 *Suppose the processes* $S^1, S^2 \ldots S^d$ *are locally bounded and that for any sequence of simple strategies* $f^n \in \mathbb{S}^d$ *and* $0 < T < \infty$, *the condition*

$$P(V_T(f^n) \geq -\frac{1}{n}) = 1, \quad \forall n \geq 1 \tag{10.7.3}$$

implies that for all $\epsilon > 0$

$$P(|V_T(f^n)| \geq \epsilon) \to 0 \text{ as } n \to \infty. \tag{10.7.4}$$

Then S^j is a semimartingale for each j.

The condition (10.7.3) $\Rightarrow$ (10.7.4) is essentially ruling out approximate arbitrage and has been called NFLVR—*No Free Lunch with Vanishing Risk* by Delbaen–Schachermayer.

Thus we now assume that $S^1, S^2 \ldots S^d$ are semimartingales. $\mathbb{L}_v(S^1, \ldots, S^d)$ is taken as the class of trading strategies and for $f \in \mathbb{L}_v(S^1, \ldots, S^d)$, the value process for the trading strategy f is defined to be $V_t(f) = \int_0^t \langle f, dS \rangle$. A trading strategy f is said to be admissible if for some constant K, one has

$$P(\int_0^t \langle f, dS \rangle \geq -K \ \forall t) = 1 \tag{10.7.5}$$

and

$$\int_0^t \langle f, dS \rangle \text{ converges in probability (to say} V(f)) \text{ as } t \to \infty. \tag{10.7.6}$$

The following theorem for one-dimensional case was proven in [11] (Corollary 1.2). For the multidimensional case see Theorem 8.2.1 in [13]. This also follows from Theorem 10.45 below.

Theorem 10.44 *Suppose $S^1, S^2 \ldots S^d$ are locally bounded semimartingales and that for any sequence of admissible strategies f^n the condition*

$$P(V(f^n) \geq -\frac{1}{n}) = 1, \quad \forall n \geq 1 \tag{10.7.7}$$

implies that for all $\epsilon > 0$

$$P(|V(f^n)| \geq \epsilon) \to 0 \text{ as } n \to \infty. \tag{10.7.8}$$

Then there exists a probability measure $\mathbf{Q}$ equivalent to $\mathbf{P}$ such that each S^j is a local martingale on $(\Omega, \mathcal{F}, \mathbf{Q})$.

Here is the final version of the (first) *Fundamental Theorem of Asset Pricing* —Theorem 14.1.1 in [12]. Also see [30], who independently proved the result.

Theorem 10.45 *Suppose the processes $S^1, S^2 \ldots S^d$ are semimartingales and that for any sequence of admissible strategies f^n the condition*

$$P(V(f^n) \geq -\frac{1}{n}) = 1, \quad \forall n \geq 1 \tag{10.7.9}$$

implies that for all $\epsilon > 0$

$$P(|V(f^n)| \geq \epsilon) \rightarrow 0 \text{ as } n \rightarrow \infty. \tag{10.7.10}$$

Then there exists a probability measure Q equivalent to P such that each S^j is a sigma-martingale on $(\Omega, \mathcal{F}, Q)$.

We can recast this result as follows: for semimartingales $S^1, S^2 \ldots S^d$, the set $\mathbb{E}_P^\sigma$ $(S^1, S^2 \ldots S^d)$ is non-empty if and only if $S^1, S^2 \ldots S^d$ satisfy NFLVR (namely $(10.7.9) \Rightarrow (10.7.10)$).

We now come to derivative securities and the role of NA condition (and NFLVR). A derivative security, also called a contingent claim, is a type of security traded whose value is contingent upon (or depends upon) the prices of the stocks. Thus the payout ξ, say at time T, could be $\xi = g(S_T^1, S_T^2, \ldots, S_T^d)$ for a function $g : \mathbb{R}^d \mapsto \mathbb{R}$ or could be a function of the paths $\{S_t^j : 0 \leq t \leq T, 1 \leq j \leq d\}$. All we require is that ξ is $\mathcal{F}_T$ measurable so that at time T, ξ is observed or known.

For example, $\xi = (S_T^1 - K)^+$: this is called the European Call Option (on S^1 with strike price K and terminal time T). Call Options have been traded on various exchanges across the world for close to a century. It was in the context of pricing of options that Bachelier had introduced in 1900 Brownian motion as a model for stock prices.

Suppose that ξ ($\mathcal{F}_T$ measurable random variable) is a contingent claim, $x \in \mathbb{R}$ and $f \in \mathbb{L}_v(S^1, S^2 \ldots S^d)$ is a trading strategy with

$$x + \int_0^T <f, dS> = \xi \text{ a.s.} \tag{10.7.11}$$

Even if ξ is not offered for trade, an investor can always replicate it with an initial investment x following the strategy f. If $(10.7.11)$ holds, (x, f) is called replicating strategy. In such a case, the price p of the contingent claim (assuming that the market is in equilibrium), must be equal to x. For if $p > x$, an investor could sell one the contingent claim at p, keep aside $p - x$, invest x and follow the strategy f. At time T the portfolio is worth exactly what the investor has to pay for the contingent claim. Thus the investor has made a profit of $(p - x)$ without any risk; in other words, it is an arbitrage opportunity. The possibility $p < x$ can be ruled out likewise, this time the investor buys a contingent claim at p and follows strategy $(-x, -f)$.

Thus if $(10.7.11)$ holds, in other words, a replicating strategy exists for a contingent claim ξ, the price of the contingent claim equals the initial investment needed for the strategy.

The market consisting of (discounted) stocks $S^1, S^2, \ldots, S^d$ is said to be complete if for all bounded $\mathcal{F}_T$ measurable random variables ξ, $\exists x \in \mathbb{R}$ and $f \in \mathbb{L}_v$ $(S^1, S^2, \ldots, S^d)$ such that for some $K < \infty$,

$$| \int_0^t < f, dS > | \le K \ \forall t \ge 0 \ a.s. \tag{10.7.12}$$

and

$$\xi = x + \int_0^T < f, dS > \ a.s. \tag{10.7.13}$$

Here is the second fundamental theorem of asset pricing.

Theorem 10.46 *Suppose* $S^1, S^2, \ldots, S^d$ *is a semimartingale such that*

$$\mathbb{E}_P^\sigma(S^1, S^2, \ldots, S^d) \ne \phi.$$

i.e. $S^1, S^2, \ldots, S^d$ *admit a Equivalent sigma-martingale measure (ESMM).*
   Then the market is complete if and only if the ESMM is unique, i.e. $\mathbb{E}_P^\sigma(S^1, S^2, \ldots, S^d)$ *is a singleton.*

Proof Let $Q \in \mathbb{E}_P^\sigma(S^1, S^2, \ldots, S^d)$. Suppose the market is complete. Fix $T > 0$ and let $\xi \in \mathbb{L}^\infty(\Omega, \mathcal{F}_T, Q)$. Consider the contingent claim ξ. Using completeness of market, obtain x, f satisfying (10.7.12) and (10.7.13). Under Q, $S^1, S^2, \ldots, S^d$ being sigma-martingales $N_t = \int_0^t < f, dS >$ for $t \le T$ and $N_t = N_{t \wedge T}$ is also a sigma-martingale. Being bounded (in view of (10.7.12)) it follows that N is a martingale. Thus $\xi \in \mathbb{K}_T^v(S^1, S^2, \ldots, S^d)$. Thus completeness of market is same is

$$\mathbb{K}_T^v(S^1, S^2, \ldots, S^d) \supseteq \mathbb{L}^\infty(\Omega, \mathcal{F}_T, Q).$$

By Theorem 10.42, this is equivalent to $\mathbb{E}_Q^\sigma(S^1, S^2, \ldots, S^d) = Q$. $\square$

Chapter 11
Dominating Process of a Semimartingale

In Chap. 7, we saw that using random time change, any continuous semimartingale can be transformed into a amenable semimartingale, and then one can have a growth estimate on the stochastic integral similar to the one satisfied by integrals w.r.t. Brownian motion.

When it comes to r.c.l.l. semimartingales, this is impossible in view of the jumps. Here we are faced with a difficulty as the stochastic integral is essentially created via an $\mathbb{L}^2$ estimate while the integral w.r.t. a process with finite variation is essentially defined as an $\mathbb{L}^1$ object-as in Lebesgue–Stieltjes integral. The problem is compounded by the fact that not every semimartingale need be locally integrable.

Metivier–Pellaumail obtained an inequality that makes all semimartingales amenable to the $\mathbb{L}^2$ treatment. Indeed, P. A. Meyer in a private correspondence had drawn our attention to the Metivier–Pellaumail inequality when he had seen the random change technique in [34]—both have an effect of making every semimartingale amenable to the $\mathbb{L}^2$ theory. As in earlier chapters, we fix a filtration $(\mathcal{F}_\cdot)$ on a complete probability space $(\Omega, \mathcal{F}, \mathsf{P})$ and we assume that $\mathcal{F}_0$ contains all P null sets in $\mathcal{F}$.

The Metivier–Pellaumail inequality relies on predictable quadratic variation $\langle M, M \rangle$ of a square integrable martingale, which we discussed in the previous chapter. The inequality states that for a square integrable martingale M and a stopping times τ, one has

$$\mathsf{E}[\sup_{0 \le t < \tau} |M_t|^2] \le 4\mathsf{E}[\ [M, M]_{\tau-} + \langle M, M \rangle_{\tau-}]. \qquad (11.0.1)$$

Since given any adapted process A with r.c.l.l. paths, we can get stopping times τ_n such that

$$\mathsf{E}[\sup_{0 \le t < \tau_n} |A_t|^2] < \infty$$

© Springer Nature Singapore Pte Ltd. 2018

R. L. Karandikar and B. V. Rao, *Introduction to Stochastic Calculus*,
Indian Statistical Institute Series, https://doi.org/10.1007/978-981-10-8318-1_11

the estimate (11.0.1) makes it feasible to obtain an estimate on growth of stochastic integral $\int f \, dX$ for any semimartingale X as we will see in later in this chapter.

11.1   An Optimization Result

Let $\mathcal{H}$ be a sub-σ-field of $\mathcal{F}$ and ξ be a square integrable r.v. such that

$$E[\xi \mid \mathcal{H}] = 0.$$

Let $A \in \mathcal{F}$. Consider the class $\mathbb{L}$ of random variables ϕ such that $E[\phi \mid \mathcal{H}] = 0$ and $\phi 1_A = \xi 1_A$. Consider the problem of minimizing $E[\phi^2]$, $\phi \in \mathbb{L}$.

Let us examine this in a special case, where $\mathcal{H}$ is the σ-field generated by a countable partition $\{H_n : n \geq 1\}$ of Ω. Let $E_n = H_n \cap A$ and $F_n = H_n \cap A^c$. Let $p_n = P(E_n)$ and $q_n = P(F_n)$

$$a_n = E[\xi 1_{E_n}], \quad b_n = E[\xi 1_{F_n}].$$

Since $E[\xi \mid \mathcal{H}] = 0$, it follows that $a_n + b_n = 0$. It follows that for any $\phi \in \mathbb{L}$, $E[\phi 1_{F_n}] = b_n$ since $E[\phi \mid \mathcal{H}] = 0$ and $\phi 1_A = \xi 1_A$. Since there is no other restriction on ϕ, it is clear that in this case, the minimum is attained when ϕ is a constant on each F_n, equal to 0 if $p_n = 0$ or $q_n = 0$ and equal to $\frac{b_n}{q_n}$ when $q_n > 0$. Thus let $N' = \{n : p_n > 0, \, q_n > 0\}$ and

$$\psi = \xi 1_A + \sum_{n \in N'} \frac{b_n}{q_n} 1_{F_n}$$

and it follows that for $\phi \in \mathbb{L}$

$$E[\phi^2] \geq E[\psi^2].$$

We would like to get a description of ψ as well as $E[\psi^2]$ in terms of ξ, $\mathcal{H}$ and A. For this, let $\mathcal{G} = \sigma(\mathcal{H}, A)$ and $\eta = E[\xi \mid \mathcal{G}]$. Then

$$\eta = \sum_{n \in N'} \frac{a_n}{p_n} 1_{E_n} + \sum_{n \in N'} \frac{b_n}{q_n} 1_{F_n}$$

and

$$\psi = \xi 1_A + \eta 1_{A^c}.$$

Thus

$$E[\psi^2] = E[\xi^2 1_A] + \sum_{n \in N'} \frac{b_n^2}{q_n}.$$

Now it can be checked that (using $a_n^2 = b_n^2$)

$$E[\eta^2 \mid \mathcal{H}] = \sum_{n \in N'} \frac{1}{p_n + q_n} \left(\frac{a_n^2}{p_n} + \frac{b_n^2}{q_n}\right) 1_{H_n}$$

$$= \sum_{n \in N'} \frac{1}{p_n + q_n} b_n^2 \left(\frac{p_n + q_n}{p_n q_n}\right) 1_{H_n}$$

$$= \sum_{n \in N'} \frac{b_n^2}{p_n q_n} 1_{H_n}$$

It thus follows that

$$E[1_A E[\eta^2 \mid \mathcal{H}]] = \sum_{n \in N'} E[\frac{b_n^2}{p_n q_n} 1_{E_n}]$$

$$= \sum_{n \in N'} \frac{b_n^2}{q_n}$$

and hence

$$E[\psi^2] = E[\xi^2 1_A] + E[1_A E[\eta^2 \mid \mathcal{H}]].$$

We will now show that the result is true in general. The calculations done above give us a clue as to the answer.

Theorem 11.1 *Let $\mathcal{H}$ be a sub-σ-field of $\mathcal{F}$, and let ξ be a random variable with $E[\xi^2] < \infty$ such that $E[\xi \mid \mathcal{H}] = 0$. Let $A \in \mathcal{F}$ and*

$$\mathbb{L} = \{\phi : E[\phi \mid \mathcal{H}] = 0, \ \phi 1_A = \xi 1_A\}.$$

Then for $\phi \in \mathbb{L}$

$$E[\phi^2] \geq E[\psi^2]$$

where $\mathcal{G} = \sigma(\mathcal{H}, A)$, $\eta = E[\xi \mid \mathcal{G}]$ and

$$\psi = \xi 1_A + \eta 1_{A^c}.$$

Further,

$$E[\psi^2] = E[\xi^2 1_A] + E[1_A E[\eta^2 \mid \mathcal{H}]]. \tag{11.1.1}$$

Proof Let us begin by noting that $\mathcal{G} = \{(B \cap A) \cup (C \cap A^c) : B, C \in \mathcal{H}\}$. Hence

$$\mathbb{L}^2(\Omega, \mathcal{G}, P) = \{\beta 1_A + \theta 1_{A^c} : \beta, \theta \in \mathbb{L}^2(\Omega, \mathcal{H}, P)\}. \tag{11.1.2}$$

Thus, $\eta = E[\xi \mid \mathcal{G}]$ can be written as $\eta = \beta 1_A + \theta 1_{A^c}$ where β, θ are $\mathcal{H}$ measurable square integrable random variables. Note that

$$E[\phi^2] = E[(\phi - \psi)^2 + \psi^2 + 2(\phi - \psi)\psi]$$

Now $(\phi - \psi)1_A = 0$ and hence $(\phi - \psi)\psi = (\phi - \psi)1_{A^c}\psi = (\phi - \psi)1_{A^c}\theta = (\phi - \psi)\theta$ and hence

$$\begin{aligned}
E[2(\phi - \psi)\psi] &= E[E[2(\phi - \psi)\theta \mid \mathcal{H}]] \\
&= E[\theta E[2(\phi - \psi) \mid \mathcal{H}]] \\
&= 0
\end{aligned}$$

as $E[\psi \mid \mathcal{H}] = 0$ and $E[\phi \mid \mathcal{H}] = 0$. This proves the first part. For the second part, recall $\eta = \beta 1_A + \theta 1_{A^c}$. Let $\alpha = E[1_A \mid \mathcal{H}]$. Since $E[\eta \mid \mathcal{H}] = 0$, we have

$$\beta\alpha + \theta(1 - \alpha) = 0. \tag{11.1.3}$$

Thus

$$\begin{aligned}
E[\eta^2 \mid \mathcal{H}] &= E[\beta^2 1_A + \theta^2 1_{A^c} \mid \mathcal{H}] \\
&= \beta^2\alpha + \theta^2(1 - \alpha)
\end{aligned}$$

and hence

$$\begin{aligned}
E[1_A E[\eta^2 \mid \mathcal{H}]] &= E[E[\eta^2 \mid \mathcal{H}]E[1_A \mid \mathcal{H}]] \\
&= E[\beta^2\alpha + \theta^2(1 - \alpha)E[1_A \mid \mathcal{H}]] \\
&= E[\beta^2\alpha^2 + \theta^2(1 - \alpha)\alpha] \\
&= E[\theta^2(1 - \alpha)^2 + \theta^2(1 - \alpha)\alpha] \\
&= E[\theta^2(1 - \alpha)]
\end{aligned} \tag{11.1.4}$$

where we have used (11.1.3). On the other hand

$$\begin{aligned}
E[\eta^2 1_{A^c}] &= E[\theta^2 1_{A^c}] \\
&= E[\theta^2 E[1_{A^c} \mid \mathcal{H}] \\
&= E[\theta^2(1 - \alpha)]
\end{aligned} \tag{11.1.5}$$

Thus (11.1.4) and (11.1.5) yield

$$E[\eta^2 1_{A^c}] = E[1_A E[\eta^2 \mid \mathcal{H}]]. \tag{11.1.6}$$

Now, from the definition of ψ, we have

$$\begin{aligned}
E[\psi^2] &= E[\xi^2 1_A] + E[\eta^2 1_{A^c}] \\
&= E[\xi^2 1_A] + E[1_A E[\eta^2 \mid \mathcal{H}]].
\end{aligned} \tag{11.1.7}$$

where the last step follows from (11.1.6). $\square$

Let us observe that $\eta = \mathsf{E}[\xi \mid \mathcal{G}]$ and hence by Jensen's inequality, one has.

$$\mathsf{E}[1_A \mathsf{E}[\eta^2 \mid \mathcal{H}]] \leq \mathsf{E}[1_A \mathsf{E}[\xi^2 \mid \mathcal{H}]].$$

Thus the previous result leads to

Theorem 11.2 *Let ξ be a random variable such that $\mathsf{E}[\xi \mid \mathcal{H}] = 0$ and $\mathsf{E}[\xi^2] < \infty$. For $A \in \mathcal{F}$, there exists a random variable ψ such that*

  (i) $\psi 1_A = \xi 1_A$.
 (ii) $\mathsf{E}[\psi \mid \mathcal{H}] = 0$.
(iii) $\mathsf{E}[\psi^2] \leq \mathsf{E}[1_A (\xi^2 + \mathsf{E}[\xi^2 \mid \mathcal{H}])]$.

11.2   Metivier–Pellaumail Inequality

We are now in a position to prove

Theorem 11.3 (Metivier–Pellaumail inequality) *Let M be a square integrable martingale with $M_0 = 0$ and σ be a stopping time. Then we have*

$$\mathsf{E}[\sup_{t < \sigma} |M_t|^2] \leq 4[\mathsf{E}[M, M]_{\sigma-} + \mathsf{E}[\langle M, M \rangle_{\sigma-}]]. \tag{11.2.1}$$

Proof Suffices to prove it for σ bounded as the general case follows by using (11.2.1) for $\sigma \wedge m$ and taking limit over m. So we assume $\sigma \leq T$. Now we can assume that $M_t = M_{t \wedge T}$.

Let $\{\tau_k : k \geq 1\}$ be predictable stopping times as in Theorem 8.75. Let $\xi_k = (\Delta M)_{\tau_k}$, $U^k = \xi_k 1_{[\tau_k, \infty)}$, $Z = \sum_{k \in F} U^k$, $N = M - Z$. Since τ_k is predictable, U^k is a martingale. We have seen in Theorem 8.75 that $\langle N, N \rangle$ is a continuous increasing process. Moreover

$$[Z, Z]_t = \sum_{k \in F} \xi_k^2 1_{[\tau_k, \infty)}(t) \tag{11.2.2}$$

and

$$\langle Z, Z \rangle_t = \sum_{k \in F} \mathsf{E}[\xi_k^2 \mid \mathcal{F}_{\tau_k-}] 1_{[\tau_k, \infty)}(t) \tag{11.2.3}$$

For $k \in F$, let $\mathcal{G}_k = \sigma(\mathcal{F}_{\tau_k-}, \{\sigma > \tau_k\})$, $\eta_k = \mathsf{E}[\xi_k \mid \mathcal{G}_k]$ and

$$\psi_k = \xi_k 1_{\{\sigma > \tau_k\}} + \eta_k 1_{\{\sigma \leq \tau_k\}}.$$

Since $\mathsf{E}[\xi_k \mid \mathcal{F}_{\tau_k-}] = 0$ and $\mathcal{F}_{\tau_k-} \subseteq \mathcal{G}_k$, it follows that $\mathsf{E}[\psi_k \mid \mathcal{F}_{\tau_k-}] = 0$ and hence

$$V_t^k = \psi_k 1_{[\tau_k, \infty)}(t)$$

is a martingale. Since V^k, V^j do not have common jumps for $j \neq k$

$$[V^k, V^j] = 0, \quad \langle V^k, V^j \rangle = 0. \tag{11.2.4}$$

Moreover, for all $k \in F$, Theorem 11.2 implies

$$\mathsf{E}[(\psi_k)^2] \leq \mathsf{E}[1_{\{\sigma > \tau_k\}}(\xi_k^2 + \mathsf{E}[\xi_k^2 \mid \mathcal{F}_{\tau_k-}])] \tag{11.2.5}$$

This of course also gives

$$\mathsf{E}[(\psi_k)^2] \leq 2\mathsf{E}[(\xi_k)^2].$$

Let $Y = \sum_{k \in F} V^k$. If F is finite, clearly, Y is a martingale. In case F is infinite, the series converges and Y is a martingale as in the proof of Theorem 8.75. Noting that $\psi_k 1_{\{\sigma > \tau_k\}} = \xi_k 1_{\{\sigma > \tau_k\}}$, we have

$$\begin{aligned}
V_t^k 1_{\{t < \sigma\}} &= \psi_k 1_{\{t < \sigma\}} 1_{[\tau_k, \infty)}(t) \\
&= \xi_k 1_{\{t < \sigma\}} 1_{[\tau_k, \infty)}(t) \\
&= U_t^k 1_{\{t < \sigma\}}
\end{aligned} \tag{11.2.6}$$

As a consequence of (11.2.6), we get

$$Z_t 1_{\{t < \sigma\}} = Y_t 1_{\{t < \sigma\}}. \tag{11.2.7}$$

Moreover, using (11.2.5), (11.2.2) and (11.2.3) we get

$$\begin{aligned}
\mathsf{E}[[Y, Y]_\sigma] &= \mathsf{E}[\sum_{k \in F} \psi_k^2 1_{\{\sigma \geq \tau_k\}}] \\
&\leq \mathsf{E}[\sum_{k \in F} \psi_k^2] \\
&\leq \mathsf{E}[\sum_{k \in F} 1_{\{\sigma > \tau_k\}}(\xi_k^2 + \mathsf{E}[\xi_k^2 \mid \mathcal{F}_{\tau_k-}])] \\
&\leq \mathsf{E}[[Z, Z]_{\sigma-} + \langle Z, Z \rangle_{\sigma-}]
\end{aligned} \tag{11.2.8}$$

Let $X = N + Y$. Then in view of (11.2.7), $X_t 1_{\{t < \sigma\}} = M_t 1_{\{t < \sigma\}}$ and hence

$$\begin{aligned}
\mathsf{E}[\sup_{t < \sigma} |M_t|^2] &= \mathsf{E}[\sup_{t < \sigma} |X_t|^2] \\
&\leq \mathsf{E}[\sup_{t \leq \sigma} |X_t|^2] \\
&\leq 4\mathsf{E}[[X, X]_\sigma]
\end{aligned} \tag{11.2.9}$$

Finally, using (11.2.8) along with $[N, Y] = 0$ and $\mathsf{E}[[N, N]_\sigma] = \mathsf{E}[\langle N, N \rangle_\sigma]$, we get

$$\mathsf{E}[[X, X]_\sigma] = \mathsf{E}[[N, N]_\sigma] + \mathsf{E}[[Y, Y]_\sigma]$$
$$\le \mathsf{E}[\langle N, N\rangle_\sigma] + \mathsf{E}[[Z, Z]_{\sigma-}] + \mathsf{E}[\langle Z, Z\rangle_{\sigma-}] \tag{11.2.10}$$

Since $\langle N, N\rangle$ is continuous, we have $\langle N, N\rangle_{\sigma-} = \langle N, N\rangle_\sigma$. As seen in Theorem 8.75 $\langle N, Z\rangle = 0$ and hence we get

$$\langle N, N\rangle_\sigma + \langle Z, Z\rangle_{\sigma-} = \langle M, M\rangle_{\sigma-}.$$

This along with (11.2.9) and (11.2.10) implies

$$\mathsf{E}[\sup_{t<\sigma}|M_t|^2] \le 4[\mathsf{E}[[Z, Z]_{\sigma-}] + \mathsf{E}[\langle M, M\rangle_{\sigma-}]] \tag{11.2.11}$$

Finally, $[Z, Z] \le [M, M]$ implies (11.2.1). $\qquad\square$

11.3 Growth Estimate

The Metivier–Pellaumail inequality enables us to obtain a growth estimate on $\int f\, dX$ for any semimartingale X. Given a locally bounded predictable process f and a decomposition $X = M + A$ of a semimartingale X, where M is a locally square integrable martingale with $M_0 = 0$ and A is a process with finite variation paths, let $Y = \int f\, dX$, $N = \int f\, dM$ and $B = \int f\, dA$. Then $Y = N + B$, N is a locally square integrable martingale and $B \in \mathbb{V}$. Further,

$$[N, N]_t = \int_0^t f_s^2\, d[M, M]_s, \tag{11.3.1}$$

$$\langle N, N\rangle_t = \int_0^t f_s^2\, d\langle M, M\rangle_s \tag{11.3.2}$$

and thus in view of (11.2.1), we have

$$\mathsf{E}[\sup_{t<\tau}|\int_0^t f\, dM|^2] \le 4\mathsf{E}[\int_0^{\tau-} f_s^2\, d[M, M]_s + \int_0^{\tau-} f_s^2\, d\langle M, M\rangle_s]. \tag{11.3.3}$$

Writing $|A|_t = \mathrm{VAR}_{[0,t]}(A)$, we have for all t,

$$|\int_0^t |f_s|\, dA_s|^2 \le |A|_t \int_0^t |f_s^2|\, d|A|_s$$

and hence

$$\mathsf{E}[\sup_{t<\tau}|\int_0^t f\, dA|^2] \le \mathsf{E}[|A|_{\tau-} \int_0^{\tau-} |f_s^2|\, d|A|_s]. \tag{11.3.4}$$

We can combine the two estimates (11.3.3) and (11.3.4) as follows: let

$$V_t = 8(1 + [M, M]_t + \langle M, M \rangle_t + \text{VAR}_{[0,t]}(A))$$ (11.3.5)

and then we have (compare with (7.2.5) for amenable semimartingales)

$$\mathsf{E}[\sup_{t < \tau} |\int_0^t f \, dX|^2] \leq \mathsf{E}[V_{\tau-} \int_0^{\tau-} |f_s^2| d|V|_s].$$ (11.3.6)

The significant point about this estimate is that given any locally bounded predictable f and any semimartingale X, we can get a sequence of stopping times τ_n increasing to ∞ such that the expression on the right-hand side in (11.3.6) is finite. This may not be the case for the estimate

$$\mathsf{E}[\sup_{t \leq \tau} |\int_0^t f \, dX|^2] \leq \mathsf{E}[V_\tau \int_0^\tau |f_s^2| d|V|_s]$$

which indeed can be obtained without any need for the Metivier–Pellaumail inequality. The process V introduced above (modulo a constant) was called a control process of the semimartingale X by Metivier–Pellaumail.

While the control process was used to successfully deal with stochastic differential equations driven by semimartingales, the notion is not natural as even if the semimartingale is small in Emery topology, the control process may not be small. Further, if V is control process for a semimartingale X, for a constant c, cV may not be a control process for cX.

Definition 11.4 An (adapted) increasing process U is said to be a *dominating process* for a semimartingale X if there exists a decomposition $X = M + A$, with M a locally square integrable martingale with $M_0 = 0$, A a process with finite variation paths such that the process B defined by

$$B_t = U_t - 2\sqrt{2}([M, M]_t + \langle M, M \rangle_t)^{1/2} - \sqrt{2}|A|_t$$ (11.3.7)

belongs to $\mathbb{V}^+$; *i.e.* B is an increasing process with $B_0 \geq 0$.

Theorem 11.5 *Every semimartingale X admits a dominating process.*

Proof As noted in Corollary 5.60, X admits a decomposition $X = M + A$, where M is a locally square integrable martingale with $M_0 = 0$ and $A \in \mathbb{V}$. Then

$$U_t = 2\sqrt{2}([M, M]_t + \langle M, M \rangle_t)^{1/2} + \sqrt{2}|A|_t$$ (11.3.8)

is a dominating process.   □

Remark **11.6** One of the reasons that we did not define U given by (11.3.8) for some decomposition $X = M + A$ as the dominating process is that now we can have a common dominating process for finitely many semimartingales.

With this definition we have the following inequality.

Theorem 11.7 *Let X be a semimartingale and U be a dominating process for X. Then for any stopping time σ we have*

$$\mathsf{E}[\sup_{0 \le t < \sigma} |X_t|^2] \le \mathsf{E}[U_{\sigma-}^2] \tag{11.3.9}$$

Proof Let $X = M + A$ be a decomposition of semimartingale X as in Definition 11.4, with B defined by (11.3.7) being an increasing process with $B_0 \ge 0$. Then it follows that $B_t \ge 0$ for all t and hence

$$2\sqrt{2}([M, M]_t + \langle M, M \rangle_t)^{1/2} + \sqrt{2}|A|_t \le U_t$$

and as a result

$$8([M, M]_t + \langle M, M \rangle_t) + 2|A|_t^2 \le U_t^2 \tag{11.3.10}$$

On the other hand, for any stopping time σ, we have

$$\mathsf{E}[\sup_{0 \le t < \sigma} |X_t|^2] \le 2\mathsf{E}[\sup_{0 \le t < \sigma} |M_t|^2] + 2\mathsf{E}[\sup_{0 \le t < \sigma} |A_t|^2] \tag{11.3.11}$$

By Metivier–Pellaumail inequality (11.2.1) we have

$$\mathsf{E}[\sup_{t < \sigma}|M_t|^2] \le 4[\mathsf{E}[M, M]_{\sigma-} + \mathsf{E}[\langle M, M \rangle_{\sigma-}]]. \tag{11.3.12}$$

At the same time $|A_t| \le |A|_t$ (where $|A|$ is the total variation of A). As a result

$$\mathsf{E}[\sup_{t < \sigma}|A_t|^2] \le \mathsf{E}[|A|_{\sigma-}^2]. \tag{11.3.13}$$

Combining (11.3.11)–(11.3.13), we get

$$\mathsf{E}[\sup_{0 \le t < \sigma} |X_t|^2] \le \mathsf{E}[8([M, M]_{\sigma-} + \langle M, M \rangle_{\sigma-}) + 2|A|_{\sigma-}^2]. \tag{11.3.14}$$

Now the required estimate (11.3.9) follows from (11.3.10) and (11.3.14). □

Ideally, we would have liked a notion of dominating process such that if U^1, U^2 be dominating processes of semimartingale X^1, X^2, respectively, then $V = U^1 + U^2$ is a dominating process for $Y = X^1 + X^2$. While this is not quite true, we will show that Y admits a dominating process W such that $W_t \le V_t$. To prove this, we need the following result.

Lemma 11.8 *For M, N be locally square integrable martingales, let*

$$q(M, N) = [M, N] + \langle M, N \rangle.$$

Then, for all t

$$(q(M+N, M+N)_t)^{\frac{1}{2}} \le (q(M, M)_t)^{\frac{1}{2}} + (q(N, N)_t)^{\frac{1}{2}}. \qquad (11.3.15)$$

Also, if M^i are locally square integrable martingales, $i = 1, 2, \ldots, k$ then

$$(q(\sum_{i=1}^{k} M^i, \sum_{i=1}^{k} M^i)_t)^{\frac{1}{2}} \le \sum_{i=1}^{k} (q(M^i, M^i)_t)^{\frac{1}{2}}. \qquad (11.3.16)$$

Proof Since $M, N \mapsto [M, N]$ and $M, N \mapsto \langle M, N \rangle$ are bilinear maps, it follows that same is true of $M, N \mapsto q(M, N)$. Further, $q(M, M)_t \ge 0$. Then proceeding as in the proof of Theorems 4.77 and 8.59, we can conclude that

$$q(M, N)_t \le (q(M, M)_t)^{\frac{1}{2}}(q(N, N)_t)^{\frac{1}{2}} \qquad (11.3.17)$$

and as a result one has

$$q(M+N, M+N)_t = q(M, M)_t + q(N, N)_t + 2q(M, N)_t$$
$$\le q(M, M)_t + q(N, N)_t + 2(q(M, M)_t)^{\frac{1}{2}}(q(N, N)_t)^{\frac{1}{2}}.$$
$$= [(q(M, M)_t)^{\frac{1}{2}} + (q(N, N)_t)^{\frac{1}{2}}]^2$$

This proves (11.3.15). The estimate (11.3.16) is just the k variable version of the same. $\square$

Theorem 11.9 *Let X^1, X^2 be semimartingales, and let U^1, U^2 be dominating processes for X^1 and X^2, respectively. Let $Y = X^1 + X^2$. Then the semimartingale Y admits a dominating process V such that*

$$V_t \le U_t^1 + U_t^2 \quad \forall t. \qquad (11.3.18)$$

Proof Let $X^i = M^i + A^i$ $(i = 1, 2)$ be decompositions of the semimartingales with M^i being a local square integrable martingale with $M_0^i = 0$ and A^i being a process with finite variation paths such that

$$D_t^i = U_t^i - C_t^i$$

are increasing processes with $D_0^i \ge 0$ where

$$C_t^i = 2\sqrt{2}([M^i, M^i]_t + \langle M^i, M^i \rangle_t)^{1/2} + \sqrt{2}|A^i|_t.$$

Let $N = M^1 + M^2$, $B = A^1 + A^2$ and

$$V_t = 2\sqrt{2}([N, N]_t + \langle N, N \rangle_t)^{1/2} + \sqrt{2}|B|_t.$$

Then V is a dominating process for Y and to complete the proof of the first part, we need to show (11.3.18). Clearly,

$$|B|_t \leq |A^1|_t + |A^2|_t \quad \forall t. \tag{11.3.19}$$

Since $N = M^1 + M^2$, it follows from Lemma 11.8 that

$$([N, N]_t + \langle N, N \rangle_t)^{1/2}$$
$$\leq ([M^1, M^1]_t + \langle M^1, M^1 \rangle_t)^{1/2} + ([M^2, M^2]_t + \langle M^2, M^2 \rangle_t)^{1/2} \tag{11.3.20}$$

Now estimates (11.3.19) and (11.3.20) imply

$$V_t \leq C_t^1 + C_t^2.$$

Since $C_t^i \leq U_t^i$, the required result follows.   □

We now move to exploring the connection of dominating process with stochastic integral. Here are a sequence of auxiliary results that we need later.

Lemma 11.10 *Let $U, V \in \mathbb{V}^+$ be such that W defined by $W_t = V_t - U_t$ belongs to $\mathbb{V}^+$. Then Z defined by $Z_t = V_t^2 - U_t^2$ also belongs to $\mathbb{V}^+$.*

Proof Note that $U, V \in \mathbb{V}^+$ implies $U_t \geq 0$ and $V_t \geq 0$ and

$$Z_t - Z_s = (V_t - U_t)(V_t + U_t) - (V_s - U_s)(V_s + U_s)$$
$$= W_t(V_t + U_t) - W_s(V_s + U_s)$$
$$= (W_t - W_s)(V_s + U_s) + W_t(V_t - V_s) + W_t(U_t - U_s)$$

Since $W, U, V \in \mathbb{V}^+$, it follows that $Z_t - Z_s \geq 0$ and thus Z is increasing. Also, easy to see that $Z_0 \geq 0$ and so $Z \in \mathbb{V}^+$.   □

Remark **11.11** Essentially the same argument as in Lemma 11.8 (see also (4.6.21)) gives us for locally square integrable martingales $N^j, j = 1, 2, \ldots k$

$$\langle N^i, N^j \rangle_t \leq (\langle N^i, N^i \rangle_t)^{\frac{1}{2}} (\langle N^j, N^j \rangle_t)^{\frac{1}{2}} \tag{11.3.21}$$

$$[N^i, N^j]_t \leq ([N^i, N^i]_t)^{\frac{1}{2}} ([N^j, N^j]_t)^{\frac{1}{2}} \tag{11.3.22}$$

and as a consequence

$$(\langle \sum_{i=1}^{k} N^i, \sum_{j=1}^{k} N^j \rangle_t)^{\frac{1}{2}} = (\sum_{i=1}^{k} \sum_{j=1}^{k} \langle N^i, N^j \rangle_t)^{\frac{1}{2}}$$

$$\leq [\sum_{i=1}^{k} \sum_{j=1}^{k} (\langle N^i, N^i \rangle_t)^{\frac{1}{2}} (\langle N^j, N^j \rangle_t)^{\frac{1}{2}}]^{\frac{1}{2}} \qquad (11.3.23)$$

$$= \sum_{i=1}^{k} (\langle N^i, N^i \rangle_t)^{\frac{1}{2}}$$

and similarly,

$$([\sum_{i=1}^{k} N^i, \sum_{j=1}^{k} N^j]_t)^{\frac{1}{2}} \leq \sum_{i=1}^{k} ([N^i, N^i]_t)^{\frac{1}{2}} \qquad (11.3.24)$$

For a locally bounded predictable process f and an increasing process V, let $\theta_t(f, V)$ be defined by

$$\theta_t(f, V) = (\int_0^t |f_s|^2 dV_s^2)^{\frac{1}{2}} + \int_0^t |f_s| dV_s. \qquad (11.3.25)$$

Note that $\theta(f, V)$ is an increasing process,

$$\theta_t(f, V) \leq 2(\sup_{0 \leq s \leq t} |f_s|) V_t, \qquad (11.3.26)$$

$$\int_0^t |f_s|^2 dV_s^2 \leq \theta_t^2(f, V) \leq 2 \int_0^t |f_s|^2 dV_s^2 + 2(\int_0^t |f_s| dV_s)^2 \qquad (11.3.27)$$

and also

$$\int_0^t |f_s|^2 dV_s^2 \leq \theta_t^2(f, V) \leq 2(\int_0^t |f_s|^2 dV_s^2 + V_t \int_0^t |f_s|^2 dV_s). \qquad (11.3.28)$$

The following result gives interplay of this notion of dominating process with that of stochastic integral.

Theorem 11.12 *Let X be a semimartingale, U be a dominating processes for X and f be a locally bounded predictable process. Let $Y = \int f dX$. Then the semimartingale Y admits a dominating process V such that*

$$V_t \leq \theta_t(f, U) \quad \forall t. \qquad (11.3.29)$$

Proof Let $X = M + A$ be a decomposition of the semimartingale X with M being a locally square integrable martingale with $M_0 = 0$ and A being a process with finite variation paths such that $U - C - \sqrt{2}|A| \in \mathbb{V}^+$ where $C \in \mathbb{V}^+$ is given by

$$C_t = 2\sqrt{2}(\langle M, M \rangle_t + [M, M]_t)^{1/2}.$$

Let $N = \int f \, dM$, $B = \int f \, dA$. Then N is a locally square integrable martingale with $N_0 = 0$, B is a process with finite variation paths such that $Y = N + B$. Now

$$\langle N, N \rangle_t + [N, N]_t = \int_0^t |f|^2 d\langle M, M \rangle + \int_0^t |f|^2 d[M, M]$$

and $|B|_t = \int_0^t |f| d|A|$ and as a consequence, V defined below is a dominating process:

$$V_t = 2\sqrt{2}(\langle N, N \rangle_t + [N, N]_t)^{1/2} + \sqrt{2}|B|$$
$$= (\int_0^t |f|^2 dC^2)^{1/2} + \sqrt{2} \int_0^t |f| d|A| \qquad (11.3.30)$$

Since $U - C \in \mathbb{V}^+$, it follows that (using Lemma 11.10) $U^2 - C^2 \in \mathbb{V}^+$ and hence

$$\int_0^t |f|^2 dC^2 \leq \int_0^t |f|^2 dU^2 \qquad (11.3.31)$$

and $U - \sqrt{2}|A| \in \mathbb{V}^+$ implies

$$\sqrt{2} \int_0^t |f| d|A| \leq \int_0^t |f| d|U| \qquad (11.3.32)$$

Combining (11.3.30)–(11.3.32), we get

$$V_t \leq (\int_0^t |f|^2 dU^2)^{1/2} + \int_0^t |f| d|U| = \theta_t(f, U). \qquad (11.3.33)$$

This proves (11.3.29). □

Putting together Theorems 11.7 and 11.12 we now obtain an estimate on the growth of a stochastic integral.

Theorem 11.13 *Let X be semimartingale and f be a locally bounded predictable process. Let V be a dominating process for X. Then for any stopping time τ one has*

$$E[\sup_{t < \tau} |\int_0^t f \, dX|^2] \leq E[\theta_{\tau-}^2(f, V)] \qquad (11.3.34)$$

Further,

$$E[\sup_{t < \tau} |\int_0^t f \, dX|^2] \leq 4E[(\sup_{0 \leq s < \tau} |f_s^2|) V_{\tau-}^2] \qquad (11.3.35)$$

Proof As noted earlier, (11.3.34) follows from Theorems 11.7 and 11.12 and then (11.3.35) follows from (11.3.26). □

Remark **11.14** Before proceeding, we would like to stress that given a locally bounded predictable f and an increasing process $V \in \mathbb{V}^+$, one can always get bounded stopping times τ_m increasing to ∞ such that

$$E[\theta^2_{\tau_m-}(f, V)] < \infty$$

and thus the estimate (11.3.34) is meaningful for any locally bounded predictable f and any semimartingale.

We can now add to the Dellacherie–Meyer–Mokobodzky–Bichteler Theorem. Each of the seven equivalent conditions in Theorem 5.89 is equivalent to existence of a dominating process. We will list here only two out of the seven.

Theorem 11.15 *Let X be an r.c.l.l. $(\mathcal{F}_{\cdot})$ adapted process. Let J_X be defined by (4.2.1)–(4.2.2). Then the following are equivalent.*

(i) *X is a weak stochastic integrator; i.e. if $f^n \in \mathbb{S}$, $f^n \to 0$ uniformly, then $J_X(f^n)_t \to 0$ in probability $\forall t < \infty$.*

(ii) *X is a semimartingale; i.e. X admits a decomposition $X = M + A$ where M is a local martingale and A is a process with finite variation paths.*

(iii) *There exists an increasing adapted process V such that for all stopping times τ and for all $f \in \mathbb{S}$, one has*

$$E[\sup_{t < \tau}|J_X(f)_t|^2] \leq 2E[\int 1_{[0,\tau)}(s)|f_s|^2 dV_s^2 + (\int 1_{[0,\tau)}(s)|f_s|dV_s)^2]$$

$$(11.3.36)$$

Proof We have already shown that (i) and (ii) are equivalent. Using Theorem 11.13, it follows that (ii) implies (iii) follows. To see that (iii) implies (i), note that given any adapted increasing process V, $s < \infty$ and $\epsilon > 0$, we can get a stopping time τ such that $V_{\tau-}$ is bounded and $P(s < \tau) \geq (1 - \epsilon)$. See Remark 11.14. Now the result follows from Theorem 11.13 and the estimate (11.3.27).   □

11.4 Alternate Metric for Emery Topology

We will now introduce another metric on the space of semimartingales in terms of dominating process and then show that this metric is equivalent to the metric introduced earlier for the Emery topology.

Definition 11.16 For semimartingales X, Y, let

$$\mathbf{d}_{sm}(X, Y) = \inf\{\mathbf{d}_{ucp}(V, 0) : V \text{ is a dominating process for } X - Y\}.$$

It is easy to see that if V is a dominating process for $X - Y$ then V is also a dominating process for $Y - X$ and thus $\mathbf{d}_{sm}(X, Y) = \mathbf{d}_{sm}(Y, X)$. The next two results will show that $\mathbf{d}_{sm}$ is a metric.

Lemma 11.17 *Let X, Y be semimartingales such that $\mathbf{d}_{sm}(X, Y) = 0$. Then $X = Y$.*

Proof Get $V^k \in \mathbb{V}^+$ such that V^k dominates $X - Y$ and $\mathbf{d}_{ucp}(V^k, 0) \leq 2^{-k}$. Then

$$\sum_{k=1}^{\infty} \sum_{n=1}^{\infty} 2^{-n} \mathsf{E}[V_n^k \wedge 1] \leq 1$$

and as a consequence, for every n, (using Fubini's Theorem) we have

$$\mathsf{E}[\sum_{k=1}^{\infty} 2^{-n}[V_n^k \wedge 1]] \leq 1.$$

Thus (noting $V_t^k \geq 0$)

$$\sum_{k=1}^{\infty} 2^{-n}[V_n^k \wedge 1] < \infty \quad a.s.$$

and hence for every $t < \infty$

$$U_t = [\sum_{k=1}^{\infty} V_t^k] < \infty \quad a.s.$$

Now let τ_m be stopping times increasing to ∞ such that $U_{\tau_m-} \leq m$. In view of (11.3.9), we have for every k

$$\mathsf{E}[\sup_{0 \leq t < \tau_m} |X_t - Y_t|^2] \leq \mathsf{E}[(V_{\tau_m-}^k)^2]. \tag{11.4.1}$$

Now $(V_{\tau_m-}^k)^2$ converges to zero in probability as $k \to \infty$ and is dominated by m^2, and thus the right-hand side in (11.4.1) converges to zero. Thus for every m,

$$\mathsf{E}[\sup_{0 \leq t < \tau_m} |X_t - Y_t|^2] = 0$$

showing that $X = Y$. $\qquad \square$

Remains to show that $\mathbf{d}_{sm}$ satisfies triangle inequality, which we do next.

Lemma 11.18 *Let X, Y, Z be semimartingales. Then*

$$\mathbf{d}_{sm}(X, Z) \leq \mathbf{d}_{sm}(X, Y) + \mathbf{d}_{sm}(Y, Z). \tag{11.4.2}$$

Proof Given $\varepsilon > 0$, get $U, V \in \mathbb{V}^+$ such that U is a dominating process for $X - Y$, V is a dominating process for $Y - Z$ and

$$\mathbf{d}_{ucp}(U, 0) \le \mathbf{d}_{sm}(X, Y) + \varepsilon$$
$$\mathbf{d}_{ucp}(V, 0) \le \mathbf{d}_{sm}(Y, Z) + \varepsilon.$$

Since $X - Z = (X - Y) + (Y - Z)$, invoking Theorem 11.9, we can get $W \in \mathbb{V}^+$ such that W is a dominating process for $X - Z$ and $W \le U + V$. As a result, we note that

$$\mathbf{d}_{ucp}(W, 0) \le \mathbf{d}_{ucp}(U, 0) + \mathbf{d}_{ucp}(V, 0).$$

Putting these estimates together, we get

$$\begin{aligned}
\mathbf{d}_{sm}(X, Z) &\le \mathbf{d}_{ucp}(W, 0) \\
&\le \mathbf{d}_{ucp}(U, 0) + \mathbf{d}_{ucp}(V, 0) \\
&\le \mathbf{d}_{sm}(X, Y) + \mathbf{d}_{sm}(Y, Z) + 2\varepsilon.
\end{aligned}$$

Since ε is arbitrary, this proves (11.4.2). $\qquad\square$

Using the previous two results, we conclude that $\mathbf{d}_{sm}$ is a metric on the space of semimartingales. We will show that this metric also induces the Emery topology. The first step is to show that the space of semimartingales is complete in this metric.

Theorem 11.19 *Let X^n be a sequence of semimartingales that is Cauchy in $\mathbf{d}_{sm}$ metric. Then there exists a semimartingale X with $\mathbf{d}_{sm}(X^n, X) \to 0$.*

Proof By taking a subsequence if necessary, we assume that X^n is such that (writing $X^0 = 0$)

$$\mathbf{d}_{sm}(X^n, X^{n-1}) \le 2^{-n}. \tag{11.4.3}$$

For $n \ge 1$, let $V^n \in \mathbb{V}^+$ be dominating process for $X^n - X^{n-1}$ such that

$$\mathbf{d}_{ucp}(V^n, 0) \le \mathbf{d}_{sm}(X^n, X^{n-1}) + 2^{-n} \le 2.2^{-n}. \tag{11.4.4}$$

Thus there exists a decomposition $X^n - X^{n-1} = M^n + A^n$ with M^n being a locally square integrable martingale with $M_0^n = 0$, $A^n \in \mathbb{V}$ and such that U^n defined by

$$U_t^n = V_t^n - 2\sqrt{2}([M^n, M^n]_t + \langle M^n, M^n \rangle_t)^{1/2} - \sqrt{2}|A^n|_t \tag{11.4.5}$$

is an increasing process with $U_0^n \ge 0$. In particular, for all n, t

$$2\sqrt{2}([M^n, M^n]_t + \langle M^n, M^n \rangle_t)^{1/2} + \sqrt{2}|A^n|_t \le V_t^n \tag{11.4.6}$$

As in Lemma 11.17, using (11.4.4) we can conclude that

$$V_t = \sum_{n=1}^{\infty} V_t^n < \infty \quad \forall t < \infty. \tag{11.4.7}$$

Let us define $B_t^m = \sum_{n=1}^m A_t^n$ and $B_t = \sum_{n=1}^\infty A_t^n$. Then

$$|B - B^m|_t \le \sum_{n=m+1}^\infty |A^n|_t. \qquad (11.4.8)$$

In view of (11.4.6) and (11.4.7), it follows that D defined by

$$D_t = \sum_{n=1}^\infty (\langle M^n, M^n \rangle_t)^{1/2}$$

satisfies $D_t \le V_t$, is a predictable increasing process, and $D_0 = 0$. Thus D is locally bounded, and we can get stopping times σ_j increasing to infinity such that D_{σ_j} is bounded (say by c_j) for each j. Let

$$D_t^m = \sum_{n=m+1}^\infty (\langle M^n, M^n \rangle_t)^{1/2}.$$

Then $D_{\sigma_j}^m \le D_{\sigma_j} \le c_j$ and converges to zero almost surely and as a consequence

$$\lim_{m \to \infty} \mathsf{E}[D_{\sigma_j}^m] = 0 \qquad (11.4.9)$$

Let us define N^m as follows: $N^0 = 0$ and

$$N^m = \sum_{n=1}^m M^n.$$

Then N^m is also a locally square integrable martingale and $X^m = N^m + B^m$. Noting that for $m \le k$, $N^k - N^m = \sum_{n=m+1}^k M^n$ and hence using (11.3.23) we get

$$\begin{aligned}
(\langle N^k - N^m, N^k - N^m \rangle_t)^{1/2} &= (\langle \sum_{n=m+1}^k M^n, \sum_{n=m+1}^k M^n \rangle_t)^{1/2} \\
&\le \sum_{n=m+1}^k (\langle M^n, M^n \rangle_t)^{1/2} \qquad (11.4.10) \\
&\le D_t^m
\end{aligned}$$

and thus in view of (11.4.9), we have

$$\lim_{m \to \infty} \sup_{k > m} \mathsf{E}[\langle N^k - N^m, N^k - N^m \rangle_{\sigma_j}] = 0. \qquad (11.4.11)$$

In turn, using Doob's maximal inequality, we get

$$\lim_{\substack{m \to \infty \\ k > m}} \sup \mathsf{E}[\sup_{t \leq \sigma_j} |N_t^k - N_t^m|^2] = 0. \qquad (11.4.12)$$

Invoking arguments given in Lemma 2.75, it follows that N^k converges in *ucp* metric to an r.c.l.l. adapted process N. Further,

$$\lim_{m \to \infty} \mathsf{E}[\sup_{t \leq \sigma_j} |N_t - N_t^m|^2] = 0. \qquad (11.4.13)$$

Thus N is a locally square integrable martingale. Using Theorem 8.61 and arguments as in (11.4.10) we conclude

$$(\langle N - N^m, N - N^m \rangle_t + [N - N^m, N - N^m]_t)^{1/2}$$
$$= \lim_k (\langle N^k - N^m, N^k - N^m \rangle_t + [N^k - N^m, N^k - N^m]_t)^{1/2}$$
$$\leq \lim_k \sum_{n=m+1}^{k} ([M^n, M^n]_t + \langle M^n, M^n \rangle_t)^{1/2} \qquad (11.4.14)$$

Let us define $X = N + B$. Then X is a semimartingale. Further, $X - X^m = N - N^m + B - B^m$ and thus U^m defined by

$$U^m = 2\sqrt{2}([N - N^m, N - N^m]_t + \langle N - N^m, N - N^m \rangle_t)^{1/2} + \sqrt{2}|B - B^m|_t$$

is a dominating process for $X - X^m$. Using (11.4.6), (11.4.8), (11.4.14), it follows that

$$U_t^m \leq \sum_{n=m+1}^{\infty} V_t^n.$$

In view of (11.4.7), it follows that $\sum_{n=m+1}^{\infty} V_t^n$ converges to zero almost surely (as $m \to \infty$). Thus, U_t^m converges to 0 in probability. Since for each m, U^m is an increasing process, we conclude that $U^m \xrightarrow{ucp} 0$ and so $\mathbf{d}_{sm}(X, X^m)$ converges to 0 completing the proof. $\qquad \square$

The next result connects convergence in $\mathbf{d}_{sm}$ with that in $\mathbf{d}_{em}$.

Lemma 11.20 *Suppose X^n, X are semimartingales such that*

$$\sum_{n=1}^{\infty} \mathbf{d}_{sm}(X^n, X) < \infty. \qquad (11.4.15)$$

Then $\mathbf{d}_{em}(X^n, X) \to 0$.

Proof Let V^n be a dominating process for $X^n - X$ such that

$$\mathbf{d}_{ucp}(V^n, 0) \leq \mathbf{d}_{sm}(X^n, X) + 2^{-n}.$$

Then as seen in the proof of Lemma 11.17,

$$U_t = \sum_{n=1}^{\infty} V_t^n < \infty \quad \forall t.$$

Let $\sigma_j = \inf\{t > 0 : U_t \geq j \text{ or } U_{t-} \geq j\}$. Then

$$U_{\sigma_j-} = \sum_{n=1}^{\infty} V_{\sigma_j-}^n \leq j$$

and hence for each j,

$$\lim_{n\to\infty} E[(V_{\sigma_j-}^n)^2] = 0. \tag{11.4.16}$$

Now for any predictable process f bounded by 1, we have

$$E[\sup_{t<\sigma_j} |\int_0^t f\,dX^n - \int_0^t f\,dX|] \leq 4E[(V_{\sigma_j-}^n)^2] \tag{11.4.17}$$

Given $T < \infty$, $\eta > 0$ and $\varepsilon > 0$, get j such that

$$P(\sigma_j \geq T) \leq \frac{1}{2}\varepsilon \tag{11.4.18}$$

and using (11.4.16) and (11.4.17) for this fixed j, get n_0 such that for $n \geq n_0$

$$P(\sup_{t<\sigma_j} |\int_0^t f\,dX^n - \int_0^t f\,dX|] \geq \eta) \leq \frac{1}{2}\varepsilon \tag{11.4.19}$$

Recall that choice of n_0 is independent of f and thus (note: $\mathbb{S}_1$ is the set of predictable processes bounded by 1) for $n \geq n_0$ we have

$$\sup_{f\in\mathbb{S}_1} P(\sup_{t\leq T} |\int_0^t f\,dX^n - \int_0^t f\,dX|] \geq \eta)$$

$$\leq \sup_{f\in\mathbb{S}_1} P(\sup_{t<\sigma_j} |\int_0^t f\,dX^n - \int_0^t f\,dX|] \geq \eta) + P(\sigma_j \geq T)$$

$$\leq \varepsilon$$

As noted in the proof of Lemma 4.108, this shows $\mathbf{d}_{em}(X^n, X) \to 0$. □

Now we are in a position to prove that $\mathbf{d}_{sm}$ and $\mathbf{d}_{em}$ give rise to the same topology. In other words, $\mathbf{d}_{sm}$ is also a metric for the Emery topology.

Theorem 11.21 *For semimartingales* X^n, X

$$\mathbf{d}_{sm}(X^n, X) \to 0 \text{ if and only if } \mathbf{d}_{em}(X^n, X) \to 0.$$

Proof Let X^n, X be such that $\mathbf{d}_{sm}(X^n, X) \to 0$. We will show that $\mathbf{d}_{em}(X^n, X) \to 0$.
Take any subsequence $\{n^k\}$ and let $Y^k = X^{n^k}$. Since $\mathbf{d}_{sm}(Y^k, X) \to 0$, we can choose
a subsequence $\{k^m\}$ such that $Z^m = Y^{k^m}$ satisfies

$$\sum_{k=1}^{\infty} \mathbf{d}_{sm}(Z^m, X) < \infty.$$

Now Lemma 11.20 yields that $\mathbf{d}_{em}(Z^m, X) \to 0$. Thus the sequence $\{X^n\}$ satisfies
the property that given any subsequence, there exists a further subsequence that
converges to X in $\mathbf{d}_{em}$ metric. Hence $\mathbf{d}_{em}(X^n, X) \to 0$.

From the definition of the metrics $\mathbf{d}_{em}$ and $\mathbf{d}_{sm}$, it follows that the space of semimartingales is a linear topological space under each. Further, we have shown that the
space is complete under each of the metrics. The identity mapping being continuous
is then a homeomorphism in view of the open mapping theorem. See [55].   □

The following result gives a technique to prove almost sure convergence of
stochastic integrals.

Theorem 11.22 *Suppose* X^n, X *are semimartingales such that*

$$V^n \text{ dominates } (X^n - X) \tag{11.4.20}$$

and

$$\sum_{n=1}^{\infty} (V_t^n)^2 < \infty \quad \forall t < \infty. \tag{11.4.21}$$

Let f^n, f *be locally bounded predictable processes such that for all* $T < \infty$

$$\sum_{n=1}^{\infty} [\sup_{0 \le t \le T} |f_t^n - f_t|^2] < \infty \quad a.s. \tag{11.4.22}$$

Then for all $T < \infty$

$$\sum_{n=1}^{\infty} [\sup_{0 \le t \le T} |\int_0^t f^n \, dX^n - \int_0^t f \, dX|^2] < \infty \quad a.s. \tag{11.4.23}$$

and as a consequence

$$\lim_{n \to \infty} [\sup_{0 \le t \le T} |\int_0^t f^n \, dX^n - \int_0^t f \, dX|^2] = 0 \quad a.s. \tag{11.4.24}$$

Proof Let W be a dominating process for X and let

$$V_t = W_t + \sqrt{\sum_{n=1}^{\infty}(V_t^n)^2} + \sqrt{\sum_{n=1}^{\infty}[\sup_{0 \le s \le t}|f_s^n - f_s|^2]} + \sup_{0 \le s \le t}|f_s|.$$

For $j \ge 1$, let

$$\tau_j = \inf\{t \ge 0 : V_t \ge j \text{ or } V_{t-} \ge j\}.$$

Note that $\sup_{0 \le s < \tau_j}|f_s| \le j$, and hence

$$\sup_{0 \le s < \tau_j}|f_s^n| \le \sup_{0 \le s < \tau_j}|f_s| + \sup_{0 \le s < \tau_j}|f_s^n - f_s|$$

$$\le 2j.$$

Using the fact that V is a dominating process for $X^n - X$ as well as for X and that $V_{\tau_j -} \le j$, we have (invoking (11.3.35))

$$E[\sup_{0 \le t < \tau_j}|\int_0^t f^n dX^n - \int_0^t f^n dX|^2] \le 16 j^2 E[(V_{\tau_j -}^n)^2],$$

$$E[\sup_{0 \le t < \tau_j}|\int_0^t f^n dX - \int_0^t f dX|^2] \le 4 j^2 E[\sup_{0 \le s < \tau_j}|f_s^n - f_s|^2]$$

and hence

$$E[\sum_{n=1}^{\infty} \sup_{0 \le t < \tau_j}|\int_0^t f^n dX^n - \int_0^t f dX|^2]$$

$$\le 2E[\sum_{n=1}^{\infty} \sup_{0 \le t < \tau_j}|\int_0^t f^n dX^n - \int_0^t f^n dX|^2$$

$$+ \sum_{n=1}^{\infty} \sup_{0 \le t < \tau_j}|\int_0^t f^n dX - \int_0^t f dX|^2] \qquad (11.4.25)$$

$$\le 32 j^2 E[\sum_{n=1}^{\infty}(V_{\tau_j -}^n)^2 + \sum_{n=1}^{\infty} \sup_{0 \le s < \tau_j}|f_s^n - f_s|^2]]$$

$$\le 32 j^2 E[V_{\tau_j -}^2]$$

$$\le 32 j^4.$$

Thus for all j

$$\sum_{n=1}^{\infty} \sup_{0 \le t < \tau_j}|\int_0^t f^n dX^n - \int_0^t f dX|^2 < \infty \quad a.s.$$

Since τ_j increases to ∞, this implies (11.4.23) which in turn implies (11.4.24). $\square$

Chapter 12
SDE Driven by r.c.l.l. Semimartingales

In this chapter, we will consider stochastic differential equations as in Sect. 7.3 where the driving semimartingale need not be continuous.

We will consider the SDE (7.3.1), where b would be as in Sect. 7.3 but Y would be an r.c.l.l. semimartingale. We will continue to use the conventions used in that section on matrix–vector-valued processes and stochastic integrals.

Here we will use the Metivier–Pellaumail inequality and the notion of dominating process introduced earlier, and we will see that invoking these, the proofs of existence and uniqueness are essentially same as in the case of SDE's driven by Brownian motion. In Sect. 7.3 we had used random time change to achieve the same.

We begin with an analogue of the Gronwall's inequality, a key step in the study of differential equations.

12.1 Gronwall Type Inequality

We will obtain an analogue of Gronwall's inequality that would be useful in dealing with the stochastic differential equations driven by semimartingales in the next section. The first one is from Metivier [50], and the second one is essentially based on the same idea.

Theorem 12.1 *Let $A, B \in \mathbb{V}^+$ (increasing processes with $A_0 \geq 0$, $B_0 \geq 0$) and a stopping time τ be such that $B_{\tau-} \leq M$. Suppose that for all stopping times $\sigma \leq \tau$*

$$\mathsf{E}[A_{\sigma-}] \leq a + \beta \mathsf{E}[\int_{[0,\sigma)} A^- dB]. \tag{12.1.1}$$

For $\alpha > 0$ let $C(\alpha) = \sum_{j=0}^{[\alpha]} \alpha^j$. Then we have

© Springer Nature Singapore Pte Ltd. 2018
R. L. Karandikar and B. V. Rao, *Introduction to Stochastic Calculus*,
Indian Statistical Institute Series, https://doi.org/10.1007/978-981-10-8318-1_12

$$\mathsf{E}[A_{\tau-}] \leq 2aC(2\beta M) \tag{12.1.2}$$

Proof Let us define $V_t = \frac{1}{M}B_t$. Then $V_{\tau-} \leq 1$ and we have

$$\mathsf{E}[A_{\sigma-}] \leq a + \beta M \mathsf{E}[\int_{[0,\sigma)} A^- dV] \tag{12.1.3}$$

For integers $i \geq 1$, let $\tau_i = \inf\{t \geq 0 : A_t \geq i \text{ or } A_{t-} \geq i\} \wedge \tau$. Note that $A_{\tau_i-} \leq i$, $\tau_i \uparrow \tau$ and since A is increasing process, it follows that

$$\mathsf{E}[A_{\tau_i-}] \to \mathsf{E}[A_{\tau-}] \text{ as } i \to \infty. \tag{12.1.4}$$

Fix i and $\delta = \frac{1}{2\beta M}$ and let σ_k be defined inductively by $\sigma_0 = 0$ and for $k \geq 0$

$$\sigma_{k+1} = \inf\{t > \sigma_k : (V_t - V_{\sigma_k}) \geq \delta \text{ or } (V_{t-} - V_{\sigma_k}) \geq \delta\} \wedge \tau_i \tag{12.1.5}$$

If $\sigma_{k+1} < \tau_i$, then $(V_{\sigma_{k+1}} - V_{\sigma_k}) \geq \delta$ and hence

$$\sigma_N = \tau_i, \quad \text{for } N = [2\beta M] + 1. \tag{12.1.6}$$

Moreover, for all k, $(V_{\sigma_{k+1}-} - V_{\sigma_k}) \leq \delta$.

For $k \geq 0$, let $Z_k = A_{\sigma_k-}$ and $\theta_k = \mathsf{E}[Z_k]$. Then

$$\begin{aligned}
\mathsf{E}[Z_{k+1}] &\leq a + \beta M \mathsf{E}[\int_{[0,\sigma_k)} A_{s-} dV_s] + \beta M \mathsf{E}[(\int_{[\sigma_k,\sigma_{k+1})} A_{s-} dV_s] \\
&\leq a + \beta M \mathsf{E}[Z_k] + \beta M \mathsf{E}[Z_{k+1}]\delta \\
&\leq a + \beta M \mathsf{E}[Z_k] + \frac{1}{2}\mathsf{E}[Z_{k+1}]
\end{aligned}$$

Thus we have for $k \geq 1$ (note that a priori we know that $A_{\sigma_{k+1}-} \leq i$ and hence θ_{k+1} is finite)

$$\theta_{k+1} \leq 2a + 2\beta M \theta_k.$$

Likewise, we can conclude that $\theta_1 = \mathsf{E}[Z_1] \leq 2a$. Then by induction it follows that

$$\theta_{k+1} \leq 2a(1 + \sum_{j=1}^{k}(2\beta M)^j).$$

Thus invoking (12.1.6), we have $\mathsf{E}[A_{\tau_i-}] = \theta_N \leq 2aC(2\beta M)$. In view of (12.1.4), this completes the proof of (12.1.2). $\qquad\square$

The next result is an analogue of the inequality obtained in Theorem 12.1 for $(A_{\sigma-})^2$.

Theorem 12.2 *Let $A, B \in \mathbb{V}^+$ (increasing processes with $A_0 \geq 0$, $B_0 \geq 0$) and a stopping time τ be such that $B_{\tau-} \leq M$ and for all stopping times $\sigma \leq \tau$*

$$E[(A_{\sigma-})^2] \leq a + \beta E[(\theta_{\sigma-}(A^-, B))^2] \tag{12.1.7}$$

For $\alpha > 0$, let $C(\alpha) = \sum_{j=0}^{[\alpha]} \alpha^j$. Then we have

$$E[(A_{\tau-})^2] \leq 3aC(10\beta M^2) \tag{12.1.8}$$

Proof Let us define $V_t = \frac{1}{M} B_t$. Then $V_{\tau-} \leq 1$ and we have

$$E[(A_{\sigma-})^2] \leq a + \beta M^2 E[(\theta_{\sigma-}(A^-, V))^2] \tag{12.1.9}$$

For integers $i \geq 1$, let $\tau_i = \inf\{t \geq 0 : A_t \geq i \text{ or } A_{t-} \geq i\} \wedge \tau$. Note that $A_{\tau_i-} \leq i$, $\tau_i \uparrow \tau$ and since A is increasing process, it follows that

$$E[A_{\tau_i-}^2] \to E[A_{\tau-}^2] \text{ as } i \to \infty. \tag{12.1.10}$$

Fix $\delta = \frac{1}{10\beta M^2}$ and let σ_k be defined inductively by $\sigma_0 = 0$ and for $k \geq 0$

$$\sigma_{k+1} = \inf\{t > \sigma_k : (V_t - V_{\sigma_k}) \geq \delta \text{ or } (V_{t-} - V_{\sigma_k}) \geq \delta\} \wedge \tau_i \tag{12.1.11}$$

If $\sigma_{k+1} < \tau_i$, then $(V_{\sigma_{k+1}} - V_{\sigma_k}) \geq \delta$ and hence $\sigma_N = \tau_i$, for $N = [10\beta M^2] + 1$. Moreover, for all k, $(V_{\sigma_{k+1}-} - V_{\sigma_k}) \leq \delta$. Noting that for $f \geq 0$

$$(\theta_{t-}(f, V))^2 \leq 2 \int_{[0,t)} f_s^2 dV_s^2 + 2V_t \int_{[0,t)} f_s^2 dV_s$$

and using the inequality (12.1.9), we have, writing $U_t = V_t^2$ for convenience,

$$E[(A_{\sigma_{k+1}-})^2] \leq a + \beta M^2 E[(\theta_{\sigma_{k+1}}(A^-, V))^2]$$
$$\leq a + 2\beta M^2 E[(\int_{[0,\sigma_{k+1})} A_{t-}^2 dU_t + \int_{[0,\sigma_{k+1})} A_{t-}^2 dV_t)]. \tag{12.1.12}$$

For $k \geq 0$, let $Z_k = (A_{\sigma_k-})^2$ and $\theta_k = E[Z_k] = E[(A_{\sigma_k-})^2]$. Since $\sigma_k \leq \tau_i$, θ_k is finite for each k. Using (12.1.12) for $k \geq 1$

$$\theta_{k+1} \leq a + 2\beta M^2 E[(\int_{[0,\sigma_k)} A_{t-}^2 dU_t + \int_{[0,\sigma_k)} A_{t-}^2 dV_t)]$$
$$+ 2\beta M^2 E[(\int_{[\sigma_k,\sigma_{k+1})} A_{t-}^2 dU_t + \int_{[\sigma_k,\sigma_{k+1})} A_{t-}^2 dV_t)] \tag{12.1.13}$$
$$\leq a + 2\beta M^2 E[Z_k(U_{\sigma_k-} - U_0) + Z_k(V_{\sigma_k-} - V_0)]$$
$$+ 2\beta M^2 E[Z_{k+1}(U_{\sigma_{k+1}-} - U_{\sigma_k}) + Z_{k+1}(V_{\sigma_{k+1}-} - V_{\sigma_k})]$$

Using $U_t \le 1$, $V_t \le 1$ for all $t < \tau$ and $(V_{\sigma_{k+1-}} - V_{\sigma_k}) \le \delta$ and

$$(U_{\sigma_{k+1-}} - U_{\sigma_k}) = (V_{\sigma_{k+1-}} - V_{\sigma_k})(V_{\sigma_{k+1-}} + V_{\sigma_k}) \le 2\delta$$

the inequality (12.1.13) yields

$$\theta_{k+1} \le a + 2\beta M^2 (2\theta_k) + 2\beta M^2 (3\theta_{k+1})\delta$$
$$\le a + 4\beta M^2 \theta_k + \frac{6}{10}\theta_{k+1} \tag{12.1.14}$$

Thus, for $k \ge 1$

$$\theta_{k+1} \le \frac{10}{4}a + 10\beta M^2 \theta_k \le 3a + 10\beta M^2 \theta_k.$$

Same argument as above also yields $\theta_1 \le 3a$. Thus by induction it follows that

$$\theta_{k+1} \le 3a[\sum_{j=0}^{k} (10\beta M^2)^j]. \tag{12.1.15}$$

Since $\sigma_N = \tau_i$ as noted earlier, it follows that $\theta_N = \mathsf{E}[(A_{\tau_i-})^2]$. Thus (12.1.15) implies that, writing $\alpha = 10\beta M^2$,

$$\mathsf{E}[(A_{\tau_i-})^2 \le 3a[\sum_{j=0}^{[\alpha]} \alpha^j]$$

This proves the required estimate in view of (12.1.10).   □

12.2   Stochastic Differential Equations

Let $Y^1, Y^2, \ldots Y^m$ be r.c.l.l. semimartingales w.r.t. the filtration $(\mathcal{F}_.)$. Here we will consider an SDE
$$dU_t = b(t, \cdot, U)dY_t, \quad t \ge 0, \quad U_0 = \xi_0 \tag{12.2.1}$$

where the functional b is given as follows. Recall that $\mathbb{D}_d = \mathbb{D}([0, \infty), \mathbb{R}^d)$. Let $\mathcal{B}(\mathbb{D}_d)$ be the smallest σ-field on $\mathbb{D}_d$ under which the coordinate mappings are measurable. Let
$$a : [0, \infty) \times \Omega \times \mathbb{D}_d \to \mathsf{L}(d, m) \tag{12.2.2}$$

be such that for all $t \in [0, \infty)$,

$$(\omega, \gamma) \mapsto a(t, \omega, \gamma) \text{ is } \mathcal{F}_t \otimes \mathcal{B}(\mathbb{D}_d) \text{ measurable,} \qquad (12.2.3)$$

for all $(\omega, \gamma) \in \Omega \times \mathbb{D}_d$

$$t \mapsto a(t, \omega, \gamma) \text{ is an r.c.l.l. mapping} \qquad (12.2.4)$$

and suppose that there is an increasing r.c.l.l. adapted process K such that for all $\gamma, \gamma_1, \gamma_2 \in \mathbb{D}_d$,

$$\sup_{0 \le s \le t} \|a(s, \omega, \gamma)\| \le K_t(\omega) \sup_{0 \le s \le t} (1 + |\gamma(s)|) \qquad (12.2.5)$$

$$\sup_{0 \le s \le t} \|a(s, \omega, \gamma_2) - a(s, \omega, \gamma_1)\| \le K_t(\omega) \sup_{0 \le s \le t} |\gamma_2(s) - \gamma_1(s)|. \qquad (12.2.6)$$

Let $b : [0, \infty) \times \Omega \times \mathbb{D}_d \to \mathbb{L}(d, m)$ be given by

$$b(s, \omega, \gamma) = a(s-, \omega, \gamma). \qquad (12.2.7)$$

Note that (12.2.6) implies

$$\sup_{0 \le s \le t} \|b(s, \omega, \gamma_2) - b(s, \omega, \gamma_1)\| \le K_{t-}(\omega) \sup_{0 \le s < t} |\gamma_2(s) - \gamma_1(s)|. \qquad (12.2.8)$$

Lemma 12.3 *Suppose a satisfies (12.2.2)–(12.2.6). Then we have*

(i) For an r.c.l.l. $(\mathcal{F}_.)$ adapted process V, Z defined by $Z_t = b(t, \cdot, V)$ (i.e. $Z_t(\omega) = a(t, \omega, V(\omega))$) is an r.c.l.l. $(\mathcal{F}_.)$ adapted process.
(ii) For any stopping time τ,

$$(\omega, \zeta) \mapsto a(\tau(\omega), \omega, \zeta) \text{ is } \mathcal{F}_\tau \otimes \mathcal{B}(\mathbb{C}_d) \text{ measurable} \qquad (12.2.9)$$

Proof For part (i), let us define a process V^t by $V_s^t = V_{s \wedge t}$. Note that in view of (12.2.6), $Z_t = a(t, \cdot, V^t)$. The fact that $\omega \mapsto V^t(\omega)$ is $\mathcal{F}_t$ measurable along with (12.2.3) implies that Z_t is also $\mathcal{F}_t$ measurable. For part (ii), when τ is a simple stopping time, (12.2.9) follows from (12.2.3). For a general bounded stopping time τ, the conclusion (12.2.9) follows by approximating τ from above by simple stopping times and using right continuity of $a(t, \omega, \zeta)$. For a general stopping time τ, (12.2.9) follows by approximating τ by $\tau \wedge n$. $\qquad \square$

Recall that we had introduced matrix–vector-valued processes and stochastic integral $\int f dX$ where f, X are matrix–vector-valued while dealing with SDEs driven by continuous semimartingales in Sect. 7.6. We will continue to use the same notation. As in the case of continuous semimartingales, here too, an r.c.l.l. ($\mathbb{R}^d$-valued) adapted process U is said to be a solution to the Eq. (12.2.1) if

$$U_t = U_0 + \int_0^t b(s, \cdot, U) dY_s \qquad (12.2.10)$$

i.e. for $1 \leq j \leq d$,

$$U_t^j = U_0^j + \sum_{k=1}^{m} \int_{0+}^{t} b_{jk}(s, \cdot, U) dY_s^k$$

where $U = (U^1, \ldots, U^d)$ and $b = (b_{jk})$.

Let us recast the growth estimate Theorem 11.13 in matrix–vector form for later use:

Lemma 12.4 *Let $X = (X^1, X^2, \ldots X^m)$, where X^j is a semimartingale for each j, $1 \leq j \leq m$. Suppose V is a dominating process for each of X^j, $1 \leq j \leq m$. Then for any locally bounded $L(d, m)$-valued predictable f, and a stopping time σ, one has*

$$E[\sup_{0 \leq s < \sigma} | \int_{0+}^{s} f dX|^2] \leq dm^2 E[\theta_{\sigma-}^2(\|f\|, V)]. \tag{12.2.11}$$

Proof Fix $T < \infty$. Then

$$E[\sup_{0 \leq s < \sigma \wedge T} | \int_{0+}^{s} f dX|^2] = \sum_{j=1}^{d} E[\sup_{0 \leq s < \sigma \wedge T} | \sum_{k=1}^{m} \int_{0+}^{s} f_{jk} dX^k|^2]$$

$$\leq m \sum_{j=1}^{d} \sum_{k=1}^{m} E[\sup_{0 \leq s < \sigma \wedge T} | \int_{0+}^{s} f_{jk} dX^k|^2] \tag{12.2.12}$$

$$\leq m \sum_{j=1}^{d} \sum_{k=1}^{m} E[\theta_{(\sigma \wedge T)-}^2(f_{jk}, V)].]$$

$$\leq dm^2 E[\theta_{\sigma-}^2(\|f\|, V)].$$

The result follows taking limit as $T \uparrow \infty$ in (12.2.12). $\qquad \square$

We are now in a position to prove uniqueness of solution to the SDE (12.2.1). The proof is essentially the same as for SDE driven by Brownian motion or by a continuous semimartingale. Here we use the growth estimate (7.3.11) in place of (3.4.4) or (7.2.5) for a continuous semimartingale satisfying (7.2.2). The technique of time change used for continuous semimartingale is replaced here by the notion of dominating process and the estimate in Theorem 12.2 replacing Gronwall's lemma—Lemma 3.27.

Lemma 12.5 *Let $Y^1, Y^2, \ldots Y^m$ be r.c.l.l. semimartingales w.r.t. the filtration $(\mathcal{F}_\cdot)$. Let a satisfy (12.2.2)–(12.2.6) and let b be defined by (12.2.7). Suppose H and G be r.c.l.l. adapted processes and let X and Z satisfy*

$$X_t = H_t + \int_{0+}^{t} b(s, \cdot, X) dY_s \tag{12.2.13}$$

$$Z_t = G_t + \int_{0+}^{t} b(s, \cdot, Z) dY_s. \tag{12.2.14}$$

Let V be a (common) dominating process for Y^j, $1 \leq j \leq m$ and let σ be a stopping time such that $V_{\sigma-} \leq M$ and $K_{\sigma-} \leq \beta$. Then

$$E[\sup_{0 \leq s < \sigma} |X_s - Z_s|^2] \leq 6E[\sup_{0 \leq s < \sigma} |H_s - G_s|^2]C(2dm^2\beta^2M^2) \qquad (12.2.15)$$

where $C(\alpha)$ is as in Theorem 12.2.

Proof Using (12.2.13) and (12.2.14), we have

$$E[\sup_{0 \leq s < \sigma} |X_s - Z_s|^2] \leq 2E[\sup_{0 \leq s < \sigma} |H_s - G_s|^2]$$

$$+ 2E[\sup_{0 \leq s < \sigma} |\int_0^s (b(u, \cdot, Z) - b(u, \cdot, X))dY_u|^2]$$

Using the Lipschitz condition (12.2.6), the fact that $K_{\sigma-} \leq \beta$, it follows that for $s < \sigma$

$$\|(b(s, \cdot, Z) - b(s, \cdot, X))\| \leq \beta \sup_{0 \leq t < s} |X_t - Z_t|.$$

Thus writing $A_s = \sup_{0 \leq t \leq s} |X_t - Z_t|$, we get for any stopping time $\tau \leq \sigma$, using (12.2.11),

$$E[A_{\tau-}^2] \leq 2E[\sup_{0 \leq s < \tau} |H_s - G_s|^2] + 2\beta^2 dm^2 E[\theta_{\tau-}^2(A^-, V)].$$

Now $V_{\sigma-} \leq M$ and Theorem 12.2 together imply the required estimate (12.2.15). $\square$

This result immediately leads to:

Theorem 12.6 *Let $Y^1, Y^2, \ldots Y^m$ be r.c.l.l. semimartingales w.r.t. the filtration $(\mathcal{F}.)$. Let a satisfy (12.2.2)–(12.2.6) and let b be defined by (12.2.7). Let H be an adapted r.c.l.l. process. Suppose X and Z satisfy*

$$X_t = H_t + \int_{0+}^t b(s, \cdot, X)dY_s \qquad (12.2.16)$$

$$Z_t = H_t + \int_{0+}^t b(s, \cdot, Z)dY_s \qquad (12.2.17)$$

Then $X = Z$.

Proof Let V be a common dominating process for Y^j, $1 \leq j \leq m$. Let $U_t = V_t + K_t$ and σ_n be defined by

$$\sigma_n = \inf\{t \geq 0 : U_t \geq n \text{ or } U_{t-} \geq n\}.$$

Then σ_n increases to ∞. Since $V_{\sigma_n-} \leq n$ and $K_{\sigma_n-} \leq n$, Lemma 12.5 implies

$$\mathsf{E}[\sup_{0 \le s < \sigma_n} |X_t - Z_t|^2] = 0$$

for $n \ge 1$. This proves $X = Z$. $\square$

The same proof essentially yields the following

Theorem 12.7 *Let $Y^1, Y^2, \ldots Y^m$ be r.c.l.l. semimartingales w.r.t. the filtration $(\mathcal{F}_.)$.
Let a satisfy (12.2.2)–(12.2.6) and let b be defined by (12.2.7). Let H be an adapted
r.c.l.l. process. Suppose τ is a stopping time and X and Z satisfy*

$$X_{t \wedge \tau} = H_{t \wedge \tau} + \int_{0+}^{t \wedge \tau} b(s, \cdot, X) dY_s \tag{12.2.18}$$

$$Z_{t \wedge \tau} = H_{t \wedge \tau} + \int_{0+}^{t \wedge \tau} b(s, \cdot, Z) dY_s \tag{12.2.19}$$

Then

$$\mathsf{P}(X_{t \wedge \tau} = Z_{t \wedge \tau} \ \forall t) = 1. \tag{12.2.20}$$

Proof Let V be a common dominating process for $Y^j, 1 \le j \le m$. Let $U_t = V_t + K_t$
and σ_n be defined by

$$\sigma_n = \inf\{t \ge 0 : U_t \ge n \text{ or } U_{t-} \ge n\} \wedge \tau.$$

Then σ_n increases to τ. Since $V_{\sigma_n-} \le n$ and $K_{\sigma_n-} \le n$, Lemma 12.5 implies

$$\mathsf{E}[\sup_{0 \le t < \sigma_n} |X_t - Z_t|^2] = 0$$

for $n \ge 1$. This proves $\mathsf{P}(X_t = Z_t \ \forall t < \tau) = 1$. The required result (12.2.20) follows from this. $\square$

Having proven uniqueness of solution to the SDE (12.2.16), we now move onto
proving existence of solution to the equation. We will show this by showing that
Picard's successive approximation method converges to a solution of the equation.
The proof will be very similar to the proof in the Brownian motion case.

Theorem 12.8 *Let $Y^1, Y^2, \ldots Y^m$ be r.c.l.l. semimartingales w.r.t. the filtration $(\mathcal{F}_.)$.
Let a satisfy (12.2.2)–(12.2.6) and let b be defined by (12.2.7). Let H be an adapted
r.c.l.l. process. Then there exists an adapted r.c.l.l. process X such that (12.2.16)
holds. In other words, existence and uniqueness hold for the SDE (12.2.16).*

Proof Let $X_t^{[0]} = H_t$ for all $t \ge 0$ and for $n \ge 1$ let $X^{[n]}$ be defined inductive as
follows:

$$X_t^{[n]} = H_t + \int_{0+}^{t} b(s, \cdot, X^{[n-1]}) dY_s \tag{12.2.21}$$

Then note that

$$X_t^{[n+1]} - X_t^{[n]} = \int_{0+}^{t} (b(s, \cdot, X^{[n]}) - b(s, \cdot, X^{[n-1]}))dY_s. \tag{12.2.22}$$

As in the proof of Lemma 12.5, let V be a (common) dominating process for Y^j, $1 \leq j \leq m$. Let

$$U_t = V_t + V_t^2 + \sup_{0 \leq s \leq t} |H_s| + K_t$$

where K is as in conditions (12.2.5) and (12.2.6). Let σ_j be the stopping times defined by

$$\sigma_j = \inf\{t \geq 0 : U_t \geq j \text{ or } U_{t-} \geq j\}.$$

Note that $\sigma_j \uparrow \infty$ as $j \uparrow \infty$. For $n \geq 0$ let

$$A_t^{[n]} = \sup_{0 \leq s \leq t} |X_s^{[n+1]} - X_s^{[n]}|.$$

For $s < \sigma_j$,

$$\|(b(s, \cdot, X^{[n]}) - b(s, \cdot, X^{[n-1]}))\| \leq j A_{s-}^{[n-1]}$$

and thus using (12.2.22) along with the estimate (12.2.11), we get for any stopping time $\tau \leq \sigma_j$, for $n \geq 1$ (using (11.3.28) for the last step)

$$
\begin{aligned}
E[(A_{\tau-}^{[n]})^2] &= E[\sup_{t < \tau} |X_t^{[n+1]} - X_t^{[n]}|^2] \\
&= E[\sup_{t < \tau} |\int_0^t (b(s, \cdot, X^{[n]}) - b(s, \cdot, X^{[n-1]})dY_s|^2] \\
&\leq dm^2 j^2 E[\theta_{\tau-}^2(A^{[n-1]-}, V) \tag{12.2.23} \\
&\leq 2dm^2 j^2 E[\int_{[0,\tau)} (A_{s-}^{[n-1]})^2 dV_s^2 + V_t \int_{[0,\tau)} (A_{s-}^{[n-1]})^2 dV_s] \\
&\leq 4dm^2 j^3 E[\int_{[0,\tau)} (A_{s-}^{[n-1]})^2 dD_s]
\end{aligned}
$$

where $D_s = V_s^2 + V_s$. Hence writing $B_t^{[k]} = \sum_{n=0}^{k} 4^n (A_t^{[n]})^2$, we thus get for any stopping time $\tau \leq \sigma_j$

$$E[B_{\tau-}^{[k]}] \leq E[(A_{\tau-}^{[0]})^2] + 16dm^2 j^3 E[\int_{[0,\tau)} B_{s-}^{[k-1]} dU] \tag{12.2.24}$$

Also, recalling that $|X_t^{[0]}| = |H_t| \leq j$ for all $t < \sigma_j$, we have

$$\mathsf{E}[(A_{\tau-}^{[0]})^2] = \mathsf{E}[\sup_{t<\tau}|X_t^{[1]} - X_t^{[0]}|^2]$$

$$= \mathsf{E}[\sup_{t<\tau}|\int_0^t b(s,\cdot,X^{[0]})dY|^2$$

$$\leq dm^2 \mathsf{E}[\theta_{\tau-}^2(j,V)]$$

$$\leq 4dm^2 j^4.$$

Using (12.2.24) we get for any stopping time $\tau \leq \sigma_j$

$$\mathsf{E}[B_{\tau-}^{[k]}] \leq 4dm^2 j^4 + 16dm^2 j^3 \mathsf{E}[\int_{[0,\tau)} B_{s-}^{[k]}dU]$$

and then using the version of Gronwall inequality given in Theorem 12.1, we conclude

$$\mathsf{E}[B_{\sigma_j-}^{[k]}] \leq C_1(j,d,m)$$

where $C_1(j,d,m) = 8dm^2 j^4 C(32dm^2 j^4)$ and $C(\alpha) = \sum_{j=0}^{[\alpha]} \alpha^j$. Taking limit as $k \uparrow \infty$, we conclude

$$\mathsf{E}[B_{\sigma_j-}] \leq C_1(j,d,m).$$

Thus, for each $j \geq 1$,

$$\sum_{n=0}^{\infty} 4^n \mathsf{E}[(A_{\sigma_j-}^{[n]})^2] < \infty$$

and as a consequence, for large n, $\mathsf{E}[(A_{\sigma_j-}^{[n]})^2] \leq 4^{-n}$ and hence

$$\sum_{n=0}^{\infty} \sqrt{\mathsf{E}[(A_{\sigma_j-}^{[n]})^2]} = \sum_{n=0}^{\infty} \|[\sup_{s<\sigma_j}|X_s^{[n+1]} - X_s^{[n]}|]\|_2 < \infty \qquad (12.2.25)$$

The relation (12.2.25) implies

$$\|[\sum_{n=0}^{\infty} \sup_{s<\sigma_j}|X_s^{[n+1]} - X_s^{[n]}|]\|_2 < \infty \qquad (12.2.26)$$

as well as

$$\sup_{k\geq 1}\|[\sup_{s<\sigma_j}|X_s^{[n+k]} - X_s^{[n]}|]\|_2 \leq \sup_{k\geq 1}\|[\sum_{j=n+1}^{n+k} \sup_{s<\sigma_j}|X_s^{[j+1]} - X_s^{[j]}|]\|_2$$

$$\leq [\sum_{j=n+1}^{\infty} \|(\sup_{s<\sigma_j}|X_s^{[n+1]} - X_s^{[n]}|)\|_2] \qquad (12.2.27)$$

$$\to 0 \text{ as } n \text{ tends to } \infty.$$

Let $N = \cup_{j=1}^{\infty}\{\omega : \sum_{n=1}^{\infty} \sup_{s<\sigma_j} |X_s^{[n+1]}(\omega) - X_s^{[n]}(\omega)| = \infty\}$. Then by (12.2.26) $P(N) = 0$ and for $\omega \notin N$, $X_s^{[n]}(\omega)$ converges uniformly on $[0, \sigma_j(\omega))$ for every $j < \infty$. So let us define X as follows:

$$X_t(\omega) = \begin{cases} \lim_{n\to\infty} X_t^{[n]}(\omega) & \text{if } \omega \in N^c \\ 0 & \text{if } \omega \in N. \end{cases}$$

Since $P(N) = 0$, it follows that

$$X^{[n]} \text{ converges to } X \text{ uniformly on compact subsets of } [0, \infty) \text{ } a.s. \qquad (12.2.28)$$

In view of (12.2.27), it also follows that, for each j,

$$\lim_{n\to\infty} E[(\sup_{s<\sigma_j} |X_s - X_s^{[n]}|)^2] = 0. \qquad (12.2.29)$$

Recalling the Lipschitz condition (12.2.6) and the fact that $K_{\sigma_j-} \le j$, we have

$$\sup_{s<\sigma_j} \|(b(s, \cdot, X) - b(s, \cdot, X^{[n]}))\| \le j \sup_{s<\sigma_j} |X_s - X_s^{[n]}|. \qquad (12.2.30)$$

As a consequence, writing $f_s^n = b(s, \cdot, X) - b(s, \cdot, X^{[n]})$

$$E[\sup_{t<\sigma_j} | \int_0^t (b(s, \cdot, X)dY_s - \int_0^t b(s, \cdot, X^{[n]})dY_s|^2] \le E[\theta_{\sigma_j-}^2(f^n, V)]$$

$$\le j^2 . j^2 E[\sup_{s<\sigma_j} |X_s - X_s^{[n]}|^2]$$

$$\to 0 \text{ as } n \to \infty.$$

This along with (12.2.21) and (12.2.28) yields that

$$X_t = H_t + \int_0^t (b(s, \cdot, X)dY_s,$$

in other words X is a solution to the Eq. (12.2.16). □

By modifying the successive approximation scheme (evaluating the integral defining $X^{[n]}$ approximately) we can obtain a pathwise formula for the solution to the SDE as obtained in Sect. 7.4 for the case of SDE's driven by continuous semimartingales. However, this approximation involves an iterated limit.

12.3   Pathwise Formula for Solution to an SDE

In this section, we will consider the SDE

$$dX_t = g(t, G, X)dY \tag{12.3.1}$$

where $f, g : [0, \infty) \times \mathbb{D}_r \times \mathbb{D}_d \mapsto \mathsf{L}(d, m)$ are such that

$$\forall (\zeta, \gamma) \in \mathbb{D}_r \times \mathbb{D}_d, \ t \mapsto f(t, \zeta, \gamma) \text{ is an r.c.l.l. function,} \tag{12.3.2}$$

and g is related to f via

$$g(t, \zeta, \gamma) = f(t-, \zeta, \gamma) \tag{12.3.3}$$

and G is an $\mathbb{R}^r$-valued r.c.l.l. adapted process and Y is a semimartingale. Here for an integer k, $\mathbb{D}_k = \mathbb{D}([0, \infty), \mathbb{R}^k)$.

Recall that $\mathcal{B}(\mathbb{D}_k)$ is the σ-field generated by the coordinate mappings. We assume that

$$f \text{ is measurable w.r.t. } \mathcal{B}([0, \infty)) \otimes \mathcal{B}(\mathbb{D}_r) \otimes \mathcal{B}(\mathbb{D}_d). \tag{12.3.4}$$

For $t < \infty$, $\gamma \in \mathbb{D}_d$ and $\zeta \in \mathbb{D}_r$, let $\gamma^t(s) = \gamma(t \wedge s)$ and $\zeta^t(s) = \zeta(t \wedge s)$ and we assume that f satisfies

$$f(t, \zeta, \gamma) = f(t, \zeta^t, \gamma^t), \ \ \forall \zeta \in \mathbb{D}_r, \ \gamma \in \mathbb{D}_d, \ 0 \le t < \infty. \tag{12.3.5}$$

We also assume that there exists a function $C : [0, \infty) \times \mathbb{D}_r \mapsto \mathbb{R}$ measurable w.r.t. $\mathcal{B}([0, \infty)) \otimes \mathcal{B}(\mathbb{D}_r)$ such that $\forall \zeta \in \mathbb{D}_r, \ \gamma, \gamma_1, \gamma_2 \in \mathbb{D}_d, \ 0 \le t \le T$

$$\|f(t, \zeta, \gamma)\| \le C(t, \zeta)(1 + \sup_{0 \le s \le t} |\gamma(s)|) \tag{12.3.6}$$

$$\|f(t, \zeta, \gamma_1) - f(t, \zeta, \gamma_2)\| \le C(t, \zeta)(\sup_{0 \le s \le t} |\gamma_1(s) - \gamma_2(s)|) \tag{12.3.7}$$

and for all $\zeta \in \mathbb{D}_r$,

$$t \to C(t, \zeta) \text{ is r.c.l.l.} \tag{12.3.8}$$

As in Sect. 6.2, we will now obtain a mapping

$$\Psi : \mathbb{D}_d \times \mathbb{D}_r \times \mathbb{D}_m \mapsto \mathbb{D}([0, \infty), \mathbb{R}^d)$$

such that for adapted r.c.l.l. process H, G ($\mathbb{R}^d$, $\mathbb{R}^r$-valued, respectively) and an $\mathbb{R}^m$-valued r.c.l.l. semimartingale Y,

$$X = \Psi(H, G, Y)$$

yields the unique solution to the SDE

$$X_t = H_t + \int_0^t g(s, G, X) dY. \tag{12.3.9}$$

We will define mappings

$$\Psi^{(n)} : \mathbb{D}_d \times \mathbb{D}_r \times \mathbb{D}_m \mapsto \mathbb{D}([0, \infty), \mathbb{R}^d)$$

inductively for $n \geq 1$. Let $\Psi^{(0)}(\eta, \zeta, \gamma)(s) = \eta$ for all $s \geq 0$ and having defined $\Psi^{(0)}, \Psi^{(1)}, \ldots, \Psi^{(n-1)}$, we define $\Psi^{(n)}$ as follows. Fix n and $\eta \in \mathbb{D}_m$, $\zeta \in \mathbb{D}_r$ and $\gamma \in \mathbb{D}_d$.

Let $t_0^{(n)} = 0$ and let $\{t_j^{(n)} : j \geq 1\}$ be defined inductively as follows: ($\{t_j^{(n)} : j \geq 1\}$ are themselves functions of (η, ζ, γ), which are fixed for now, and we will suppress writing it as a function) if $t_j^{(n)} = \infty$, then $t_{j+1}^{(n)} = \infty$ and if $t_j^{(n)} < \infty$, then writing

$$\Gamma^{(n-1)}(\eta, \zeta, \gamma)(s) = f(s, \zeta, \Psi^{(n-1)}(\eta, \zeta, \gamma))$$

let

$$t_{j+1}^{(n)} = \inf\{s \geq t_j^{(n)} : \|\Gamma^{(n-1)}(\eta, \zeta, \gamma)(s) - \Gamma^{(n-1)}(\eta, \zeta, \gamma)(t_j^{(n)})\| \geq 2^{-n}$$
$$\text{or } \|\Gamma^{(n-1)}(\eta, \zeta, \gamma)(s-) - \Gamma^{(n-1)}(\eta, \zeta, \gamma)(t_j^{(n)})\| \geq 2^{-n}\}$$

since $\Gamma^{(n-1)}(\eta, \zeta, \gamma)$ is an r.c.l.l. function, $t_j^{(n)} \uparrow \infty$ as $j \uparrow \infty$. Let

$$\Psi^{(n)}(\eta, \zeta, \gamma)(s) = \eta + \sum_{j=0}^{\infty} \Gamma^{(n-1)}(\eta, \zeta, \gamma)(t_j^{(n)})(\gamma(s \wedge t_{j+1}^{(n)}) - \gamma(s \wedge t_j^{(n)})).$$

Now we define

$$\Psi(\eta, \zeta, \gamma) = \begin{cases} \lim_n \Psi^{(n)}(\eta, \zeta, \gamma) & \text{if the limit exists in ucc topology} \\ 0 & \text{otherwise.} \end{cases} \tag{12.3.10}$$

Now it can be seen that

$$a(s, \omega, \gamma) = f(s, G(\omega), \gamma), \quad b(s, \omega, \gamma) = g(s, G(\omega), \gamma)$$

satisfies (12.2.2)–(12.2.7) with $K_t(\omega) = C(t, G(\omega))$. Let

$$X(\omega) = \Psi(H(\omega), G(\omega), Y(\omega)). \tag{12.3.11}$$

Note that an ω path of X has been defined directly in terms of the ω paths of G, H, Y via the functional Ψ. We will prove

Theorem 12.9 *X defined by (12.3.11) is the (unique) solution to the SDE (12.3.9).*

Proof Let $Z_t^{(0)} = H_0$. The processes $Z^{(n)}$ are defined by induction on n. Fix n. Having defined $Z^{(0)}, \ldots, Z^{(n-1)}$, we define $Z^{(n)}$:

Let $\tau_0^{(n)} = 0$ and let $\{\tau_j^{(n)} : j \geq 1\}$ be defined inductively as follows: if $\tau_j^{(n)} = \infty$, then $\tau_{j+1}^{(n)} = \infty$ and if $\tau_j^{(n)} < \infty$, then

$$\tau_{j+1}^{(n)} = \inf\{s \geq \tau_j^{(n)} : \|f(s, G, Z^{(n-1)}) - f(\tau_j^{(n)}, G, Z^{(n-1)})\| \geq 2^{-n}$$
$$\text{or } \|f(s-, G, Z^{(n-1)}) - f(\tau_j^{(n)}, G, Z^{(n-1)})\| \geq 2^{-n}\}. \tag{12.3.12}$$

Since the process $s \mapsto f(s, G, Z^{(n-1)})$ is an adapted r.c.l.l. process, it follows that each $\tau_j^{(n)}$ is a stopping time and $\lim_{j \uparrow \infty} \tau_j^{(n)} = \infty$. Let $Z_0^{(n)} = H_0$ and for $j \geq 0$, $\tau_j^{(n)} < t \leq \tau_{j+1}^{(n)}$ let

$$Z_t^{(n)} = Z_{\tau_j^{(n)}}^{(n)} + f(\tau_j^{(n)}, G, Z^{(n-1)})(Y_t - Y_{\tau_j^{(n)}}).$$

Equivalently,

$$Z_t^{(n)} = H_t + \sum_{j=0}^{\infty} f(\tau_j^{(n)}, G, Z^{(n-1)})(Y_{t \wedge \tau_{j+1}^{(n)}} - Y_{t \wedge \tau_j^{(n)}}) \tag{12.3.13}$$

It can be seen from the respective definitions that

$$Z^{(n)}(\omega) = \Psi^{(n)}(H(\omega), G(\omega), Y(\omega)).$$

Thus to complete the proof, suffices to show that $Z^{(n)}$ converges to a solution Z of the SDE (12.3.9). Uniqueness would then imply that $Z = X$.

For $n \geq 1$, let us define W^n and S^n by

$$S_t^{(n)} = \sum_{j=0}^{\infty} f(\tau_j^{(n)}, G, Z^{(n-1)}) 1_{[\tau_j^{(n)}, \tau_{j+1}^{(n)})}(t) \tag{12.3.14}$$

$$W_t^{(n)} = H_t + \int_0^t f(s-, G, Z^{(n-1)}) dY_s. \tag{12.3.15}$$

Let us note that

$$Z_t^{(n)} = H_t + \int_0^t S_{s-}^{(n)} dY_s \tag{12.3.16}$$

Noting that by definition of $\{\tau_j^{(n)} : j \geq 1\}$,

$$\|S_t - f(t, G, Z^{(n-1)})\| \leq 2^{-n}. \tag{12.3.17}$$

As in the proof of Theorem 12.8, let V be a (common) dominating process for Y^j, $1 \leq j \leq m$. Let

$$U_t = V_t + V_t^2 + \sup_{0 \leq s \leq t} |H_s| + K_t$$

where $K_t(\omega) = C(t, G(\omega))$ and C is as in the Lipschitz and growth conditions (12.3.6)–(12.3.7), $D_s = V_s^2 + V_s$ and let σ_j be the stopping times defined by

$$\sigma_j = \inf\{t \geq 0 : U_t \geq j \text{ or } U_{t-} \geq j\}.$$

Note that $\sigma_j \uparrow \infty$ as $j \uparrow \infty$. Using (12.3.15)–(12.3.17) and the fact that $V_{\sigma_j-} \leq j$ along with the fact that V is a common dominating process for Y^j, $1 \leq j \leq m$ we get

$$E[\sup_{0 \leq s < \sigma_j} |W_t^{(n)} - Z_t^{(n)}|^2] \leq dm^2 j^2 2^{-2n} \tag{12.3.18}$$

For $n \geq 0$ let

$$A_t^{[n]} = \sup_{0 \leq s \leq t} |Z_s^{[n+1]} - Z_s^{[n]}|.$$

For any stopping time $\tau \leq \sigma_j$, for $n \geq 1$ (using (11.3.28) for the last step)

$$E[(A_{\tau-}^{[n]})^2] = E[\sup_{t < \tau} |Z_t^{[n+1]} - Z_t^{[n]}|^2]$$

$$\leq 3E[\sup_{0 \leq s < \sigma_j} |W_t^{(n+1)} - Z_t^{(n+1)}|^2] + 3E[\sup_{0 \leq s < \sigma_j} |W_t^{(n)} - Z_t^{(n)}|^2]$$

$$+ 3E[\sup_{t < \tau} |W_t^{[n+1]} - W_t^{[n]}|^2]$$

$$\leq 6dm^2 j^2 2^{-2n} + 3E[\sup_{t < \tau} |\int_0^t (g(s, G, Z^{[n]}) - g(s, G, Z_s^{[n-1]}))dY_s|^2]$$

$$\leq 6dm^2 j^2 2^{-2n} + 3dm^2 j^2 E[\theta_{\tau-}^2 (A^{[n-1]-}, V)$$

$$\leq 6dm^2 j^2 2^{-2n} + 6dm^2 j^2 E[\int_{[0,\tau)} (A_{s-}^{[n-1]})^2 dV_s^2$$

$$+ V_t \int_{[0,\tau)} (A_{s-}^{[n-1]})^2 dV_s]$$

$$\leq 6dm^2 j^2 2^{-2n} + 12dm^2 j^3 E[\int_{[0,\tau)} (A_{s-}^{[n-1]})^2 dD_s]$$

$$\tag{12.3.19}$$

Hence writing $B_t = \sum_{n=0}^{\infty} 2^n (A_t^{[n]})^2$, we thus get for any stopping time $\tau \leq \sigma_j$

$$E[B_{\tau-}] \leq E[A_{\tau-}^{[0]}] + \sum_{n=0}^{\infty} 2^n 6dm^2 j^2 2^{-2n} + 24dm^2 j^3 E[\int_{[0,\tau)} B_{s-} dU] \tag{12.3.20}$$

As in the proof of Theorem 12.8, it follows that

$$\mathsf{E}[A_{\tau-}^{[0]}] \le dm^2 j^4$$

and hence that

$$\mathsf{E}[(B_{\tau-})^2] \le dm^2 j^4 + 6dm^2 j^2 + 12dm^2 j^3 \mathsf{E}[\int_{[0,\tau)} (B_{s-})^2 dD_s]. \tag{12.3.21}$$

Now proceeding exactly as in the proof of Theorem 12.8, we can conclude that $Z^{(n)}$ converges to a solution Z of Eq. (12.3.9). Since $Z^n(\omega) = \Psi^n(H(\omega), G(\omega), Y(\omega))$, and Z^n converges to Z, it follows that $Z(\omega) = \Psi(H(\omega), G(\omega), Y(\omega))$ completing the proof.

12.4 Euler–Peano Approximations

We are going to show that Euler–Peano approximations (for the solution to the SDE (12.3.9) converge to the solution and indeed converge almost surely, and this yields a pathwise formula for the solution. In the formula given in Sect. 12.3, the approximation $\Psi^{(n)}$ depended upon $\Psi^{(n-1)}$, whereas in the approximation constructed in this section, the approximation $\tilde{\Psi}^{(n)}$ is defined directly in terms of the coefficients and thus is preferable from computational point of view as compared to the formula (12.3.10). These results were obtained in [40]. The formulation given here is taken from [41]. We need this auxiliary lemma later.

Lemma 12.10 *Let* $0 = \tau_0 \le \tau_1 \le \ldots \le \tau_i \le \ldots$ *be an increasing sequence of stopping times. For an r.c.l.l. adapted processes* U, *let* S *be defined by*

$$S_t = \sum_{i=0}^{\infty} U_{\tau_i} 1_{[\tau_i, \tau_{i+1})}(t). \tag{12.4.1}$$

Then S *is an r.c.l.l. adapted process.*

Proof Let $\sigma = \lim_{i \to \infty} \tau_i$. Fix $T < \sigma(\omega)$. For $t \in [0, T]$, $S_t(\omega)$ is a finite sum of r.c.l.l. functions and hence is r.c.l.l.

If $\sigma(\omega) < \infty$, then for $t \ge \sigma(\omega)$, $S_t(\omega) = 0$ and thus $S.(\omega)$ is a right continuous function. Remains to show that when $\sigma(\omega) < \infty$, the left limit of S at $\sigma(\omega)$ exists. Fix ω such that $a = \sigma(\omega) < \infty$. If $\sigma(\omega) = \tau_i(\omega)$ for some i, then the claim is obvious. In the other case $U_{\tau_i(\omega)}(\omega) \to U_{a-}(\omega)$ and if $t_n \uparrow a$ with $t_n < a$, then $S_{t_n}(\omega)$ is a subsequence of $U_{\tau_i(\omega)}(\omega)$ and hence left limit of $S.(\omega)$ at a exists and equals $U_{a-}(\omega)$. Since each summand is adapted, so is S. $\qquad\square$

We will consider the framework as in Sect. 12.2. Let $Y^1, Y^2, \ldots Y^m$ be r.c.l.l. semimartingales w.r.t. the filtration $(\mathcal{F}.)$, H be an r.c.l.l. adapted process. Consider the SDE

$$U_t = H_t + \int_{0+}^{t} b(s, \cdot, U) dY_s, \tag{12.4.2}$$

where the functional b is given as follows. Let

$$a : [0, \infty) \times \Omega \times \mathbb{D}_d \to \mathsf{L}(d, m) \qquad (12.4.3)$$

be such that for all $t \in [0, \infty)$

$$(\omega, \gamma) \mapsto a(t, \omega, \gamma) \text{ is } \mathcal{F}_t \otimes \mathcal{B}(\mathbb{D}_d) \text{ measurable}, \qquad (12.4.4)$$

for all $(\omega, \gamma) \in \Omega \times \mathbb{D}_d$

$$t \mapsto a(t, \omega, \gamma) \text{ is an r.c.l.l. mapping} \qquad (12.4.5)$$

and suppose that there is an increasing r.c.l.l. adapted process K such that for all $\gamma, \gamma_1, \gamma_2 \in \mathbb{D}_d$,

$$\sup_{0 \leq s \leq t} \|a(s, \omega, \gamma)\| \leq K_t(\omega) \sup_{0 \leq s \leq t} (1 + |\gamma(s)|) \qquad (12.4.6)$$

$$\sup_{0 \leq s \leq t} \|a(s, \omega, \gamma_2) - a(s, \omega, \gamma_1)\| \leq K_t(\omega) \sup_{0 \leq s \leq t} |\gamma_2(s) - \gamma_1(s)|. \qquad (12.4.7)$$

Let $b : [0, \infty) \times \Omega \times \mathbb{D}_d \to \mathsf{L}(d, m)$ be given by

$$b(s, \omega, \gamma) = a(s-, \omega, \gamma). \qquad (12.4.8)$$

As proved in Theorem 12.6, the SDE (12.4.2) admits a unique solution X under the conditions (12.4.3)–(12.4.8).

Let us fix $\varepsilon > 0$, and we will construct an ε-approximation $Z = Z^\varepsilon$ to the solution X of the SDE. We will drop ε from the notation here and in what follows till the next theorem, where we will give an estimate on $X - Z = X - Z^\varepsilon$.

For $i \geq 0$, let stopping times τ_i and processes W^i be defined inductively by:

$$\tau_0 = 0 \text{ and } W_t^0 \equiv H_0$$

and having defined τ_j, W^j for $j \leq i$, if $\tau_i < \infty$ let

$$A_t^{i+1} = (H_t - H_{\tau_i} + a(\tau_i, \cdot, W^i)(Y_t - Y_{\tau_i})) 1_{[\tau_i, \infty)}(t)$$
$$B_t^{i+1} = (a(t, \cdot, W^i) - a(\tau_i, \cdot, W^i)) 1_{[\tau_i, \infty)}(t)$$
$$U_t^{i+1} = A_t^{i+1}(1 + K_t)$$
$$\tau_{i+1} = \inf\{t > \tau_i : |U_t^{i+1}| \geq \varepsilon \text{ or } |U_{t-}^{i+1}| \geq \varepsilon \text{ or } |B_t^{i+1}| \geq 4\varepsilon \text{ or } |B_{t-}^{i+1}| \geq 4\varepsilon\}$$
$$W_t^{i+1} = H_0 + \sum_{j=1}^{i+1} A_{\tau_j}^j 1_{[\tau_j, \infty)}(t)$$

$$(12.4.9)$$

and if $\tau_i = \infty$, then $\tau_{i+1} = \infty$, $A^{i+1} = 0$, $B^{i+1} = 0$, $U^{i+1} = 0$ and $W^{i+1} = W^i$. Note that for $i < k$, W^i and W^k agree on $[0, \tau_i]$ by definition and as a consequence, we have

$$a(\tau_i, \cdot, W^i) = a(\tau_i, \cdot, W^k). \tag{12.4.10}$$

For $i \geq 0$ define Z^{i+1} by $Z_0^{i+1} = H_0$ and

$$Z_t^{i+1} = \begin{cases} W_{\tau_j}^{i+1} + A_t^{j+1} & \text{for } \tau_j \leq t < \tau_{j+1}, \; j \leq i \\ W_{\tau_{i+1}}^{i+1} & \text{for } t \geq \tau_{i+1}. \end{cases}$$

Thus, by the choice of $\{\tau_j : j \geq 1\}$, we have

$$\sup_t \; |W_t^j - Z_t^j| \leq \varepsilon \tag{12.4.11}$$

and

$$\sup_t \; K_t |W_t^j - Z_t^j| \leq \varepsilon \tag{12.4.12}$$

for all $j \geq 1$. As a consequence of the Lipschitz condition on a we also have

$$\sup_t |a(t, \cdot, W^j) - a(t, \cdot, Z^j)| \leq \varepsilon. \tag{12.4.13}$$

We can now check that

$$Z_t^k = H_{t \wedge \tau_k} + \sum_{i=0}^{k-1} a(\tau_i, \cdot, W^k)(Y_{t \wedge \tau_{i+1}} - Y_{\tau_i}) \tag{12.4.14}$$

and

$$A_{\tau_{i+1}}^{i+1} = Z_{\tau_{i+1}}^{i+1} - Z_{\tau_i}^{i+1}. \tag{12.4.15}$$

Let us define a mapping $T : \Omega \times \mathbb{D}_d \to \mathbb{D}_d$ as follows:

$$T(\omega, \gamma)(t) = \gamma(\tau_i(\omega)) \text{ for } \tau_i(\omega) \leq t < \tau_i(\omega).$$

Lemma 12.10 ensures that $T(\omega, \gamma)$ is an r.c.l.l. function. We now define mapping $\mathcal{J}$ that maps r.c.l.l. adapted processes into r.c.l.l. adapted processes by

$$\mathcal{J}(U(\omega)) = T(\omega, U(\omega))$$

or equivalently,

$$\mathcal{J}(U) = \sum_{i=0}^{\infty} U_{\tau_i} 1_{[\tau_i, \tau_{i+1})}. \tag{12.4.16}$$

Let us note that for all $k \geq 1$, by definition of Z^k, W^k we have

$$\mathcal{J}(Z^k) = W^k. \tag{12.4.17}$$

Let us define $\tilde{a} : [0, \infty) \times \Omega \times \mathbb{D}_d \to \mathsf{L}(d, m)$ as follows:

$$\tilde{a}(t, \omega, \gamma) = T(\omega, a(\cdot, \omega, T(\omega, \gamma))).$$

Easy to check that $\tilde{a}$ satisfies (12.2.2)–(12.2.6) and hence defining

$$\tilde{b}(t, \omega, \gamma) = \tilde{a}(t-, \omega, \gamma),$$

it follows from Theorem 12.6 that the SDE

$$Z_t = H_t + \int_{0+}^{t} \tilde{b}(s, \cdot, Z) dY_s \tag{12.4.18}$$

admits a unique solution.

We can check (using (12.4.17)) that

$$\tilde{a}(t, \cdot, Z^k) = \sum_{i=0}^{\infty} a(\tau_i, \cdot, W^k) 1_{[\tau_i, \tau_{i+1})}(t)$$

and so

$$\tilde{b}(t, \cdot, Z^k) = \sum_{i=0}^{\infty} a(\tau_i, \cdot, W^k) 1_{(\tau_i, \tau_{i+1}]}(t).$$

Hence it follows from (12.4.14) that

$$Z_t^k = H_{t \wedge T_k} + \int_0^{t \wedge T_k} \tilde{b}(s, \cdot, Z^k) dY_s \tag{12.4.19}$$

Then invoking Theorem 12.7, we conclude

$$P(Z_{t \wedge T_k}^k = Z_{t \wedge T_k} \ \forall t \geq 0) = 1. \tag{12.4.20}$$

Lemma 12.11

$$\lim_{i \to \infty} \tau_i = \infty \ \ a.s.$$

Proof We will show that for ω such that

$$Z_{t \wedge T_k}^k(\omega) = Z_{t \wedge T_k}(\omega) \ \ \forall t \geq 0, \ \forall k \geq 1, \tag{12.4.21}$$

$\lim_{i \to \infty} \tau_i(\omega) = \infty$. Note that if $\tau_{i+1}(\omega) < \infty$, then

$$U_{\tau_{i+1}}^{i+1}(\omega) = (1 + K_{\tau_{i+1}}(\omega))(Z_{\tau_{i+1}}(\omega) - Z_{\tau_i}(\omega)),$$

$$U_{\tau_{i+1}-}^{i+1}(\omega) = (1 + K_{\tau_{i+1}}(\omega))(Z_{\tau_{i+1}-}(\omega) - Z_{\tau_i}(\omega)),$$

$$a(\tau_i, \cdot, W^i)(\omega) = a(\tau_i, \cdot, T(Z))(\omega),$$

$$a(\tau_{i+1}-, \cdot, W^i)(\omega) = a(\tau_{i+1}-, \cdot, T(Z))(\omega).$$

Suppose there exists an ω such that (12.4.21) holds and such that $\theta = \lim_{i \to \infty} \tau_i(\omega) < \infty$. Then from the definition of the sequence $\{\tau_j : j \geq 1\}$ it follows that at least one of the following four inequalities

$$(1 + K_{\tau_{i+1}}(\omega))(Z_{\tau_{i+1}}(\omega) - Z_{\tau_i}(\omega)) \geq \varepsilon, \tag{12.4.22}$$

$$(1 + K_{\tau_{i+1}}(\omega))(Z_{\tau_{i+1}-}(\omega) - Z_{\tau_i}(\omega)) \geq \varepsilon, \tag{12.4.23}$$

$$|a(\tau_{i+1}, \cdot, W^i)(\omega) - a(\tau_i, \cdot, T(Z))(\omega)| \geq 2\varepsilon, \tag{12.4.24}$$

$$|a(\tau_{i+1}-, \cdot, T(Z))(\omega) - a(\tau_i, \cdot, T(Z))(\omega)| \geq 2\varepsilon \tag{12.4.25}$$

must be satisfied for countably many i. Note that in general $a(\tau_{i+1}, \cdot, W^i))(\omega)$ may not be equal to $a(\tau_{i+1}, \cdot, T(Z))(\omega) = a(\tau_{i+1}, \cdot, W^{i+1}))(\omega)$.

If (12.4.22) or (12.4.23) holds for countably many i, then the left limit $Z_{\theta-}(\omega)$ at θ cannot exist—a contradiction. Likewise, if (12.4.25) holds for countably many i, then the left limit $a(\theta-, \cdot, T(Z))(\omega)$ cannot exist, again a contradiction, since $a(t, \cdot, \alpha)$ is r.c.l.l. for all α.

Now for i such that (12.4.22) does not hold,

$$|a(\tau_{i+1}, \cdot, W^i))(\omega) - a(\tau_{i+1}, \cdot, T(Z)))(\omega)|$$
$$= |a(\tau_{i+1}, \cdot, W^i))(\omega) - a(\tau_{i+1}, \cdot, W^{i+1}))(\omega)|$$
$$\leq K_t(\omega) \sup_t |W_t^i - W_t^{i+1}|$$
$$= K_t(\omega)|Z_{\tau_{i+1}}(\omega) - Z_{\tau_i}(\omega)|$$
$$\leq \varepsilon$$

Thus for i such that (12.4.22) does not hold but (12.4.24) holds,

$$|a(\tau_{i+1}, \cdot, T(Z))(\omega) - a(\tau_i, \cdot, T(Z))(\omega)|$$
$$\geq |a(\tau_{i+1}, \cdot, W^i)(\omega) - a(\tau_i, \cdot, T(Z))(\omega)|$$
$$\quad - |a(\tau_{i+1}, \cdot, W^{i+1}))(\omega) - a(\tau_{i+1}, \cdot, W^i))(\omega)| \tag{12.4.26}$$
$$\geq 2\varepsilon - \varepsilon = \varepsilon$$

Thus if (12.4.24) holds for countably many i (and since we have already shown that (12.4.22) holds at most finitely many times), it follows that (12.4.26) holds countably many times thus $a(t, \cdot, T(Z))$ cannot have a left limit at θ—again a contradiction.

It thus follows that $\theta = \infty$. □

We have suppressed ε from notations but $\tilde{b}$ depends on $\{\tau_i\}$ which in turn depends upon ε and the process Z also depends upon ε. Note that

$$\tilde{b}(s, \cdot, Z) = \sum_{i=0}^{\infty} a(\tau_i, \cdot, W^{i+1}) 1_{(\tau_i, \tau_{i+1}]}(s)$$

and hence using the Lipschitz condition (12.4.7), definition of $\{\tau_j\}$ we get

$$\|b(s, \cdot, Z) - \tilde{b}(s, \cdot, Z)\| = \sum_{i=0}^{\infty} \|a(s-, \cdot, Z) - a(\tau_i, \cdot, W^{i+1})\| 1_{(\tau_i, \tau_{i+1}]}(s)$$

$$= \sum_{i=0}^{\infty} \|a(s-, \cdot, Z^{i+1}) - a(\tau_i, \cdot, W^{i+1})\| 1_{(\tau_i, \tau_{i+1}]}(s)$$

$$= \sum_{i=0}^{\infty} \|a(s-, \cdot, Z^{i+1}) - a(s-, \cdot, W^{i+1})\| 1_{(\tau_i, \tau_{i+1}]}(s)$$

$$+ \sum_{i=0}^{\infty} \|a(s-, \cdot, W^{i+1}) - a(\tau_i, \cdot, W^{i+1})\| 1_{(\tau_i, \tau_{i+1}]}(s)$$

$$\leq K_s \sum_{i=0}^{\infty} |Z^{i+1} - W^{i+1}| 1_{(\tau_i, \tau_{i+1}]}(s) + \varepsilon$$

$$\leq (K_s + 1)\varepsilon \tag{12.4.27}$$

Lemma 12.12 *Let X be the solution to (12.4.2) and $Z \equiv Z^{\varepsilon}$ be as defined in preceding paragraphs (satisfying (12.4.27)) for fixed ε. Let V be a common dominating process for Y^j, $1 \leq j \leq m$. Let $U_t = V_t + K_t$ (where K appears in condition (12.4.7) on a) and for $j \geq 1$ let σ_j be defined by*

$$\sigma_j = \inf\{t \geq 0 : U_t \geq j \text{ or } U_{t-} \geq j\}.$$

Then there exists a constant $k(j, d, m)$ depending only on j, d, m such that

$$E[\sup_{0 \leq s < \sigma_j} |X_t - Z_t^{\varepsilon}|^2] \leq \varepsilon^2 k(j, d, m) \tag{12.4.28}$$

Proof Let us define (dropping the suffix ε on Z)

$$A_t = \sup_{0 \leq s \leq t} |X_t - Z_t|$$

Now, for any $\tau \leq \sigma_j$

$$E[|A_{\tau-}|^2] \leq E[\sup_{0 \leq t < \sigma} |\int_{0+}^{t} (b(s, \cdot, X) - \tilde{b}(s, \cdot, Z))dY|^2]$$

$$\leq 2E[\sup_{0 \leq t < \sigma} |\int_{0+}^{t} (b(s, \cdot, X) - b(s, \cdot, Z))dY|^2] \qquad (12.4.29)$$

$$+ E[\sup_{0 \leq t < \sigma} |\int_{0+}^{t} (b(s, \cdot, Z) - \tilde{b}(s, \cdot, Z))dY|^2]$$

$$\leq 2j^2 dm^2 E[\theta_{\tau-}^2(A^-, V)] + 2(j+1)^2 dm^2 \varepsilon^2$$

Using Theorem 12.1, it now follows that

$$E[|A_{\sigma_j-}|^2] \leq 4(j+1)^2 dm^2 \varepsilon^2 C(4j^3 dm^2) \qquad (12.4.30)$$

where $C(\alpha) = \sum_{j=0}^{[\alpha]} \alpha^j$. Thus (12.4.28) holds with

$$k(j, d, m) = 4(j+1)^2 dm^2 C(4j^3 dm^2).$$

$\square$

We have thus proved

Theorem 12.13 *Let X^n denotes the approximation Z^ε for $\varepsilon = 2^{-n}$ constructed in the preceding paragraphs. Then X^n converges almost surely to the solution X of the SDE (12.4.2).*

This in turn helps us obtain a pathwise formula for solution to the SDE (12.4.2).

We will consider the framework from Sect. 12.3 and obtain a pathwise formula involving a single limit rather than an iterative limit. Let $f, g : [0, \infty) \times \mathbb{D}_r \times \mathbb{D}_d \mapsto L(d, m)$ be such that

$$\forall (\zeta, \gamma) \in \mathbb{D}_r \times \mathbb{D}_d, \ t \mapsto f(t, \zeta, \gamma) \text{ is an r.c.l.l. function,} \qquad (12.4.31)$$

and g is related to f via

$$g(t, \zeta, \gamma) = f(t-, \zeta, \gamma) \qquad (12.4.32)$$

and G is an $\mathbb{R}^r$-valued r.c.l.l. adapted process and X is a semimartingale. Suppose

$$f \text{ is measurable w.r.t. } \mathcal{B}([0, \infty)) \otimes \mathcal{B}(\mathbb{D}_r) \otimes \mathcal{B}(\mathbb{D}_d). \qquad (12.4.33)$$

For $t < \infty$, $\gamma \in \mathbb{D}_d$ and $\zeta \in \mathbb{D}_r$, let $\gamma^t(s) = \gamma(t \wedge s)$ and $\zeta^t(s) = \zeta(t \wedge s)$ and we assume that f satisfies

$$f(t, \zeta, \gamma) = f(t, \zeta^t, \gamma^t), \ \forall \zeta \in \mathbb{D}_r, \ \gamma \in \mathbb{D}_d, \ 0 \leq t < \infty. \qquad (12.4.34)$$

We also assume that there exists a function $C : [0, \infty) \times \mathbb{D}_r \mapsto \mathbb{R}$ measurable w.r.t. $\mathcal{B}([0, \infty)) \otimes \mathcal{B}(\mathbb{D}_r)$ such that $\forall \zeta \in \mathbb{D}_r$, $\gamma, \gamma_1, \gamma_2 \in \mathbb{D}_d$, $0 \leq t \leq T$

$$\|f(t, \zeta, \gamma)\| \le C(t, \zeta)(\sup_{0 \le s < t} |\gamma(s)|) \tag{12.4.35}$$

$$\|f(t, \zeta, \gamma_1) - f(t, \zeta, \gamma_2)\| \le C(t, \zeta)(\sup_{0 \le s < t} |\gamma_1(s) - \gamma_2(s)|) \tag{12.4.36}$$

and for all $\zeta \in \mathbb{D}_r$,

$$t \to C(t, \zeta) \text{ is r.c.l.l.} \tag{12.4.37}$$

For $n \ge 1$, we define

$$\tilde{\Psi}^{(n)} : \mathbb{D}_d \times \mathbb{D}_r \times \mathbb{D}_m \mapsto \mathbb{D}([0, \infty), \mathbb{R}^d)$$

as follows: for $\eta \in \mathbb{D}_m$, $\zeta \in \mathbb{D}_r$ and $\gamma \in \mathbb{D}_d$ (fixed) let $t_0 = 0$ and let $\{t_j : j \ge 1\}$ and $\{\alpha^j : j \ge 1\}$, $\{\beta^j : j \ge 1\}$, $\{\xi^j : j \ge 1\}$ be defined inductively as follows: (these are themselves functions of n, (η, ζ, γ), which are fixed for now and we will suppress writing these as a function) if $t_j = \infty$, then $t_{j+1} = \infty$ and if $t_j < \infty$, then

$$\alpha_t^i = (\eta_t - \eta_{t_i} + f(t_i, \zeta, \xi^i)(\gamma_t - \gamma_{t_i}))1_{[t_i, \infty)}(t)$$
$$\beta_t^i = (f(t, \zeta, \xi^i) - f(t_i \zeta, \xi^i))1_{[t_i, \infty)}(t) \tag{12.4.38}$$

$$t_{i+1} = \inf\{t > t_i : |\alpha_t^i| \ge 2^{-n} \text{ or } |\alpha_{t-}^i| \ge 2^{-n} \text{ or } \|\beta_t^i\| \ge 2^{-n} \text{ or } \|\beta_{t-}^i\| \ge 2^{-n}\}$$

and

$$\xi_t^{i+1} = \begin{cases} \xi_t^i & \text{for } t < t_{i+1} \\ \xi_t^i + \alpha_{t_{i+1}}^i & \text{for } t \ge t_{i+1}. \end{cases}$$

Thus, ξ^{i+1} is a function that has jumps at $t_1, ..., t_{i+1}$ and is constant on the intervals

$$[0, t_1), \ldots, [t_j, t_{j+1}), \ldots [t_i, t_{i+1}), [t_{i+1}, \infty).$$

Also ξ^i and ξ^{i+1} agree on $[0, t_{i+1})$ by definition.

We finally define

$$\tilde{\Psi}^{(n)}(\eta, \zeta, \gamma)(t) = \eta_t + \sum_{i=0}^{\infty} f(t \wedge t_i, \zeta, \xi^i)(\gamma_{t \wedge t_{i+1}} - \gamma_{t \wedge t_i}) \tag{12.4.39}$$

and for $\eta \in \mathbb{D}_d$, $\zeta \in \mathbb{D}_r$ and $\gamma \in \mathbb{D}_m$ we define

$$\tilde{\Psi}(\eta, \zeta, \gamma) = \begin{cases} \lim_n \tilde{\Psi}^{(n)}(\eta, \zeta, \gamma) & \text{if the limit exists in ucc topology} \\ 0 & \text{otherwise.} \end{cases} \tag{12.4.40}$$

As in Sects. 6.2 and 12.3, it should be noted that the mapping

$$\tilde{\Psi} : D_d \times D_r \times D_m \mapsto D([0, \infty), \mathbb{R}^d)$$

has been defined without any reference to a probability measure or any semimartingale. As in Theorem 12.9, this also yields a pathwise formula. This one is preferable
from computation point of view as here, in order to construct nth approximation, we
do not need the $(n − 1)$th approximation.

Theorem 12.14 *Let* f, g *satisfy conditions* (12.4.31)–(12.4.37). *Let* Y *be a semimartingale w.r.t. a filtration* $(\mathcal{F}_{.})$ *and let* H, G *be r.c.l.l.* $(\mathcal{F}_{.})$ *adapted processes
taking values in* $\mathbb{R}^d$, $\mathbb{R}^r$, *respectively. Let* $\tilde{\Psi}$ *be as defined in* (12.4.40) *and let*

$$\tilde{X} = \tilde{\Psi}(H, G, Y).$$

Then X *satisfies the SDE*

$$\tilde{X}_t = H_t + \int_{0+}^{t} g(t, G, \tilde{X})dY. \tag{12.4.41}$$

The proof follows from observing that

$$a(t, \omega, \gamma) = f(t, G(\omega), \gamma), \ b(t, \omega, \gamma) = g(t, G(\omega), \gamma)$$

satisfy (12.3.2)–(12.3.8) and further,

$$\tilde{\Psi}^{(n)}(H, G, Y) = X^n$$

where X^n is the 2^{-n} approximation constructed in this section earlier. It now follows
that

$$\tilde{\Psi}(H, G, Y) = \tilde{X}$$

is the unique solution to the Eq. (12.4.41). $\square$

12.5 Matrix-Valued Semimartingales

In this section, we will consider matrix-valued r.c.l.l. semimartingales. We will use the
notations introduced in Sect. 7.6. Recall that $\mathsf{L}(m, k)$ is the set of all $m \times k$ matrices,
and $L_0(d)$ denotes the set of non-singular $d \times d$ matrices.

Recall that when $X = (X^{pq})$ is an $\mathsf{L}(m, k)$-valued semimartingale and $f = (f^{ij})$
is an $\mathsf{L}(d, m)$-valued predictable process such that $f^{ij} \in \mathbb{L}(X^{jq})$ (for all i, j, q), then
$Y = \int f dX$ is defined as an $\mathsf{L}(d, k)$-valued semimartingale as follows: $Y = (Y^{iq})$
where

$$Y^{iq} = \sum_{j=1}^{m} \int f^{ij} dX^{jq}$$

and that for $\mathsf{L}(d, d)$-valued semimartingales X, Y let $[X, Y] = ([X, Y]^{ij})$ be the $\mathsf{L}(d, d)$-valued process defined by

$$[X, Y]_t^{ij} = \sum_{k=1}^{d} [X^{ik}, Y^{kj}].$$

We can consider an analogue of the SDE (12.2.1)

$$dU_t = b(t, \cdot, U) dY_t, \quad t \geq 0, \quad U_0 = \xi_0 \tag{12.5.1}$$

where now Y is an $\mathsf{L}(m, k)$-valued continuous semimartingale, U is an $\mathsf{L}(d, k)$-valued process, ξ_0 is $\mathsf{L}(d, k)$-valued random variable and here

$$b : [0, \infty) \times \Omega \times \mathbb{D}([0, \infty), \mathsf{L}(d, k)) \to \mathsf{L}(d, m).$$

Exercise 12.15 Formulate and prove analogues of Theorems 12.6, 12.8 and 12.14 for Eq. (12.5.1).

Exercise 12.16 Let X be an $\mathsf{L}(d, d)$-valued semimartingale with $X(0) = 0$ and let I denote the $d \times d$ identity matrix. Show that the equations

$$Y_t = I + \int_0^t Y_{s-} dX_s \tag{12.5.2}$$

and

$$Z_t = I + \int_0^t (dX_s) Z_{s-} \tag{12.5.3}$$

admit unique solutions.

The solutions Y, Z are denoted, respectively, by $e(X)$ and $e'(X)$ and are the left and right exponential of X.

Exercise 12.17 Let X be an $\mathsf{L}(d, d)$-valued semimartingale with $X(0) = 0$ and let $Y = e(X)$ and $Z = e'(X)$. Show that

(i) If Y and Y^- are $L_0(d)$-valued, then $(I + \Delta X)$ is $L_0(d)$-valued.
(ii) If Z and Z^- are $L_0(d)$-valued, then $(I + \Delta X)$ is $L_0(d)$-valued.

For a matrix $A \in \mathbb{L}(d, d)$ we will denote (only in this section) the Hilbert–Schmidt norm of A by $\|A\|$. The following facts are standard. The norm is defined as

$$\|A\|^2 = \sum_{i,j=1}^{d} (a_{ij})^2.$$

If $\|A\| < 1$, then $B = (I + A)$ belongs to $L_0(d)$. Further, for $\|A\| \leq \alpha < 1$, one has

$$\|(I + A)^{-1} - I + A\| \leq \frac{1}{1 - \alpha}\|A\|^2. \tag{12.5.4}$$

Exercise 12.18 For an L(d, d)-valued semimartingale X, show that

$$\sum_{0 < s \leq t} \|(\Delta X)_s\|^2 \leq Trace([X, X]_t).$$

Exercise 12.19 Let X be an L(d, d)-valued semimartingale with $X(0) = 0$ such that $(I + \Delta X)$ is $L_0(d)$-valued. Then

(i) Show that $W_t = \sum_{0 < s \leq t}[\{(I + \Delta X)^{-1}\} - I + (\Delta X) + (\Delta X)^2]$ is well defined.

(ii) Show that $\Delta W = \{(I + \Delta X)^{-1}\} - I + (\Delta X) - (\Delta X)^2$.

(iii) Let $U = -X + [X, X] + W$. Show that

$$X + U + [X, U] = 0 \tag{12.5.5}$$

(iv) Show that

$$\mathfrak{e}(X)\mathfrak{e}'(U) = I \tag{12.5.6}$$

and

$$\mathfrak{e}(U)\mathfrak{e}'(X) = I. \tag{12.5.7}$$

(v) Let $Y = \mathfrak{e}(X)$ and $Z = \mathfrak{e}'(X)$. Show that Y, Y^-, Z and Z^- are $L_0(d)$-valued.

HINT: For (i), separate jumps bigger than half, these are finitely many. For the rest of the jumps, use estimate (12.5.4). For (iv) use integration by parts formula, (7.6.1).

For a L(d, d)-valued semimartingale Y such that $Y_0 = I$ and such that Y and Y^- are $L_0(d)$-valued, let

$$\mathfrak{log}(Y)_t = \int_{0+}^t (Y^-)^{-1} dY$$

and

$$\mathfrak{log}'(Y) = \int_{0+}^t (dY)(Y^-)^{-1}.$$

The next exercise is to show that $\mathfrak{e}$ and $\mathfrak{log}$ are inverses of each other. We will say that a matrix-valued process is a local martingale (or a process with finite variation) if each of its components is so.

Exercise 12.20 Let X be an $L(d, d)$-valued semimartingale with $X(0) = 0$ such that $(I + \Delta X)$ is $L_0(d)$-valued and let Y be a $L(d, d)$-valued semimartingale such that Y and Y^- are $L_0(d)$-valued. Then show that

(i) $e(\log(Y)) = Y$, $e'(\log'(Y)) = Y$.
(ii) $\log(e(X)) = X$, $\log'(e'(X)) = X$.
(iii) $X \in \mathbb{M}_{\text{loc}}$ if and only if $e(X) \in \mathbb{M}_{\text{loc}}$.
(iv) $X \in \mathbb{V}$ if and only if $e(X) \in \mathbb{V}$.
(v) $Y \in \mathbb{M}_{\text{loc}}$ if and only if $\log(Y) \in \mathbb{M}_{\text{loc}}$.
(vi) $Y \in \mathbb{V}$ if and only if $\log'(Y) \in \mathbb{V}$.

Exercise 12.21 Let X^i be $L(d, d)$-valued semimartingale with $X^i(0) = 0$ such that $(I + \Delta X^i)$ is $L_0(d)$-valued, for $i = 1, 2$. Let $Y = e(X^2)$ and $U^1 = \int Y^-(dX^1)$ $(Y^-)^{-1}$. Then show that

$$e(X^1 + X^2 + [X^1, X^2]) = e(U^1)e(X^2) \tag{12.5.8}$$

The formula (12.5.8) has an important consequence. Given a $L(d, d)$-valued semimartingale Y such that Y and Y^- are $L_0(d)$-valued, let $X = e(Y)$. If we can write $X = M + A + [M, A]$ such that $M \in \mathbb{M}_{\text{loc}}$ and $A \in \mathbb{V}$ with $(I + \Delta M)$, $(I + \Delta A)$ are $L_0(d)$-valued, then it would follow that

$$Y = NB$$

where $N = e(M) \in \mathbb{M}_{\text{loc}}$ and $B = e(A) \in \mathbb{V}$ yielding a multiplicative decomposition of Y. The next exercise is about this.

Exercise 12.22 Let Y be a $L(d, d)$-valued semimartingale such that $Y_0 = I$ with Y and Y^- being $L_0(d)$-valued. Let $X = e(Y)$. Let

$$D_t = \sum_{0 < s \leq t} (\Delta X)_s 1_{\{\|(\Delta X)_s\| \geq \frac{1}{3}\}}$$

$$Z_t = X_t - D_t.$$

(i) Show that

(a) $P(\|(\Delta Z)_t\| \leq \frac{1}{3} \ \forall t) = 1$.
(b) Z is locally integrable (i.e. each component is locally integrable).

(ii) Let $Z = M + A$ be the decomposition with $M \in \mathbb{M}_{\text{loc}}$ and $A \in \mathbb{V}$, $Z_0 = 0$ and A being predictable. Show that

(a) $P(\|(\Delta A)_t\| \leq \frac{1}{3} \ \forall t) = 1$.
(b) $P(\|(\Delta M)_t\| \leq \frac{2}{3} \ \forall t) = 1$.
(c) $(I + \Delta M)$ is $\mathbb{L}_0(d)$-valued.

(iii) Let $B_t = A_t + D_t$ so that $X = M + B$. Let $C_t = B_t - \sum_{0 \leq s \leq t}(I + (\Delta M)_s)^{-1}$ $(\Delta B)_s$. Show that

(a) $B = C + [M, C]$.
(b) $X_t = M_t + C_t + [M, C]_t$.
(c) $(I + \Delta X) = (I + \Delta M)(I + \Delta C)$.
(d) $(I + \Delta M)$ and $(I + \Delta C)$ are $L_0(d)$-valued.

Let $H = \varepsilon(C)$, $N = \int H^-(dM)(H^-)^{-1}$ and $R = \varepsilon(N)$. Show that

(a) $R \in \mathbb{M}_{\text{loc}}$ and $H \in \mathbb{V}$ and

$$X = RH. \tag{12.5.9}$$

Chapter 13
Girsanov Theorem

In this chapter, we will obtain Girsanov Theorem and its generalizations by Meyer. Let M be a martingale on $(\Omega, \mathcal{F}, \mathsf{P})$ and let Q be another probability measure on $(\Omega, \mathcal{F})$, absolutely continuous w.r.t. P. Then as noted in Remark 4.26, M is a semimartingale on $(\Omega, \mathcal{F}, \mathsf{Q})$. We will obtain a decomposition of M into N and B, where N is a Q-martingale. This result for Brownian motion was due to Girsanov, and we will also present the generalizations due to Meyer.

13.1 Preliminaries

Let $(\Omega, \mathcal{F}, \mathsf{P})$ be a complete probability space and $(\mathcal{F}_\cdot)$ be a filtration such that $\mathcal{F}_0$ contains all P null sets. Let Q be a probability measure on $(\Omega, \mathcal{F})$ such that P and Q are equivalent; i.e., for $A \in \mathcal{F}, \mathsf{P}(A) = 0$ if and only if $\mathsf{Q}(A) = 0$. In such a case, P and Q are also called mutually absolutely continuous. Let $\xi = \frac{d\mathsf{Q}}{d\mathsf{P}}$. Let Z be the r.c.l.l. version of the martingale $\mathsf{E}_\mathsf{P}[\xi \mid \mathcal{F}_t^+]$. Of course Z is a uniformly integrable martingale with $\mathsf{E}_\mathsf{P}[Z_t] = 1$. Also, for $A \in \mathcal{F}_t^+$

$$Q(A) = \int_A Z_t \, d\mathsf{P}. \tag{13.1.1}$$

Here is a simple observation on Z.

Lemma 13.1 Z *is a* $(0, \infty)$ *valued process, i.e.* $\mathsf{P}(Z_t > 0 \ \forall t \geq 0) = 1$.

Proof Since P and Q are equivalent, $\mathsf{P}(\xi > 0) = 1$ and $\eta = \xi^{-1}$ is the Radon–Nikodym derivative of P w.r.t. Q. Let Y be the r.c.l.l. version of the martingale $\mathsf{E}_\mathsf{Q}[\eta \mid \mathcal{F}_t^+]$. It follows that $Z_t Y_t = 1$ almost surely for each t and since the two processes are r.c.l.l. it follows that $\mathsf{P}(Z_t Y_t = 1 \ \forall t \geq 0) = 1$. The result follows. $\square$

© Springer Nature Singapore Pte Ltd. 2018
R. L. Karandikar and B. V. Rao, *Introduction to Stochastic Calculus*,
Indian Statistical Institute Series, https://doi.org/10.1007/978-981-10-8318-1_13

Lemma 13.2 *Let* Q, ξ, Z *be as above. Suppose* Z *is* $(\mathcal{F}_.)$ *adapted. Let* M *be an adapted process. Then*

 (i) *M is a* Q-*martingale if and only if* MZ *is a* P-*martingale.*
(ii) *M is a* Q-*local martingale if and only if* MZ *is a* P-*local martingale.*

Proof For a stopping time σ, let η be a non-negative bounded $\mathcal{F}_\sigma$ measurable random variable. Then

$$\mathsf{E}_\mathsf{Q}[\eta] = \mathsf{E}_\mathsf{P}[\eta Z] = \mathsf{E}_\mathsf{P}[\eta \mathsf{E}[Z \mid \mathcal{F}_\sigma]] = \mathsf{E}_\mathsf{P}[\eta Z_\sigma].$$

Thus M_s is Q integrable if and only if $M_s Z_s$ is P-integrable. Further, for any bounded stopping time σ,

$$\mathsf{E}_\mathsf{Q}[M_\sigma] = \mathsf{E}_\mathsf{P}[M_\sigma Z_\sigma]. \tag{13.1.2}$$

Thus (i) follows from Theorem 2.57. For (ii), if M is a Q-local martingale, then get stopping times $\tau_n \uparrow \infty$ such that for each n, $M_{t \wedge \tau_n}$ is a Q-martingale. Then for any bounded stopping time σ, we have

$$\mathsf{E}_\mathsf{Q}[M_{\sigma \wedge \tau_n}] = \mathsf{E}_\mathsf{P}[M_{\sigma \wedge \tau_n} Z_{\sigma \wedge \tau_n}]. \tag{13.1.3}$$

Thus $M_{t \wedge \tau_n} Z_{t \wedge \tau_n}$ is a P-martingale, and thus MZ is a P-local martingale. The converse follows similarly. $\qquad\square$

Remark **13.3** Note that $Z_t = \mathsf{E}_\mathsf{Q}[\eta \mid \mathcal{F}_t^+]$ is $(\mathcal{F}_.)$ adapted if the filtration $(\mathcal{F}_.)$ is right continuous or Z is a continuous process.

13.2 Cameron–Martin Formula

Let $\tilde{\Omega} = \mathbb{C}_d = \mathbb{C}([0, \infty), \mathbb{R}^d)$, $\tilde{\mathcal{F}} = \mathcal{B}(\mathbb{C}_d)$. Let X_t be defined by

$$X_t(\zeta) = \zeta(t), \quad \zeta \in \mathbb{C}_d. \tag{13.2.1}$$

Let μ_w be the Wiener measure on $(\tilde{\Omega}, \tilde{\mathcal{F}})$ so that X is a Brownian motion on $(\tilde{\Omega}, \tilde{\mathcal{F}}, \mu_w)$. Let $\theta \in \mathbb{C}_d$ be fixed such that $\theta(0) = 0$. Consider the mapping

$$T_\theta : \tilde{\Omega} \mapsto \tilde{\Omega}$$

given by

$$T_\theta(\zeta) = \zeta + \theta.$$

Let $\mathsf{Q}_\theta = \mu_w \circ T_\theta^{-1}$. Note that for $A \in \mathcal{B}(\mathbb{C}_d)$,

$$\mathsf{Q}_\theta(A) = \mu_w(\zeta : T_\theta(\zeta) \in A) = \mu_w(A - \theta) \tag{13.2.2}$$

where $A - \theta = \{\zeta : \zeta + \theta \in A\}$. The next result gives conditions under which μ_w and Q_θ are equivalent. In what follows the filtration is taken to be $(\mathcal{F}_.) = (\mathcal{F}_.^X)$. Recall that an absolutely continuous function θ is differentiable almost everywhere w.r.t. Lebesgue measure, and we will denote the derivative by $\dot{\theta}$. When θ is $\mathbb{R}^d$-valued, absolute continuity and the derivative are interpreted coordinatewise.

Theorem 13.4 *Let* T_θ, Q_θ *be as above. Then* Q_θ *is equivalent to* μ_w *if and only if* θ *is absolutely continuous and*

$$\alpha = \sum_{i=1}^{d} \int_0^\infty (\dot{\theta}_s^i)^2 ds < \infty. \tag{13.2.3}$$

Further,

$$\theta_t^i = \int_0^t \dot{\theta}_s^i ds \;\; \forall t < \infty. \tag{13.2.4}$$

If (13.2.3) *holds, then* Z *defined by*

$$Z_t = \exp\{\sum_{i=1}^{d} \int_0^t \dot{\theta}_s^i dX_s^i - \frac{1}{2} \sum_{i=1}^{d} \int_0^t (\dot{\theta}_s^i)^2 ds\}$$

is a uniformly integrable martingale and for $A \in \mathcal{F}_t$,

$$\int_A Z_t dP = Q_\theta(A). \tag{13.2.5}$$

Further,

$$Z_t \text{ converges to } \frac{dQ_\theta}{d\mu_w} \text{ in } \mathbb{L}^1(\mu_w). \tag{13.2.6}$$

Proof Suppose that θ is absolutely continuous and (13.2.3) and (13.2.4) hold. Then it follows that $M_t = \sum_{i=1}^{d} \int_0^t \dot{\theta}_s^i dX_s^i$ is a continuous square integrable martingale and

$$[M, M]_t = \sum_{i=1}^{d} \int_0^t (\dot{\theta}_s^i)^2 ds.$$

Thus, $Z_t = \exp(M_t - \frac{1}{2}[M, M]_t)$. Further,

$$Z_t = 1 + \int_0^t Z dM \tag{13.2.7}$$

(see Exercise 4.101) and is thus a local martingale. As seen in Exercise 3.26, for each t, M_t has normal distribution with mean zero and $E[M_t^2] = [M, M]_t$. As a consequence,

$$E[Z_t^2] = E[\exp(2M_t - [M, M]_t)]$$
$$= \exp([M, M]_t)$$
$$\leq \exp(\alpha).$$

Now invoking Corollary 5.22 we conclude that Z is a square integrable martingale and further, being $\mathbb{L}^2$ bounded, it is uniformly integrable. Thus by martingale convergence theorem, Z_t converges in $\mathbb{L}^1$ and almost surely to, say, ξ and for $t < \infty$, $E[\xi \mid \mathcal{F}_t] = Z_t$. Clearly, $E[\xi] = 1$ and $\xi \geq 0$ almost surely. Since

$$E[(M_t - M_s)^2] = [M, M]_t - [M, M]_s$$

and $[M, M]_t$ converges to α, it follows that M_t converges in $\mathbb{L}^2$ to say η. So $\xi = \exp(\eta - \frac{1}{2}\alpha)$ and thus $\mu_w(\xi > 0) = 1$. So $\tilde{Q}$ is equivalent to μ_w.

Let $\tilde{Q}$ be defined by

$$\tilde{Q}(A) = \int_A \xi \, d\mu_w.$$

Then for $A \in \mathcal{F}_t$, using $E[\xi \mid \mathcal{F}_t] = Z_t$, we also have

$$\tilde{Q}(A) = \int_A Z_t \, d\mu_w. \tag{13.2.8}$$

Let $W_t = X_t - \theta_t$. Fix $(\lambda^1, \lambda^2, \ldots, \lambda^d) \in \mathbb{R}^d$ such that $\sum_{i=1}^{d}(\lambda^i)^2 = 1$ and let $U_t = \sum_{i=1}^{d} \lambda^i X_t^i$, $Y_t = \sum_{i=1}^{d} \lambda^i W_t^i$, $V_t = \sum_{i=1}^{d} \lambda^i \theta_t^i$ and $\phi_t = \sum_{i=1}^{d} \lambda^i \dot{\theta}_t^i$. Observe that $Y_t = U_t - V_t$ and

$$(U_t - V_t)Z_t = \int_0^t (U_s - V_s)dZ_s + \int_0^t Z_s dU_s - \int_0^t Z_s\phi_s ds + [U - V, Z]_t. \tag{13.2.9}$$

Since $\tilde{Q}$ is equivalent to μ_w, the quadratic variation of a semimartingale is the same under μ_w and $\tilde{Q}$. Now

$$[U - V, Z]_t = [U, Z]_t$$

$$= \sum_{i=1}^{d} \lambda^i [X^i, Z]_t$$

$$= \sum_{i=1}^{d} \lambda^i \int_0^t Z_s d[X^i, M]_s \tag{13.2.10}$$

$$= \sum_{i=1}^{d} \lambda^i \int_0^t Z_s \dot{\theta}_s^i ds$$

$$= \int_0^t Z_s \phi_s ds.$$

Using (13.2.9) and (13.2.10) it follows that $Y_t Z_t = (U_t - V_t) Z_t$ is a μ_w-local martingale, being a sum of stochastic integrals w.r.t. continuous martingales. Using Lemma 13.2, it follows that Y_t is a $\tilde{Q}$-local martingale. Since V is a process with finite variation paths, it follows that

$$
\begin{aligned}
[\textstyle\sum_{i=1}^d \lambda^i W^i, \sum_{i=1}^d \lambda^i W^i]_t &= [Y, Y]_t \\
&= [U, U]_t \\
&= [\textstyle\sum_{i=1}^d \lambda^i X^i, \sum_{i=1}^d \lambda^i X^i]_t \\
&= \sum_{i=1}^d (\lambda^i)^2 t.
\end{aligned}
$$

Invoking Lévy's characterization, Theorem 3.8, we conclude that W is a d-dimensional Brownian motion under $\tilde{Q}$. For $A \in \mathcal{B}(\mathbb{C}_d)$, we have

$$
\begin{aligned}
\tilde{Q}(A) &= \tilde{Q}(X - \theta \in A - \theta) \\
&= \tilde{Q}(W \in A - \theta) \\
&= \mu_w(A - \theta) \\
&= Q_\theta(A)
\end{aligned}
$$

where the last step was noted in (13.2.2). Thus $\tilde{Q} = Q_\theta$.

This proves one part. For the other part, let us assume that Q_θ is equivalent to μ_w. Now the process X is a semimartingale under μ_w and hence under Q_θ. On the other hand, under Q_θ, $W = X - \theta$ is a Brownian motion and hence a Q_θ-semimartingale. Thus θ considered as a stochastic process is a semimartingale. Thus for each i, θ^i is a function with finite variation on $[0, T]$ for every $T < \infty$ (see Exercise 5.62).

We will show that θ must satisfy (13.2.3) and (13.2.4). Let ξ be the Radon–Nikodym derivative $\frac{dQ_\theta}{d\mu_w}$, and let Z be the martingale

$$
Z_t = E_{\mu_w}[\xi | \mathcal{F}_t^+].
$$

Since $\theta(0) = 0$, it follows that $Z_0 = 1$. Since all $(\mathcal{F}_{\cdot}^+)$-martingales on $(\Omega, \mathcal{F}, \mu_w)$ admit a stochastic integral representation w.r.t. X (Theorem 10.20), we can get predictable processes $f^j \in \mathbb{L}(X^j)$, $1 \le j \le d$ such that

$$
Z_t = 1 + \sum_{j=1}^d \int_0^t f_s^j \, dX_s^j. \tag{13.2.11}
$$

It follows that Z is continuous and hence $Z_t = E_{\mu_w}[\xi | \mathcal{F}_t]$. As noted in the previous section, $\mu_w(Z_t > 0 \; \forall t) = 1$. Let

$$M_t = \int_0^t Z_s^{-1} dZ_s. \tag{13.2.12}$$

Then M is a local martingale, and writing $g_s^j = (Z_s)^{-1} f_s^j$, it follows that

$$M_t = \sum_{j=1}^d \int_0^t g_s^j dX_s^j. \tag{13.2.13}$$

Let $V_t^j = \int_0^t g_s^j ds$, for $1 \le j \le d$. We will next show that $X_t^j - V_t^j$ is a $\mathbf{Q}_\theta$ martingale for each j. For this, using integration by parts formula, we get

$$(X_t^j - V_t^j)Z_t = \int_0^t (X_s^j - V_s^j)dZ_s + \int_0^t Z_s dX_s^j - \int_0^t Z_s g_s^j ds + [X^j, Z]_t. \tag{13.2.14}$$

Now, $[X^j, Z]_t = \sum_{k=1}^d \int_0^t f_s^k d[X^j, X^k]$. Since $[X^j, X^k] = 0$ if $j \ne k$ and $[X^j, X^j]_t = t$, it follows that

$$[X^j, Z]_t = \int_0^t f_s^j ds = \int_0^t Z_s g_s^j ds.$$

Hence, using (13.2.14), we conclude

$$(X_t^j - V_t^j)Z_t = \int_0^t (X_s^j - V_s^j)dZ_s + \int_0^t Z_s dX_s^j$$

and is thus a μ_w-local martingale, being a sum of stochastic integrals w.r.t. continuous martingales. Hence by Lemma 13.2, $(X^j - V^j)$ is a $\mathbf{Q}_\theta$-local martingale. As noted earlier, $X^j - \theta^j$ is a Brownian motion under $\mathbf{Q}_\theta$ and thus $(V^j - \theta^j)$ is itself a continuous local martingale under $\mathbf{Q}_\theta$. But we have noted that θ^j is a function with finite variation and by definition V^j is a process with finite variation paths. Thus invoking Theorem 5.24, we conclude $\mu_w(V_t^j - \theta_t^j = 0 \ \forall t) = 1$. Thus

$$\mu_w\left(\int_0^t g_s^j ds = \theta_t^j \ \forall t\right) = 1.$$

This proves θ is absolutely continuous and (13.2.4) holds. Remains to show that θ satisfies (13.2.3) to complete the proof. Now we have

$$Z_t = \exp(M_t - \frac{1}{2}[M, M]_t)$$

where $M_t = \sum_{j=1}^d \int_0^t \dot\theta_s^j dX_s^j$ and $[M, M]_t = \sum_{j=1}^d \int_0^t (\dot\theta_s^j)^2 ds$. Further, Z_t converges in $\mathbb{L}^1(\mu_w)$ to ξ. We are to show that $[M, M]_t \to \alpha < \infty$. Suppose not, i.e. $[M, M]_t \uparrow \infty$. Get $t_n \uparrow \infty$ such that $[M, M]_{t_n} = n$. Then

$$|Z_{t_{n+1}} - Z_{t_n}| = Z_{t_n} |\exp((M_{t_{n+1}} - M_{t_n}) - \frac{1}{2}) - 1|.$$

Using that Z_{t_n} and $(M_{t_{n+1}} - M_{t_n})$ are independent, it follows that

$$E[|Z_{t_{n+1}} - Z_{t_n}|] = E[Z_{t_n}]E[|\exp((M_{t_{n+1}} - M_{t_n}) - \tfrac{1}{2}) - 1|]$$
$$= E[|\exp((M_{t_{n+1}} - M_{t_n}) - \tfrac{1}{2}) - 1|]$$
$$= E[|\exp(\eta - \tfrac{1}{2}) - 1|].$$

where η is a random variable with standard normal distribution. As a result, $E[|Z_{t_{n+1}} - Z_{t_n}|]$ does not converge to 0 contradicting $L^1(\mu_w)$ convergence of Z_t. Hence (13.2.3) holds completing the proof. □

Lemma 13.5 *Let M be a continuous local martingale and let f be a predictable process such that*

$$\int_0^\infty f_s^2 d[M, M]_s < \infty \quad a.s. \tag{13.2.15}$$

Then $N_t = \int_0^t f\, dM$ converges in probability as $t \to \infty$.

Proof For $k \geq 1$, let

$$\tau_k = \inf\{t : \int_0^t f_s^2 d[M, M]_s \geq k\}.$$

Note that the assumption (13.2.15) yields

$$\mu_w(\tau_k < \infty) \to 0 \text{ as } k \to \infty. \tag{13.2.16}$$

Then continuity of $[N, N]_t$ implies that

$$[N, N]_{t \wedge \tau_k} \leq k. \tag{13.2.17}$$

Since $[N, N]$ is increasing, this yields

$$\lim_{T \to \infty} \sup_{s,t \geq T} E[\,|[N, N]_{t \wedge \tau_k}] - E[[N, N]_{s \wedge \tau_k}|] = 0. \tag{13.2.18}$$

Observe that for any k,

$$\lim_{T \to \infty} \sup_{s,t \geq T} \mu_w(|N_s - N_t| \geq \varepsilon)$$

$$\leq \lim_{T \to \infty} \sup_{s,t \geq T} \mu_w(|N_{s \wedge \tau_k} - N_{t \wedge \tau_k}| \geq \varepsilon) + \mu_w(\tau_k < \infty)$$

$$\leq \lim_{T \to \infty} \sup_{s,t \geq T} \frac{1}{\varepsilon^2} E[|N_{s \wedge \tau_k} - N_{t \wedge \tau_k}|^2] + \mu_w(\tau_k < \infty)$$

$$= \lim_{T \to \infty} \sup_{s,t \geq T} \frac{1}{\varepsilon^2} E[\,|[N, N]_{t \wedge \tau_k}] - E[[N, N]_{s \wedge \tau_k}|] + \mu_w(\tau_k < \infty).$$

In view of (13.2.16) and (13.2.18), we conclude that N_t is Cauchy in probability and hence converges in probability. □

Remark **13.6** For a semimartingale X and $f \in \mathbb{L}(X)$, if $\int_0^t f \, dX$ converges in probability as $t \to \infty$, the limit is denoted by

$$\int_0^\infty f \, dX.$$

The conclusion of Theorem 13.4 can be restated as

$$\frac{d\mathbf{Q}_\theta}{d\mu_w} = \exp\{\sum_{i=1}^d \int_0^\infty \dot{\theta}_s^i \, dX_s^i - \frac{1}{2} \sum_{i=1}^d \int_0^\infty (\dot{\theta}_s^i)^2 ds\}. \qquad (13.2.19)$$

This is known as the Cameron–Martin formula.

13.3   Girsanov Theorem

Girsanov [19] generalized the Cameron–Martin formula to the case when the Brownian motion is translated by an adapted process. We begin with a simple observation.

Lemma 13.7 *Let N be a continuous local martingale such that $N_0 = 0$. Let $Y_t = \exp(N_t - \frac{1}{2}[N, N]_t)$. Then Y is a supermartingale, and for all $\tau \in \mathbb{T}_b$, $\mathsf{E}[Y_\tau] \le 1$.*

Proof Using Ito's formula, it follows that

$$Y_t = 1 + \int_0^t Y \, dN.$$

Thus Y is a local martingale. The rest follows from Lemma 5.7.   □

Let Z be a d-dimensional Brownian motion adapted to a filtration $(\mathcal{F}_\cdot)$ such that $(Z_t, \mathcal{F}_t)_{\{t \ge 0\}}$ is a Wiener martingale. Let $f = (f^j)$ be an $\mathbb{R}^d$-valued predictable process such that $f^j \in \mathbb{L}(Z^j)$. Suppose that

$$\alpha = \sum_{j=1}^d \int_0^\infty (f_s^j)^2 ds < \infty \ \ a.s. \qquad (13.3.1)$$

$$M_t = \sum_{j=1}^d \int_0^t f_s^j \, dZ_s^j. \qquad (13.3.2)$$

Then M is a local martingale, and in view of the assumption (13.3.1), it follows from Lemma 13.5 that

$$M_t \to \eta \ \text{ in probability}$$

where

$$\eta = \sum_{j=1}^{d} \int_0^\infty f_s^j \, dZ_s^j.$$ (13.3.3)

Theorem 13.8 (Girsanov Theorem) *Let $f^1, \ldots, f^d$ be predictable processes satisfying* (13.3.1). *Let*

$$\xi = \exp(\sum_{j=1}^{d} \int_0^\infty f_s^j \, dZ_s^j - \frac{1}{2} \sum_{j=1}^{d} \int_0^\infty (f_s^j)^2 ds).$$ (13.3.4)

Suppose

$$E[\xi] = 1.$$ (13.3.5)

Let Q be the probability measure defined by

$$\frac{dQ}{dP} = \xi.$$

Then the process $Y = (Y^1, \ldots, Y^d)$ defined by

$$Y_t^j = Z_t^j - \int_0^t f_s^j \, ds$$

is a d-dimensional Brownian motion, and $(Y_t, \mathcal{F}_t)_{\{t \geq 0\}}$ is a Wiener martingale under Q.

Proof Let

$$U_t = \exp(\sum_{j=1}^{d} \int_0^t f_s^j \, dZ_s^j - \frac{1}{2} \sum_{j=1}^{d} \int_0^t (f_s^j)^2 ds)$$ (13.3.6)

As noted earlier, (13.3.1) implies that U_t converges to ξ in probability. From Lemma 13.7 it follows that U is a supermartingale. Thus, for $A \in \mathcal{F}_s$, $t \to E[U_t 1_A]$ is a decreasing function for $t \in [s, \infty)$. Choosing a sequence $\{t_n : n \geq 1\} \subseteq [s, \infty)$ increasing to ∞ such that U_{t_n} converges to ξ almost surely, we conclude using Fatou's lemma that

$$E[\xi 1_A] \leq E[U_s 1_A].$$ (13.3.7)

In particular,

$$1 = E[\xi] \leq E[U_s] \leq 1.$$

Thus, $E[U_t] = 1$ for all t and as a consequence, U is a martingale. Moreover, since (13.3.7) holds for all $A \in \mathcal{F}_s$ and $E[U_s] = 1$ it follows that equality holds in (13.3.7) and hence

$$E_P[\xi \mid \mathcal{F}_s] = U_s.$$ (13.3.8)

Thus for $A \in \mathcal{F}_s$,

$$Q(A) = \int_A \xi dP = \int_A U_t dP \tag{13.3.9}$$

Note that by definition $[Y^i, Y^j] = [Z^i, Z^j]$ and that the quadratic variation under P is same as that under Q. Thus, in view of the Levy's characterization of Brownian motion, all one needs to show is that Y^j is a Q-local martingale.

Let us observe that $U_t = 1 + \int_0^t U_s dM_s$ and so $[Y^j, U]_t = [Z^j, U]_t = \int_0^t U_s d$ $[Z^j, M]_s = \int_0^t U_s f_s^j ds$. Thus

$$\begin{aligned}
Y_t^j U_t &= \int_0^t Y_s^j dU_s + \int_0^t U_s dY_s^j + [Y^j, U]_t \\
&= \int_0^t Y_s^j dU_s + \int_0^t U_s dZ_s^j - \int_0^t U_s f_s^j ds + [Y^j, U]_t \\
&= \int_0^t Y_s^j dU_s + \int_0^t U_s dZ_s^j
\end{aligned}$$

Since U and Z^j are martingales under P, it follows that $Y^j U$ is P-local martingale. Invoking Lemma 13.2 we conclude that Y^j is Q local martingale. As noted above, this implies Y is an $\mathbb{R}^d$-valued Brownian motion and $(Y_t, \mathcal{F}_t)_{\{t \geq 0\}}$ is a Wiener martingale. $\square$

A natural question that arises is: *given $f^1, f^2, \ldots, f^d$, such that (13.3.1) is true when does (13.3.5) hold?* What are known as sufficient conditions, but no necessary and sufficient condition is known. We now give sufficient conditions, due to Novikov and Kazamaki.

Theorem 13.9 *Let M be a continuous local martingale such that $M_0 = 0$. Suppose that*

$$\sup_{\tau \in \mathbb{T}_b} E[\exp(\tfrac{1}{2} M_\tau)] = K < \infty. \tag{13.3.10}$$

Then $U_t = \exp(M_t - \tfrac{1}{2}[M, M]_t)$ is a uniformly integrable martingale.

Further, $[M, M]_t \to \eta$, $M_t \to \phi$ and $U_t \to \xi = \exp(\phi - \tfrac{1}{2}\eta)$ in probability as $t \uparrow \infty$ and

$$E[\xi] = 1. \tag{13.3.11}$$

Proof Since for any $\alpha \in [0, \tfrac{1}{2}]$, $\exp(\alpha x) \leq (1 + \exp(\tfrac{1}{2} x))$, it follows that for any $\alpha \in [0, \tfrac{1}{2}]$ and $\tau \in \mathbb{T}_b$, we have

$$E[\exp(\alpha M_\tau)] \leq (1 + K). \tag{13.3.12}$$

Fix $\lambda \in (0, 1)$ and let

$$U_t^\lambda = \exp(\lambda M_t - \tfrac{1}{2}\lambda^2 [M, M]_t).$$

For $\tau \in \mathbb{T}_b$, $p > 1$ and $a > 0$ (to be chosen later) let us write

$$(U_\tau^\lambda)^p = \exp(a\lambda M_\tau - \tfrac{1}{2}\lambda^2 p[M, M]_\tau) \exp(\lambda(p - a)M_\tau).$$

For $b > 1$, $c > 1$ such that $\tfrac{1}{b} + \tfrac{1}{c} = 1$, using Holder's inequality, we conclude

$$\mathsf{E}[(U_\tau^\lambda)^p] \le (\mathsf{E}[\exp(ab\lambda M_\tau - \tfrac{1}{2}\lambda^2 pb[M, M]_\tau)])^{\frac{1}{b}} (\mathsf{E}[\exp(c\lambda(p - a)M_\tau)])^{\frac{1}{c}}.$$

We now choose $a = \frac{\sqrt{p}}{\sqrt{b}}$ so that $a^2 b^2 = pb$, and hence the first factor on right-hand side above is ≤ 1 in view of Lemma 13.7. We thus get

$$\mathsf{E}[(U_\tau^\lambda)^p] \le (\mathsf{E}[\exp(c\lambda(p - a)M_\tau)])^{\frac{1}{c}}. \tag{13.3.13}$$

For $\delta > 0$, take $p = (1 + \delta^2)^2$ and $b = (1 + \delta)^2$. Since $a = \frac{\sqrt{p}}{\sqrt{b}}$, we get $a = \frac{(1+\delta^2)}{(1+\delta)}$. Also $c = \frac{b}{b-1} = \frac{(1+\delta)^2}{2\delta+\delta^2}$. Thus,

$$\begin{aligned}
c\lambda(p - a) &= \frac{(1 + \delta)^2}{2\delta + \delta^2}\lambda((1 + \delta^2)^2 - \frac{(1 + \delta^2)}{(1 + \delta)}) \\
&= \frac{(1 + \delta)}{2\delta + \delta^2}\lambda(1 + \delta^2)((1 + \delta)(1 + \delta^2) - 1) \tag{13.3.14} \\
&= \lambda\frac{(1 + \delta)}{2 + \delta}(1 + \delta^2)(1 + \delta + \delta^2)
\end{aligned}$$

Since $\lambda < 1$, in view of (13.3.14), we can choose $\delta > 0$ such that $c\lambda(p - a) \le \tfrac{1}{2}$, and as a result we have by (13.3.12)

$$\mathsf{E}[(U_\tau^\lambda)^p] \le (1 + K)^{\frac{1}{c}} \tag{13.3.15}$$

where $p > 1$, a and c are as chosen above and K is as in (13.3.10). This shows

$$\sup_{\tau \in \mathbb{T}_b} \mathsf{E}[(U_\tau^\lambda)^p] < \infty.$$

Invoking Lemma 5.6 we conclude that for $0 < \lambda < 1$, U^λ is uniformly integrable martingale and hence by Theorem 2.25 U_t^λ converges in probability for each such λ. Using this for distinct values of λ, say $\tfrac{1}{2}$ and $\tfrac{1}{4}$ we can conclude that $[M, M]_t$ and M_t converge in probability say to η and ϕ and then that

$$U_t^\lambda \to \exp(\lambda\phi - \frac{1}{2}\lambda^2\eta).$$

Further it follows that and that for any $\lambda < 1$

$$E[\exp(\lambda\phi - \tfrac{1}{2}\lambda^2\eta)] = 1. \tag{13.3.16}$$

Thus it follows that U_t converges in probability to $\xi = \exp(\phi - \tfrac{1}{2}\eta)$. Remains to show (13.3.11).

Since $M_t \to \phi$, using (13.3.12) and Fatou's lemma (along a sequence $t_n \uparrow \infty$ such that U_{t_n} converges almost surely) we conclude that for any $\alpha \in [0, \tfrac{1}{2}]$, we have

$$E[\exp(\alpha\phi)] \le (1 + K). \tag{13.3.17}$$

By Lemma 13.7, for all t we have

$$E[\exp(M_t - \tfrac{1}{2}[M, M]_t] \le 1$$

and hence again using Fatou's lemma we conclude

$$E[\exp(\phi - \tfrac{1}{2}\eta)] \le 1. \tag{13.3.18}$$

Note that

$$\exp(\lambda\phi - \tfrac{1}{2}\lambda^2\eta) = \exp(\lambda^2\phi - \tfrac{1}{2}\lambda^2\eta)\exp(\lambda(1 - \lambda)\phi).$$

Now using this relation along with Holder's inequality with $p = \frac{1}{\lambda^2}$ and $q = \frac{1}{(1-\lambda^2)}$, we get

$$
\begin{aligned}
1 &= E[\exp(\lambda\phi - \tfrac{1}{2}\lambda^2\eta)] \\
&= E[\exp(\lambda^2\phi - \tfrac{1}{2}\lambda^2\eta)\exp(\lambda(1 - \lambda)\phi)] \\
&\le (E[\exp(\phi - \tfrac{1}{2}\eta)])^{\lambda^2}(E[\exp(\frac{\lambda}{(1 + \lambda)}\phi)])^{(1-\lambda^2)} \\
&\le (E[\exp(\phi - \tfrac{1}{2}\eta)])^{\lambda^2}(1 + K)^{(1-\lambda^2)}
\end{aligned}
\tag{13.3.19}
$$

where the first equality is (13.3.16), and in the last step we have used (13.3.17). Now taking limit as $\lambda \uparrow 1$ in (13.3.19), we conclude

$$1 \le E[\exp(\phi - \tfrac{1}{2}\eta)].$$

In view of (13.3.18), this shows

$$E[\exp(\phi - \tfrac{1}{2}\eta)] = 1.$$

$\square$

The condition (13.3.10) is due to Kazamaki [44]. Earlier, a slightly stronger condition (13.3.20) was proposed by Novikov, which we give below. In practice, the Novikov condition may be easier to check than the Kazamaki condition.

Theorem 13.10 *Let M be a continuous local martingale such that $M_0 = 0$. Suppose that*

$$\sup_{T < \infty} \mathsf{E}[\exp(\tfrac{1}{2}[M, M]_T)] = K < \infty. \tag{13.3.20}$$

Then $U_t = \exp(M_t - \tfrac{1}{2}[M, M]_t)$ is a uniformly integrable martingale. Further, $[M, M]_t \to \eta$, $M_t \to \beta$ and $U_t \to \xi = \exp(\beta - \tfrac{1}{2}\eta)$ in probability with

$$\mathsf{E}[\xi] = 1. \tag{13.3.21}$$

Proof We will show that (13.3.20) implies (13.3.10). For any bounded stopping time τ, bounded by T observe that

$$
\begin{aligned}
\mathsf{E}[\exp(\tfrac{1}{2}M_\tau)] &= \mathsf{E}[\exp(\tfrac{1}{2}M_\tau - \tfrac{1}{4}[M, M]_\tau)\exp(\tfrac{1}{4}[M, M]_\tau)] \\
&\leq (\mathsf{E}[\exp(M_\tau - \tfrac{1}{2}[M, M]_\tau)])^{\frac{1}{2}}(\mathsf{E}[\exp(\tfrac{1}{2}[M, M]_\tau)])^{\frac{1}{2}} \\
&\leq (\mathsf{E}[\exp(\tfrac{1}{2}[M, M]_\tau)])^{\frac{1}{2}} \tag{13.3.22} \\
&\leq (\mathsf{E}[\exp(\tfrac{1}{2}[M, M]_T)])^{\frac{1}{2}} \\
&\leq \sqrt{K}.
\end{aligned}
$$

where we have used Lemma 13.7 and Cauchy–Schwarz inequality. Taking supremum over $\tau \in \mathbb{T}_b$ on LHS, the result follows, namely that (13.3.20) implies (13.3.10). $\square$

The results given above lead to the generalization of the Cameron–Martin formula by Girsanov to the case when the Brownian motion is translated by a possibly nonlinear predictable functional g.

In this section, we continue to denote the coordinate process on $\mathbb{C}_d$ by $X = (X_t)$. For $t \geq 0$ let $\mathcal{G}_t = \sigma(X_u : u \leq t)$. Let

$$g : [0, \infty) \times \mathbb{C}_d \mapsto \mathbb{R}^d \text{ be } (\mathcal{G}_.)\text{-predictable.} \tag{13.3.23}$$

Suppose on some probability space $(\Omega, \mathcal{F}, \mathsf{P})$, we have a filtration $(\mathcal{F}_.)$, a Brownian motion W such that $(W_t, \mathcal{F}_t)_{\{t \geq 0\}}$ is a Wiener martingale. Note that if Y is any continuous $(\mathcal{F}_.)$ adapted process, then $U_t = g(t, Y)$ is a $(\mathcal{F}_.)$ predictable process. Suppose Y is a solution of the stochastic differential equation

$$dY_t = dW_t + g(t, Y)dt \tag{13.3.24}$$

i.e. Y is an adapted continuous process such that

$$Y_t^j = W_t^j + \int_0^t g^j(s, Y)ds. \tag{13.3.25}$$

Let $\nu = P \circ Y^{-1}$ be the distribution of Y—thus ν is a probability measure on $(\mathbb{C}_d, \mathcal{B}(\mathbb{C}_d))$. Let μ_w denote the Wiener measure on $(\tilde{\Omega}, \tilde{\mathcal{F}}) = (\mathbb{C}_d, \mathcal{B}(\mathbb{C}_d))$. Thus X is Brownian motion under μ_w.

The next result, due to Girsanov, shows that under some conditions, ν is absolutely continuous w.r.t. μ_w and gives a formula for the Radon–Nikodym derivative. Recall that $\mathbb{T}_b$ denotes the class of all bounded stopping times w.r.t. the filtration under consideration.

Theorem 13.11 *Let* $g = (g^1, \ldots, g^d)$ *satisfy* (13.3.23). *Suppose Y is a solution to the SDE* (13.3.25) *where W is a Brownian motion. Assume that*

$$\sum_{j=1}^{d} \int_0^\infty (g^j(s, Y_s))^2 ds < \infty \quad a.s. \ \mathsf{P}. \tag{13.3.26}$$

and

$$\sup_{\tau \in \mathbb{T}_b} \mathsf{E}_\mathsf{P}[\exp(-\frac{1}{2} \sum_{j=1}^{d} \int_0^\tau g^j(s, Y) dW_s^j)] < \infty. \tag{13.3.27}$$

Then $\nu = \mathsf{P} \circ Y^{-1}$ is absolutely continuous w.r.t. μ_w and

$$\frac{d\nu}{d\mu_w} = \exp(\sum_{j=1}^{d} \int_0^\infty g^j(s, X) dX_s^j - \frac{1}{2} \sum_{j=1}^{d} \int_0^\infty (g^j(s, X))^2 ds). \tag{13.3.28}$$

Thus, uniqueness of weak solution to the SDE (13.3.25) *holds in the class of solutions satisfying* (13.3.27). *Moreover, for $T < \infty$*

$$\frac{d\nu}{d\mu_w}|_{\mathcal{F}_T} = \exp(\sum_{j=1}^{d} \int_0^T g^j(s, X) dX_s^j - \frac{1}{2} \sum_{j=1}^{d} \int_0^T (g^j(s, X))^2 ds) \tag{13.3.29}$$

i.e. for $A \in \mathcal{G}_T$

$$\nu(A) = \int_A \exp(\sum_{j=1}^{d} \int_0^T g^j(s, X) dX_s^j - \frac{1}{2} \sum_{j=1}^{d} \int_0^T (g^j(s, X))^2 ds) d\mu_w. \tag{13.3.30}$$

Proof First note that the condition (13.3.27) is same as

$$\sup_{\tau \in \mathbb{T}_b} \mathsf{E}_\mathsf{P}[\exp(-\frac{1}{2} \sum_{j=1}^{d} \int_0^\tau g^j(s, Y) dY_s^j + \frac{1}{2} \sum_{j=1}^{d} \int_0^\tau (g^j(s, Y))^2 ds)] < \infty. \tag{13.3.31}$$

Let a process Z be defined on $(\mathbb{C}_d, \mathcal{B}(\mathbb{C}_d))$ via

$$Z_t^j = X_t^j - \int_0^t g^j(s, X)ds. \tag{13.3.32}$$

Since $\nu = P \circ Y^{-1}$ it follows that distribution of Z under ν is the same as that of W under P, and in other words Z is a Brownian motion under ν.

Let $\tilde{\mathbb{T}}_b$ denote the class of bounded $(\mathcal{F}_\cdot^X)$ stopping times. The condition (13.3.31) implies

$$\sup_{\sigma \in \tilde{\mathbb{T}}_b} E_\nu[\exp(-\frac{1}{2}\sum_{j=1}^d \int_0^\sigma g^j(s, X)dX_s^j + \frac{1}{2}\sum_{j=1}^d \int_0^\sigma (g^j(s, X))^2 ds)] < \infty.$$

$$\tag{13.3.33}$$

Let $M_t = -\sum_{j=1}^d \int_0^t g^j(s, X)dZ_s^j$. Then M is a ν-local martingale. Noting that

$$M_t = -\sum_{j=1}^d \int_0^t g^j(s, X)dX_s^j + \int_0^t (g^j(s, X))^2 ds$$

The relation (13.3.33) yields

$$\sup_{\sigma \in \tilde{\mathbb{T}}_b} E_\nu[\exp(\frac{1}{2}M_\tau)] < \infty \tag{13.3.34}$$

Thus invoking Theorem 13.9 we conclude that

$$U_t = \exp(M_t - \frac{1}{2}[M, M]_t)$$

is a uniformly integrable martingale and U_t converges to ξ in $\mathbb{L}^1(\nu)$ with $\nu(\xi > 0) = 1$ and $E_\nu[\xi] = 1$ where

$$\xi = \exp(-\sum_{j=1}^d \int_0^\infty g^j(s, X)dZ_s^j - \frac{1}{2}\sum_{j=1}^d \int_0^\infty (g^j(s, X))^2 ds). \tag{13.3.35}$$

Note that the assumption (13.3.26) along with Lemma 13.5 ensures that for each j, $\int_0^t g^j(s, X)dZ_s^j$ converges to $\int_0^\infty g^j(s, X)dZ_s^j$ in ν probability. Let us define a probability measure $\tilde{\mu}$ on $(\mathbb{C}_d, \mathcal{B}(\mathbb{C}_d))$ by

$$\frac{d\tilde{\mu}}{d\nu} = \xi.$$

By Theorem 13.8, it follows that $X_t^j = Z_t^j + \int_0^t g^j(s, X)ds$ is a d-dimensional Brownian motion on $(\mathbb{C}_d, \mathcal{B}(\mathbb{C}_d), \tilde{\mu})$. Recalling that X is the coordinate process on $\mathbb{C}_d$, we conclude $\tilde{\mu} = \mu_w$. Since $\nu(\xi > 0) = 1$, it follows that ν and μ_w are equivalent and

$$\frac{d\nu}{d\mu_w} = \xi^{-1}.$$

Now

$$\xi^{-1} = \exp(\sum_{j=1}^{d} \int_0^\infty g^j(s, X) dZ_s^j + \frac{1}{2} \sum_{j=1}^{d} \int_0^\infty (g^j(s, X))^2 ds)$$

$$= \exp(\sum_{j=1}^{d} \int_0^\infty g^j(s, X) dX_s^j - \frac{1}{2} \sum_{j=1}^{d} \int_0^\infty (g^j(s, X))^2 ds)$$

where in the second step we have used

$$\int_0^\infty g^j(s, X) dZ_s^j = \int_0^\infty g^j(s, X) dX_s^j - \int_0^\infty (g^j(s, X))^2 ds.$$

Since the distribution $\nu = \mathsf{P} \circ Y^{-1}$ for any solution Y to the SDE (13.3.25) satisfies (13.3.28), uniqueness of weak solution to SDE (13.3.25) follows. Observing that

$$V_t = \exp(\sum_{j=1}^{d} \int_0^T g^j(s, X) dX_s^j - \frac{1}{2} \sum_{j=1}^{d} \int_0^T (g^j(s, X))^2 ds)$$

is a martingale under μ_w, we can conclude that (13.3.30) holds. This completes the proof. □

Let us note that the condition (13.3.33) can be recast as follows in terms of integral w.r.t. the Wiener measure.

$$\mathsf{E}_\nu[\exp(-\frac{1}{2} \sum_{j=1}^{d} \int_0^\sigma g^j(s, X) dX_s^j + \frac{1}{2} \sum_{j=1}^{d} \int_0^\sigma (g^j(s, X))^2 ds)]$$

$$= \mathsf{E}_{\mu_w} \xi^{-1}[\exp(-\frac{1}{2} \sum_{j=1}^{d} \int_0^\sigma g^j(s, X) dX_s^j + \frac{1}{2} \sum_{j=1}^{d} \int_0^\sigma (g^j(s, X))^2 ds)]$$

$$= \mathsf{E}_{\mu_w}[\exp(\frac{1}{2} \sum_{j=1}^{d} \int_0^\sigma g^j(s, X) dX_s^j)]$$

Thus the condition (13.3.27) implies

$$\sup_{\sigma \in \tilde{\mathbb{T}}_b} \mathsf{E}_{\mu_w}[\exp(\frac{1}{2} \sum_{j=1}^{d} \int_0^\sigma g^j(s, X) dX_s^j)] < \infty. \qquad (13.3.36)$$

Indeed, if in (13.3.27), we take the underlying filtration to be $(\mathcal{F}_\cdot^Y)$, then (13.3.27) is equivalent to (13.3.36). The advantage of the condition (13.3.36) is that it only involves integrals w.r.t. the Wiener measure. Having proven the uniqueness, we will now show existence under suitable conditions.

Theorem 13.12 *Suppose that* $g = (g^1, \ldots, g^d)$ *satisfies* (13.3.23),

$$\sum_{j=1}^{d} \int_0^\infty (g^j(s, X_s))^2 ds < \infty \quad a.s. \ \mu_w \tag{13.3.37}$$

and (13.3.36). *Then there exists a probability space* $(\widehat{\Omega}, \widehat{\mathcal{F}}, \widehat{\mathsf{P}})$ *with filtration* $(\widehat{\mathcal{F}}_\cdot)$, *a Brownian motion* $\widehat{W}$ *such that* $(\widehat{W}_t, \widehat{\mathcal{F}}_t)_{\{t \geq 0\}}$ *is a Wiener martingale, and an adapted process* $\widehat{Y}$ *satisfying*

$$\sum_{j=1}^{d} \int_0^\infty (g^j(s, \widehat{Y}_s))^2 ds < \infty \quad a.s. \ \widehat{\mathsf{P}}, \tag{13.3.38}$$

$$\sup_{\tau \in \widehat{\mathbb{T}}_b} \mathsf{E}_{\widehat{\mathsf{P}}}[\exp(-\frac{1}{2} \sum_{j=1}^{d} \int_0^\tau g^j(s, \widehat{Y}) d\widehat{W}_s^j)] < \infty \tag{13.3.39}$$

and

$$\widehat{Y}_t^j = \widehat{W}_t^j + \int_0^t g^j(s, \widehat{Y}) ds, \quad 1 \leq j \leq d. \tag{13.3.40}$$

Here $\widehat{\mathbb{T}}_b$ *is the class of bounded stopping times on* $(\widehat{\Omega}, \widehat{\mathcal{F}}, \widehat{\mathsf{P}})$ *w.r.t. the filtration* $(\widehat{\mathcal{F}}_\cdot)$.

Proof Let us define a measure ν on $(\mathbb{C}_d, \mathcal{B}(\mathbb{C}_d))$ by (13.3.28). The assumption (13.3.37) implies that ν is a probability measure and Theorem 13.8 then implies that Z defined by (13.3.32) is a d-dimensional Brownian motion and

$$dX_t = dZ_t + g(t, X) dt.$$

Let us take $(\widehat{\Omega}, \widehat{\mathcal{F}}, \widehat{\mathsf{P}}) = (\mathbb{C}_d, \mathcal{B}(\mathbb{C}_d), \nu)$, $(\widehat{\mathcal{F}}_\cdot) = (\mathcal{F}_\cdot^X)$, $\widehat{W} = Z$, $\widehat{Y} = X$. It follows that (13.3.38) and (13.3.40) hold. Retracing steps in the proof of Theorem 13.11, we can verify that (13.3.36) implies (13.3.39). $\square$

In other words, if g satisfies (13.3.23), (13.3.37) then existence and uniqueness of weak solution to (13.3.40) holds in the class of solutions satisfying (13.3.38) and (13.3.39).

We now briefly consider analogues of the results in this section for solutions to stochastic differential equations driven by Brownian motion. Let us fix $\sigma : [0, \infty) \times \mathbb{C}_d \mapsto L(d, d)$ and $h : [0, \infty) \times \mathbb{C}_d \mapsto \mathbb{R}^d$ satisfying conditions (7.5.1), (7.5.2), (7.5.9)–(7.5.12). Also let us fix $y_0 \in \mathbb{R}^d$. We have seen in Theorem 7.26 that the SDE,

$$Y_t^j = y_0 + \sum_{k=1}^d \int_0^t \sigma^{jk}(s, Y) dW_s^k + \int_0^t h^j(s, Y) ds, \qquad (13.3.41)$$

where W is a Brownian motion, admits a unique strong solution Y and that $P \circ Y^{-1}$ is uniquely determined; i.e., the SDE (13.3.41) has a unique weak solution. we denote $P \circ Y^{-1} = \mu_*$.

We will continue to denote by X the coordinate process on $\mathbb{C}_d$ defined by (13.2.1). Let $C_t = \sigma(X_u : u \le t)$

Here is a result extending existence of weak solutions to equation of the type (13.3.41).

Theorem 13.13 *Suppose W is a Brownian motion on some probability space $(\Omega, \mathcal{F}, P)$ and Y is a solution to the SDE (13.3.41). Let $\phi : [0, \infty) \times \mathbb{C}_d \mapsto \mathbb{R}^d$ be predictable (for the filtration $(C_.)$). Suppose*

$$\sum_{j=1}^d \int_0^\infty (\phi^j(s, Y))^2 ds < \infty \quad a.s. \ P. \qquad (13.3.42)$$

$$\phi = \phi 1_{\{|\sigma| \ne 0\}} \qquad (13.3.43)$$

and

$$\sup_{\tau \in \mathbb{T}_b} E_P[\exp(\frac{1}{2} \sum_{j=1}^d \int_0^\tau \phi^j(s, Y) dW_s^j)] < \infty. \qquad (13.3.44)$$

Let

$$f = h + \sigma \phi \qquad (13.3.45)$$

i.e. $f^j = h^j + \sum_{k=1}^d \sigma^{jk} \phi^k$. Then the SDE,

$$V_t^j = y_0 + \sum_{k=1}^d \int_0^t \sigma^{jk}(s, V) dU_s^k + \int_0^t f^j(s, V) ds, \qquad (13.3.46)$$

where U denotes a Brownian motion, admits a weak solution. Further, for such a solution V defined on $(\widehat{\Omega}, \widehat{\mathcal{F}}, \widehat{P})$,

$$\sum_{j=1}^d \int_0^\infty (\phi^j(s, V_s))^2 ds < \infty \quad a.s. \ \widehat{P} \qquad (13.3.47)$$

and

$$\sup_{\tau \in \widetilde{\mathbb{T}}_b} E_{\widehat{P}}[\exp(-\frac{1}{2} \sum_{j=1}^d \int_0^\tau \phi^j(s, V) dU_s^j)] < \infty. \qquad (13.3.48)$$

Proof Defining Q on $(\Omega, \mathcal{F})$ by

$$\frac{dQ}{dP} = \exp(\sum_{j=1}^{d} \int_0^\infty \phi^j(s, Y)dW_s^j - \frac{1}{2}\sum_{j=1}^{d} \int_0^\infty (\phi^j(s, Y))^2 ds)$$

it follows from Theorem 13.9 that Q is a probability measure and invoking Theorem 13.8 it follows that U defined by

$$U_t^j = W_t^j - \int_0^t \phi^j(s, Y)ds \qquad (13.3.49)$$

is a Brownian motion under Q. Clearly, using the definition of U and f in terms of W and σ, ϕ, h we can deduce that

$$Y_t^j = y_0 + \sum_{k=1}^{d} \int_0^t \sigma^{jk}(s, Y)dU_s^k + \int_0^t f^j(s, Y)ds. \qquad (13.3.50)$$

Noting that

$$-\frac{1}{2}\sum_{j=1}^{d} \int_0^T \phi^j(s, Y)dU_s^j + \sum_{j=1}^{d} \int_0^T \phi^j(s, Y)dW_s^j - \frac{1}{2}\sum_{j=1}^{d} \int_0^T (\phi^j(s, Y))^2 ds$$

$$= \frac{1}{2}\sum_{j=1}^{d} \int_0^T \phi^j(s, Y)dW_s^j$$

it follows that

$$E_Q[\exp(-\frac{1}{2}\sum_{j=1}^{d} \int_0^T \phi^j(s, Y)dU_s^j)] = E_P[\frac{dQ}{dP}\exp(-\frac{1}{2}\sum_{j=1}^{d} \int_0^T \phi^j(s, Y)dU_s^j)]$$

$$= E_P[\exp(\frac{1}{2}\sum_{j=1}^{d} \int_0^T \phi^j(s, Y)dW_s^j)].$$

Thus the condition (13.3.44) implies (13.3.48) with $\widehat{P} = Q$, $V = Y$. Since Q is absolutely continuous w.r.t. P, (13.3.47) holds with $\widehat{P} = Q$, $V = Y$. $\square$

Having proved existence, we will now show uniqueness of the weak solution by identifying its distribution. Recall $\mu_* = P \circ Y^{-1}$, where Y is the unique strong solution to the SDE

$$Y_t^j = y_0 + \sum_{k=1}^{d} \int_0^t \sigma^{jk}(s, Y)dW_s^k + \int_0^t h^j(s, Y)ds$$

with W being a Brownian motion.

Theorem 13.14 *Let U be a Brownian motion on $(\widehat{\Omega}, \widehat{\mathcal{F}}, \widehat{Q})$ adapted to a filtration $(\widehat{\mathcal{F}_t})$ such that $(U_t, \widehat{\mathcal{F}_t})_{\{t \geq 0\}}$ is a Wiener martingale and let V be an $(\widehat{\mathcal{F}_t})$ adapted continuous process satisfying the SDE (13.3.46), where f is defined by (13.3.45). Further, suppose ϕ, V satisfy (13.3.47) and (13.3.48). Let $\nu = \widehat{Q} \circ V^{-1}$ be the distribution of V. Then ν is absolutely continuous w.r.t. μ_* and for $T < \infty$*

$$\frac{d\nu}{d\mu_*}\Big|_{\mathcal{F}_T} = \exp(L) \tag{13.3.51}$$

where

$$L = \sum_{j=1}^{d} \int_0^T \psi^j(s, X) dX_s^j - \sum_{j=1}^{d} \int_0^T h_s^j \psi^j(s, X) ds - \frac{1}{2} \sum_{j=1}^{d} \int_0^T (\phi^j(s, X))^2 ds,$$

X is the coordinate process on $\mathbb{C}_d$ and $\psi = \phi \sigma^{-1} 1_{\{|\sigma| \neq 0\}}$. As a consequence, weak solution to the SDE (13.3.46) is unique.

Proof Let us define a measure $\widehat{P}$ on $(\widehat{\Omega}, \widehat{\mathcal{F}})$ by

$$\frac{d\widehat{P}}{d\widehat{Q}} = \exp(-\sum_{j=1}^{d} \int_0^\infty \phi^j(s, V) dU_s^j - \frac{1}{2} \sum_{j=1}^{d} \int_0^\infty (\phi^j(s, V))^2 ds). \tag{13.3.52}$$

In view of the assumption (13.3.48), it follows from Theorem 13.9 that $\widehat{P}$ is a probability measure and from Theorem 13.9 that Z defined by

$$Z_t^j = U_t^j + \int_0^t \phi^j(s, V) ds. \tag{13.3.53}$$

is a Brownian motion under $\widehat{P}$. Now recalling that V satisfies (13.3.46) and that h, f and ϕ are related via (13.3.45), we note that

$$V_t^j = y_0 + \sum_{k=1}^{d} \int_0^t \sigma^{jk}(s, V) dZ_s^k + \int_0^t h^j(s, V) ds. \tag{13.3.54}$$

Since Z is a Brownian motion under $\widehat{P}$, it follows that V is a weak solution to the SDE (13.3.41) on $(\widehat{\Omega}, \widehat{\mathcal{F}}, \widehat{P})$ and as a consequence, we have

$$\widehat{P} \circ \widehat{V}^{-1} = \mu_*. \tag{13.3.55}$$

Since $\widehat{P}$ and $\widehat{Q}$ are mutually absolutely continuous, it follows that μ_* and ν are mutually absolutely continuous.

Let

$$S_t = \exp(-\sum_{j=1}^{d} \int_0^t \phi^j(s, V)dU_s^j - \frac{1}{2}\sum_{j=1}^{d} \int_0^t (\phi^j(s, V))^2 ds).$$

It then follows that S is a Q-martingale and for $A \in \widehat{\mathcal{F}}_t$

$$P(A) = \int_A S_t dQ. \qquad (13.3.56)$$

Let

$$R_t = \exp(\sum_{j=1}^{d} \int_0^t \phi^j(s, V)dZ_s^j - \frac{1}{2}\sum_{j=1}^{d} \int_0^t (\phi^j(s, V))^2 ds). \qquad (13.3.57)$$

Note that

$$
\begin{aligned}
S_t^{-1} &= \exp(\sum_{j=1}^{d} \int_0^t \phi^j(s, V)dU_s^j + \frac{1}{2}\sum_{j=1}^{d} \int_0^t (\phi^j(s, V))^2 ds) \\
&= \exp(\sum_{j=1}^{d} \int_0^t \phi^j(s, V)dZ_s^j - \frac{1}{2}\sum_{j=1}^{d} \int_0^t (\phi^j(s, V))^2 ds) \\
&= R_t.
\end{aligned}
\qquad (13.3.58)
$$

Using (13.3.56) it follows that

$$\widehat{Q}(A) = \int_A R_t d\widehat{P} \quad \forall A \in \widehat{\mathcal{F}}_t, \; \forall t \qquad (13.3.59)$$

and thus R is a P-martingale. Using (13.3.43) and (13.3.54) and $\psi = \phi\sigma^{-1}1_{\{|\sigma|\neq 0\}}$ it follows that

$$
\begin{aligned}
\sum_{j=1}^{d} \int_0^t \phi^j(s, V)dZ_s^j &= \sum_{j=1}^{d} \int_0^t \phi^j(s, V)1_{\{|\sigma|\neq 0\}}dZ_s^j \\
&= \sum_{j=1}^{d} \int_0^t \psi^j(s, V)dV_s^j - \sum_{j=1}^{d} \int_0^t h^j(s, V)\psi^j(s, V)ds
\end{aligned}
$$

Thus,

$$R_t = \exp(\zeta)$$

where

$$\zeta = \sum_{j=1}^{d} \int_0^t \psi^j(s, V) dV_s^j - \sum_{j=1}^{d} \int_0^t h^j(s, V) \psi^j(s, V) ds - \frac{1}{2} \sum_{j=1}^{d} \int_0^t (\phi^j(s, V))^2 ds$$

Let us define α_t on $(\mathbb{C}_d, \mathcal{B}(\mathbb{C}_d), \mu_w)$ by

$$\alpha_t = \exp(\eta)$$

where

$$\eta = \sum_{j=1}^{d} \int_0^t \psi^j(s, X) dX_s^j - \sum_{j=1}^{d} \int_0^t h^j(s, X) \psi^j(s, X) ds - \frac{1}{2} \sum_{j=1}^{d} \int_0^t (\phi^j(s, X))^2 ds$$

We would like to show that for $B \in \mathcal{F}_t^X$,

$$\int_{\{V \in B\}} R_t d\widehat{\mathsf{P}} = \int_{\{X \in B\}} \alpha_t d\mu_*. \tag{13.3.60}$$

If ψ^j were continuous, the stochastic integral could be expressed pathwise using Theorem 6.2 and then (13.3.60) would follow from the usual change of variable formula. For the general case, (13.3.60) follows using Exercise 4.54.

$$\nu(B) = \mathsf{Q}(V \in B)$$

$$= \int_{\{V \in B\}} R_t d\widehat{\mathsf{P}} \tag{13.3.61}$$

$$= \int_B \alpha_t d\mu_*.$$

Thus, ν is uniquely determined on $\mathcal{F}_t^X$ and since $\cup_{t>0} \mathcal{F}_t^X$ is a field that generates $\mathcal{B}(\mathbb{C}_d)$, it follows that ν is uniquely determined. Thus weak uniqueness holds for solution to the SDE (13.3.46). $\square$

13.4 The Girsanov–Meyer Theorem

The following result is due to Meyer building upon the idea by Girsanov in the context of a Wiener process. We return to the framework of Sect. 13.1. Recall Q is a probability measure equivalent to P, $\xi = \frac{d\mathsf{Q}}{d\mathsf{P}}$, Z is the (r.c.l.l. version) of the martingale $\mathsf{E}_\mathsf{P}[\xi \mid \mathcal{F}_t^+]$. We assume that Z is $(\mathcal{F}_\cdot)$ adapted. Of course, if the filtration $(\mathcal{F}_\cdot)$ is right continuous, this assumption is always satisfied.

Theorem 13.15 (Girsanov–Meyer) *Let M be a* P-*local martingale. Then*

$$N_t = M_t - \int_0^t (Z_s)^{-1} d[M, Z]_s \qquad (13.4.1)$$

is a Q-*local martingale.*

Proof Let $U_t = \int_0^t (Z_s)^{-1} d[M, Z]$. Then

$$N_t Z_t = M_t Z_t - U_t Z_t$$

$$= (M_t Z_t - [M, Z]_t) + [M, Z]_t - \left(\int_0^t U_{s-} dZ_s + \int_0^t Z_{s-} dU_s + [U, Z]_t \right)$$

where we have used integration by parts formula, (4.6.7) along with $U_0 = 0$. Now $(M_t Z_t - [M, Z]_t)$ is P-local martingale (see Theorem 9.30). Further, $\int_0^t U_{s-} dZ_s$ is a P-local martingale (see Corollary 9.15). It thus follows that

$$N_t Z_t = L_t + [M, Z]_t - \int_0^t Z_{s-} dU_s - [U, Z]_t \qquad (13.4.2)$$

where $L_t = (M_t Z_t - [M, Z]_t) - \int_0^t U_{s-} dZ_s$ and thus L is a P-local martingale. Since $U \in \mathbb{V}$ is a process with finite variation paths, $[U, Z]_t = \sum_{0 < s \leq t} (\Delta U)_s (\Delta Z)_s$ and as a consequence

$$\int_0^t Z_{s-} dU_s + [U, Z]_t = \int_0^t Z_s dU_s. \qquad (13.4.3)$$

From the definition of U, it follows that

$$\int_0^t Z_s dU_s = \int_0^t Z_s (Z_s)^{-1} d[M, Z] = [M, Z]. \qquad (13.4.4)$$

Thus using (13.4.2)–(13.4.4) it follows that

$$N_t Z_t = L_t$$

and thus NZ is a P-local martingale. Hence N is a Q-local martingale.   □

Here is the predictable version of the Girsanov–Meyer Theorem.

Theorem 13.16 (Girsanov–Meyer) *Let M be a* P-*local martingale. Further suppose that MZ is locally integrable. Then*

$$L_t = M_t - \int_0^t (Z_{s-})^{-1} d\langle M, Z \rangle_s \qquad (13.4.5)$$

is a Q-*local martingale.*

Proof Once again we need to show that LZ is a P-local martingale. We have observed that (see Remark 9.34), $M_t Z_t - \langle M, Z \rangle_t$ is a local martingale. So in order to show that LZ is a P-local martingale, it suffices to show that N defined by

$$N_t = \langle M, Z \rangle_t - Z_t \int_0^t (Z_{s-})^{-1} d\langle M, Z \rangle_s \tag{13.4.6}$$

is a P-local martingale. Let $V_t = \int_0^t (Z_{s-})^{-1} d\langle M, Z \rangle_s$ so that

$$N_t = \langle M, Z \rangle_t - Z_t V_t.$$

First we observe that V has a jump at a stopping time σ if and only if $\langle M, Z \rangle$ has a jump at σ and then

$$(\Delta V)_\sigma = (Z_{\sigma-})^{-1} (\Delta \langle M, Z \rangle)_\sigma.$$

Predictability of $\langle M, Z \rangle$ implies that for a predictable stopping time σ, $V_{\sigma-}$ is $\mathcal{F}_{\sigma-}$ measurable. It follows using Lemma 8.25 that V is predictable. Then we have (note $V_0 = 0$)

$$Z_t V_t = \int_0^t Z_{s-} dV_s + \int_0^t V_{s-} dZ_s + [V, Z]_t. \tag{13.4.7}$$

Since $\int_0^t Z_{s-} dV_s = \langle M, Z \rangle_t$, we conclude

$$N_t = - \int_0^t V_{s-} dZ_s - [V, Z]_t.$$

Now $\int_0^t V_{s-} dZ_s$ is a P-local martingale by Corollary 9.15 and $[V, Z]_t$ is a P-local martingale by Theorem 9.32. Thus N is a P-local martingale. As noted earlier, this completes the proof that L is Q-local martingale. $\square$

Bibliography

1. Bachelier, L. *Theorie de la Speculation*, Ann. Sci. Ecole Norm. Sup., 17, (1900), 21–86. English translation in: The Random Character of stock market prices (P. Cootner, editor), (1964), MIT Press.
2. Bhatt, A. G. and Karandikar, R. L. *On the Fundamental Theorem of Asset Pricing*, Communications on Stochastic Analysis 9, (2015), 251–265.
3. Bichteler, K. *Stochastic integration and L^p -theory of semi martingales*, Annals of Probability, 9, (1981), 49–89.
4. Billingsley, P. *Probability and Measure*, (1995), John Wiley, New York.
5. Breiman, L. *Probability*, (1968), Addison-Wesley, Reading, Mass.
6. Burkholder, D. L. *A Sharp Inequality for Martingale Transforms*, Annals of Probability, 7, (1979), 858–863.
7. Cherny, A. S. and Shiryaev, A. N. *Vector stochastic integrals and the fundamental theorems of asset pricing*, Proceedings of the Steklov Institute of Mathematics, 237, (2002), 6–49.
8. Chou, C. S. *Characterization dune classe de semimartingales*, Seminaire de Probabilites XIII, Lecture Notes in Mathematics, 721 (1977), Springer-Verlag, 250–252.
9. Cinlar, E., Jacod, J., Protter, P., and Sharpe, M. J. *Semimartingales and Markov processes* Z. Wahrscheinlichkeitstheorie verw. Gebiete 54, (1980), 161–219.
10. Clark, J. M. C. *The representation of functionals of Brownian motion as stochastic integrals*, Annals of Math. Stat., 41, (1979), 1282–1295; correction 1778.
11. Delbaen, F. and Schachermayer, W. *A general version of the fundamental theorem of asset pricing*, Mathematische Annalen 300, (1994), 463–520.
12. Delbaen, F. and Schachermayer, W. *The Fundamental Theorem of Asset Pricing for Unbounded Stochastic Processes*, Mathematische Annalen, 312, (1998), 215–250.
13. Delbaen, F. and Schachermayer, W. *The Mathematics of Arbitrage*, (2006), Springer-Verlag.
14. Davis, B. *On the integrability of the martingale square function*, Israel J. Math. 8, (1970), 187–190.
15. Davis, M. H. A. and Varaiya, P. *On the multiplicity of an increasing family of sigma-fields*. Annals of Probability, 2, (1974), 958–963.
16. Emery, M. *Une topologie sur l'espace des semi-martingales*, Seminaire de Probabilites XIII, Lecture Notes in Mathematics, 721, (1979), 260–280, Springer-Verlag.
17. Emery, M. *Compensation de processus a variation finie non localement integrables*. Seminaire de Probabilites XIV, Lecture Notes in Mathematics, 784, (1980), 152–160, Springer-Verlag.
18. Ethier, S. N. and Kurtz, T. G. *Markov processes: Characterization and Convergence*. Wiley, New York. (1986).

© Springer Nature Singapore Pte Ltd. 2018
R. L. Karandikar and B. V. Rao, *Introduction to Stochastic Calculus*,
Indian Statistical Institute Series, https://doi.org/10.1007/978-981-10-8318-1

19. Girsanov, I. V. *On Transforming a Certain Class of Stochastic Processes by Absolutely Continuous Substitution of Measures.* Theory of Probability and Its Applications. 5 (1960), 285–301.

20. Graf, S. A measurable selection theorem for compact-valued maps. *Manuscripta Math.* 27 (1979), 341–352.

21. Harrison, J. M, Kreps, D. M. *Martingales and Arbitrage in Multi-period Securities Markets,* (1979), Journal of Economic Theory 20, 381–408.

22. Harrison, J. M. and Pliska, S. R. *Martingales and Stochastic Integrals in the theory of continuous trading,* Stochastic Processes and Applications, 11 (1981), 215–260.

23. Harrison, J. M. and Pliska, S. R. *A stochastic calculus model of continuous trading: Complete markets,* Stochastic Processes and their Applications 15, (1983), 313–316.

24. Ikeda, N. and Watanabe, S. *Stochastic differential equations and diffusion processes,* (1981), North Holland.

25. Ito, K. *Lectures on stochastic processes,* Tata Institute of Fundamental Research, Bombay, (1961).

26. Jacod, J. *Calcul Stochastique et Problemes de Martingales.* Lecture Notes in Mathematics, 714, (1979), Springer-Verlag.

27. Jacod, J. *Integrales stochastiques par rapport a une semi-martingale vectorielle et changements de filtration,* Lecture Notes in Mathematics, 784, (1980), 161–172, Springer-Verlag.

28. Jacod, J. *Grossissement initial, hypothese (H') et theoreme de Girsanov* In: Grossissements de filtrations: exemples et applications, Jeulin and Yor (eds.), Lecture Notes in Mathematics 1118, (1985), 15–35, Springer-Verlag.

29. Jacod, J. and Yor, M. Etude des solutions extremales et representation intégrale des solutions pour certains problemes de martingales, Z. Wahrscheinlichkeitstheorie ver. Geb., 125, (1977), 38–83.

30. Kabanov, Y. M. *On the FTAP of Kreps-Delbaen-Schachermayer,* (English). Y. M. Kabanov (ed.) et al., Statistics and control of stochastic processes. The Liptser Festschrift. Papers from the Steklov seminar held in Moscow, Russia, 1995–1996, 191–203, World Scientific, Singapore.

31. Kallianpur, G. *Stochastic Filtering Theory,* (1980), Springer-Verlag.

32. Kallianpur, G. and Karandikar, R. L. *Introduction to option pricing theory.* (2000), Birkhauser.

33. Karandikar, R. L. *Pathwise stochastic calculus of continuous semimartingales,* Ph.D. Thesis, Indian Statistical Institute, 1981.

34. Karandikar, R. L. *Pathwise solution of stochastic differential equatios.* Sankhya A, 43, (1981), 121–132.

35. Karandikar, R. L. *On quadratic variation process of a continuous martingale,* Illinois journal of Mathematics, 27, (1983), 178–181.

36. Karandikar, R. L. *A. S. approximation results for multiplicative stochastic integrals,* Seminaire de Probabilites XVI, Lecture notes in Mathematics, 920, (1982), 384–391, Springer-Verlag.

37. Karandikar, R. L. *Multiplicative decomposition of non-singular matrix valued continuous semimartingales,* The Annals of Probability, 10, (1982), 1088–1091.

38. Karandikar, R. L. *On Metivier-Pellaumail inequality, Emery toplogy and Pathwise formuale in Stochastic calculus.* Sankhya A, 51, (1989), 121–143.

39. Karandikar, R. L. *Multiplicative decomposition of non-singular matrix valued semimartingales,* Seminaire de Probabilites XVII, Lecture notes in Mathematics Vol. 1485, (1991), 262–269, Springer-Verlag.

40. Karandikar, R. L. *On a. s. convergence of modified Euler-Peano approximations to the solution of a stochastic differential equation,* Seminaire de Probabilites XVII, Lecture notes in Mathematics, 1485, (1991), 113–120, Springer-Verlag.

41. Karandikar, R. L. *On pathwise stochastic integration,* Stochastic processes and their applications, 57, (1995), 11–18.

42. Karandikar, R. L. and Rao, B. V. *On Quadratic Variation of Martingales.* Proc. Indian Academy of Sciences, 124, (2014), 457–469.

43. Karatzas, I. and Shreve, S. E. *Brownian Motion and Stochastic Calculus,* (1988), Springer-Verlag.

44. Kazamaki, N. *On a problem of Girsanov.* Tohoku Math. Journal, 29, (1977), 597–600.

45. Kreps, D. M. *Arbitrage and Equilibrium in Economics with infinitely many Commodities*, (1981), Journal of Mathematical Economics, 8, 15–35.
46. Kunita, H. and Watanabe, S. *On square integrable martingales*, Nagoya Math. J., 30, (1967), 209–245.
47. McKean, H. P. *Stochastic Integrals*, (1969), Academic Press, New York.
48. J. Memin *Espaces de semimartingales et changement de probabilites*. Z. Wahrscheinlichkeit-stheorie verw. Geb, Vol 52, 1980, pp 9–39.
49. Merton, R. C. *The theory of rational option pricing*, Bell J. Econ. Manag. Sci. 4, (1973), 141–183.
50. Metivier, M. *Semimartingales*, (1982), de Gruyter, Berlin.
51. Meyer, P. A. *Un cours sur les integrales stochastiques*, Seminaire de Probabilites X, Lecture Notes in Mathematics, 511, (1976), 245–400, Springer-Verlag.
52. Protter, P. *Stochastic Integration and Differential Equations*, (1980), Springer-Verlag.
53. Revuz, D. and Yor, M. *Continuous Martingales and Brownian Motion*, Grundlehren der Mathematischen Wissenschaften, 293, (1991), Springer-Verlag.
54. Ross, S. *The arbitrage theory of capital asset pricing*, J. Econ. Theor. 13, (1976), 341–360.
55. Rudin, W. *Functional Analysis*, (1950), McGrew-Hill.
56. Samuelson, P.A. *Proof that properly anticipated prices fluctuate randomly*, (1965), Industrial Management Review 6, 41–50.
57. Srivastava, S. M. *A Course on Borel Sets*, (1998), Springer-Verlag.
58. Stricker, Ch. *Arbitrage et Lois de Martingale*, Annales de l'Institut Henri Poincare - Probabilites et Statistiques 26, (1990), 451–460.
59. Williams, D. *Probability with Martingales*, (1991), Cambridge University Press.
60. Stroock, D. W. and Varadhan, S. R. S. *Multidimensional Diffusion Processes*, (1979), Springer-Verlag.
61. Yan, J.A. *Caracterisation d' une classe d'ensembles convexes de L^1 ou H^1*, Seminaire de Probabilites XIV, Lecture Notes in Mathematics, 784, (1980), 220–222, Springer-Verlag.
62. Yor, M. *Sous-espaces denses dans L^1 ou H^1 et representation des martingales* Seminaire de Probabilites XII, Lecture Notes in Mathematics, 649, (1978), 265–309, Springer-Verlag.

Index

© Springer Nature Singapore Pte Ltd. 2018

R. L. Karandikar and B. V. Rao, *Introduction to Stochastic Calculus*,
Indian Statistical Institute Series, https://doi.org/10.1007/978-981-10-8318-1